Infant Chimpanzee and Human Child

SERIES IN AFFECTIVE SCIENCE CLASSICS

Series Editors
Richard J. Davidson
Paul Ekman
Klaus Scherer

Infant Chimpanzee and Human Child: A Classic 1935
Comparative Study of Ape Emotions and Intelligence
By N. N. Ladygina-Kohts
Edited by Frans B. M. de Waal

Infant Chimpanzee and Human Child

A Classic 1935 Comparative Study of Ape Emotions and Intelligence

N. N. LADYGINA-KOHTS

Edited by Frans B. M. de Waal
Translated by Boris Vekker
Introduction by Allen and Beatrix Gardner
The publication of this volume was sponsored by
Living Links, Yerkes Regional Research Center at Emory University.

OXFORD
UNIVERSITY PRESS

2002

OXFORD

UNIVERSITY PRESS

Oxford   New York

Athens   Auckland   Bangkok   Bogotá   Buenos Aires   Calcutta   Cape Town
Dar es Salaam   Delhi   Florence   Hong Kong   Istanbul   Karachi
Kolkata   Kuala Lumpur   Madrid   Melbourne   Mexico City   Mumbai   Nairobi
Paris   São Paulo   Shanghai   Singapore   Taipei   Tokyo   Toronto   Warsaw

and associated companies in
Berlin   Ibadan

Copyright © 2002 by Oxford University Press, Inc.

This volume was originally published in 1935, by the Museum Darwinianum,
as Volume III in their series, Scientific Memoirs of the
Museum Darwinianum in Moscow.

Published by Oxford University Press, Inc.
198 Madison Avenue, New York, New York 10016

Oxford is a registered trademark of Oxford University Press.

Library of Congress Cataloging-in-Publication Data
Ladygina-Kohts, N. N. (Nadezhda Nikolaevna), 1890–1963.
Infant chimpanzee and human child : a classic 1935 comparative study of
ape emotions and intelligence / N. N. Ladygina-Kohts ; translated by Boris Vekker ;
edited by Frans B. M. de Waal.
p.   cm. — (Series in affective science)
"The publication of this volume was sponsored by Living Links,
Yerkes Regional Primate Research Center at Emory University."
Includes bibliographical references and index.
ISBN 0-19-513565-2
1. Chimpanzees—Psychology.   2. Animal intelligence.   3. Behavioral assessment of
children.   4. Child development—Testing.   I. Waal, F. B. M. de (Frans B. M.), 1948–
II. Title.   III. Series.
QL785 .L28   2000
599.885′139—dc21   00-028546

65.95

1 3 5 7 9 8 6 4 2

Printed in the United States of America
on acid-free paper

CONTENTS

FOREWORD

A primatological masterpiece, first published in 1935, that many readers may recognize from isolated photographs reproduced elsewhere, finally makes its richly illustrated debut in an English translation. Nadezhda Ladygina-Kohts produced a remarkably thorough and insightful study given her status as a relatively isolated pioneer in Stalinist Moscow. The book confirms how far we have come in our understanding of chimpanzees, yet reminds us how it all began.

That Kohts amassed so much detailed information on her young chimpanzee, Joni, shows that she was fully aware that every drop of knowledge would be appreciated. We can only marvel at how her closeness to the subject of study helped produce insights and how many current academic debates are foreshadowed in her perceptive observations. Of the various studies reported by Kohts, it is probably the research on facial expressions—appropriately conducted at an institute named after Charles Darwin, author of *The Expression of the Emotions in Man and Animals*—that makes the deepest impression. In text and photographs, Kohts makes a convincing case that the chimpanzee's face, too, is the mirror of its soul. But, her work on perceptual abilities, emotions, and cognition in general is also informative and inspiring.

The relative isolation of Kohts had its advantages. At the same time that American behaviorists were closing the door resolutely on the mind and representing animals as robots devoid of thoughts and feelings, Kohts opened her heart and eyes to every nuance of sensitivity, empathy, and intelligence in her young charge. Working before the rise of ethology in continental Europe, she followed the same style as the early ethologists—a style adopted from comparative zoology and anatomy—documenting every behavior in the fullest possible detail as a goal in and of itself. Description, and more description, was the early etholo-

gists' motto and provided them with a starting point broader than the precon-
ceived notions of the behaviorists. Kohts thus was as much a pioneer of primate
ethology as of cognitive psychology.

Robert Yerkes, the great American expert on ape temperament, promoted
Kohts's works in the English-speaking world, devoting many pages of *The Great
Apes* (Yerkes & Yerkes, 1929) to her work. It is only appropriate, therefore, that
Living Links, part of the Yerkes Regional Primate Research Center, contributes
to the appearance of this important work. Together with the fine studies of Wolf-
gang Köhler and Yerkes, Kohts's classic belongs on the bookshelf of any student
of primate behavior and cognition. The book is introduced capably by the two
scientists who first taught American Sign Language to apes, Allen and the late
Beatrix Gardner. Kohts's text is followed by a review by Lisa Parr, Signe Preuschoft, and myself of current knowledge about primate facial communication.

We obviously are presenting this historic document in a faithful translation,
without alterations or omissions. Precisely because of this, it is necessary to point
out to the modern reader that the text before us is, at times, as disturbing for
the experts as it must be for the lay reader. The ethical awareness of scientists,
especially those working with nonhuman primates, has undergone a radical transformation over the last few decades. In some circles, there is even talk of equal
rights for apes and humans, but even those who do not want to go this far now
question invasive research more than used to be the case and consider apes as
susceptible to psychological abuse as humans. One simply cannot maintain that
apes are a lot like us while at the same time denying their potential for suffering.

Kohts does not seem to have reached the same conclusions, however. She
affectionately calls Joni "our little prisoner" and describes how he is whipped as
punishment for his misdeeds. He is extremely attached to Kohts, taking food
only from her hands, sucking on parts of her body, and acting jealously if others
pay attention to her. All of this is normal for a chimpanzee of his age, and it is
not surprising, therefore, that every time she leaves him alone he goes into a
depression. In defense of Kohts, it should be noted that she may not have realized
how many years chimpanzees remain dependent on their caregivers: A chimpanzee of Joni's age normally still is nursed by its mother. As also noted in the
introduction by the Gardners, Kohts seems to have suffered almost as much as
the ape from the rough treatment. The book reveals no intentional cruelty or
neglect, but rather heart-breaking ignorance and a misconception of what is best
for this kind of animal.

Even though the Yerkes Primate Center has kept apes in captivity for nearly
80 years, we do not look favorably on ownership of pet apes and consider the
care for apes a grave responsibility that requires appropriate nutrition, round-
the-clock veterinary care, and companionship by members of the same species.
On all three counts, Joni seems to have been deprived. Perhaps as a result of
malnutrition and lack of calcium—he ate plaster from the walls—combined with
loneliness, he died at a very young age.

It should be realized, however, that this fate befell many young primates at
zoos during this period, and that corporal punishment of children was common
practice at the time as well. This was also a period in which natural historians

thought that the best way to study a bird up close was to shoot it out of a tree. It is hard to believe how far we have come in ethical and environmental awareness in less than a century, and Kohts's honest and revealing account brings this home in a way that may disturb, but should not surprise.

At times, Kohts complains that she has become a willing slave to her young "despot," showing that she had a strong urge to satisfy Joni's every need and demand. In a context of great affection and mutual attachment, she gathered information that was appreciated before behaviorism came along and is appreciated once more now that behaviorism is in its dying days. The doors to cognition are open again, and Kohts's free discussion of Joni's artistic impulses, imitations, deceptions, and social emotions makes the book a very modern read indeed. The tone of her writing conveys the wonder felt by all of us who work with creatures so similar to us and yet so different at the same time.

Atlanta, Georgia   Frans B. M. de Waal
January 2000

INTRODUCTION TO THE ENGLISH EDITION

Allen and Beatrix Gardner

Nadezhda Nikolaevna Ladygina-Kohts (1889–1963) received her degree from Moscow University in 1917, specializing in comparative psychology. In 1913, while a student at the Moscow Advanced Courses for Women, she opened the Psychological Laboratory of the Darwin Museum with her studies of the infant chimpanzee, Joni. Kohts's husband to be, Alexander Fiodorovich Kohts, founded the Darwin Museum in 1907, and he was also very involved in the work with Joni. Kohts's research with Joni and also with macaques is well known and widely cited throughout the world, and her distinguished career in Moscow earned her a series of honors in the Soviet Union.

Kohts's major objective was to test Joni's perceptual and conceptual abilities. She published the results in Russian, German, and French in the early 1920s, but English-speaking psychologists mainly know this work from contemporary summaries published by Yerkes and others. Kohts's methods with Joni were highly inventive for any time. In one series of tests, for example, Joni reached into a bag and selected a required object from several objects concealed in the bag. This method avoided the possibility that Kohts could be cuing Joni with her eyes; at the same time, it tested Joni's haptic discrimination. Kohts's truly pioneering work was a primary source of information about the intelligence of chimpanzees for many years.

Joni's origins were unknown to Kohts and probably also to the Moscow animal traders who sold him to her. He died of a respiratory ailment in 1916 after about 2½ years of study. The best contemporary estimates of his age placed him at roughly 4 years old in 1916.

In 1925, Kohts's own son, Roody, was born; around 1929, she began work on a comparative study based on the detailed descriptions in her diary records, of

Joni and Roody. In 1935, the Darwin Museum published this study in a lavish two-volume edition. One volume consisted entirely of fine photographs of Joni and Roody standing, sitting, walking, climbing, gesturing, manipulating objects, making faces, and engaging in all sorts of childish activities. The main volume ended with a generous English summary of the Russian text and a supplementary section of English captions for the photographs. For most of us, the bulk of this truly classic comparative work remained as opaque as the Cyrillic script in which it was written.

Choice descriptions of Joni's play, curiosity, and fears scattered throughout the summary only sharpened our appetites for the complete work. We longed to know much more about the chimpanzee, the child, and the lively young woman in the photographs, who Joni and Roody hugged and led about by the hand and whose ingenious experiments reveal such a keen scientific mind. Until now, we thought we would always have to wonder about the treasures hidden in that Cyrillic script. We at last have this fine English translation, and it is as exciting as we always hoped it would be.

The value of this book survives so well because faithful description survives the passage of time so much better than fashionable theory. Kohts describes Joni's curiosity. Even when he was desperately evading her, she could catch him by peering intently into her cupped hands. Joni could not resist peeking for himself, even at the cost of capture. Kohts describes Joni's empathy. She could lure him back from a wicked escapade on the roof by crying out in pain and distress from inside her window as an assistant pretended to attack her. Kohts describes Joni's sensitivity. A harsh word about his test performance reduced him to a whimpering, clinging creature who was impossible to retest for nearly an hour.

Joni was very young. Infant chimpanzees of that age are helpless and dependent. Under natural conditions in Africa, they cannot survive if their mother dies before they are 3 years old, even when older siblings attempt to care for them. Weaning only begins when they are between 4 and 5 years old, and the change from milk teeth to adult dentition only begins at about 5 years of age. Until weaning, a chimpanzee infant stays close enough to touch her mother most of the time and rarely leaves her sight. In Africa, young chimpanzees usually live with their mothers until they are 7 years old; females often stay with their mothers until the age of 10 or 11. Menarche occurs when wild females are 10 or 11, and the first infant is born when the mothers are between 12 and 15 years old. Captive chimpanzees have remained vigorously alive, taking tests and solving experimental problems at an age verified as more than 50 years.

It was only in the 1960s that Jane Goodall's field reports revealed the length of infant immaturity and dependence in chimpanzees. Taking our cue from Goodall's work and also from the Hayes's descriptions of their cross-fostering study of Viki around 1950, we never left our infant chimpanzees alone. Like human infants, Washoe, Moja, Pili, Tatu, and Dar stayed close to their human foster families throughout the waking day and lived in a world filled with human objects, activities, and events. Joni, by contrast, spent most of his waking hours in solitary confinement in an enclosure furnished with a couple of benches, a wooden bed with sheets, and a hanging trapeze.

Without the benefit of Goodall's reports, Kohts was sensitive enough to see how immature and dependent Joni was. She describes Joni's misery at each parting. She needed many ruses to bring Joni within reach because she had to catch him and lock him up before she could leave him alone. At least once, Kohts stayed to watch Joni from an outside hiding place; she saw that he remained huddled in depression for hours. The bond between Joni and Kohts was reciprocal; she also suffered with each parting. Yet, as clearly as Kohts could appreciate Joni's deprivation, she plainly failed to appreciate the profound effect this must have had on his intellectual and emotional development.

We have seen most of Joni's traits in other infant chimpanzees, but Kohts often reports traits that are distinct traits of Joni rather than general chimpanzee traits. Because she only knew Joni, it was easy to slip into the habit of attributing all of his personal traits to chimpanzees in general. We fell into this same trap when the only chimpanzee we knew was Washoe. As we enlarged our circle of familiar infant chimpanzees, we soon learned that two chimpanzees can be as distinctly different from each other as two human children.

We were surprised by Kohts's reports that Joni resisted all attempts to clothe him or wrap him in blankets. Although Joni did drape ribbons and strings around his neck as if decorating himself, he rejected anything that covered his arms, even a shawl on a cold day. Very likely, Joni would have accepted clothing as readily as our cross-fosterlings if he had been introduced to clothing early enough. Gua, who was cross-fostered by the Kelloggs around 1930, and Viki, who was cross-fostered by the Hayeses around 1950, also accepted clothing readily. Washoe, Moja, and Tatu amused themselves by dressing up in assorted human clothes that we kept for them in a special box. They preened before mirrors from the time they were toddlers, and Moja in her 20s continues to dress up in human clothes.

We were even more surprised to read that Joni was extremely jealous of his food and refused to share any, even with Kohts. By contrast, Washoe, Moja, Pili, Tatu, and Dar eagerly shared food with their good friends. They were insistent about pushing food to our mouths; sham eating never satisfied them. Believe us, it could be very difficult to eat food out of those often filthy hands, which might have been anywhere since their last good washing.

These and most of the other things that Joni did poorly or not at all, such as bipedal walking and eating with spoons and forks, are things that cross-fostered chimps typically do with ease. We can now see that virtually all of Joni's failures reflect his deprived, caged environment, which was as different from the life of a wild chimpanzee raised with his own mother and siblings as it was from the life of a child like Roody. Joni's lively intelligence and social sensitivity are evidence of remarkable chimpanzee resilience.

Joni frequently suffered from respiratory infections and died of pneumonia, a very common fate for captive infant chimpanzees to this day. In 1966, when we started our experiments in cross-fostering, experienced colleagues repeatedly warned us about high infant mortality and advised us to use older subjects. We found instead that infant chimpanzees that are kept like human children are at least as hardy as human children. Now, we suspect that deprived environments are responsible for the high rate of infant chimpanzee mortality in laboratories. Similarly, writers used to puzzle over the high mortality rate of human infants in orphanages; these deaths occurred in spite of professional standards of nutrition and medical attention.

Inevitably, the sections of the book on Roody suffer by comparison with the sections on Joni. Roody seems less vivid, but perhaps Roody's life and doings are so humanly familiar that they make rather bland reading next to those of Joni. Kohts herself says that, as she wrote, her images of the infant Joni were more vivid for her than the images of the infant Roody. Nevertheless, the sections on Roody are valuable because they put the Joni story into much clearer perspective. The contrast between Roody's world and that of Joni was very great, indeed.

Perceptive as she was, Kohts recognized the dramatic difference between Roody's life and Joni's. She comments on this in several places. In her view, however, Roody's wider intellectual and social world served as a wider arena in which he could show his human superiority. Kohts seems to have overlooked the reciprocal effect, that Roody's wider world stimulated his human growth and made him more human. Worse still, she seems to have overlooked the negative effect of Joni's narrow world, which constrained his growth and made him less of a chimpanzee.

This was the conventional wisdom of her day, of course. Infants were thought to develop according to an inexorable species-specific plan. Provided with sufficient food, water, and shelter, Joni must develop into a typical chimpanzee, Roody into a typical human. Not long after Kohts published this book in Stalin's Moscow, B. F. Skinner recommended a small, sterile, climatically controlled chamber as the best place to raise human infants. Even though everybody hated Skinner for this proposal, they never questioned its scientific and practical merit.

Kohts gives us many details that set her work in its time and fill in the background with historical and local color. Joni rode in horse-drawn vehicles. In the streets of Moscow, Roody watched political demonstrations with marchers who wore red badges and waved red flags. Roody repeated a story that a retired cavalryman told him about an army horse. When Joni insisted on wiping his nose with his forearm instead of his handkerchief or when he enjoyed having his head groomed, he reminded Kohts of peasants in old Russia. All these precious details give us a sense of time and place that draws us into the story.

The descriptive passages in this book are timeless. They will always instruct and entertain. At last, there is an English version to study and to savor. The theoretical passages have some historical interest because they suggest the trend of thought of colleagues in Moscow in those days. More recent work shows clearly that Joni's life, while remarkably enriched for a laboratory chimpanzee, precluded any valid comparison with normal human children.

One historical lesson is that absence of evidence is never evidence of absence. Later work tends to fill in the most significant gaps in earlier studies. Another historical lesson is that each fresh set of findings seems to evoke new theories based on gaps in the new evidence. The less one knows, the easier it is to construct an Aristotelian theory of logic-tight divisions by drawing boundaries wherever the evidence ends.

On the positive side, each generation seems to produce fresh investigators who are challenged by limits in the evidence. There will always be those who see the gaps not as battlefields on which to fight to the death for the sacred difference between man and beast, but rather as unknown lands to be explored and mapped. The adventures that Kohts describes so well should inspire future adventurers.

Infant Chimpanzee and Human Child

This volume is dedicated to
all of those working in the
Museum Darwinianum.

PREFACE

This research is based on my own observations of a chimpanzee, Joni, from 1913 to 1916, and of my son, Roody, from 1925 to 1929, from his birth to when he was 4 years of age.[1] Since accuracy in age estimates was very important for the purpose of parallel comparison of the two infants, we had to be absolutely certain about the chimpanzee's age. It was not easy to figure out, especially because both the question of anthropoids' life span and the principles of its calculation have been radically reviewed lately.

Professor Brandes of Dresden[2] concludes that many mistakes have been made in determining the age of anthropoids in captivity. I originally had estimated that my Joni was 7 years old based on the time when his teeth changed from baby teeth to permanent ones; however, I did not take into account proper body and weight measurements.

Professor Brandes, who was particularly interested in the question of Joni's age and who had requested from me a series of pictures of Joni's dental and skull structure, discovered that I (like many other authors) overestimated the animal's age twofold.[3] Consequently, Joni at the time of his death was no more than 3 years old. On the other hand, Professor Yerkes (United States), who knew about Professor Brandes' calculations and had all my factual data for solving this difficult problem, concluded that Professor Brandes had underestimated the chimpanzee's age. Professor Yerkes tended to think that Joni died at the age of 4 to 5 years.

Three scientists at the Anthropoid Station in Florida, deducing from the data I submitted, showed a discrepancy of 3 to 5 years in determining Joni's age. In light of this discrepancy, Professor Yerkes proposed taking the proportional average of four age estimates, which brought us to the age 4 years and 3 months.

On my part, I also included Professor Brandes' data and took the proportional average of three measurements, coming up with 4 years as an estimate for Joni's age. Subtracting from that number the 2½ years of Joni's life in our home, I figured that Joni had come to us at the age of 1½ years; hence, my 2½-year observations had been of an animal 1½ to 4 years old.

Processing of the data from my observations of the chimpanzee and my son, collected over a 7-year period, has taken 5 years.[4]

All the illustrative materials in this study are original photographs (most of which are being published for the first time) or graphic sketches of the live animal made by artist V. A. Vatagin with my direct participation. Only the sketches of chimpanzees' heads are designs reproduced on the basis of my studies of their facial expressions, and they were made by combining a series of photographs.

The notes describing the chimpanzee's behavior on the one hand and that of my son on the other were separated by a 12-year period. The infants were developing and being observed at different times; therefore, they could not influence each other. I was intentionally trying not to train Joni and to teach him as few human habits as I could; the goal was to observe his natural and more spontaneous behavior. The notes about the child were made regardless of the goals of comparative psychology research, which came about post factum and considerably later than 1929.

To the skeptic with doubts about the validity of an observation of *one* chimpanzee and *one* child as a basis for generalization, I have to mention the following: The central part of my theme that gives the material for the conclusions—the comparison of instincts, emotions, and expressive movements—deals with biopsychological features characteristic of the genus. It is highly improbable that greater individual variations of these features could be found in species of *Homo sapiens* and chimpanzee (*Pan troglodytes*) of the same age and gender.

In addition, I had extensive autopsy material on anthropoids (chimpanzees and orangutans kept in the zoos of Moscow, Berlin, and London), which has also been taken into account. These results did not in any way contradict my conclusions; quite the opposite, they confirmed them. As to the data on my child, they were complemented if possible with the observations of other children; these observations are used widely in the narrative.

The seeming singularity of the observed subjects (chimpanzee and human) strikes the eye because of the originality and vividness of the illustrative material and because the observations of Roody are precise chronological documentations; the observed facts are well known to every biopsychologist or pedologist, which is not to say that they are banal.

The notes on the child in this research were a typical mother's diary in which new patterns of behavior of a first and only child were logged with precision, joy, and love every day; it is no wonder that this diary was richer in facts than the notes on the chimpanzee.

It was remarkable that Roody's behavior prompted so many associations with the chimpanzee, and so often restored in my memory the forgotten details of Joni's behavior. Eventually, it motivated me to conduct a factual comparison of the behavior of both infants. This impression increased even more when I was working on the notes about the chimpanzee in connection with the theme, infant chimpanzee in his plays, instincts, habits, emotions, and expressive movements.

The last chapter of this work was planned to include the comparison between a chimpanzee's psyche and that of a human child of the same age. But, the comparison and analysis of the notes on Joni and Roody presented me, quite unexpectedly, with such an enormous amount of extremely diverging analogies that I was forced to embark on a more comprehensive, broad, and profound consideration of all the notes, numbering thousands of pages. The end of the theme, which seemed to be so close at hand, was extending into the uncertain future.

I found myself in the position of a traveler who had just reached, with great difficulty, the peak of a very tall mountain.[5] The person had been climbing the mountain for 3 years, day after day, hour after hour, from early in the morning 'til late at night; the traveler has been navigating the stony path persistently and unyieldingly.

Sometimes it was tough; his weary heart beat wildly, his mind fogged, and his energy abandoned him. But the thought that from this peak he would see the farthest horizons instilled in him a new stream of energy; momentary weakness had been overcome, and he found the enthusiasm for a new drive upward.

When the end seemed to be two or three steps away, another, even more unsurmountable, peak suddenly appeared before his eyes; the point that he had reached was only at the foot of this new mountain.[5] He could not help but flinch from this new predicament. How can he not let himself down? Where can he find the strength for overcoming new difficulties.

But what is beyond humans—the untamed, powerful, compelling spirit of exploration—was urging him to rise higher and higher up that mountain, so that from the new scientific horizons and a new perspective, he could view what was left behind.

So, I had to climb a new, steeper, and hardly passable path for 2 more years. Far in the background of my consciousness sounded a constant warning: "Do not let yourself slide and fall down!"

Even now, when my task has been completed[6] and I am looking at it from new heights, can I really say that I have seen everything that I have intended? I have seen more than before, but not everything. Why? Because above me I see new and higher mountaintops[7] from which unmeasurable, greater spaces can be seen. I am striving with all my energy to go forward and up to the new peaks. May these peaks come up to the sun itself!

Speaking about the history of this book, I ought to emphasize that this work, in its present form, could not have been published before 1917; only the October Revolution, initiating the work of the Darwin Museum on its current scale, made it possible to create a framework for my research that I could not have dreamed of 20 years ago.

I feel obligated to acknowledge the help of the State Darwin Museum, which created exceptionally favorable conditions for my years of steady and secure scientific work and provided me with both the financial support and the opportunity to publish my results. The publication of this book is timed with the 30th anniversary of the Darwin Museum (1905–1935).

I would like to acknowledge those who helped me conduct this work, turn a dream into words, and materialize the words into deed. To whom do I owe mention? Could 10 years of work, thousands of pages of notes, hundreds of pages of text, many dozens of photographs and drawings possibly be a one person job? Could it possibly be done with one pair of hands, one pair of eyes, and one mind?

How many of them? Those famous and those unknown, close and more distant men and women, who encouraged me when I was tired, gave advice in the moments of scientific speculations, prompted my thoughts to soar, fixed by a camera, pencil, or brush what was available to the eye. . . . I had dozens of helping hands and eyes, three or four faithful hearts, and one inspiring image!

I decided to acquire the chimpanzee Joni 20 years ago, and I said with the enthusiasm of an adolescent: "I would love to buy an anthropoid for a scientific study!" But where would I find the money? Who responded to this pleading cry of burning scientific passion? Who helped out at a difficult time with the hundreds of rubles needed for the purchase? It was not fellow scientists, not scientific institutions; they could hardly give an advance to a young beginner and put their trust in the creativity of her plans. It was F. E. Fedulov, the oldest and most devoted associate of the Darwin Museum, whose penetrating mind always had held science in great esteem, whose affectionate heart believed in the sincerity and earnestness of my scientific ideas. Fedulov's subsequent help with the complicated procedure of photographing Joni (and then Roody) was invaluable and indispensable as well.

A different kind of help was no less timely. Now that Joni had been acquired and was under supervision, he had to be given a space to act and move freely both inside and outside. Who would create such conditions? Who would agree to associate himself with a restless, anxious, and unusual creature? At that moment, two generous hearts responded, and loving hands were extended, those of Nataliya Nikolayevna Engelman and Mikhail Nikolayevich Engelman; they provided everything that was necessary for the continuation of my work.

I was watching Joni and recording his behavior in detail. How valuable would these records be without the photographs that not only documented events that my eyes had seen, but also brought to my knowledge details so fleeting and minute that I was not able to discern them? On hot and humid summer days, under the scorching sun, filling his vacation days with hard work, straining under the burden of a 5-pound reflex camera, standing for hours before the white screen, or following Joni in search of "a good moment"—he, my constant, faithful, and bright companion who came onto the orbit of my life—A. F. Kohts made hundreds of pictures that are part of the foundation of this research, pictures considered unsurpassed even by the American press.[8]

When I analyzed the observations and examined the pictures of Joni, I found that they did not depict everything that needed to be depicted or the way I wanted. Some of them abounded in photographic details that needed to be generalized to make the most important features clear; others required special printing and processing. Who would fix the mechanical eye of a camera and the defects of the negatives? Two men, two names emerge from my memory: One is still alive and incessantly active, unsurpassed in his field, our famous artist-animalist V. A. Vatagin; the other is A. T. Trofimov, an old photographer who died unknown.

It has been 12 years since my observations of Joni; a new animal has arrived, and there are new working conditions, new problems. In addition to my three old and faithful colleagues A. F. Kohts, F. E. Fedulov, and V. A. Vatagin, new ones have appeared.

Records of Roody's behavior and rough copies of the text describing Joni were typed by Yu. A. Polakova, always friendly and calm; my illegible, obscure, hastily made (and therefore sloppy) notes were rendered lucid, easy to read, and ready for the subsequent analysis and final typing.

At different times, two other photographers were assigned to take pictures of Roody. Each had his own strengths: L. M. Sytin's strength was in the field of artistic photography; M. A. Sirotkin's was an amazingly quick comprehension of the finest biopsychological phenomena and artistic processing of the negatives, which provided exceptional phototypical rendition of pictures (he took upon himself the main part of this job).

Three new young, talented artists worked on the drawings: N. N. Kondakov, who masterfully created thousands of exquisite watercolor sketches illustrating the strengthening of the elementary mental processes in the child; M. M. Potapov, who brilliantly painted Roody's portrait; and V. V. Trofimov, who prepared excellent graphic drawings.

A great number of people donated their time and leisure for the observation of Roody, taking care of the child during my work hours! Most clearly, I can see the tender and caring eyes of the grandmother (E. A. Kohts) who unselfishly gave her last reserves of energy to Roody. She did not live to experience the joy of this book coming off the press.

I would like to mention the participation of two physicians who treated my son: Dr. A. N. Khanevsky and our indefatigable Dr. V. V. Postnikov, whose timely and knowledgeable advice sustained the bodily welfare and well-being of my examinee, saved him not only from any grave illness, but also from typical "childhood" diseases, which gave me the opportunity of uninterrupted observation of the boy for the first 7 years of his life.

I want to emphasize my appreciation of the attention and scientific and bibliographical support by Professor R. Yerkes (director of the Anthropoid Station in Florida) and the following foreign colleagues: Professors Osborne, Gregory, Fernberger, and Gezell (United States); Professors Huxley and Keith (England); Professors Brandes, Stern, Lippman, and W. Köhler (Germany); Professor Révész (Holland); and Professor Klapared (Switzerland); they expressed great interest in my work while it was still a manuscript and stimulated intensive and extensive research.

With his usual stylistic proficiency, S. S. Tolstoy translated the summary of my work into English.

The story of this book would be incomplete if I failed to mention that, while I was watching Roody and Joni, often writing notes about the infants' behavior in their presence in order to document an accurate picture, two pairs of small eyes, the brown of Joni's and greenish gray of Roody's, looked at me with gloom and sadness. Someone who has and loves children, reading these words, will understand the depth of emotions with which I am flooded at these recollections.

Indeed, Roody and Joni were children, and children always want to have fun; but suddenly they would notice that, for some reason, I was looking at them so seriously and with such a concentration. Children always like to be entertained by adults present; seeing me engaged in my own activity was incomprehensible and strange to them.

For both of them, I was the closest and most desired person; it was me with whom they so joyfully played. And now I seemed to be brushing them aside, sitting motionlessly and writing for long periods of time. I feel and I see how both of them were trying by any means to make me forget for a moment about science and to remind me that they were real, live children. Baby Joni pulled at my dress, inviting me to play, snatching from me the horrible pencil and notebook. Roody, my son, whined pleadingly while getting onto my lap, "When are you going to finish?"

Once again, about the leading participant, the founder of the Darwin Museum—A. F. Kohts: Many times during these long 10 years I was overcome by moments of moral exhaustion, by stretches of mental stagnation. But every time I saw before me his always enthusiastic image and heard his inspiring words, the spark of my scientific drive, which had almost faded out, again gleamed and burned.

So, when I begin browsing in my thoughts, I can clearly see how relatively small my own share in this work has been.

What actually has come my way? Long, purposeful scientific research, the joyful work of observation, patient analysis, tentative synthesis, and the wish to cry out: "This published work in its entirety is not my work. This is the labor, patience, endurance, skills, energy, and inspiration of all those who have been and are beside me. Some have been a part of the group for only a short time, but left an indelible mark here, on these pages."

To all of them—close or distant, alive or dead, famous or unknown, mentioned or unmentioned—I am dedicating this work of mine.

INTRODUCTION

"Παντα ρετν." These two words have come to us from the depth of past centuries, reflecting an entire philosophy. "Everything flows, everything moves, you cannot step into the same stream of water twice—it has already gone" (Heraclitus).

As in a historical chronicle, for which the dates—years, months, even hours, minutes, and seconds—do not repeat, the mental condition of an individual does not remain the same for a single moment, and the same mental experience does not return. Are we able to define accurately and measure such a short period when, in the mind of a human or an animal, a certain mental balance settles in and stays unchanged? It is easy to realize how remarkably difficult it is to portray truly, accurately, and distinctly this endlessly billowing element of mental conditions and how distorted this reflection is in the experimenter's nets, ingeniously woven for the catch.

Squeezed into the framework of an experiment, mental activity is like a free bird, accustomed to unhampered flights and unlimited spaces, that has been caught and put into a small cage. Like a live bird, the mental activity bumps into the walls of the experiment, unwilling to be confined by them, struggles to get out, and breaks the nets of theoretical expectations and the experimenter's plans. A contemplative observer who grants total freedom of expression to this kaleidoscopical change of mental conditions will be more likely to watch its development, follow it wherever it goes, catch it, and describe it more easily.

The modern observer has elaborated the methods of observation. The natural senses are broadened, specified, and perfected by auxiliary technical devices: the addition to the eyes of a photo camera, a telephoto lens, and a movie camera; to the ears the sensitive membrane of a phonograph; and to the writing hand the registering lever of a kymograph.

Making these totally submissive, machinelike, mechanistic coworkers reproduce, repeatedly and extensively, the outward expressions of mental conditions, the modern observer undoubtedly has a right and more to compete in research methods with the modern experimenter. The former has a clear advantage over the latter when choosing the course of a "natural experiment": not curbing the freedom of the observed animal, not interfering with the flow of mental conditions, only changing external circumstances. When the former has the opportunity, extensive registration of observed facts and their objective interpretation are performed. But, the method of observation has a vulnerability: Its merits vary, depending on the field and the object of its application.

In the sphere of comparative psychology, by limiting the flow of mental phenomena, the experimental method preserves its main scientific qualities even when applied to enormously wide fields of an entire hierarchy of live creatures (from amoeba to humans) and all kinds of mental phenomena (from tropism to the highest intellectual processes). At the same time, the observation method, which needs an open space for a phenomenon to evolve, is limited in its applicability: It shows its most valuable qualities only when the mental activity is molded into expressive and distinct outward shapes.

It is not difficult to notice that this method is especially meaningful in applications to a human—considering the extremely rich human expressions, especially of a human child—with the uninhibited human ways of behavior, and to an anthropoid ape as apes surpass humans in expressiveness of movements and directness of manners.

The face is the mirror of the soul. This truth is as trivial as it is easily proved. This is strikingly evident in the recall of the face of a genius and that of an idiot. But, the facial substance is much richer: It reveals to us not only the level of intellect, but also temperament and emotional endowment. These revelations are relatively more noticeable: an asthenic, athletic, or pyknic individual has a certain physical and mental structure. A melancholic, sanguine, or choleric person has a personal facial type. Our every mood finds an outward expression first on the face. On one's face is registered the main tone of one's real emotions; to a degree, one's past spiritual life is imprinted on it. We do not read and do not decipher it only because we do not know how or do not want to do it, although we have the needed ability. For example, we can notice, by the facial expressions of our close friends and relatives, every nuance of changing mood, every shade of mental state, but this is registered as though outside our will, unconsciously, almost intuitively; we cannot clearly comprehend what this bodily change has been.

For the primeval people, expressive gestures and body movements were part of not only emotional, but also intellectual, communications. Gestures sometimes were used instead of images and acts and even instead of concepts and judgments. Or, look closely at a sad and moving act—the wordless, noiseless conversation of deaf-mutes: The fine, exquisite symbolism of the lip and finger movements are accentuated with expressive movements of eyebrows, eyes, hands, and even the entire body.

Experienced psychologists, poets, and writers long ago intuitively appreciated the emotional impact of gestures and expressive movements. Therefore, in emotionally charged scenes, they make their heros act, not speak, and show emotions not with words, but with vivid images to render the emotions visually obvious. "The people are silent!" are the final words of Pushkin's poem, "Boris Godunov." The silent scene is also at the end of Gogol's play, "The Inspector."

A writer or virtuoso actor understands that, at moments of maximal emotional inspiration, words must give way to a more effective kind of expression, an image, which is in full accordance with the fact that at a time of high emotional stress or great moral crisis, people understand each other without the use of spoken words. Words and more words flow only when the path of mutual trust and understanding has been lost. "The thought uttered is a lie," says Tutchev, clearly realizing how difficult it is to verbalize an inscrutable state of mind.

More eloquent than words is the language of those in love; a single glance is enough for two loving hearts ("Was vom Herzen geht, geht zum Herzen"); they do not need a method of expression by proxy (verbal language). A word acquires its special meaning outside the emotional sphere in a world of intellectual activity for which accuracy and subtle differentiation of events, mental perceptions, images, notions, and such are required.

Let us look back at our childhood and infancy. At that time, the language of expression and gesture was our only language because we did not have another one, articulate speech. We all know that a child understands the facial expressions of people around the child much earlier than the meaning of words and communicates with people first using nonverbal expression.

When articulate speech appears, it replaces this wordless communication. Words begin to convey our thoughts so perfectly, and intonation adds emotional flavor, without which the thought and the words themselves are lifeless. Growing up, we come to trust our auditory perceptions so much that we often entirely forget about the accompanying language of expressive movements and expressions. We recall it only when we do not trust words, do not understand them; then, we examine the face and eyes and can often uncover the true sense of the words. Faces rarely can conceal what words conceal with no difficulty.

The expression "eyes don't lie" is no less banal. Animals do not understand our speech; they understand gestures, poses, and sharp changes in human facial expressions much easier than words. In turn, when we communicate with animals, we understand their desires and feelings mainly on the basis of our visual perception of their expressive movements. The following is clear: The mouths of animals are "sealed": they cannot communicate words about themselves, and their sounds are undifferentiated, very few, and so far investigated so insufficiently that they become understandable to us only in connection with the actions of the animals at the time of the sound. The facial expressions, poses, gestures, and body movements unveil the meaning of sounds emitted by the animals and disclose the emotions of the speechless creatures.

Expressive movements, accompanied by instinctive sounds, are the animals' means of communicating with their kin; these movements uncover for the observer the entire complicated hidden mechanism of an animal's mental life, and the animal becomes accessible to observation and study. Animals, contrary to people, cannot hide their emotions.

This natural, unconditional language of their expressive movements is direct, straightforward and truthful; it has an unequivocal advantage over totally conditional, flexible, ambiguous human language, which is often used by many to conceal their true, secret thoughts.

It is far from unintentional that the entirely rational Western intellectual culture, by withering minds and hearts and extinguishing flickers of unrestrained and vivid expression of feelings and thoughts, created an ideal for teaching "gentlemanly" manners, that is, a statuelike immobile body, a face reminiscent of a mask, and limited gesticulation cramped as though by paralysis.

It was as if these manners of a "civilized" society were specially created so that it would be impossible during the communications process for a person involuntarily to bring to the periphery of his bodily image the hidden thoughts and feelings that might divulge the subjective attitude toward people and events. Conversely, the winning charm of animals, like that of children, is in the naive unpretentiousness of their behavior, which generates trust and warms hearts to them.

The sincerity and openness of the outward manifestation (inadequately understood at the present time and as though symbolic and enciphered) of this way of revealing the inner world of a wordless animal shows, due to its truthfulness and spontaneity, the easiest and the most proper road to a comprehensive understanding of an animal's mental structure, especially the dominating emotional sphere.

The language of expressive movements of anthropoids is particularly colorful, eloquent, and rich. This is especially so for the chimpanzee.

In fact, the reticent, uncommunicative, melancholic gorilla, the phlegmatic orangutan, and the unexpressive gibbon are far from being as articulate as the communicative, vivid, vivacious, and sanguine chimpanzee, the possessor of unusually complicated facial muscles. By comparison of appearance to that of other anthropoids, the chimpanzee has the greatest number of anthropoid features.

To acquaint the reader with this most unusual creature, I take the liberty of introducing my 1½-year-old male chimpanzee Joni (Pan chimpanse Meyer—*Anthropopithecus troglodytes*). Let us look closely at our Joni with such attention and love as I did 20 years ago, sitting him up before me and watching his remarkably expressive face—this complicated keyboard on which the primeval, wild, exotic symphony of his emotional life has been played so emphatically and so vividly.

We need to examine this face in all its minute details, much as the composer, to achieve the desired expressiveness in creating a piano piece, needs to be sure in advance of the precise location of every key of the instrument and to remember the tone and the volume of sounds emitted by every part of the keyboard so that they can be made to sound or be kept silent at any moment. We also must be familiar with the details of a chimpanzee's face in stasis so that we can reveal more easily and distinctly its change in motion.

PART I

Descriptive

Behavior of Infant Chimpanzee

Chapter 1

Description of the Chimpanzee

Chimpanzee's Face   The head of our chimpanzee is elongated and a wide
in Stasis   oval at the crown; it starts to narrow from the upper
edge of the ears and keeps narrowing gradually to the chin.

Forehead. Looking at the chimpanzee *en face*, one can see that the forehead
part of his head has a broad, flattened spherical shape; it is covered with black
hair and stands in sharp contrast to the distinguishable light-color, above-the-eye
arcs and to the face proper, which is tightly enveloped in dark, flesh-color skin
that is almost without hair.

A small part of the forehead, right above the point at which the arcs converge,
displays the tendency for balding[1]; it is covered with hair so short and thin that
the light skin seems transparent. Every time the animal looks up, the skin forms
five narrow wrinkles that run parallel to each other and to the arcs[2] [see group of
furrows 1, including wrinkles 1 to 5 in figure 1(1)].

The three lower wrinkles are longer than the upper ones. When the animal
delicately knits his brow, the three lower wrinkles on both sides of the face almost
touch one another in the middle of the forehead; in a more intense facial expression, the wrinkles converge at an angle and form a promontory that comes down
above the bridge of the nose. The two upper wrinkles are so short that they
never reach the middle of the forehead, and one never touches the other.

The hair on the balding part of the forehead and on the adjacent parts is short
and soft, but it gets longer, thicker, and harder as it approaches the crown; at
the forehead sides, near the upper edge of the ears, the hair is especially long
and hard. It sticks out, partially covers the ears, descends at the side of the cheeks
in the form of side-whiskers and converges under the chin, with the hair fringe
framing the light face of the animal.

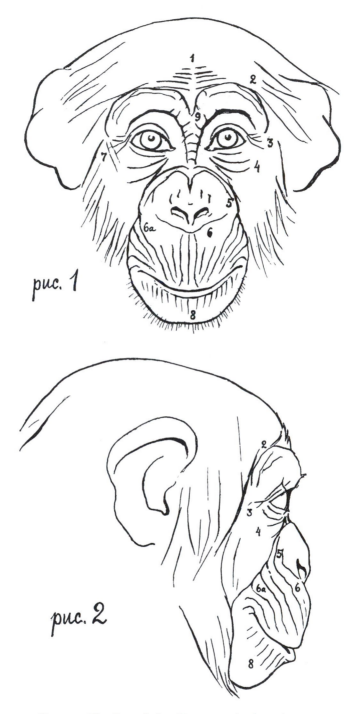

рис. 1

рис. 2

Figure 1. The face of the chimpanzee in the quiet state:
(1) en face; (2) profile.

The raven black hairy forehead of the chimpanzee, as well as the side-whiskers adjacent to it, quite sharply set off the rest of the face, which is light and bare (this particular part, if looked at very briefly, in fact may be thought of as the *face*). However, the forehead, as such, blends in color with the crown and seems to be a part of the area of the back of the head.

Above-the-eye arcs and bridge of the nose. The face proper begins with the above-the-eye arcs. The arcs tower conspicuously over the upper forehead, as well as over the lower eyelids and eyes; they have the shape of flattened rolls, are broadly outlined, and are uniformly wide (1.5 to 2 cm) along the entire length of the arc. Where they come down in the middle of the forehead, they become very thin, merging at the narrow bony core of the bridge of the nose. The above-the-eye arcs are separated from the forehead by sharp furrows, which run parallel to the arcs and form a promontory where the arcs meet [figure 1(1), 1(2)].

The above-the-eye arcs are covered with dark, flesh-color, somewhat un-smooth, rugged skin that has brown pigmentation in the form of blurred stains near the outer ends of the arcs. Skin at the arcs is very mobile; it looks as if it is capable of rolling over the arcs, up and down. When the animal looks up, the skin rises, thickens, and forms three shallow, elongated furrows (second group of furrows, No. 6 to 8), at the upper and rear edges of the arcs, parallel to the outline of the arc. When the animal looks down (or during sleep), the skin slips down so noticeably that the dark, hairy edge of the forehead brims over to the upper edges of the arcs. The lower edge of the arcs slides down almost onto the eyelids; the border furrows move over onto the arcs and, forming a promontory, come down to the bridge of the nose. Skin at the front part of the arcs gathers in multiple short, deep, transverse wrinkles that are accentuated strongly where both arcs converge and at the bridge of the nose where they suddenly get closer and intersect at a certain angle (figure 1(1)].

Skin at the above-the-eye arcs, after thorough examination, does not appear totally bare. In fact, it is covered with thick, white, fluffy, silky hair visible only in transmitted light. Nevertheless, even after a brief inspection, one undoubtedly can see hard, long (2 to 2.5 cm), thin, setaceous hair sticking out irregularly scattered along the arcs—a semblance of eyebrows. It seems appropriate to attri-bute the term *eyebrows* to the lightest above-the-eye arcs that clearly are set off against the dark background of the forehead and face.

The bridge of the nose is covered tightly with somewhat pigmented skin. Above the point where the arcs converge, the bridge of the nose is split by the narrow strip of hair that descends from the forehead and reaches the base of the bridge, as though indicating the place where the bridge begins.

The bridge is covered with rare dark hair that is shorter than hair on the above-the-eye arcs. Also, it is covered with down consisting of soft, light hair. The bridge is furrowed by three transverse wrinkles and by two tilted ones that near each other when the nose wrinkles; the ninth group of furrows includes wrinkles 46–50 [figure 1(1)].

The light above-the-eye arcs and the bridge of the nose are seen clearly against the surrounding background due to the adjacent darkly pigmented skin. The wide pigmented spot beginning under the lower edge of the inner end of the arcs extends parallel to the arcs, almost disappearing in the middle part of the upper eyelid and increasing before the outer corner of the eye, where it merges with the adjacent pigmented spot at the outer ends of the arcs. At the lower front edge of the arc, the pigmentation starts with a narrow strip close to the outer edge of the bridge, runs parallel to the bridge axis, substantially broadens at the level of the axis end, crawls to the sides under the lower eyelids, and continues into the upper parts of the cheeks, where it gradually fades out. The middle and back parts of the upper eyelids and the tiny spots at the front and back corners do not have pigmentation.

The part of the face containing the eye orbits, cheeks, and bridge of the nose is depressed noticeably compared with the forehead part and the highly distinguishable arcs. The deepest part of this depression is in the vicinity of the bridge, where the protruding cartilaginous nose begins, and in the lower parts of the cheeks, which look like hollows compared to the hilly, protruding jaw part of the face.

Cheeks. The cheek part of the face is dark and sharply separated by an almost linear boundary from both the lower eyelids and the jaw part of the face. The dark color of the cheek part depends on both the pigmentation and, to a greater degree, on the dark hair covering it. This hair, appearing on the upper parts of the cheeks (under the eyelids where the pigmented spots end), in the beginning is very rare, then it becomes longer, thickens, and continues across the entire surface of the middle and the back parts of the cheeks, merging at the sides with the long side-whiskers, framing the face.

Eyelids. The eyes of chimpanzee's are framed with the eyelids. The upper eyelids (similar to those of a human) are considerably wider than the lower ones and end with a roll-like rim.

The upper eyelid, viewed in a half-closed position along its whole length (from the inner to outer corner), are cut across with four thin furrows. When the eyelid is raised partially, these furrows gather close to one another and form, above the roll-like rim of the eyelid, a double crease that splits at the outer end of the eyelid into two wrinkles. When the eyelid is raised fully, all four creases merge into one deep, thick crease that hangs over the roll-like rim of the eyelid and also divides near the outer corner of the eye into three or four parallel wrinkles, so-called goose paws [the third group of wrinkles, includes 9–12; see figure 1 (1)].

The lower eyelid is represented by a narrow, flat, thick strip. The lower border is more or less visible, depending on the lighting and on the position of the animal's eyes.

The upper eyelid is equipped with four straight, soft, rather dense eyelashes, all the same size (approximately 3 mm). The eyelashes of the lower eyelid, compared to those of the upper eyelid, are thinner and two times shorter.

Wrinkles under eyelids. Under the lower eyelids, separated by a shallow depression that seems to follow the eyelid outline, there is a more or less distinct skin swelling, which is furrowed by four deep creases that fan out from the front corner of the eye [fourth group of furrows, 13–16; see figure 1(1)].

The first of these creases, the thinnest, starts almost from the middle of the lower eyelid and is almost parallel with its outline. The second and the third creases, which are deeper, start together from the front corner of the eye and diverge (to 5 mm); then, as they depart from the corner of the eye, they continue to diverge. The fourth crease, which branches off from the upper third of the third crease, is extremely distinct and is two times shorter than the third one. Further along its course, the fourth crease departs from the third crease and descends, but not to the side, absolutely parallel with the protruding jaw part of the face. The fairly sharp 17th tilted crease [the first from the fifth group of wrinkles; see figure 1(1)] cuts across the light-color jaw part of the face and along the lower, lateral parts of the cheeks. It goes from the base of the nasal ridge, precisely parallel with the converging furrows that separate the cartilaginous nose.

Speaking of the wrinkles of this particular part of the face, we have to mention two other long, thin wrinkles that come from the outer edge of the above-the-eye arcs, cross the wrinkles of the outer corner of the eyelids, travel at an angle and inward along the middle part of the cheeks, and vanish in the dark hair [the seventh group of furrows, including 35 and 36; see figure 1(1)].

Eyes. The upper and the lower eyelids frame the almond-shaped eye orifice. The wide end of this orifice is the outer corner of the eye and is raised, while the front end (the inner corner of the eye) is pulled down a little; therefore, the eyes are positioned somewhat obliquely.

The iris of the eye is dark brown, highly pigmented. This pigmentation increases at the iris edges, where it forms a darker ring with blurred outer configuration. Both the front corner of the eye and the mucous primary eyelid are darker; compared with those, the adjacent part of the white is lighter and in some places whitishly translucent.

The iris itself is perfectly round and usually is shaded from above and below by the eyelids. It reveals its real outline while the eyelids are raised or lowered (i.e., when chimpanzee is looking up or down). The iris is light golden-brown and is distinguished clearly against the background of the white. The pupil of the eye is round and small, but it enlarges in the dusk so significantly that only a narrow strip of the iris remains visible.

Bridge of the nose and the jaw part of the face. The bony nasal core, the bridge proper, begins from the point at which the above-the-eye arcs converge. At its end, beside the base of the protruding cartilaginous nose near the border of the jaw part of the face, the bridge broadens swordlike and infringes, in the shape of a promontory, on the cartilaginous nose, bulging in the form of a heart [we have already mentioned that in this region, the ninth group of furrows is located, 46–50; see figure 1(1)].

The jaw part of the face is defined distinctly and is distinguished clearly due to the strong protrusion of the cartilaginous nose and expansion of the jaw bones and also because this part is lighter (lack of brown pigmentation and dark hair). The jaw is oval-round *en face*; it begins with an arched cavity near the sword-shape part of the nasal core and encompasses the base of the cartilaginous nose.

It is separated from the cheeks by the tilted, sharp 17th furrow; it comes down, somewhat departing from the wings of the nose, proceeds to the outer corners of the mouth, and then narrows and finishes at the chin.

The jaw part of the face can be divided quite naturally into two unequal parts: the part containing the cartilaginous nose and the jaw part, which includes the wide upper and lower lips. These parts are separated from each other by the fairly deep, arclike 20th furrow (from the 5th group), which frames the cartilaginous nose [figure 1(1)].

Nose. The cartilaginous nose is a small, heart-shape knob that gently slopes in the beginning and comes down at the sides (at the wings of the nose) and in the front, where it ends with a short, thin shaft that completes the nasal partition.

The small elongated nostrils are almost entirely hidden as they go downward by the overhanging parts of the nose and are supplied with dense, light-colored, fluffy hair.

In profile, the cartilaginous nose of the chimpanzee strongly protrudes compared with the impressed bridge of the nose. Because of this, the outline of the chimpanzee's profile is reminiscent of that of a syphilitic [figure 1(2); plates 1(2), 2(1)].

The configurations of the nose itself are quite blurred, and they merge unnoticeably with the adjacent parts. The nose shape is fairly complex. In the form of a promontorylike knob, it is dissected in the middle by a deep furrow (19) that comes from the base of the bridge of the nose and cuts across its upper third to the highest point of the nose. At the sides and in the lower part of the nose, slightly off its edges, there is an unevenly deep arched impression. The lower dome of this arch is located exactly above the nasal partition, and its upper edges are above the nostrils. In three places, the arch has deeper impressions: where the arcs converge, forming a dome, and at the arch right and left ends (at the sides of the nose). Consequently, there are four cross-shape pits located near the edge of the upper part of the nose: at the top,[3] at the bottom, on the right and on the left. These pits are hardly visible on a calm face; however, even at a slight raising of the nose (which follows the raising of the upper lip), the upper and the side impressions deepen to such an extent that they make the nose's configuration more complicated. This is especially true for the upper part, where the nose takes the shape of a coil positioned across the whole upper nose part [figure 2(a) and plate 3(5)].

At the border of the nose and the jaw, there is a roundly outlined, dark-brown rim that slightly overlaps the 20th furrow, dividing the nose and jaw. This rim broadens at the outer edges; proceeds under the nose; bends up at the level of the nostrils; envelops the nostrils from outside and from the sides with two bending, narrow, pigmented strips that come up to the nasal mound; and then has a sharp arc to the upper part of the nose and to the upper ends of the arched impression. These pigmented strips then become even thinner and converge into an arch on the median line at the very top of the nose, forming a pigmented arc as if following the outlines of the arched impression. A thin pigmented strip also runs along the 19th furrow; it comes from the bony shaft of the bridge of the nose and cuts across the nose along the median line.

(a)

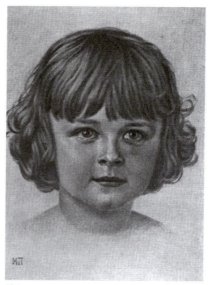

(b)

Figure 2. The faces of the infant chimpanzee and the human child: (a) my chimpanzee Joni (age 4 years); (b) my son Roody (age 4 years).

Lips. The jaw includes the lips, which swarm with wrinkles and therefore are exceptionally mobile. The lips frame a long, arched mouth.

The upper lip does not have a visible thick roll at the edge; it gently comes down to the mouth cleavage. However, after thorough examination, we can see that the lower edge of the upper lip differs somewhat from the adjacent parts. First, a thin (no wider than 1 mm) strip at the edge of the lips is covered by a slightly reddish mucous membrane and differs in color from the outside of the lips and particularly from the inner, considerably lighter, side of the lips. Second, this whole reddish strip is cut across by tiny transverse dots and has no hair (in contrast to the upper parts, which are covered with soft, white, short hair and with harder, dark hair becoming rare by the middle of the lip and dense by the mouth corners). Third, all the bigger wrinkles that cut across the lip along its entire length and bring it into motion stop short at the beginning of the mucous strip at the edge. Therefore, at a slight contraction of the lip, its very edge, because it does not participate in this movement, bulges more or less and forms a semblance of a lip roll. This roll thickens as soon as the lip stops moving [figure 2(a), static lips without the roll; plate 4(3), lips in motion with the lip roll highly visible].

As we have mentioned already, the whole upper lip from the beginning of 20th arc-shape furrow under the nose and up to the mucous rim, is cut by extremely rare, visible, longitudinal wrinkles. Three vertical wrinkles, converging at the top and diverging at the bottom, come from the corner of the under-the-nose furrow. At some distance from the under-the-nose furrow, two shorter wrinkles (less distinct in the upper parts of the lip, more distinct in the lower parts) from each side are parallel to each other and to the lower ends of the previous wrinkles. These wrinkles are from the sixth group of furrows (21–24). Eventually, at an angle to these last wrinkles, five parallel wrinkles come off the lower parts of the cheeks to form a tilted line in the direction of the outer corners of the mouth [from group 6a, furrows 25–29; see figure 1(1)].

The upper lip is cut by 17 wrinkles from groups 6 and 6a, 21–29. One of the wrinkles is central (21); it comes off the corner of the arc-shape under-the-nose furrow and proceeds straight down. Six other wrinkles are parallel to that central one (22–24). The rest of the 10 tilted wrinkles are parallel to each other (25–29), but are positioned at an angle to the previous wrinkles.

The entire upper lip (with the exception of the mucous strip at the edge) is covered with white, hairy down. Along its edges, the lip is framed with longer, harder, lighter hair that is straight or bent and has ends that more often angle down, but sometimes are sideways. In some areas in the midst of this hair, blacker and harder hairs are blend in haphazardly.

The lower lip, compared to the upper one, is shorter and wider. When the mouth is not moving, the lower lip looks as if it picks up the upper one and comes forth ahead of it [figures 1(1), 1(2)]. In contrast with the upper lip, the lower one contains, even in stasis, a conspicuous roll-like thickening. This lip roll is covered with a mucous membrane completely without hair and rife with transverse dots.

Compared with the mucous edge of the upper lip, the lip roll is wider and a more intense red. The entire lower lip, starting from the lip roll where it merges with the chin, is covered with white, soft hair. The hair adjacent to the lip roll is shorter and rarer. Nearer the chin, it becomes denser and bigger, forming a gray beard at the chin itself. This beard stands out against the contrasting black background of the adjacent side-whiskers and dark hairy cover of the chest (see portrait of chimpanzee, figure 2).

The skin of the lower lip and the chin, which usually appears through the not very dense thicket of hair, is bumpy. At the slightest forward movement of the lower lip, it forms parallel creases in the same direction as the creases of the upper lip. Therefore, 17 creases can be determined to come from the lip roll and across the lower lip. The creases are shorter near the outer corners of the mouth (they stop short of them) and become longer in the vicinity of the median line of the lips and chin (the eighth group of furrows, 37–45). As can be concluded from our description, the chimpanzee's lips differ from those of European humans; the structure of the lower lip is more similar to that of a human than the upper lip. To conclude our analysis of the chimpanzee's furrows, they are presented in the following list:

List of the Group of Furrows on Chimpanzee's Face

1st group   5 furrows on the forehead: 1, 2, 3, 4, 5
2nd group   3 furrows above the eyebrow: 6, 7, 8
3rd group   4 furrows at the outer corners of the eyes: 9, 10, 11, 12

4th group	4 furrows under the eyelids: 13, 14, 15, 16
5th group	4 furrows in the nose area: 17, 18, 19, 20[4]
6th group	4 furrows on the central upper lip: 21,[5] 22, 23, 24
6a group	5 furrows on the side upper lip: 25, 26, 27, 28, 29
6b[6] group	5 furrows framing the corners of the mouth: 30, 31, 32, 33, 34
7th group	2[7] furrows on the upper cheek: 35, 36
8th group	9 furrows on the lower lip: 37,[8] 38, 39, 40, 41, 42, 43, 44, 45
9th group	% furrows on the bridge of the nose[9]: 46, 47, 48, 49, 50

Thus, we found 50 furrows on the chimpanzee's face, of which 5 are unsymmetric (18–21 and 37) and 45 are symmetric, that is, there are 90 symmetrical and 5 unsymmetrical furrows (95 furrows total) [figures 1(1), 1(2)].

Ear. The first look at the face of our chimpanzee calls our attention to his enormous, light-color, semihard, thin-cartilage ears, which stand out clearly against the black background of the facial and head hair [plates 2(1), 2(4), 2(7), 2(8), 5(1), 5(2)]. The ears are bare; they are the same color as the chimpanzee's face. Only along the upper inner edge under the outer turn of the helix are the ears covered with fairly long, soft, dense black hair, which disappears at the lower edge of the ears. The outer edge of the pinna is completely without hair (if we do not count the white down that is visible only in transparent light). The length of the ears is somewhat greater than their width.

The outer acoustic duct is located at the same horizontal level as the base of the cartilaginous nose. The upper border of the pinna is slightly higher than the level of the above-the-eye arcs [plate 6(2)].

Chimpanzee's ear does not have a lobe (lobulus auriculae). The outer coil of the ear (helix) is outlined visibly only at the upper part of the ear; in the middle of the pinna, it suddenly becomes thinner and then fades. At the point at which the edge of the pinna becomes thinner, the pinna makes a sharp turn outward. As a result, the inner side of the ear also bulges outward, and the entire outer edge of the ear seems broken. In the inner part of the pinna, we find the strongly distinguished tragus and antitragus; deep incisura intertragica; very noticeable crus helicis, which is clearly separated by the inner roll (antenelix); cavum conchae; and crura antehelicis.

Usually, the ears are covered partially with the hair of the side-whiskers; they are disguised almost up to the outer edges when the side-whiskers are pushed to the sides. This happens when the chimpanzee is worried and therefore bristling. The ears are usually immobile and cannot move independently; however, during a run, they shake somewhat. The ears move when the chimpanzee is chewing energetically and sometimes when he is listening intently.

Chimpanzee's Arms and Hands   Our Joni's arm is about twice as long as his leg. Of the three parts that constitute the arm, the hand is the shortest; the upper arm is longer than the hand, and the forearm is longer than the upper arm. If the chimpanzee is in the utmost vertical straight position, his hands come down considerably lower than his knees [plates 7(1), 7(2)]; his fingertips reach to the middle of the shins.

The chimpanzee's arm is covered almost along its entire length with fairly dense, hard, pitch-black hair, which has different direction, length, and density on different parts of the arm.

On the chimpanzee's upper arm the hair goes down, and it is generally denser and longer then the hair on the forearm and hand. There is more hair on the back of the upper arm than on its inner side, which has translucent, light skin. There is almost no hair in the armpits.

On the forearm, the hair goes up; again, it is denser and longer than hair on the hands. On the inner side of the forearm, especially near the elbow joint and at the base of the hand, it is much rarer than on the outer side.

On the back of the hand, the hair almost reaches the second phalange of the fingers; the inner side of the hand is totally without hair and is covered with skin that is much darker than the skin of the face [plates 8(1), 8(3)]. The hand is very long: Its length is almost three times greater than its width; its wrist part is slightly longer than its phalange part. The palm is long and narrow; its length is greater by one third than its width.[10]

Fingers. The fingers are long, strong, thick, and appear inflated; they get somewhat narrower at the distal ends. The main phalanges are more subtle than the middle ones; the distal phalanges are smaller, shorter, narrower, and thinner than the main ones. If the fingers were arranged in diminishing order by the length, the order would be third, fourth, second, fifth, and first.

Examining the fingers from the back, we notice that all of them are covered with thick, rugged skin covered with hair only on the main phalanges. At the borders of the main and middle phalanges, there are significant swellings of the skin, which form soft, calluslike thickenings. The swellings between the middle and the distal phalanges are much smaller. The terminal phalanges end with small, shiny, slightly bulging dark-brown nails framed at the outer edge with a narrow dark strip.

In a healthy animal, the fingernail rim hardly runs over the flesh of the distal phalange; grown nails become gnawed off before long. Only in ill animals can be seen excessively long nails.

Lines of the hands. Compared with the hand of a young female chimpanzee described by Schlaginhaufen,[11] the line structure of our Joni's palm is more complicated[12] [figure 3(1), plate 8(3)].

The first horizontal line (1st or aa,)[13] is clear, distinct, and in the position and the shape shown on the diagrams, but becomes more complicated with additional branches. Soon after it comes off the ulna part of the hand (at its intersection with a vertical line V across from the fifth finger), the first line gives off a spur (1a) that proceeds to the base of the inner edge of the 2nd finger phalange and stops at the first line crossing at the finger's base.

The second horizontal line (2d or bb,) has an initial part that is located 1 cm more proximal than the first line; it starts with a small fork from vertical line V. This fork converges into one line that, at the point at which it crosses the vertical line III, slants sharply in the direction of the first horizontal line at the point of

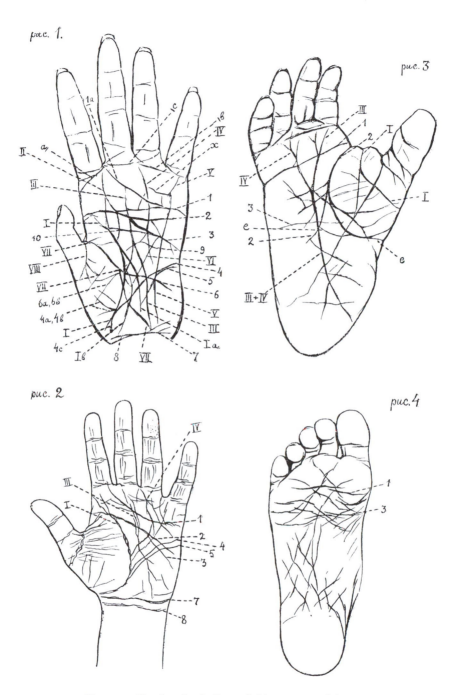

*Figure 3. Hand and sole lines of chimpanzee and human:
(1) hand lines of the chimpanzee Joni; (2) hand lines of the
human infant; (3) sole lines of the chimpanzee Joni; (4)
sole lines of the human infant.*

its intersection with vertical line II (dd$_1$), situated across from the axis of the index finger.

The initial part of the third horizontal line (3d or cc$_1$) is more proximal than the second line and starts from the very edge of the ulna part of the hand. Along its whole length, it has the tendency to ascend. At the intersections with vertical lines V and IV, the divergence of the third line from the second one is only 1 cm. At the intersection with vertical line III, the third and second lines merge completely. Incidentally, the beginning of the third line at the ulna edge of the hand includes a short horizontal branch. In the middle of its run (in the center of the palm), it breaks, and the 10th horizontal line (a detailed description is given below) can be considered its continuation.

The other, bigger, transverse lines of the palm, are discussed next. The fourth line (4th or gg$_1$) begins at the ulna part of the palm at the point at which the third horizontal line comes off and goes obliquely down to the first line (or FF$_1$); it crosses the first line, and forms three small branches. Two of these branches (4a, 4b) diverge, roll-like, at the base of the thumb's mound; the other one (4c) comes down to the seventh and eighth (ii$_1$) lines of the wrist.

Almost next to the initial part of the fourth line, parallel to it, there is a furrow; the fifth horizontal line of this furrow (at the point of its intersection with vertical line V) descends at an angle, crosses vertical line III, and almost reaches the first spur (1a) of vertical line 1.

The sixth horizontal line begins 1 cm lower than the previous one, goes almost horizontally and somewhat up, and ends soon after it crosses (at the point of intersection of the sixth line with line VII) the two weak branches 6a and 6b.

The seventh horizontal line (7th or hh$_1$) is located at the base of the hand and has two small branches that are at an angle and ascend the lowest part of the little finger's mound. The eighth horizontal line (8th or ii$_1$) is short and weak; it almost blends with the previous line, but is located lower and is more radia. The ninth horizontal line is hardly distinguishable; it passes right in the middle of the palm and is 1 cm more proximal than a segment of the 10th horizontal line.

The tenth horizontal line, which is located at the top and in the middle of the palm, is parallel to the second horizontal line (bb$_1$), and in its middle part (between vertical lines IV and II), it passes 1 cm from the previous line. The 10th line is, in my opinion,[14] a spur of the third line (cc$_1$).

Turning to the lines that cut across the palm vertically and at an angle, we have to mention the following. Vertical line 1 (FF$_1$) starts at a first crossing line (1 or at aa$_1$) 1 cm from the radial edge of the hand and, as it encircles the thumb mound with a wide arc, descends almost to the wrist line (hh$_1$).

In the area of the central part of the hand, vertical line 1 branches off. The first branch, 1a, departs at the level of its upper third, almost across from the ninth weak, transverse line, and slants inward to the medial part of the palm. Vertical line 1 crosses the fourth and sixth horizontal lines. The second branch (1b) of the vertical line 1 comes off 2 mm lower than line (1a), goes in almost the same direction, but ends slightly lower than 1a, reaching the seventh and eighth wrist lines (hh$_1$, ii$_1$) as if cutting them.

Sharp furrow VII, the most distinct of all the palm lines, starts from vertical line I inward, from a recess near the thumb. This line, which encircles the thumb's mound from above with a wide arc, crosses lines 1a and 1b (FF$_1$) somewhat lower than their middle part and descends obliquely down, reaches the seventh wrist lines, and crosses lines 4 (gg$_1$) and 1b.

Of the other more or less distinguishable vertical lines, four have to be mentioned. Short line II (ee$_1$ according to Schlaginhaufen) is situated in the upper fourth of the hand and goes in the direction of the second finger's axis. It begins between the second and third fingers and travels straight down, merging at its lower end with line I (FF$_1$) at the point at which a segment of the 10th horizontal line comes up to line I.

Line III is one of the longest lines on the palm (dd$_1$ according to Schlaginhaufen). It starts from the top in the form of a not very distinct furrow across from the middle finger's axis, cuts a spur of the first transverse line (aa$_1$), and sharply crosses lines 1 and 2 (at the point at which the second and third line merge). Then, it crosses lines 9 and 10, diverges from the ulna part of the hand, passes across the point at which lines 4 and 6 intersect, descends further, and crosses the ends of line 5 and of a branch of the seventh horizontal line, reaching the seventh wrist line.

Vertical line IV (kk$_1$ according to Schlaginhaufen), situated across from the fourth finger's axis, begins in the form of a weak furrow (visible only in good lighting conditions), comes off between the third and the fourth fingers, and makes a straight descent. This line becomes more noticeable right under the second line. Descending, vertical line IV sequentially crosses the third and ninth horizontal lines and imperceptibly fades, stopping short of the fifth horizontal line.

Vertical line V, the longest of all vertical lines, is located across from the fifth finger's axis; it begins from a transverse line at its base. Then, it descends, crossing transverse lines 1–6, and meets the tilted lines that come off the seventh line on the wrist.

On the upper part of the hand, above line 1 (aa$_1$), a small horizontal bridge x (between vertical lines IV and V) can be seen under a good lighting.

As to other clear lines of the hand, we have to mention long tilted line VI, which cuts across the lower part of the hand. This line begins from the lower branch of the second line, slants down to the points at which it crosses lines 1a, 1b, and the sixth horizontal line, and descends further to the point of its merger with 1b; it then goes to the seventh wrist line.

Now, we describe the lines at the base of the fingers. At the thumb's base, we find two tilted, diverging lines, VII and VIII, which meet at the big recess of the hand. From the lower of these two (i.e., line VII, which encircles the thumb), four radially diverging smaller lines descend. They are crossed in the middle of the thumb's mound by a thin transverse crease; the higher of these lines has already been described.

At the base of the index and the little fingers, we find three lines each. These lines begin separately at the outer edge of the fingers and converge near the inner corners between the fingers. Slightly above the bases of the middle and the fourth fingers, we find single transverse lines.

Apart from these lines, we find three additional arc-shape lines that pair the fingers as follows: second with third (a), fourth with fifth (b), and third with fourth (c).

1. An arc-shape line (a) goes from the outer edge of the second finger in the direction of the inner edge of the third finger; this line ascends to a transverse line at the latter finger's base.
2. An arc-shape line (b) goes from the outer edge of the fifth finger (particularly, from the medial transverse line of its base) in the direction of the inner edge of the fourth finger; this line comes up to the transverse line of the latter finger's base.
3. An arc-shape line (c) unites the bases of the third and the fourth fingers. Coming out of the corner between the second and the third fingers, it goes in the direction of the corner between the fourth and the fifth fingers (particularly to the transverse line at the base of the fourth finger).

We find double parallel lines by the bases of the second phalanges of the fingers (from the second to the fifth).

By the bases of all the nail phalanges of the fingers (from the first to the fifth), we again can notice transverse lines.

Hence, our Joni's palm, especially centrally, is furrowed by a fine net of 8 vertical and 10 horizontal lines discovered only in the course of a thorough and detailed analysis.[15]

The relief of our Joni's palm is considerably more complex not only when compared with the hand of a young female chimpanzee presented by Schlaginhaufen (this chimpanzee had 10 main lines at the most), but in comparison with the drawings[16] in my possession of the hands of other chimpanzees: a young male chimpanzee who lived from 1913 at the Moscow Zoo (judging by his appearance, he was somewhat younger than Joni) [figure 4(8)], an 8-year-old female chimpanzee nicknamed Mimoza [figures 4(3), 4(5)], and 8-year-old Petya [figures 4(1), 4(2)] kept (in 1931) at the Moscow Zoo.[17] In all these cases, as the drawings show, the total number of the main lines is not higher than 10.

Even a brief study of all these hands shows that, despite great variations in the relief of the palm; fading of certain lines and shifted positions of other lines; and despite the difference in drawings between right and left hands of the same individual [figures 4(1) and 4(2), figures 4(3) and 4(5), respectively], we can determine without any difficulty, using analogy, the identification of all the lines.

On all five prints, the position of the first horizontal transverse line (aa,) is the most undisputed and constant. The second horizontal line merges at its end with the first line [figures 4(1), 4(8)] or becomes independent [Schlaginhaufen diagram, figures 4(3) and 4(5)], or extends only a branch to the first horizontal line [figure 4(2)].

The third horizontal line varies more than the first and second by size [compare figures 4(5) and 4(8) with all others] and by location. While in figures 4(1), 4(3), 4(5), and 4(8), it occupies a quite autonomous position (in the last case, it

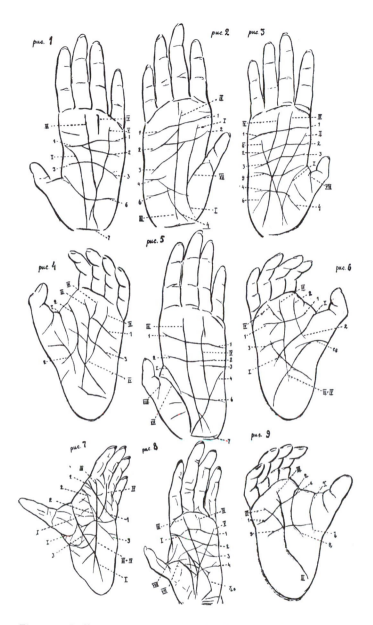

Figure 4. Difference in pattern of hand and sole lines for chimpanzees of different sex and age: (1) lines of left hand of chimpanzee Peter, 8 years old; (2) lines of right hand of Peter; (3) line of right hand of female chimpanzee Mimosa, 8 years old; (4) lines of sole of left foot of Mimosa; (5) lines of palm of left hand of Mimosa; (6) lines of sole of right foot of Mimosa; (7) lines of sole of left foot of a female chimpanzee, 3 years old; (8) lines of palm of left hand of the same ape as in (7); (9) lines of sole of right foot of chimpanzee Peter.

has only a weak twig going upward); in figure 4(2) (as well as for Joni), it joins the second horizontal line, totally merging with it in the radial part of the hand.

The fourth horizontal line, very conspicuous in Joni, is also quite distinct in figure 4(5). In figures 4(2) and 4(8), we draw only an approximate analogy based on the direction from the little finger's mound to the lower part of the thumb's mound and also based on the triple branching (we cannot exclude the possibility of confusing it with the fifth and sixth horizontal lines). This last transverse line 6 can be located unmistakenly and accurately in figures 4(1) and 4(5) as having exactly the same position and direction as in Joni. In figures 4(2) and 4(3), we are inclined to locate only its initial upward segment on the little finger's mound.

Staying with the horizontal lines on the drawings, we have to mention the lines located by the base of the wrist. Many of these lines are seen in figure 4(8); in figures 4(1)–4(3), we can find only a few. Another such line is the ninth line, which passes across the middle of the palm and is represented only once in the five cases [namely, in figure 4(3)].

Turning to the vertical lines of the hands, we have to point out that all of them are easily located on the basis of analogy, topographical position and mutual relations with the described lines of hands, although the details show certain differences from those in Joni.

Line 1 has the most constant position [as in figures 4(1), 4(2), 4(8)]. We can see in figures 4(3) and 4(5) that this line shortens and has the tendency to get near to [figure 4(5)], or perhaps merge with line VII [figure 4(3)].

Among other vertical lines clearly distinguishable are line III (represented in all the drawings and only sometimes deviating from its usual position across from the third finger's axis) and live V, which go toward the little finger.

Contrary to the case for Joni, line V in three cases does not retain its constant position (across from the fifth finger's axis), but goes in the direction of VI as though merging with it, accommodating segments of other vertical lines (IV, III, II, I). It is clearly visible in figures 4(3) and 4(8) and partially visible in figure 4(1). In two cases [figures 4(2) and 4(5)], line V is totally absent.

Vertical line IV, with one exception [figure 4(1)], is present, but it varies considerably by size and shape. It is either very short [figures 4(1) and 4(8)], broken and long [figure 4(5)], or diverges sharply from its usual position across from the fourth finger's axis [figure 4(3)]. Line II was observed to go to the index finger in only one case [figure 4(3)].

Chimpanzee's Legs and Feet

Whereas the hand and the foot of a human have profound structural differences due to the differentiation of their functions (walking and gripping), in apes, particularly in chimpanzees, the difference in structure of the extremities is considerably less. The hand, apart from its main function of gripping, participates in movement on the ground; the foot also takes part in gripping [plates 4(1)–4(4)].

The chimpanzee's leg is much shorter, thinner, and structurally more subtle than the arm [plates 7(1)–7(4)]. The thigh section of the leg is somewhat longer than the shin section; the shin section is twice as long as the foot section.

Joni's entire leg, to the very base of the toes, is covered with black, shiny, rather short hair that is rarer and shorter on the inner sides of the thigh and in the groin. On the outer sides of the thighs, the hair goes down and to the back; on the inner sides of the thighs, the hair goes up and to the front. On the entire shin, the hair is of the same length and density. Hair on the foot is rarer, especially along its edges and near the base of the fingers. On the sole and at the sides, the foot is totally bare. The part of the foot covered with hair and the hairless part are separated by an almost linear border; hair on the edge looks as though it has been cut. The main phalanges of the toes bear rare tufts of dark hair.

Let us now describe the toes [plates 9(1), 9(3)]. The first toe is considerably thicker than the thumb, but their lengths are almost equal; it stands widely apart from the other toes and is shorter and thicker. The rest of the toes do not differ much from each other; all of them are fairly thin (considerably thinner than the fingers; the littlest is the thinnest) and almost equally long. The third toe is 0.5 cm longer than the others; the second and the third toes are almost the same length; the fifth toe is 1 cm shorter than the fourth. The toes end with dark-brown, somewhat convex, relatively miniature nails. The nail of the big toe, accordingly, is larger.

The sole is also furrowed with lines despite the fact that the animal rests on it; more than that, these lines are deeper and more distinct than the lines of the hand. Comparing the relief of the palm and the sole, there are the following, very sharp, differences.

Lines of the sole. While the palm is almost evenly covered with lines (the thickest net is located in the central part of the palm), on the sole we can see an entirely different pattern. Lines are most abundant in the area near the big toe; the half of the sole beyond the line separating this mound is covered with thicker, tighter, smooth skin that is almost without furrows, except for a few short lines coming off a tilted line that passes along the entire sole (joint branches of vertical lines III and IV) and ends near its radial edge [figures 3(1), 3(3)].

Comparing the relief of Joni's sole with those presented by Schlaginhaufen and in three chimpanzees of the Moscow Zoo [figures 4(4), 4(6), 4(7), 4(9)] we find in our chimpanzee more lines and more complex lines than in other observed individuals. The main eight lines of the sole, denoted by Schlaginhaufen, can be located easily in all five cases despite certain variation in their shapes and interrelations.

As we observed in our analysis of lines of the hands, the right and the left soles of the same individual show considerable difference in the configuration and location of certain lines [figures 4(4) and 4(6)].[18] Other lines have fairly constant appearance and topographic position.

Among these lines, we first mention the first transverse line (1 or aa_1) situated in the front part approximately 2 cm from the bases of the second to fourth toes. In three cases [figures 4(7) and 4(9)], this straight line cuts across the sole from edge to edge; in Mimoza [female chimpanzee, figures 4(4) and 4(6)], it is shorter and does not reach the edge.

Less certain is the location of the second transverse line (bb_1), which comes off at a greater[19] or a smaller[20] distance from the first transverse line or even going way up and crosses line I [figures 4(6) and 4(9)]. The second transverse line takes different shapes and directions. It is (as in Schlaginhaufen's diagram) an almost straight line down to the center of the sole and almost touches an end of vertical line II at the point at which line II crosses the third horizontal line cc_i [as in figure 4(7)]. Or, it crosses the first vertical and first horizontal lines [figure 4(6)], or as for Joni, it forms strong branches that give a series of transverse bridges to a long line that passes across the entire sole and consists of two merged vertical lines, III and IV. Alternatively, it is independent in a vertical direction and crosses the first and third transverse lines [figure 4(9)].

The third transverse line (cc_i) often has a more constant position, with its central part either at the point of intersection of the second horizontal line (bb_i) and vertical lines III and IV [as in Schlaginhaufen's diagram and in figures 4(7) and 4(9)] or at its meeting point with the second line (bb_i) and with a joint branch formed by lines III and IV (as for Joni). Mimoza's case [figures 4(4) and 4(6)] differs from the cases described above. The third line is somewhat shifted to the outer edge of the sole, and although it is located in the central segment of the sole, it is sometimes above the intersection of lines III and IV (4) and sometimes below this intersection [figure 4(6)].

A weak horizontal line (ee_i according to Schlaginhaufen) can be found in our study with great difficulty. This is a first line from the base of the big toe, and it travels some distance from line FF_i, encircling the mound of the first toe [figure 4(6)]. This line is clear in Joni. It merges with a branch coming off the second line (bb_i and separates (by a wide semicircle) a cushionlike swelling by the base of the big toe, sharply distinguishing this toe from the rest of the sole.

Turning to the vertical lines of the sole, we find an almost perfect analogy with the pattern presented by Schlaginhaufen. In all diagrams, we can see clearly vertical lines III and IV and their joint spur, which passes almost across the entire sole. In four cases, the form and topographic position of these lines are almost identical with those on Schlaginhaufen's diagram. In figure 4(7) only, we find a certain shift of these lines to the inner edge of the sole. Vertical line I, which separates the big toe's mound, is always present, but its form constantly changes. In Joni, as in Schlaginhaufen's young female chimpanzee, this line is a semicircle, and it encircles the big toe apart from the second line (bb_i) inward. In other chimpanzees [figures 4(4), 4(9)], it goes straight down; in two cases [figures 4(6), 4(7)], it deviates from its usual course and goes deeply inward to the outer edge of the sole, crossing the long branch of joint lines III and IV.

To summarize the line patterns on the chimpanzee's soles, we have to point out that the lines on the sole are less developed than those on the palms. We marked 10 main lines on the palm, but we could find no more than 7 on the soles. On the hands, these lines are concentrated mainly in the middle and the lower parts of the palms. The lower part of the sole is almost without lines, and one part of the heel is completely smooth.

There are individual variations in the number, shape, and boldness of the lines. The greatest number of sole lines was for Joni and a young (died in 1917) female chimpanzee [figure 4(7)]; against the background of seven main lines, we found a host of additional small branches. In the cases of a female chimpanzee presented by Schlaginhaufen and both male and female chimpanzees [figures 4(4), 4(6), 4(9)] from the Moscow Zoo (Mimoza and Petya),²¹ we found bold, almost sketch-like, lines.

Chimpanzee's Body in Stasis   1. Sitting Positions. Let us describe now the quite usual changes in the chimpanzee's body positions and in the movement of the extremities. We have to emphasize that, in normal conditions, an infant chimpanzee is a very mobile creature and can be observed walking, running, or climbing more often than standing or sitting.

The poses of a sitting chimpanzee are fairly diverse. When sitting on an even surface, the most characteristic positions are the following [plates 10(1)–10(3)]. The chimpanzee bends his knees at a sharp angle, presses his thighs to his trunk, and rests on his feet. In this case, his arms often hang freely at the sides of his legs [plate 10(3)] or are folded on his bent knees [plate 10(2)]. Sometimes, his hands, with the fingers bent, rest firmly on the ground [plate 10(1)]. Fingers 2 to 5 are bent predominantly in the first joint and rest on the ground on the back sides of their second phalanges. The third phalanges are bent inward somewhat and raised; the first finger does not reach the ground. Sitting on an even surface the chimpanzee often has his hands firmly resting on the ground; they are strained as though ready to help him rise from the ground and change his passive position to a more lively one.

A second sitting position [plate 6(2)] is less characteristic of a chimpanzee. Sitting on an even spot, he bends his legs in a different way: one at a sharp angle, the other at a blunt angle. His first leg is pressed to his body and rests entirely on his foot. His second leg is somewhat apart from his body and in a tilted position. Therefore, he cannot press the whole foot to the floor and rests only on its heel or on the outer edge of the foot. Sometimes, he puts one leg, bent at the knee, in a vertical or tilted position (now resting on its foot, now only on its heel), lays his other leg on the floor, moves it off the body, and then touches the floor with the outer edge of its foot. In all these cases, one of his hands always rests firmly on its bent fingers, the other one is in a freer position: It either hangs down to the floor or lies on a bent knee [plate 6(2)].

The chimpanzee's much rarer pose is when he sits with his legs apart from his body and pressed flatly to the ground [plate 10(4)]. In this case, his knees are slightly bent, his soles are open, and he touches the floor only with the outer edges of his feet. Most rarely, the chimpanzee sits, extending his almost totally straightened legs parallel to each other on the floor [plate 1(2)].

In the last two cases [plates 1(2), 10(4)], the toes either are bent or are pressed into a firm fist; the hands hang freely and touch the ground. The fingers on one of the hands are bent, as usual, at the first joint; they rest with their back side on the ground and are ready in case the animal wants to stand up.

In all the examples described, as illustrated by the original photographs of animals, the variety of poses of a sitting chimpanzee was represented mainly by change in positions of the legs. Positions of the hands also vary considerably; this diversity often determines the expressiveness of the pose [figures 5(1)–5(6)].

If the chimpanzee is totally calm and prepared to sit still, he snugly folds his arms on his raised or lowered knees [figures 5(1), 5(2)]. Sometimes, the chimpanzee rests his folded arms on his closed and bent knees and leans his head to the side; therefore, he looks melancholy and as though compassionately sad [figure 5(4)]. Sometimes (very rarely), the chimpanzee assumes a very unusual pose: With one of his arms bent at the elbow, the hand, the wrist, and all three phalanges and with his knee resting on his bent leg, the chimpanzee rests the side of his chin on the back part of the bend of the hand deflected outward. He lays his other elbow on the knee of his firm right leg, moving his other hand under the elbow of his first arm [figure 5(4)]; his trunk leans to one side, and the axis of his body slants. When looked at briefly, in this pose he resembles a meditating human; however, a human is known to rest the chin more comfortably and firmly on the inner palm part of the hand, enclosing the face in the fingers. Therefore, this pose of the chimpanzee looks exaggerated, artificial, even comical.

No less peculiar is the pose of the chimpanzee when he sits, like a dummy, with his arms and legs in perfect symmetry; his arms, bent at the elbows and hands, lie on his knees. He looks mummylike and totally immobile; he usually contemplates impassively. However, this pose is rare, unsteady, and uncharacteristic [figure 5(5)].

As already noted, his usual pose is more potentially dynamic. He sits, resting on bent fingers and with his arms maximally extended in front of him. If, leaning his trunk and head, he slightly draws forward, you can be sure that he is worried and may take off at any moment [figure 5(6)].

Among other more artificial poses of the chimpanzee, we have to mention the two following: when the chimpanzee sits on the lap of a human and on a bench [plates 10(5) and 11(5)]. In both these cases, as can be seen on the photographs, the chimpanzee's poses correspond to those of a human child, and only the slouchiness of the infant chimpanzee, which is determined by the way his head is attached to the trunk, tells us that this is a different offspring of *ordinis primatum*.

We also have to underline that the chimpanzee's leg is extremely mobile, but not as much in the knee as in the hip joint. A sitting chimpanzee easily can move his leg, bent at the knee, so high that the sole is at the level of his shoulder [figure 6(3)]. The chimpanzee, standing on his one completely straightened leg, can raise the other leg so high that the raised leg forms a blunt angle with the straight one. However, to stay in this position for a while, the chimpanzee must hitch his leg to a firm object and change his vertical position into a tilted one.

2. Standing.   Of the more natural poses of a *standing* chimpanzee, we first have to mention his most usual and stereotyped pose. The chimpanzee stands and rests his soles, the spread toes of his semistraightened legs, and the bent (at the second phalanges) fingers of his straightened arms on the ground [plates 4(1), 4(3), 4(4)]. At the same time, his trunk slants, his legs are placed apart arcwise at a distance from the middle axis of his body, and his arms are symmetrical to that axis and are positioned in front of his legs. This pose is extremely firm, but

рис. 1

рис. 2

рис. 3

рис. 4

рис. 5

рис. 6

Figure 5. Sitting postures of the chimpanzee: (1)–(5) static postures of the sitting chimpanzee; (6) potentially dynamic posture of the sitting chimpanzee.

Figure 6. The chimpanzee's arm tenacity and leg mobility: (1) the chimpanzee climbing to bar with his hands; (2) the chimpanzee ready to swing; (3) leg mobility of sitting chimpanzee; (4) leg mobility of standing chimpanzee.

it is less comfortable for such an energetic and sanguine animal as the infant chimpanzee; it is too restrictive for him.

First, it is not comfortable because, slouchy as he is, the chimpanzee stoops even more; his head falls so low that its highest point is at the level of the beginning of his nape and his eyes are at the shoulder level, so that his field of vision narrows and becomes limited to what is immediately before him. When, as though hampered by this pose, the chimpanzee wants to broaden the sphere of his visual observation, he raises his eyes; it is not a mere coincidence that in both photographs illustrating this pose, he has this slanting look [plates 4(3), 4(4)].

When even minimally alert, the chimpanzee does not limit himself to raising his eyes, but straightens his entire body into a more vertical position, takes his hands off the ground, raises his head, and points his eyes upward [plates 7(1)– 7(3)]. At this moment, he has to bend his knees to remain steady, but even then he cannot stay in such a position for a long time and must resort to the use of at least one of his arms. He often places a lowered arm, slightly bent at the elbow, in the middle between his feet and slants his trunk again; then, resting on his bent fingers, he can stand for a while, freeing his other hand for touching, gripping, extending, and the like [plate 4(2)].

Only in an unusual situation and for a very short time can the chimpanzee stand in a completely vertical position, not touching the ground with his hands, not using the support of his arms, resting almost entirely on his arc-shape and pulled-apart legs, and firmly touching the ground with his sole, with a spread big toe and with the tips of the other four toes (2nd to 5th) slightly hunched [plates 7(1)–7(3); figure 7(2)].

This pose is characterized by such unsteady balance that the chimpanzee is ready to fall onto his hands or to lurch any second. To remain standing at least for a short time in this obviously uncomfortable and artificial pose, like a human unused to standing on tiptoe, he has to balance himself with his arms and trunk so he does not fall down [plates 7(2)–7(4)]. For the sake of greater firmness, he often transfers his center of gravity to the outer part of one of his legs; for this purpose, he has to take his big toe, and sometimes the entire inner part of the foot, off the ground [plate 7(2)]. Now, his vertical pose is more steady; he can straighten himself, raise his head higher, and broaden his field of visual observation [plates 7(1), 7(2)].

Chimpanzee's Body in Motion

1. Walking. Standing in the position just discussed, the chimpanzee sometimes tries to walk [plates 7(3), 7(4)], but he can hardly take two or three steps, then he loses his balance and has to rest on his arms so he does not fall. A vertical, free, long walk is impossible for the chimpanzee.

Even if the chimpanzee is held by the hand and supported by a human, he cannot walk using only his legs as a human child of the same age would do easily; instead, he must rest on one of his arms [plate 13(3)].

рис. 1.

рис. 2.

Figure 7. Standing postures of the chimpanzee: (1) the excited chimpanzee stops his walk; (2) the excited chimpanzee after arising erect.

Prior to walking, the chimpanzee stands on all for extremities [plate 12(2)]. Without changing the position of his head and trunk and only moving his arms and legs, the chimpanzee can walk with his body at an angle to the ground, resting on the back sides of the second phalanges of his fingers, on his feet, and on his spread toes; also, he usually bends his knees more than when he stands motionlessly [plates 12(1), 12(2)].

Stepping slowly, chimpanzee bends less and places his hands and feet at a fairly short distance, putting forward his one hand (for example, right) and the opposite (left) foot simultaneously. At his next step, the other pair (left hand and right foot), positioned crosswise and so far motionless, come into action [plate 12; figure 7(1)].

2. Running.   While walking fast or running, the chimpanzee bends closer to the ground; his trunk assumes an almost vertical position, and his head falls lower. For example, extending his left arm and right hand forward, the chimpanzee rests on the bent fingers of his extended left arm and on the heel of his extended right leg. At this moment, his right arm stands near or even somewhat behind his right leg, and its bent fingers rest on the ground; his entire left foot is pressed to the ground. Thus, his right extremities are maximally close; his left ones are maximally spread [plate 12(3)].

At the next moment, retaining the same position of his left arm, the chimpanzee eases the pressure on his right hand, which hardly touches the ground [plate 12(4)]. Then, he transfers the support of his body onto the middle part of the sole and onto the toes of his right foot, taking his left foot off the ground, raising the heel and the middle part of the sole, and resting only on the toes [plate 12(4)].

At the following, third, moment, the chimpanzee takes his right hand [bending it at the wrist; plates 12(5), 12(6)], and his left foot (touching the ground only with the tips of its toes) completely off the ground, as though getting ready for a few steps forward. He positions his extremities at an angle so that later—at the fourth moment of walking—he can extend forward, as far as possible, the freed hand and the leg just removed from the ground [plate 12(6)].

As we noted already, the chimpanzee is a very mobile creature; slow walking is quite uncharacteristic. He moves at a steady pace only in unfamiliar areas, or when he is alert and has to turn around often to observe the surroundings. In usual and familiar circumstances, a chimpanzee, like a playful child, prefers running to walking. In the speed of his movements, he undoubtedly exceeds not only a human child of the same age, but also a human adult; even a good runner cannot keep up with a chimpanzee.

3. Climbing.   The very high mobility of the chimpanzee is manifested not only in running, but also in climbing. In this kind of movement, the chimpanzee by far exceeds the most skilled child, which is no wonder, considering the chimpanzee's natural tools for climbing: his hooks for hitching; his powerful, muscular arms and hands with long, strong, tenacious fingers covered with thick skin; and his feet with the outstanding first toe. These provide a clear advantage over the

child's weak arms, legs, hands, feet, fingers, and toes, which are covered with thin skin. Hitched with his hands and feet to any object and, hanging with his trunk in the air, the chimpanzee can stay in such a position for a long time, evidently not feeling uncomfortable [plate 14(1)].

For the chimpanzee, climbing is a usual and desired pastime, the same as running. Naturally, the chimpanzee can climb high stairs extremely well, easily ascending on all fours and descending as easily. Even in captivity, the chimpanzee uses every opportunity to do a little climbing. He hitches one or both hands to every plank and hangs and swings on it for a long time with his legs pulled up [figures 6(1), 6(2)]. He climbs easily and freely the smooth trunk of a tree, embracing it with his hands and toes [plates 14(2), 14(3)]. The first time I helped him to a tree, he became scared and began crying [plate 14(2)], but very soon he calmed down and started quickly and fearlessly to climb up the tree, which was not a very high one; a happy smile did not leave his face [plate 14(3)].

During tree climbing, the chimpanzee's arms and legs embrace the trunk from all sides, which guarantees the firmness of attachment and the speed of movement. One arm encircles the trunk from behind; the other encircles from the front. One leg holds the trunk from the right; the other holds from the left [plate 14(3). Climbing is mainly done with the hands: The chimpanzee uses his hands first; they bear the bulk of the weight of his body. Moving his hands along the trunk, shifting them higher and higher, the chimpanzee can climb remarkably fast; in this case, the function of the legs is only auxiliary. The opposite is true for the descent. Here, the most active role belongs to the legs [plate 15(2)].

Using his reliable hooks, the chimpanzee easily travels up wooden structures. He quickly climbs smooth poles that lead to high roofs [plate 15(1)], rises up their slopes to the very top, and carefully walks down the sharp ridge of the roof. Near the edge, the chimpanzee looks down apprehensively and even more carefully comes down the steep slopes to the fences; moves along the upper edge of the fence, nimbly avoiding the sharp sticking nails; gets onto the upper planks of the gates; and comes down the poles, garden gates, or anything else that is available [plate 15(2)].

Chimpanzee's untiring mobility and constant change in his facial expression reflect his sanguine, lively temperament.

Chimpanzee's Face in Motion

The chimpanzee's face is rarely passive; it changes every moment, and sometimes these changes are quite expressive. We can distinguish eight most evident changes in the facial expressions reflecting eight various mental conditions of the animal: general excitability, sadness, joy, anger, fear, repulsion, surprise, and attention [plates 3(1)–3(8)]. His expression of general excitability (excitement) is the most spectacular and original [plates 3(1), 5(1)–5(6); figure 8(2)].

1. General Excitability.   In this case, the chimpanzee's lips contract in the corners, extend forward, and open at the end in the form of a funnel. Side-whiskers

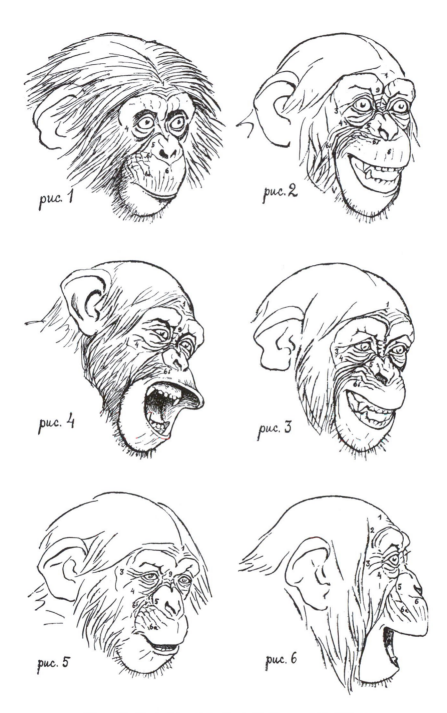

Figure 8. Typical location of wrinkle furrows with different facial expressions: (1) attention; (2) excitement; (3) narrow smile; (4) broad inviting grin; (5) laughing; (6) crying.

come up sharply and frame the cheeks with a rim of the hair that sticks out. The chimpanzee's eyes are open wide, and his intense, fixed glance is riveted to an object that stimulates his anxiety. The 17 facial furrows along the lips and weakly accentuated on the chimpanzee's calm face become deeper and longer and merge with the wrinkles coming off the eyelids, cheeks, or nose. These furrows cut across the cheeks and the upper lip in the form of parallel creases, which are deeper and wider on the cheeks and thinner and denser on the contracted upper lip.

These creases have the following pattern. Three come off the corner of the under-the-nose furrow (as on the calm face of the chimpanzee) and go toward the edge of the upper lip's funnel, only this time closer to each other. The next two creases[22] come from the point under the eyelid at which the fourth and fifth furrows end[23]; they enclose the nasal mound with a semicircle and, as they descend onto the upper lip, lie parallel to the three previous creases. In the upper area of the cheeks, they advance so far that they almost reach and seem to merge with the two wrinkles that come off the outer corners of the eyes and go across the cheeks at an angle (35 and 36). The next three creases are located outward from the previous ones and are parallel to them.[24] Outward from the last two wrinkles, at some distance from them in the very corners of the mouth, two weak wrinkles fan out; they disappear in the lateral, dark hairy parts of the cheeks.

The 17 furrows of the lower lip gather into the densely located creases near the lower half of the funnel; these furrows somewhat diverge at the sides of the chin and near the corners of the mouth.

2. Sadness.   The chimpanzee's face presents quite a different sight when he is sad or crying [plate 3(7)]. When he cries, he tosses back his head, opens his mouth wide, and closes his eyes completely [plates 16(5), 16(6)]. In this case, his facial expressions change beyond recognition compared with what has been described for a calm or worried animal. Now, it seems that not a single feature of the face remains idle; from top to bottom, everything moves and shifts from its usual position. The partially bald skin of the forehead, usually touching the above-the-eye arcs, now hangs over these arcs so that a sharp line that separates the forehead from the arcs splits them with a deep furrow, and the arcs themselves appear flattened. While the skin of the forehead and above-the-eye arcs slips down, the nose jerks up so high that the nasal mound sometimes[25] almost reaches the lower eyelids. The part of the face enclosed between the eyes and the nose (the bridge of the nose and the upper part of the cheeks) becomes so squeezed that all its wrinkles shift closer to each other, forming deep creases that cover this entire area.

However, despite being so close, the following furrows can be singled out. The bridge of the nose area is cut by five cone-shape creases.[26] The wrinkles of the closed upper eyelid gather in a double round crease that follows the eyelid outline and descends at the eye corner in the form of two parallel straight strips. Four main wrinkles under the lower eyelid become deeper and closer, give out additional (as though intermediate) thinner branches, embrace the upper part of the cheeks with wide semicircles, come down to the lower jaw, and become thin and fade out in the dark hair of the side-whiskers [plate 16(5); figure 8(6)]. In the front of these five wrinkles there are five more; these are located on the light part of the face. These wrinkles come off the sides of the nasal mound and are parallel to each other in the direction of the chin, somewhat apart from the corners of the mouth.

When the chimpanzee cries loudly, his mouth opens so widely that it assumes the shape of a large oval.[27] His gaping gullet opens so deep that it uncovers the entire inner mouth cavity. The pharynx usually is not visible because the tongue is pulled into the gullet. The tongue, positioned at an angle, rests its end on the center of the soft bottom of the mouth, and its middle rests on the had palate [plates 3(7), 16(6)].

The upper lip is jerked up to the nose, the lower one pulled to the chin, so that not only teeth, but also the gums[28] become totally exposed [plate 3(7)]. Both lips are stretched to the limit; they cover the gums so tightly that they seem on the verge of bursting. All the lip wrinkles vanish. In the middle of the upper lip, on the smooth skin, a light patch even can be found under a-certain lighting. On both lips, the hairy mounds distinguish themselves; their hair bristles. The long hair of the head and the side-whiskers is let down and pressed to the body.

Sometimes, the upper lip covers the nose so that the nasal core seems to fall into a hollow formed by the overlapping lip. The edges of this hollow merge with the nasal mound. The nose is compressed by the raised upper lip from below and the descending creases of the bridge of the nose from above. These creases push the cutting furrow on the nose so hard that the upper part of the nose and almost half of its length become divided into two parts and contain V-shaped hollows.

The chimpanzee never sheds tears when crying.

3. Joy.   The chimpanzee's joyful mood is characterized by an expression that corresponds to that of the human laugh [plates 2(7), 2(8), 3(8); figure 8(5)]. In this case, as when chimpanzee cries, the head often is tossed up, and the mouth is open wide; the similarity ends there. In all other details, the face of a laughing chimpanzee differs considerably from that of a crying animal.

Now, the chimpanzee's eyes are open and especially bright, but the look is indefinite and unsteady. The upper part of the face is smoothened; the first three under-the-lid wrinkles have their usual positions. The mouth is open, but not as tensely and widely as when the chimpanzee cries. It is particularly interesting that the corners of the mouth are stretched not to the sides, but somewhat up-ward, and the teeth are almost invisible because they are covered partially with the lips [plate 2(8); figure 8(6)].

The near-the-mouth wrinkles (16, 17, and the five lip wrinkles), encircling the corners of the mouth, are less tense and compressed than during crying; these wrinkles have different directions compared with crying. In a crying chimpanzee, the near-the-mouth wrinkles, slant down almost immediately as they come off the bridge of the nose. Here, the 16th and 17th wrinkles, bypassing the nasal mound, make a wavy turn to the corners of the mouth. From the light part of the face, almost from the middle of the upper part of the upper lip, they are abutted by five semicircular wrinkles that indefinitely and widely depart from the side of the under-the-nose furrow. These wrinkles disappear at the sides of the lower jaw and in the dark hair of the side-whiskers. The lower lip is stretched out, and all the furrows are smoothened. The lip edge forms a wide, arc-shape turn that determines a new shape of the mouth, a laughing expression; the edge of the upper lip participates, only insignificantly, in this change of configuration.

4. Anger.   The anger expression is entirely different from that of crying and laughing; the dissimilarities can be established easily [plate 3(5); figure 9(3)]. Now, the chimpanzee brings the front corners of the skin of the above-the-eye arcs closer to each other so that a furrow between the arcs becomes deeper and infringes, in the form of a promontory, on the base of the bridge of the nose. The chimpanzee's eyes are open; the inner corner of the eye is compressed strongly and turned down, and his look is fixed steadily [plate 17(2)]. The corners of the mouth are stretched tensely upward; the teeth and gums are exposed partially. It is remarkable that, while in the crying condition, the gums are exposed mostly in the incisors' area, yet in this case, the lips come down on the incisors, and the gums are uncovered mostly in the area of the canines (particularly the upper canines).

The bridge of the nose is not wrinkled, but the promontory at its base (at the root of the nasal mound) is jerked up and pressed into the under-the-lid wrinkles, so that second and third wrinkles under the inner corner of the eye also bend in the shape of a promontory [figure 9(3)]. The coil at the upper part of the nose is distinguished clearly. The under-the-nose furrow becomes deeper, turns up, and sharply separates the nasal mound, which takes the shape of a diamond. The lower corner of that diamond coincides with the nasal partition. The upper corner (truncated by a V-shaped cutout) coincides with the lower end of the bridge of the nose; the right and the left corner coincides with the point at which the first of the five semicircular near-the-mouth wrinkles departs. The under-the-nose furrow, which turns at an angle at the sides of the nasal mound, slants down in the direction of the corners of the mouth; as it forms the second arc-shaped bend, it descends to the lower jaw [plate 17(2)].

The last furrow becomes part of seven deep, long, semicircular wrinkles (group 6b) that embrace the corners of the mouth, with its position in the anger expression somewhat different from that in crying and laughing [figure 9(3)]. In this case, two wrinkles (16 and 17), squeezed at the point of their departure from the bridge of the nose by the 14th and 15th under-the-nose upper thick wrinkles, deviate from the sides of the nasal mound, go toward the sides of the cheeks, and cut across the dark part of the cheeks in a deep semicircle. In the upper part of the cheeks, they pass close to the ends of the 14th and 15th under-the-nose parallel wrinkles, which now move up to the nose, mask the beginning of the lower 16th and 17th wrinkles, and bend in an angle above the V-shape protrusions of the nose.

Four near-the-mouth wrinkles (group 6b) located below the under-the-nose furrow are in parallel rows, almost following the course of that furrow, beginning somewhat separately from the sides and from the bottom of the furrow and

Figure 9. Typical changes in the location of face furrows with different facial expressions of the chimpanzee: (1) timidness; (2) fear or terror; (3) anger; (4) rage; (5) disgust; (6) astonishment.

almost reaching the middle of the central promontory-shape bend. Then, they go up a little, turn at an angle under the side corners of the mouth and above the upper canines, proceed down, bending, become thinner near the corners of the mouth, and run down the chin in the form of four thinner creases that abut on the three previous near-the-mouth wrinkles.

Rage. When anger develops into fits of rage [plate 18(3); figure 9(4)], the upper part of the face remains almost unchanged; only the eyes become rather narrower than in anger. However, the mouth and jaw area change drastically. The jaws open wider, the mouth cavity is much larger, the lip corners approach the canines, and the mouth stretches less in width and more in height.

The lower lip and particularly the upper lip bulge forward so much that they expose not only the teeth, but also the gums almost down to the base. The upper lip hangs, like a visor, over the mouth, demonstrating the chimpanzee's offense weapons—his teeth and his vigorous assault—as though ready to be used. Interestingly, the hair on his head and face does not bristle even slightly.

5. Fear.

Timidity. The fear expression of the chimpanzee is very peculiar. Every symptom of timidity in the chimpanzee is accompanied immediately by the fluffing up of the hair of his face, of the crown of his head, and especially of the side-whiskers. The head appears almost one-and-a-half times bigger and is framed by the crown of sticking out hair, which almost covers his enormous protruding ears. The chimpanzee's eyes usually are open wide [plate 19(2)], and his look is aimed in the direction of the object that poses a threat; his eyes are literally riveted to it [figure 8(1)].

The configuration of the lower part of the face is especially original. The chimpanzee closes his teeth shut, pushes them slightly forward, and draws the corners of the mouth into the mouth. His upper lip becomes so tight in the transverse, circular direction (along a line about 2 mm from the lip edge) and so pressed inward that it bulges like a hump, making the under-the-nose furrow deeper [plate 20(1)]. This swelling of the upper lip is so significant that, looking at the chimpanzee in profile, one can watch his bulging lip sticking forward more than his cartilaginous nose.

Among the lip creases, we can recognize three central ones that come off the corner of the under-the-nose furrow and remain unchanged up to the upper edge of the lip. The rest of the longitudinal furrows that pass down the lip are as though cut across by transverse dots along the entire length of the furrows. These dots become visible due to the semicircular transverse near-the-mouth wrinkles perpendicular to the longitudinal creases.

This expression is characteristic of the initial stage of fear that we call timidity; maximal fear (our term is *horror*) is expressed much more intensely on the chimpanzee's face.

Horror. The horror expression of the chimpanzee is partly close to what has been said about anger [plate 17(2)], but under a more thorough examination, we can find sharp dissimilarities. The partial similarity is limited to the lower part of the face (it is associated with the shape of the mouth and the appearance of the teeth); the differences pertain to the eyes, eyelids, forehead, and nose.

In the horror expression [plate 3(6)], we observe the same tense stretching of the corners of the mouth to the sides and somewhat upward and the same exposure of the teeth and gums as in anger. But compared to anger and to the sadness and joy emotions described above, there is an interesting detail in the lip stretching. While laughing, the gums were completely covered, and the canines hardly visible from under the overhanging lips; in crying, the gums (and the canines) were open evenly and widely; in anger, we observed greater exposure of the upper gums (especially in the upper canine area). In fear, we see the greatest exposure in the area of the lower gums and lower canines. Thus, while the upper lip along its entire length envelopes the gums evenly and has a straight edge, the lower lip outline is curved. It slants from the corners of the mouth down to the canine, forming a blunt angle under the lower canine; in the middle of the mouth, it again comes near the base of the teeth. Five near-the-mouth wrinkles of the lip clearly are visible, but the lowest wrinkle does not reach the middle of the lip as it does in anger [plate 3(5)]. All these wrinkles do not form angled bends above the angled lateral curve of the under-the-nose furrow [compare figures 9(2) and 9(3)].

The nose is jerked up much more than in anger, making the upper lip bulge more and the under-the-nose furrow take a different course, with its middle curve round, not angled.

The upper part of the face undergoes the greatest change, especially in the bridge of the nose area, which is attacked from all four sides by the adjacent moving parts. It is pressed from below by the raised nose; from above, it is invaded by the skin of the above-the-eye arcs; from the right and left, it is approached closely by the second and third under-the-nose wrinkles and by the inner corners of the eyes. Therefore, looking briefly, one can see only the roll-like, thickened, shortened core of the bridge of the nose covered with transversally wrinkled skin.

The chimpanzee's eyes are especially expressive. While in a crying chimpanzee they are tightly closed, in a laughing animal, they are most often half-open. In a worried chimpanzee, the eyes are open widely and freely; in an angry chimpanzee, the eyes are open, but the eyelids in the inner corners of the eyes are closed tightly and are moved nearer to the bridge of the nose so that the eyes seem smaller, closer to each other in position and shape. In maximal fear, horror, the chimpanzee's eyes open widely and tensely in all directions, and his look is fixed frozenly on the source of his fear. Because of this, the eyes appear bigger and rounder than in anger. The inner corners of the eyes are also compressed and moved so close to each other that they almost converge at the median line of the bridge of the nose, joining its transverse creases [plate 3(6)].

Above the upper eyelids, following their outline, a big thick overhanging crease of skin gets thinner, fades out near the outer corner of the eyes, and grows near the inner corners of the eyes, running down to the squeezed corners of the eyelids. This bold framing of the chimpanzee's widely open eyes with the round creases formed by the closed eyelids and squeezed corners of the eyes, as well as with the transverse strips on the bridge of the nose, gives the impression that the chimpanzee is wearing a pince-nez because these enlarged eyes are outlined so clearly against the background of his face and so unusual are these creased strips connecting the eyes in the middle.

In horror, contrary to what we observed in moderate fear, the hair of the head and face is pressed completely to the body, and the side-whiskers stick out to the sides.

6. Repulsion.   The expression of repulsion is partially similar to the above-described expressions of anger and fear [plate 3(4)]. In this case, the upper third of the chimpanzee's face (the area of the above-the-eye arcs and bridge of the nose) is wrinkled as in anger, and the lip configuration is partially reminiscent of the initial stage of fear [plate 3(4)]. The chimpanzee's eyes look as usual.[29]

The corners of the mouth are squeezed and drawn inside, pressed to the gums even more than in timidity, but contrary to the timidity, the lips are not compressed, and the mouth is open a little. The upper lip, inflated in the form of a hump, evenly covers the upper teeth. The lower lip is stretched down in the middle part of the mouth and exposes the edges of the lower incisors, so the mouth assumes an almost rectangular shape.

As a result of the swelling of the upper lip and of the enveloping of it by the middle of the upper jaw, all the furrows on the lip and the under-the-nose furrow become smooth and appear only on the sides of the light jaw part of the face. They are patterned here in a network of rhombs formed by the intersection of the longitudinal and transverse near-the-mouth creases. Since the repulsion motion often includes elements of uneasiness and fear, we observe fluffing of the side-whiskers, but not as substantially as in typical fear [compare plates 3(4), 19(1), and 20(1)–20(4)].

Two other distinct expressions are worth mentioning: surprise and attention.

7. Surprise.   In the case of surprise [plate 3(3)], the chimpanzee's lower jaw hangs down and the mouth opens widely, but the lips are not drawn upward; therefore, all the teeth (even the canines) are covered absolutely with the lips. It is natural that, in such a calm position of the lips, near-the-mouth wrinkles do not form [plate 19(6)]. The upper lip is completely smooth in the front and slightly stretched to the sides; therefore, the under-the-nose furrow loses its sharp outline.

At the sides of the face where the jaw is light, there are four tilted wrinkles that go from the dark hairy cheeks toward the upper edge of the lip, but they do not reach the end of the edge. The upper part of the face does not change. The eyebrows, nose, and eyes look as usual; the chimpanzee's glance concentrates on the object that caused surprise.

Four under-the-nose wrinkles are in their normal positions, with the fourth wrinkle the boldest and most visible along a large stretch of its length. Three wrinkles in the eye corner are also bold (two upper wrinkles are extensions of the above-the-eyelid wrinkles; one lower wrinkle is an extension of one of the first under-the-eyelid wrinkles).

8. Attention.   The attention expression of the chimpanzee's face is entirely different [plate 3(2); figure 8(1)]. The upper part of the face almost does not change, but the lip configuration changes greatly. The appearance of the lower part of the face differs considerably from what we observed for other emotional conditions of the chimpanzee.

In this case, both lips are compressed tensely in the corners and at the sides of the mouth and are extended forward promontory-like; they slightly diverge at the end. The upper lip shows a small inflection that coincides with the beginning of the promontorylike extension of the lips forward and a little upward. The long creases along both lips cut across the entire lip surface, but do not reach the very edge of the lips.

On the lower lip, it is easy to locate the 17 furrows discussed above. The 17 furrows are clearly noticeable on the upper lip; these furrows were noted in the expression of general excitability, but in this case, they are bolder, and their outlines are a little different. In general excitability, the creases on the cheeks are thicker, bigger, and farther apart, and the upper lip creases are smaller and tightly compressed. In the attention expression, all these creases are almost evenly deep and thick [compare plates 3(1) and 3(2)]. In general excitability, the bold creases that encircle the nasal mound[30] make a sharp U-turn on the upper lip; this turn is followed, to a lesser extent, by all other adjacent creases. In the attention expression, the creases that encompass the nasal mound are broken and less noticeable; they approach the upper lip in even, tilted rows.

In this case, we can clearly see two upper cheek wrinkles (35 and 36, from group 7), located under the 16th and 17th under-the-eyelid wrinkles, which move upward to the very corners of the eyes. Here, we hardly notice the two wrinkles in the mouth corners that were so conspicuous in the excitability expression [figures 9(1), 9(2)]

In opposition to the excited animal, the hair of the side-whiskers is pressed to his lowered head; his eyes are open in the usual way and are fixed on the object commanding his attention.

Dual Facial Expressions

Facial expressions that we observe are rarely unequivocal and clear enough to exclude an indefinite or dual explanation of their meaning. Very often, we also find that the chimpanzee has mixed facial expressions in registering mixed feelings.

For example, Joni looks attentively at a horse that has appeared not far from him; at the same time, he is afraid of that animal. We watch his *curious attention* accompanied by *fear*, which manifests itself not only in his fixed enlarged eyes and his extended tightly closed lips (as in typical attention), but also in the rising of his head hair and sticking out of his side-whiskers [plate 19(2)] as in fear.

In surprise, the chimpanzee calmly would open his mouth wide [plate 3(3)], but if the surprise object (for instance, a live creature) at the same time provoked a joyful and lively feeling in Joni, he would pull down his lower jaw and expose his lower teeth, while his upper lip would cover all the upper teeth [plate 19(6)].

Playing with small animals with his hands, Joni pulls aside the corners of his lips in a smile and narrows his eyes [plate 2(4)]. If at this moment the animal begins to resist Joni in some way, angry feelings begin to appear and then prevail in Joni's joyous disposition. This immediately affects his facial expression, which now reflects the chimpanzee's vigor [plate 19(4)]. With his mouth in a grin, Joni exposes his teeth, but his gums remain covered completely. The upper part of the chimpanzee's face is not wrinkled, and he does not jerk up his nose as in typical anger [plate 3(5)].

An angry feeling accompanying sadness imparts a different shade to the crying expression of the chimpanzee. The chimpanzee in despair, expressing his mood in a wild crying frenzy, has the characteristic look shown in plate 3(7). If the chimpanzee is in a stage preceding the sadness spell and protests angrily against an unpleasant stimulus, he usually screams and roars, but does not open his mouth widely, does not close his eyes, and does not wrinkle the upper part of his face; his stare is tense and steady, and only the configuration of his jaws, mouth, and lips changes [plate 19(5)].

His mouth opens more or less depending on the degree of sadness, but it always is stretched moderately, and his lips are not pulled to the extreme sides as in typical crying. It is especially characteristic that, although both rows of teeth are exposed, his gums are covered almost entirely with his lips.

Loud bursts of screaming and crying do not leave any doubts as to whether or not the chimpanzee is in a somber mood. Examination of the outline of the lateral parts of the upper lip discloses circular wrinkles with angled bends, which are typical of the expression for anger; in this case, these wrinkles are broken by the segments of longitudinal wrinkles.

It is difficult for an inexperienced eye to decipher the complex expression associated with the timid, vigorously playful disposition of the chimpanzee after we showed a dead hare to him [plates 21(1), 21(2)]. Having recovered from the initial fear at the sight of the hare and not noticing any aggressiveness toward him, Joni chose to attack the animal right away; with his fingers folded into a fist, he was prepared to hit the hare.

In the end, his attack was fairly amiable, and the expression of his mouth clearly reflected this. The corners of his mouth were stretched into a smile, as in joy; his mouth was half-open, but characteristically, his lower teeth and canines were exposed, as in the initial phases of the angry mood. Attacking in such a manner, Joni probably was not quite sure of his safety, and these elements of fear the reason why the configuration of his upper lip was reminiscent of a position characteristic of the initial phases of fear, particularly of timidity [figure 9(1)]. The chimpanzee's upper lip was inflated in the shape of a hump and hung over the upper gums like a beak. This lip did not look similar to its shape in joy or anger, for which it always tightly enveloped the upper gums [plates 3(5), 3(8)].

No less complex is the combined expression associated with the triple mental condition of *repulsion*, *fear*, and *anger* [plate 22(4)]. A live hen was used as a stimulus for this complicated mental reaction. I was imposing the hen onto Joni; it was cackling desperately, trying to break loose, and beating off with its legs; although the hen was not giving Joni real trouble, it evidently was bringing out aggressive and fearful feelings in him. Joni was waving the hen off with his hand, hitting it with his fist [plate 22(3)], moving away from it, shuffling its feathers, and pinching at its beak if the hen happened to touch him; then, smelling his hand, he was anxious to wipe it off with anything to get rid of the last traces of the scent.

How was this mental condition expressed in the expressions of his face? The upper part of his face was arranged almost in the same way as in typical anger (or repulsion), but the configuration of the lips and the shape of the mouth were different. His upper lip, in the form of a wide visor, hung over the upper teeth; its upper segment protruded a little and stood apart from the gums. The upper segment of the lip was covered with the usual clearly noticeable longitudinal parallel wrinkles that stop short at the lip rim. The upper lip was as though getting ready to roll out forward, as in rage [plate 18(3)].

The chimpanzee's lower lip was broken in angles and pulled down, especially in the canine area, where it bared not only the teeth, but also the gums (as in fear). The chimpanzee's mouth assumed the shape of a trapezium: The upper lip edge was the base of that trapezium, with the middle edge of the lower lip its apex, and the sides of the lower lip, rising to the squeezed corner of his mouth, the trapezium's sides. This wide, rectangular mouth cavity, as well as the impression of the lip corners, resembled the configuration of the mouth in the repulsion expression [plate 3(4)].

The chimpanzee's side-whiskers were fluffed and bristling, as they were when the timidity and the fear elements were part of his mood.

Hence, all three constituents of his complex mental conditions—fear, anger, and repulsion—found objective expression. As always, the lower part of the chimpanzee's face—his lips—were more expressive than the upper part of his face.

Chapter 2

Chimpanzee's Emotions:
Their Expressions and Their Stimuli

In chapter 1, we described the most distinct and expressive changes in the chimpanzee's expression as it appears for eight emotional conditions of the animal. Here, we discuss in more detail the emotions general excitability, joy, and sadness and consider their stimuli.

General Excitability   The expressions of general excitability are fleeting and correspond to the maximum point in emotional development: Each of these expressions is preceded and followed by a series of rapid stages in the developing or waning of an expression associated with the increasing or fading of the emotion. It took hundreds of repeated observations and thousands of snapshots to fix this kaleidoscopic change of expressive movement on paper.

We present a description for the chimpanzee of the developing emotion of anxiety or general excitability as we call it. Let us review the expressive movements that accompany this emotion at the various phases of its flow.

General excitability is apparently the undifferentiated feeling that appears as an immediate response to any unexpected impression. Depending on the magnitude of the stimulus and of the animal's subjective reaction to the stimulus, this reaction develops to different degrees of external expressiveness. Often, general excitability presents itself only in the form of anxiety that slowly or rapidly increases to a limit and then fades. Sooner or later, it discharges into some feeling (sadness, joy, anger, fear) and can be considered a stage that precedes the appearance of this feeling.

In its most universal and pure form, without a specific emotional overtone, general excitability is extremely expressive; it is accompanied by a change in the position of hair, facial expressions, gestures, and movements and by peculiar sounds. This change is especially visible when the looks of calm and worried chimpanzees are compared directly.

When calm, the chimpanzee sits motionless, his extremities down and positioned snugly around his body, his fingers and toes relaxed, his hair tightly pressed to his body, and his facial wrinkles hardly visible; his look absentmindedly wanders along the surrounding objects. As soon as the chimpanzee begins to worry, the excitability condition comes into being, and the animal's appearance changes beyond recognition [plates 5–23, 24].

First, the chimpanzee fluffs up [plates 3(1), 23(1)]. The long hair of the lateral parts of his face (side-whiskers) rises quickly, sticks out to the sides, stands on end, and as the excitement increases, sticks out more perpendicular to the face, so that it covers his enormous ears [plate 23(2)]. This rising of hair increases and spreads from the sides of the face to the sides of the crown and to the nape; the entire face becomes framed with a wreath of hair that sticks out [plate 23(2)].

Then, the fluffing spreads to other body parts; the trunk hair rises like thorns at the chest and at the abdomen. Where hair is especially thin, the body outline transpires; the thorns of hair seem inserted into the body like closely placed black needles. Frequently, excitation is accompanied by rising of the hair on the arms and even on the legs. This fluffing spreads very fast; you can hardly notice the body parts being conquered by it one after another. At a brief glance, you can only see the chimpanzee getting bigger, increasing his size in all directions, right before your eyes [plates 20(4), 23(1)–23(6)].

The changing expressions of the chimpanzee's face enhance the peculiarity of his appearance. His lips extend rapidly forward and up [plates 5(1), 5(2)], form a funnel at the end [plates 5(3)–5(6)], and become furrowed with deep, long, longitudinal wrinkles. These wrinkles are in parallel rows along the upper lip from the sides of the nose and the cheeks. The rows get closer and narrower at the lip edge. At the lower lip, these wrinkles start close to the mouth edges, join the upper lip wrinkles near the mouth corners, and, broadening, fan out across the entire chin[1]; [plates 5(5), 5(6); figure 9(2)].

Extending of the lips is accompanied by a characteristic drawling, hooting sound that is a six-time interchange of higher and lower tones taken in every other third of an octave of the lower register. In any of six thirds, the highest tone sounds first, then a lower tone; any of the first four thirds sounds higher than the previous one, and only the last three thirds are represented by a precise three-time repetition of two sounds.

In the first three thirds, the higher tone, which sounds "uu," is more drawling and accentuated than the second, lower, tone—a drawling "khu." In the similar last three thirds, the higher tone is pronounced "ua," and the lower tone is pronounced "khu." Both tones of these last thirds are shorter than in the first three thirds and also are sharper and louder. Sometimes, they make an impression of barking sounds; sometimes, they turn into a real barking like that of a dog. Transforming this hooting into notes, we can present it by the following musical phrase:

u khu	uu khu	uuu khu
E-flat C	G-flat E-flat	A-flat G-flat

ua khu		ua khu		ua khu	
C	A	C	A	C	A

When emitting this hooting sound, the chimpanzee tosses his head further and further back with each higher sound and as if in time with it. His upper lip extends and jerks up; his lower lip is pulled down, and the funnel at the mouth edge rolls out wider, deeper, and fuller, like the corolla of a tulip in the early morning.

During weak stimulation or small or slowly increasing excitation, one can even discern sequential stages of the configuration change of a worried chimpanzee's lips. In the first stage, the chimpanzee extends his lips forward; they are closed at the edges, and opened a crack in the very middle of the mouth [plate 5(1)]. In the second stage, the extension of both lips forward increases; the upper lip slightly humps, the cleavage in the middle of the mouth opens more, and the lip edges split and separate [plate 5(2)]. In the third stage, a hump on the upper lip flattens, and both extended lips separate at the edge for a fairly large distance [plate 5(3)]. In the fourth stage, both lips extend further. The upper and lower lips shrink at their base and become pressed into the jaws; the lips keep separating from each other at their ends, so that the funnel in the middle of the mouth enlarges [plate 5(4)]. In the fifth stage, both lips become thinner at the funnel area and are pulled further apart; the upper lip bends upward, and the lower part bends downward [plate 5(5)]. In the sixth stage, the lips extend forward and lengthwise, and the funnel in the middle of the mouth opens to the limit. The mouth cleavage is the largest possible, with the lip edges thin and somewhat drooped at the corners of the mouth [plate 5(6)].

Along with the facial muscles, the rest of the chimpanzee's muscles become more active. The eye muscles are strained, and the widely open eyes are fixed on the stimulating object. The arms and the legs are pressed tightly to the body with the toes clench into firm fists. An erection often is observed [plate 23(1)].

However, this tense condition does not last long. In a few seconds, it discharges into a series of movements. The extremities are pulled from the body and tossed outward. The toes straighten. One arm rises and appears to halt in the air with the fingers either spread or cramped. The second arm rests on the ground and is ready to be raised [plate 23(2)]. Resting on this arm (most often the left arm), the chimpanzee slowly rises to his feet and stands up almost vertically, holding his second arm (bent arclike at the elbow, wrist, hand, and fingers) in front of him with his head and eyelids down [plate 23(3)]. Then, he repeatedly sits up and down, increasing the tempo with every movement [plate 23(4)]. He stands using one of his hands for support; or, standing only on his legs and quickly raising his head and eyes, he extends his arm forward in the direction of the stimulating object [plate 23(5)] or upward as if waving it in a threatening gesture [plate 23(6)].

However, in most cases, the external expressions of general excitability far from reach the highest point of development. These expressions appear in a given sequence, often partially or completely devoid of certain stages. If the stimulus is not strong enough or is too short, general excitability is expressed by extension of the lips forward and by emission of sounds; however, it is not accompanied by fluffing. In other cases, general excitability is limited to the rising of the chimpanzee's hair, but his face remains calm and the lips are tightly closed.

For example, Joni looks out the window. If common looking people walk by, he quietly watches them. Suddenly, a man appears carrying a heavy load (a big sack or case). Joni immediately extends his lips and gives out an abrupt grunt. He gives out the same grunt if he sees an unusual looking man, for example, one with a long beard or wearing a big black hat.

Joni almost always hails arrival of unfamiliar people with a grunt. Sometimes, he fluffs up, tries to smell the new arrival, fearfully extends an arm in their direction, and attempts to touch them lightly; then, making sure that they are totally harmless, he starts a closer and more personal communication. Only under special conditions does the chimpanzee become very excited and demonstrate to the viewers original gesticulation and the peculiar movements of sitting up and down, bending and straightening. These conditions are an unexpected or original stimulus; an intense and sustained stimulus; a stimulus that is not directed to the chimpanzee and, consequently, causes no specific subjective reaction from the animal. The concrete examples of general excitability explain more vividly the conditions under which it appears.

The chimpanzee was outside in the yard, surrounded by the people he knew. Suddenly, a brawl between two strange peasants was staged in his view. Armed with huge logs, two bearded, sturdy men began waving threateningly at each other, now getting closer, now breaking apart, coming and going, mimicking a struggle filled with abrupt movements, sharp cries, and bangs.

The chimpanzee immediately became alert and kept staring at this live picture full of motion; he fluffed up, raised himself, and hooted. As the struggle progressed, he became more and more excited and passed through all the poses, gestures, and body movements characteristic of general excitability. The chimpanzee calmed down shortly after this ruffle ended.

Although the rousing stimulus was for him absolutely unexpected and unusual, it was not aimed directly at him. The chimpanzee followed attentively all the events unfolding before him as a passive viewer, not as an active participant (current or prospective); he did not want to take on a more serious role.

A similar reaction was observed when Joni was at some distance from me (at the top of the cage or on the roof), and somebody was playing out an attack on me. When the aggressiveness toward me was not too intense, Joni usually was reluctant to leave his post, and his attention was limited to making expressive movements that reflected his anxiety.

Sometimes, the chimpanzee, whatever his wishes, is not able to take a more active part in the event because the scene is too far away from him; in this case, we observe a typical reaction of general excitability. We see such reaction, for example, in an experiment with the chimpanzee placed in a country house; a flock of sheep or cows moves past the windows at a distance of 100–150 feet. Or, Joni could be on the balcony and a crowd of people approach the house. The reaction could occur when Joni watches from the window and sees workers running down from the scaffolding of a building across the street.

The chimpanzee imitates all signs of excitement when he sees a worried look on people's faces and cannot see the object causing the anxiety. For example, I pretend to be crying, cover my face with my hands, make a shrieking appeal to the chimpanzee, and call him by name. The little chimpanzee restlessly looks around him, looks for somebody to defend me from; not finding anybody, he reproduces very expressively and fully all the characteristic movements of maximal general excitability: He stands up, fluffs up, extends his lips forward, rises, crouches, spreads his arms, and repeats these manipulations until I stop moaning.

As soon as any strange person arrives at the scene, the general excitability reaction in the chimpanzee is followed by an aggressive reaction toward the newcomer. In a broader sense, in the development of general excitability, there is always danger that the external stimulus will cause a specific emotional response since, in the absolute majority of cases, general excitability is like a base, a background for the initiation of fear, anger, and sadness. General excitability precedes these affects; from the start, it either has a specific emotional overtone to it, or when it reaches a limit, it suddenly discharges in a powerful affect.

Certainly, more active emotions associated with elevated spirit (for example, rage and joy) are preceded most often by maximal general excitability. More passive emotions (sadness and fear), which suppress or stifle the spirit, are preceded by only a low degree of excitability.

For example, when the animal suddenly becomes upset, the excitability that precedes sadness is limited to hardly noticeable fluffing of the body hair, extension of the lips forward, and emission of a drawling, single hooting sound. This sound, although somewhat resembling the one in the initial stages of general excitability, has sadder and more distressing inflection. The chimpanzee's gestures only complement the picture. In these cases, the chimpanzee extends his arm forward with his palm up, as when he asks for something to be given to him. At the slightest delay in satisfying of this request, he falls into real sadness with all its characteristic attributes of moaning, crying, and the like.

The following examples better explain the initiating conditions of the anxiety emotion that evolves into sadness.

Joni was in the hot sun for 3 hours during photographic shootings. He was given some water to drink. He extended his arm toward the bowl, but at the very last moment the bowl was withdrawn; the chimpanzee immediately extended his lips and emitted a moaning sound.

The same reaction appeared when, after long and lively communication with the chimpanzee, I suddenly arose and left the room or for some reason put him in the cage. In both cases, the chimpanzee was deprived of an entertainment session and was doomed to depressing isolation. A sudden and unpleasant stimulus immediately caused anxiety that evolved into sadness.

The chimpanzee expresses anxiety that precedes fear by raising his hair, straining his facial muscles, extending his lips, and emitting a single, drawling, hooting sound.[2] At the first moment coincident with fright, the chimpanzee does not demonstrate any characteristic gestures or body movements, but seems to freeze in the pose in which he was caught when the frightening stimulus oc-curred; he stares at the object and does not move [plate 24(4)]. At the slightest hint of real danger, Joni displays all the signs of fear, making a series of body movements to disguise himself and become as inconspicuous as possible to get away from the danger. In 2 or 3 seconds, in the absence of a real threat on the part of the object, the hair on the chimpanzee's body comes down, his lips are back in place, and he starts looking his usual, normal self.

The photograph to illustrate excitability followed by fear was taken under the following circumstances. The chimpanzee was outside in the yard, sitting on a bench; he distractedly stared before him, resting his bent arms on his knees [plate 24(4). He had just moved the finger on his left hand toward his head to scratch himself, but at that moment, not far from him and beyond the fence, a few cows appeared. Usually, cows frighten Joni, and he flees; but now, having spotted them, he only extended his lips forward; fluffed his hair; released a single, short hooting sound; and began looking tensely into the distance with his eyes open wide. He remained motionless, forgetting to scratch himself and with his index finger still in the air, not taking his eyes off the animals and not calming down until the cows disappeared. Only when the danger increases (for example, the cows get closer to the chimpanzee), does he take off, throw himself onto my lap, press his whole body to me, shake all over, roar loudly and run after me if I attempt to leave him, and calm down only when he feels safe and completely out of danger.

The same reaction is observed in the chimpanzee in the presence of auditory or olfactory frightening stimuli. In the first case, it can be the sudden sound of a horn; in the second, it occurs when the chimpanzee smells raw meat. In both these cases, after typically expressed anxiety, Joni shows signs of fear and the movements of self-preservation and defense (see more details in chapter 3, Self-Preservation Instinct). Usually, the degree of fluffing of the body hair is propor-tional to the amount of fear and to the strength of the frightening stimulus.

Anxiety that evolves into fear once was seen very clearly during a movie dem-onstration. The chimpanzee was taken out of his usual surroundings and put into a big room illuminated with very bright lights and filled with a crowd of strange people. Joni immediately developed a vigorous reaction of general excitability: He fluffed up, hooted, stood up, crouched, and looked scared. He turned around all the time, peered into the darkness of the next room, flinched at the slightest strange noise, momentarily came down from the table, hid under the table, or looked for shelter in a quiet place when he heard the sound of a lamp being turned on. Joni calmed down somewhat only when I took him on my lap, but even then, his whole body kept shaking for some time.

Anxiety that evolves into anger differs in its appearance from anxiety that precedes other emotions by a sharp, short, grunting sound and by a waving, threatening gesture toward the stimulating object. This condition invariably can be brought about at any moment and with any object [plate 24(2)]. You only have to impose on an infant chimpanzee something that he has been flinging energetically away from himself, and he immediately and with increasing anger will keep throwing it away, accompanying this action by aggressive gestures and snappy hooting.

Sometimes, Joni imitates someone's angry anxiety. For example, I intentionally emit shrieking cries, make threatening gestures, or throw objects away from me. Joni gets excited, fluffs up, tosses and turns, and repeatedly hoots, extending his lips forward. As soon as I start knocking on the table, the chimpanzee already is worried, fluffing up his hair; he takes the knock as a signal of insecurity, starts striking himself, assumes aggressive poses, and uses aggressive gestures.

The excitability that precedes intense aggressiveness usually develops to its highest point. The chimpanzee stands up vertically, extends his lips forward, fluffs up his hair, emits a long shrieking and hooting sound that evolves into barking, and then, standing on one foot, repeatedly crouches and straightens up, increasing the degree and frequency of the bending and straightening with every movement. Then, Joni straightens up and seems to fling his arm toward the stimulating object.

At the next stage of stronger anger, the chimpanzee stands up and crouches many times; this is followed by a vigorous offensive. If the rousing stimulus is near Joni, he bends to the ground, rests on all four extremities, now bends forward, now tosses back, and changes his support from arms to legs and back again, as though getting ready to make his best leap forward; he makes this leap, throwing his entire body on the stimulating object and digging his teeth into it.

If Joni is confined to the cage and does not have physical access to the object of his concern, he expresses his anger by running nervously around the cage; he throws himself onto the grate and presses his body to it. He gnaws at the bars so vigorously that the whole cage sways. He knocks with his hand several times on the cage walls and grabs various objects, bangs on the floor with them and throws them at the walls, creating enormous noise and seems to take his angry mood out on these inanimate objects, giving vent to his disagreeable, angry feelings.

Anxiety that precedes joy corresponds with the strength of the stimulus. It can be limited to weak fluffing up and be accompanied by a recurring, short, grouching sound; or, it can reach the limit of his development and be expressed in the raising of his hair, extending of his lips, and long intensive hooting that ends with loud triple barking, followed by gestures and body movements that reveal the chimpanzee's joyful mood [plate 24(3)].

For example, any of my unexpected arrivals at the chimpanzee's room suggests to him that entertainment is coming. This prompts his joyful expectations: He meets me fluffed up as though in greeting, extends his arm toward me, incessantly hoots, then runs up to me, grabs me with his hands, presses his body to mine, sometimes touches me with his open mouth, becomes breathless and extremely vivacious, is ready to play with me all the time, and is depressed when I do not go along with his wishes. The same reaction is observed when I put him into other rooms, into the crowd, allow him out of the cage, or when he opens the door and runs from upstairs down to the terrace.

After an initial stage of maximal general excitability with its characteristic poses, gestures, and sounds, the chimpanzee shows all the signs of joy: He smiles, runs from one place to another and from one person to another, pulls at every-body's dresses, urges people to play with him, and taps on various objects with his hands. He performs a series of useless movements, which attests to his ele-vated, joyous mood.

With happy hooting followed by loud barking, the chimpanzee hails the arrival of tasty food, for example, an orange. While the orange is being peeled, he hoots endlessly as if looking forward to the pleasure of eating his favorite fruit and enjoying the very process of eating.

With the same reaction, the chimpanzee meets the sound of the elevator, the precursor of our returning home after a long absence. Riding the elevator to the fourth floor and then banging the elevator door shut, I would hear this peculiar hooting of the chimpanzee coming through the apartment door; he was impa-tiently waiting for us and related the elevator sound to our homecoming.

Less intense and shorter grunting could be heard every day after his early morning awakening in the isolation of his room when he heard our voices in the next room. In the evenings, when he went to bed, Joni usually preferred to have somebody beside him while he was going to sleep. For example, when he was in bed and I went out of the room, leaving him alone, he immediately crawled out of the bed. If I stayed with him until he went to sleep, he emitted a grunt of content, perhaps indicating that his apprehension about my not being with him had been resolved favorably.

Joy   With the exception of the emotion of general excit-ability—an emotional experience sui generis that was uncovered only after numer-ous observations in different circumstances and that neither in humans nor in animals with lower organization has full analogs due to its unusual bodily expres-sion[3]—all subsequent emotions are comprehended easily. Among these emotions, *joy*, accompanied by a smile and laugh, has the external expression closest to that of a human. Even when the chimpanzee simply is having a good time, in periods of mental and physical contentedness, his *bien-etre* is reflected clearly on his face.

Usually, the chimpanzee's lips are closed tightly, framing the linear shape of his mouth [see figure 2(a) and plates 10(1)–10(5)]. When the chimpanzee is in a good mood, his lips open; the corners of the mouth are pulled to the sides and bent up so that the mouth assumes a round, semispherical form [plates 2(1), 2(2)].

At the same time, the chimpanzee's eyes become narrower, and in the outer corners of the eyes, two or three sharp, diverging wrinkles can be seen. These wrinkles are the continuation of the under-the-eyelid wrinkles, which correspond to the so-called goose paws of a smiling man. But, it is not yet a smile, although it takes only a little to bring out a real smile and laugh in the chimpanzee.

If you tickle the chimpanzee a little under his neck and in the lower abdomen, his facial expression becomes even more jovial. The mouth opens wider, and his teeth show [plate 2(3)]; near the corners of the mouth, five round wrinkles appear that frame the bent up corners of the lip [figures 8(3), 8(4)]. These semicircular, transverse wrinkles on the lateral part of the upper lip cross longitudinal parallel wrinkles along the upper lip and furrowing this part of the face with a network of slanting tiny rhombs [plates 2(3), 25(1)].

One more minute of tickling, and the narrow smile of the chimpanzee transforms into a wider one. His mouth opens wider, and the opening of the mouth takes the shape of a sickle. Both lips extend to the sides and envelop the gums so tightly that the wrinkles on the lips become almost smooth [plates 2(4), 25(2)]. Although the mouth is open quite wide, the teeth are not visible because they are covered almost fully with the lips. The narrowed eyes and the head, which slants somewhat to the side, are in harmony with the chimpanzee's smile; they cause an involuntary smile in everybody who watches him and thereby reveal the essence of this expression even to the lay viewer [plate 26(5)].

At the next stage, the mouth opens yet wider and becomes almost crescent shape; the corners of the mouth bend up more sharply [plate 2(5)]. The edge of the upper lip acquires an almost spherical shape and completely covers the upper row of teeth; the upper lip edges across from the canines form a blunt angle, depart from the gums in the lateral parts of the mouth, and uncover the upper parts of the crowns of five lateral teeth.[4] In the cleavage of the semiopen mouth, one can see the tongue, which is pulled inward and has its tip resting on the bottom of the mouth.

The eyes of a smiling chimpanzee, narrowed or open wide (in the latter case, the eyes look as if they are covered with oil), are so bright, lively, and sparkling that no one can have any doubts that the chimpanzee is in a joyous mood.

This broad smile of the chimpanzee is totally soundless and often is followed by a semblance of a laugh. In the rare moments of laughing (artificially induced by a physiological stimulus, tickling), the chimpanzee opens his mouth yet wider, tosses his head back, and breathes more rapidly and with a sound.

Since the mouth opens widely by the sagging lower jaw, the whole configuration of the mouth and relative positions of the lips sharply change. Now, in the laugh, the crescent cleavage of the mouth becomes semispherical because the mouth corners are blunt, and they bend upward less steeply [plates 2(6)–2(8)]. Compared with the wide smile, the upper lip at the corners of the mouth does not rise; therefore, its edges become flatter and more even and envelop the upper gums less tightly.

The lower lip edges are pulled not so much up and to the sides as down and forward, so four wrinkles that encircle the corners of the mouth are not so close as in a smile. These edges do not form sharp angles across from the corners of the mouth and come down in wide semicircles. In a loud laugh, only the tips of the upper canines and partially the lower teeth can be seen in the widely open mouth [plate 2(8)].[5] The tongue is pulled inward; the visible part of it is greater than before. The upper part of the chimpanzee's face remains unchanged. His body movements confirm our guess of his jovial disposition.

Compared with that of a human, the chimpanzee's laugh is almost soundless and therefore is less expressive, as if the laughing sounds are dimmed and represented only by rapid breathing. This impossibility to discharge his emotion of joy in the sounds of his own voice seems to urge the chimpanzee to express it in a different form—in the violent movements of his body and in all the sounds he is capable of reproducing.

In a joyous, lively chimpanzee, the facial expressions, arms, and legs, are in endless movement [plates 25(1)–25(4)]; he does not stay in one pose even for a second. Even in a sitting position, the chimpanzee varies poses all the time without appearing repetitive. As the mobile, colorful pieces of glass in a toy mosaic stereoscope give a charming, unexpected, and unique color mosaic at every turn, so the mobile face of the chimpanzee and his incessantly moving levers—arms and legs—every second produce a fleeting, bright, and distinct sketch that illustrates his joy.

However, committing this image to paper is almost impossible for the viewer because the joyous chimpanzee constantly changes his position. He takes off; runs around the room; slaps with his hands at surrounding objects; gets on chairs, sofas, and tables; and jumps down to the floor. Stopping for a second, he tap-dances and dashes around as if stung, trying to knock something over or at least touch it with his hand. Sometimes, running up to a solid object (for example, a door or wardrobe), the chimpanzee taps with his folded fingers on the object and runs away in an unstoppable storm as if insane; he then runs randomly around the room, possessed with "the movement spirit," begs for a response movement, and puts everything that comes his way into motion.

What can make a chimpanzee joyous? His mouth puts on a smile the moment someone he likes enters the room or when people assemble around him; when he is shown a new thing, taken into a new place, allowed out of the cage, given freedom and space for his activities, entertained; and certainly when he is given a tasty food. Physical well-being, tasty food, petting, freedom, new impressions, communication, play—all these sources produce an inexhaustible flow of pleasant stimuli that bring out chimpanzee bubbling, vigorous joy accompanied by a smile, a laugh, and peculiar sounds.

The facts are plenty, an entire chain of them; they are in my memory and appear in my logs, from which I give random examples.

1. I am bringing oranges to my little chimpanzee, his favorite treat. Spotting them already from far away, he smiles and grunts loudly, as if looking forward to the sweetness of his taste perceptions; he rushes to meet me, grabs one of the oranges, runs away from me and into his corner, jumps up to his bunk,[6] climbs the swing, takes a single swing, rushes down to the bunk again, bangs his feet on the floor, and then quietly sits down and starts peeling the orange, grunting endlessly and loudly. During eating, as the chimpanzee indulges in the process, this grunting transforms into loud coughing and even barking. The chimpanzee especially enjoys the meal when one of his acquaintances is sitting beside him; in this case, he grunts more frequently and strongly. From time to time, he reaches out to touch one person or another as if confirming his goodwill. If I start repeating after him in imitation of his grunting sound, his grunting becomes louder, longer, and more drawling and turns into some kind of moaning; he does not stay quietly where he is, but moves closer to me, touches my face with his, and hugs me either under the chin or at the neck. He presses his widely open

mouth to me; then, getting his lips closer, he nibbles my cheek between his lips.

Sometimes, suddenly abandoning me, he throws his arms around some-body else's neck and, as with me, touches their face with his, slightly nib-bling them with his teeth or lips. This is the chimpanzee's highest joy ac-companied by tender cuddling; this touching with the mouth can be considered the equivalent of kissing (see more details in chapter 3, Social Service).

2. The chimpanzee's mouth immediately turns into a smile when someone close to him enters the room. His smile is soundless, but it is supported by additional roaring sounds; he whisks up, reaches the hanging trapezes, and for some time produces such a deafening noise with them that you need to stop this vigorous expression of joy as soon as possible.

This joy is often expressed in a more tender and delicate way toward people to whom the chimpanzee is attached strongly, for example, me. My every visit is an enjoyment for him; my every departure is a distress. I enter his room, sit down to write (the unequivocal sign for him that I am staying for a long time); he smiles, grunts intermittently, runs to meet me, tries to touch me with his hands,[7] gets onto my lap, and touches my hands with his, strongly pressing the skin with his fingers.

Sometimes, he touches my face with his lips or my neck with his widely open mouth; or, pressing his lips together, he sucks at the skin somewhat and clenches his teeth, nibbling at the neck or at the face firmly but not strongly, at the same time breathing feverishly. With his whole body shak-ing, the chimpanzee nibbles my skin more strongly and breathes through his nose; sometimes, he breathes through his slightly opened mouth, in-creasing the speed of breathing.

The more intense his joy, the more strongly he digs his teeth into you and the longer he is not willing to let you go. Apparently, his joy in this case is combined with feelings of tenderness, desire for cuddling, and elements of sexual feelings.

The same bodily expression of joy, directed toward people he knows, is observed in Joni when he is let out of the cage, introduced to a big group, shown new things to play with, given some attention, entertained, thrown up into the air, rocked on the swing, or tickled.

It has been noticed that the chimpanzee expresses his joy toward his acquaintances more vigorously and directly than toward strange people. In the latter case, this expression is limited most often to touching with his hands and less often to touching with his lips or open mouth.

Thus, for the infant chimpanzee, as well as for the human child, it does not take much to make him entirely happy. But, as for the child, the sun of happiness becomes dimmed instantly when a tiny cloud emerges, and the reasons for his sadness are endless.

Sadness   Each of the following factors can be a sufficient rea-son for the chimpanzee's sadness: illness or tiredness; incomplete satisfaction of physiological needs; even insignificant variations in his routine (food, water, sleep); barring him from communication with people; restraining his freedom by locking him in a cage; departure of his guardians; limiting his destructive tendencies; containing his unrestrained movements; resisting his calls for play; and above all, his fear of real or imaginary danger.

Even a small worry, a weak concern, brings out in the chimpanzee an expressive emotion of sadness; his facial expression changes completely and passes a series of stages according to the development of the sad feeling.

Stage 1. In Stage 1 [plate 27(2)], the face of a saddened chimpanzee appears considerably longer than normal [plate 27(1)] because the upper part of his face (skin of the above-the-eyelid arcs, eyelids, and eyes) rises; the lower part of his face and the tightly closed lips somewhat extend forward and down.

The upper lip is furrowed with numerous (17) long, sharp, longitudinal wrinkles that come off in large creases from the sides of the nose and from the upper and lower parts of the cheeks; the wrinkles narrow near the corners of the lip.

Four central wrinkles encircle the sides of the nasal mound and make a wavy inflection in the middle of the lip; they cut the lip edge with wrinkles that fan out. Four peripheral wrinkles come off the sides of the cheeks and become smaller as they depart from the center; near the outer corners of the mouth, they form slanting creases that appear most distinct in the corners of the mouth. The entire lower lip also is furrowed with longitudinal creases, but these creases are hardly visible because of the dense gray hair covering them.

The chimpanzee stares up as if entreating; his hand is raised and forward in a begging gesture [plate 27(2)]. Often, the penile erection is observed.

Stage 2. In Stage 2 [plate 27(3)], a slight delay in satisfying chimpanzee's wishes in elimination of the distressing stimulus, and the lip configuration changes considerably. The lip edges are closed in the corners of the mouth and bent up sharply, but in the middle of the mouth, they diverge, forming a hardly visible funnel. The wavy inflection of four wrinkles that encircle the nose becomes sharper and steeper, so that the lateral parts of the upper lip converge in its center and look like two hollows. The background of these hollows sets off the central part of the lip and forms the funnel proper, which is furrowed with diverging wrinkles that start from the corner of the under-the-nose furrow. The lower lip hardly participates in forming the funnel. Then, the chimpanzee's eyes look up and retain this entreating expression, which is emphasized by a soft, moaning sound that is intermittent and frequent. Apparently, he still cherishes the hope of eliminating the unpleasant stimulus and believes in the possibility of returning to a quiet condition [plate 27(3)].

Stage 3. In Stage 3 [plate 27(4)], the chimpanzee's head is somewhat thrown back, and the eyes look up imploringly. His arm extends forward and up in the direction of his glance and freezes in the air in an appealing gesture. The mouth funnel opens wider and is directed up and forward. The funnel is formed by both lips; the lip configuration substantially changes, with the middle part of the upper lip contracted and pressed inward to such a degree that the upper part of the lip under the under-the-nose furrow (furrow 20) bulges like a small hump [plates 27(5), 27(6)].

Stage 4. In Stage 4 [plate 16(1)], the chimpanzee's head is bent down, the eyelids and glance are lowered, the mouth is open, and the funnel disappears. The straightened upper lip comes forward and completely covers the teeth and hangs over the dropped lower jaw like a visor. The lower lip is pulled down and uncovers the front row of the teeth to the level of the canines. You can hear a loud and drawling moan; the chimpanzee's mood is declining to despair.

Stage 5. For Stage 5 [plate 16(3)], the mouth opens even wider, with the lower teeth uncovered more and the upper teeth (canines) partially visible because the upper lip becomes somewhat detached from the gums and pulled to the sides.

While the skin of the lower lip extends almost to the limit and tightly envelops the chin and the lower gums so that its natural bumpiness becomes almost smooth and gray hairs of the beard stick out like horns, the skin of the upper lip is cut across by shallow longitudinal wrinkles that are abutted (at the border between the light and the dark parts of the face) by transverse, circular wrinkles near the mouth that are hardly noticeable under the eyelids and near the nasal mound. The chimpanzee's eyes look straight at you; his arm reaches out to you—he is the embodiment of unpleasant anxiety and tense suspense.[8]

The chimpanzee starts to yell intermittently in a drawling, loud, and rattling cry; he seems angry in making demanding, imploring attempts to attracting your attention to his suffering to help him overcome and reject the unpleasant stimulus. If you do not meet his halfway, his despair is obvious.

Stage 6. For Stage 6 [plate 16(4)], the chimpanzee bends his head down again, but does not look at you because he does not expect anything from you. His eyes look down blankly, apparently not seeing anything; his arm is withdrawn and seems to stick in the air. Both lips are extended to the sides so that their front longitudinal wrinkles become completely smooth. The upper lip jerks up somewhat, moves over the nose, and seems to shift the nose upward to the eyes. Two large circular wrinkles near the mouth start from the nasal mound and go to the corners of the mouth [plate 16(4)]. At this moment, the chimpanzee gives out such a deafening roar that in a few moments, he gasps for breath, loses his wind and his voice, is silent for a couple of seconds, and then resumes roaring with the same force and energy. Even during the chimpanzee's most vigorous crying fits, there are no tears in his eyes, and our skeptical mind harbors doubts as to whether the chimpanzee really is suffering. But, another more sensitive, responsive, trusting, and fair judge, our heart, starts beating compassionately at the sight of the scrawny figure of the crouched roaring chimpanzee and urges us to comfort the little animal.

You do not want to miss this chance to comfort him compassionately. In a second or two, the sadness already has overpowered the chimpanzee so completely that he remains deaf and blind to anything unfolding before him.

Stage 7. In the final stage [plates 16(5), 16(6)], the chimpanzee throws his head back, closes his eyes, and opens his mouth so wide that he entirely displays the his mouth cavity with its two rows of enormous teeth; not only are the teeth visible, but also the gums and the tongue, which is drawn back toward the gullet [figure 8(6)]. While the lower part of the face, the jaws and the lips, are stretched to the limit, the upper part is wrinkled and shifted because the above-the-eye-brow creases have slid down, the nasal and under-the-eyelid creases have moved up, and the eyes are closed so tightly that only a narrow strip of the sticking out eyelashes can be seen.

The color of chimpanzee's face somewhat darkens; he gives out an earsplitting roar, so loud that it can be heard through two or three stories of a brick house or at a horizontal distance of 50–60 meters. Intermittent thunderbolts of this roar recur with a new strength, and seem to be endless. You may also observe toes clenched in fists and an erection [plates 28(1), 28(2), 28(5), 29(1), 29(3)].

The facial expression of a crying chimpanzee is set off visibly by diverse movements of the arms. At an initial stage of sadness, the chimpanzee extends his arm in a pleading gesture toward a person from whom he expects help, satisfaction of his wishes, or comfort; later, as sadness increases, he extends both hands forward, palms up, as though getting mentally and physically ready to accept this help [plate 28(2)]. He cries and roars, but still follows your slightest movement, apparently hoping for a favorable end to this matter.

If you reject him or delay in comforting him, his cry increases, he closes his eyes and keeps crying, and his arms do not reach out to you any more, but they do not withdraw either. He throws his straightened or bent arms to the sides or slaps himself on the legs or on the abdomen [plate 28(3)].

We often can observe the chimpanzee stretching out his arms in the air [plates 28(4), 28(5)]. One straightened arm points up; the other, also straightened to the tip of the fingers, points down. Sometimes, both arms are raised, bent at the elbow. In other cases, one arm is stretched up, and the other is bent and pressed to the body [plate 29(3)]. Sometimes the chimpanzee simultaneously raises both arms very high and, bending them at the elbows, closes them over his head as though making an appeal for help not to someone near him, but to someone who is higher, farther away, and nobody knows where [plate 29(2)]. Sometimes, usually after a long cry, as if losing hope for help and in total despair, Joni places his joined hands onto his eyes, crosses his arms above the crown of his head as though to close his senses, his eyes and ears, to external impressions to protect instinctively his most valuable body part, his head, from harmful influences [plates 30(1)–30(3)]. If this also does not help, Joni falls into despair and starts beating himself severely on the head with intermittent and quick blows of both hands, once in a while pulling his hands away from his head and raising his eyes in the hope that someone will have mercy on him and comfort him.

Such a gradually increasing effect always occurs when the chimpanzee walks outside with some of us, his guardians, and he sees animals (for example, cows), he considers a threat. It is as if he is waiting for our help in case of closer contact with the animals. If we are unyielding, continue on our way, and do not pay any attention to him, his despair is limitless; he either fiercely cries and roars, to the point of grunting or loss of voice, or stands still, silently opening and closing his mouth with his head thrown back and only his mouth cavity looking forward. As soon as he notices somebody's arms extended to him, his whole body rushes to his human guardian; he is still shaking from the recent experience and is sweating. But, now he calms immediately, presses his body tightly against his defender, and is ready to go anywhere under this human protection.

A similar reaction of gradually increasing despair is observed when Joni is shown an especially dreadful object (for example, a stuffed turtle or the hide of a panther) which brings out panicked terror in him; the appearance of this object causes him to look for a refuge with the people he knows well.

The chimpanzee expresses his sadness not only with his face or with hand gestures, but also sometimes with his pose by noticeably changing the position of his body. For example, Joni pleads energetically to be allowed out of the cage; I say no, but he tries to open the door himself. I wave my finger at him; he turns his back to me and whimpers. In a little while, he turns only his head to me, not changing his pose, and stares at me and extends to me one of his arms as a sign of a plea. I say no again; he turns his back to me again and keeps sitting motionlessly for a long time. Using analogy to human expressive movements and their interpretation in psychological terms, we would have to say that the chimpanzee was distressed and offended.

For a second example, Joni was taken from his previous owners and brought to us, his new owners; he was about to be left alone in strange surroundings. What a vigorous resistance the chimpanzee put out in response to these new, distressing, and fearful conditions! This resistance was accompanied by sad emotions that reached the point of despair.

At our first attempt to leave, he started to worry, fluffed up, extended his lips forward, restlessly ran around those leaving, and give out a snappy, short moan. If you did not pay attention and continued to the door, he stood up vertically and extended his arms out to you in a pleading gesture, clearly expressing his wish of getting onto your lap or simply being with you. If you still did not take him and got closer to the door, he followed you, staying in the same vertical position; he ran around you, not taking his eyes off you, following your slightest movements and increasing his protesting moans [figure 7(2)].

If he did not see on your part any reassuring movement, he often assumed different poses: Suddenly he bent his trunk, rested his head on the floor, and stood in this position for some time as if getting ready to tumble over his head. At the same time, he emitted an earsplitting cry, which did not prevent him from rising to his feet and looking at you from time to time, as though to check the firmness of your intention of leaving and to guess your final decision.

If you stayed and came over to Joni, he immediately calmed down; if you directed your steps to the door again, he retreated to the farthest corner of the cage. Once again, he bent the front part of his body to the ground, rested on his bent elbows and semistraightened legs, and raised the back part of his body above the front.

When your intention to leave was close to realization and you were almost at the threshold of the room, near the door, it did not mean that he gave up his claims on you and that soon you would be able to leave. As soon as he noticed that you were close to the door, he quickly took off to make it to the door faster than you. He grabbed your dress with his strong, tenacious hands; caught your legs; reached out to you; and tried by all means to hold you beside him.

You hardly could shake his natural hook—his hand—off you, and he already was attached to you with three other hooks (the second hand and two feet), which came to the rescue of the first one and hitched to the same place with new strength. As soon as you freed your feet from these tenacious bonds, he wriggled his body with amazing speed and hung on your neck, on your shoulders; again and again you became the victim and became imprisoned by his little body.

Caresses, threats, punishment did not help the matter. Caresses and persuasion only enhanced the chimpanzee's wish to remain beside you; harsh screams caused even bigger confusion. He held you with amplified persistence and energy. At the time of strong mental excitation, Joni remained totally indifferent to corporal punishment.

Now, you could escape from his prison only if you resorted to a trick: You pretended that you gave up your intentions and were not going to leave. He immediately calmed down, but at the same time kept watching the door.

By waiting until he was calmed fully, you managed to go to the door and open it a crack; this did not mean that you could leave him because, like lightning, he ran to the door, put his hand into the crack, caught the edge of the door, and would not let you close or open it.

If you attempted to close the door or put pressure on his hand, you ran the risk of breaking his fingers, yet would not achieve the desired goal because he became completely insensitive to pain in the course of these events. Therefore, you stopped competing with him in strength and agility and counted on distraction and luck in opening the door.

Suddenly, through wonders of luck you managed to sneak out of the room and closed the door tightly. But, now you were not glad you were free and wanted to return to your "house arrest" inflicted on you by your little despot because you were followed by such a boisterous roar that your heart could not endure this expression of suffering of a little creature separated from his dear world. You were ready to sacrifice your freedom, your leisure, and your work and do everything in your power to comfort him.

Indeed, who would not shudder to see as the chimpanzee, left alone (but accessible to observation through glass) and totally confused, rose to his feet to a vertical position, ran around the room, stretched his arms toward the door separating him from you. Not getting an answer, he banged with his fists on the door, knocked with a hanging padlock, made noises with the doorknob, stuck his fingers under the door in trying to open it, and kept roaring desperately to the point of losing his voice.[9]

He threw himself to the floor, rested his head there, tumbled over his head, and again made a series of entreating or haphazard, confused, hopeless gestures with his arms, demonstrating his utter despair.

Mental depression. The most pitiful, most distressing sight is not when the chimpanzee is protesting and raging, but when, not believing in your help anymore, he becomes exhausted and emaciated. He puts up with his fate and falls into a state of full mental depression. In this case, he usually would retire to the farthest and darkest corner of the cage, sit down on the rug and keep sitting there for hours, rolled in a dark, tiny ball, covered with a piece of cloth. Or, he would lie down, enveloped in a blanket up to his eyes, give up everything, as though not wishing and not hoping for anything; he would lie there until somebody entered the cage.

There are four sketches [figures 10(1)–10(4)] and a photograph that convey the poses and expressions of the chimpanzee in a state of full mental prostration better than words can [plates 31(3)–31(6)].

1. Joni bends down, drops his arms symmetrically on the ground, leans his head to his chest, and extending his lips down and forward, raises his eyes and eyebrows imploringly, fluffs up his hair, as though waiting desperately and silently and asking for something [plates 31(4), 31(6); figure 10(2)].

2. Joni braces himself, bends his knees sharply, closes his legs, drops his head lower, and forms a hump. His lips touch his hand, which rests on his lap; he stares blankly before him and sits motionlessly and mournfully as if he has lost the hope of fulfilling his wishes and has given up everything [plates 31(4), 31(6); figure 10(2)].

3. It is as though he does not want to come out of this state of mental stupor, but his arms and legs are tired and numb, and he folds them in a more comfortable manner. He stretches his legs; his feet are joined and embraced by his lowered left hand so that they do not spread apart [figure 10(3)].

   His right arm, as earlier, serves as a support for his bent head, but the arm is tired already and does not rest on the sharp angles of his knee; his hand grips his left elbow. The chimpanzee's lips contract somewhat, and he stares before him; a state of dull repose has come. In time, this condition transforms into hopeless torpid tranquility [figure 10(4)].

4. The feet are positioned flat on the ground and pulled to the sides; the long arms intertwine like snakes and rest on the legs. The head drops so low that the chin is almost hidden; the lip skin flattens and moves to the sides. Now, the chimpanzee not only looks slouchy, but appears to be an absolute monster, like some octopus without a neck or a trunk but with only a huge head and dilated, desperate eyes. An indefinite number of wriggling curved sprouts, his arms and legs, come off the sides of the head in various directions.

To conclude this chapter describing sadness and its related stimuli, we need to point out another characteristic that differentiates the infant chimpanzee from the human child. During my 3-year observation, I never noticed that the chimpanzee's sadness or cry was caused by physical pain. Whether the chimpanzee fell on the floor from the swing, hurting himself badly; smashed himself against the poles of a trapeze; bumped into doorposts or furniture in his unhampered run; got a splinter in his leg; or scratched off the skin on his elbows, no matter how severely he was hurt, he never screamed or cried. When he crashed or fell, he looked stunned. Then, he calmed down, and in a second, continued his play as though nothing happened.

Once, during the study of Joni's physical endurance, a very strong man attempted, using all his strength, to squeeze the chimpanzee's finger. Joni did not

рис. 1 рис. 2

рис. 3 рис. 4

Figure 10. Postures connected with depression of the chimpanzee: (1) sad; (2) The chimpanzee in low spirts; (3)
slightly depressed; (4) depressed.

flinch a bit; he only looked at the man attentively and did not even try to remove his finger. The same person for the same purpose tried to pinch Joni as strongly as he could; the chimpanzee also remained indifferent.

When punished for a misdemeanor, Joni never cried, although he obviously got angry. One time, after he had bitten a boy, the chimpanzee was punished by whipping; he was beaten so harshly that the whip literally swished in the air. Nevertheless, he remained motionless and did not even show any intent to flee; he only curved his lips and scratched himself at the most painful spots sometimes.

Yet more characteristic of the chimpanzee is that a slight verbal reprimand on the part of a person close to him, even an insignificant expression of dissatisfaction from someone to whom chimpanzee is especially attached, often caused his bitter and desperate cry.

A couple of scenes are still fresh in my memory from in the framework of our experiments with Joni selecting a certain object from a group of objects. In reaction to Joni's selection of a wrong object, if I as much as pushed his arm away from me, he immediately started to scream and moan and was so worried and confused that he could not continue with the experiment until I had comforted him.

It took only a yell in response to Joni's disobedience for him to start crying; he reached out to me, asked to be taken on my lap, and stroked my face and my chin with his hands as though trying to mellow me. However, if I was angry with his boisterous behavior and was going to leave him, if I slapped him slightly and waved him away from me when he tried to follow me, he gave out an awful roar and ran over to me with his arms extended, his whole body shaking. He asked to get onto my lap and expressed such immeasurable despair that I was quick to relieve his punishment and restore the lost entente.

What are the natural surroundings and the forms of behavior in which a chimpanzee's mental activity develops? The behavior of an infant chimpanzee, as well as that of a human child, is expressed fully in his instincts, play, and habits. His instincts give him an impetus for action and determine the direction and development of his actions up to the limits indispensable and useful for his species. The play broadens the sphere of his activity and enriches him as an individual with new experiences. The ability to form habits (to keep and utilize experience) reveals his individual gifts.

We consider three groups of instincts that govern the life of a young chimpanzee: the self-sustaining instinct that regulates the main physiological needs of an organism (food, water, sleep, and self-care) and is aimed at attaining physical well-being; the self-preservation instinct (defensive and offensive) protecting the chimpanzee from harmful influences; and the communication instinct, which determines the chimpanzee's relationship with other live creatures in his environment.

Chapter 3

Chimpanzee's Instincts

Self-Sustaining Instinct   The difference between good health and illness in
(in the Healthy and   the chimpanzee's behavior and in the consistency
Sick Chimpanzee)   of his actions is most obvious. Even an inexperienced observer can determine easily by the chimpanzee's appearance whether the
physical or mental condition of the chimpanzee is good or bad [plates 31(1),
31(2)].

When in good condition, the chimpanzee's entire muscle structure is ener-
gized. His legs are actively bent; they rest firmly on the ground and can lift
him quickly and easily. His arms are positioned asymmetrically: The right, most
frequently used, arm is free and can be put onto action immediately. His raven-
black facial and body hair is shiny, as if it were oiled, and is pressed tightly to
the body.

The chimpanzee's head leans back somewhat, so that his back does not appear
as slouch as it usually does. The chin is raised, the head assumes a more brachy-
cephalic outline, and the face seems rotund and fleshy [plate 31(1)].

The lips are closed lightly, the corners of the mouth are pulled up loosely,
and the mouth has a wide spherical outline. All the lip wrinkles are smooth since
both lips are inflated a little.

The upper part of the face in the area of the bridge of the nose and the cheek
is cut across by numerous wrinkles because the skin of the above-the-eye arcs
slides down, the eyes narrow, and the wrinkles near the eyelids gather in small
dense creases that encircle the eyes from below. In the corners of the eyes, the
"goose paws" (the fanning out, diverging wrinkles) are particularly visible.

Judging by the chimpanzee's external appearance, one definitely can say that
he looks friendly, or one can assume that the chimpanzee is in a good disposition.

When the chimpanzee is sick, his face and figure present quite a different sight. His facial features and limbs become elongated and flaccid [plate 31(2)]. His arms hang passively and symmetrically and seem to have dropped onto the bent knees. His head leans forward as though falling on his chest; it rests on his arms and therefore seems to sink into the shoulders. The chimpanzee slouches more. His face appears to become leaner and elongated. The chin drops and shrinks, and the tightly closed lips (in the form of a promontory) extend forward and are cut across by a series of longitudinal wrinkles.

The wrinkles of the upper part of the face become smoother as the skin of the above-the-eye arcs shifts up. The eyes open widely and tensely; the glance is fixed, sullen, and aimed upward.

The chimpanzee's facial color fades. His hair seems to have tarnished; it haphazardly fluffs up, sticks out from the body, and even bristles in some places. The chimpanzee looks miserable and helpless.

The chimpanzee acts in full accordance with his appearance. A healthy chimpanzee is vivacious and mobile; he will not sit still for a minute. When he is in the cage, on the swing, on the trapezes, or is put into a room, he runs all over the place, avoiding no corner, paying attention to everything, and wishing to involve all his surroundings in play.

When the chimpanzee is sick, his behavior changes radically, and you will notice it at first sight. You enter the room, and he will not meet you with joyful cries, but will sit motionlessly with his head bent and his lips extended, or he will wallow phlegmatically on his back with his legs up in the air and his hands over the back of his head, slowly swaying from side to side.

If you start playing with him, he does not respond, but asks to get onto your lap. He presses his body to yours and sits on your lap silently, not willing to get down [plate 10(5)]. If you try to leave him, he reacts painfully and nervously. He does not let you go even for a second and holds you firmly with his hands. When you attempt to shake him off, he ardently protests, moaning, screaming, and crying.

1. Treatment of Chimpanzee.   During illness, the chimpanzee is especially quiet and tender with those to whom he is attached; he is extremely sensitive to any sign of dissatisfaction or reprimand. In time of illness, like any sick creature, he asks and demands the guardianship of the people close to him. So, he willingly offers himself for treatment as a rare child does.

If he has a runny nose, to which he is particularly prone and that comes with even a slight cold, he timidly allows his nose to be treated with boric Vaseline and invaded with cotton wads. When he has tracheitis or bronchitis, he takes, with no objection, a strong-smelling medication. During intestinal diseases (constipation), he does not gripe when given an enema.

With a very high fever, reaching 40°C (which Joni had during an intestinal disease and then later during the fatal lung disease, membranous pneumonia), he does not protest with the slightest movement when the thermometer is placed in his armpit or (more conveniently) in his groin.

In the beginning, he seems a little afraid of new objects (a tube of Vaseline and a thermometer), but if he is allowed to examine and to feel them, he calmly accepts application of the ointment and measurement of the temperature.

Once, when I was inserting a cotton wad placed on the tip of a match into Joni's nose, he accidentally jerked his head and pricked himself with the match. He immediately assaulted me with his mouth open and caught my hand with his teeth, but thought better of it. Instead of the serious bite that I had expected, Joni only slightly pressed by hand with his teeth, not even leaving teeth marks on my hand.

Another time, when Joni had a cough, he let me rub him with turpentine. Later, when he felt his skin burning, he kept touching it with his fingers and scratching. He turned around nervously, looking for a biting "enemy" under the furniture, and explored all the secret places in the room as if hoping to find the culprit spoiling his physical well-being.

After the chimpanzee's several accidental leg injuries, I had to treat the sore spots with iodine. At first, the burning wound made him pull back his leg, but then he quickly calmed down. Many times afterward, he again was ready for the same procedure.

Of all existing methods of treatment, we were unable to use the simplest one. We could not treat him with a moist compress or even a plain bandage when he needed them.[1]

The chimpanzee categorically and by all means rejected all, even partial, attempts to bandage him. He tore and unbound the bandages, became angry and nervous, bit everybody, and raged and roared so loudly that it did more harm to him than good. We eventually had to cancel this treatment.

Joni absolutely could not stand even the simplest bandaging of his fingers or toes and removed the bandages as soon as they were put on. It goes without saying that he allowed only "his own people" to do all these manipulations.

Once, when the chimpanzee had an intestinal disease, we invited the leading Moscow pediatrician as a consultant. The venerable professor was very excited about seeing such an unusual little patient, but as soon as he touched the chimpanzee's swollen abdomen, our sanguine animal suddenly gave out such a thundering and hooting sound that the famous doctor hastily retired to the faraway corner and was quick to assure the embarrassed hosts that the patient was in no danger. In all appearances, the patient was "absolutely healthy."

2. Self-Treatment.   Sometimes, the chimpanzee tries to treat himself on his own. If he gets a splinter in his foot, he limps to a well-illuminated place, sits down, gives out a couple of sad hooting sounds, gets serious, raises his foot to the level of his lips, bends his head low, extends both closed lips, and examines the sore spot very thoroughly, touching here and there, as though trying to locate the source of the pain [plate 32(1)].

If I come to the rescue, he lets me look at the splinter, but becomes more alert; he does not take his eyes off me and seems to collaborate in the examination. He sits still when I touch the splinter, take it out, and treat the sore spot with iodine.

He follows all my manipulations intensely. Even when everything has been taken care of, he is absorbed for some time in now superfluous examination.

Sometimes the chimpanzee happens to scratch, prick, or cut himself. The tiniest scratch bothers him for a long time: He looks at it attentively from one point or another, bends his head to the sides, touches the scratch with his fingers or with his lips, sucks in the blood that appears and licks it [plate 32(4)].

Rather rarely, Joni resorts to another method of examination. He takes a stick and touches a sore sport with it. Feeling the pain, he twitches, leaves the sore spot alone for a moment, and then resumes the whole procedure, starting from the beginning: looking, licking, picking, sucking, and so on. The chimpanzee may scratch his leg to the quick; these become chronic sore spots and cannot be cured because picking at it and examining it is a systematic and favorite pastime for the chimpanzee.

First, the chimpanzee examines for a long time any deviation from the normal skin condition. Second, he wants to remove the crust by any means, picking at it with his fingernails, and entertains himself with this for hours. He scrapes the coarsened skin until it bleeds, often twitching from pain, but still continues to scale the crust off. Sometimes, the hair around the cut, glued with blood, prevents him from examining the cut, but he softens the hair with his saliva [plate 33(4)], licking or gnawing it, and he completely unblocks the access to the sore spot. Feeling the callus, Joni, in time with his groping fingers, moves his extended lips as though chewing something, breathes rapidly, picks at the most sensitive spots, and yet does not stop his manipulations.

Here is an example of his persistence in removing any strange object from his body. Once, I intentionally stopped Joni's examination and his picking at a small healed scratch on his foot. I used all means of distraction that were in my possession to involve him in his favorite activities, but every time, after swift satisfaction of his curiosity, he resumed his interrupted activity—picking at the scratch. He became angry and cried when I tried to stop his picking and calmed down only when he scaled the scratch off completely and there was nothing more to be done with it. Joni was also very interested in a long and thorough inspection of people's skin and the same methods of examination used for this purpose. This persistence in his wishes and actions is all the more surprising because, in his other pursuits, the chimpanzee is rather inconsistent and inattentive.

3. Self-Maintenance (Self-Attention) Hygienic Procedures.   Apart from the chimpanzee's ailments that require his participation and attention, in his day-to-day life, he shows great curiosity toward the usual hygienic procedures he received.

He is always ready to surrender himself for washing, wiping, brushing, or cutting his hair and with tense attention followed our actions, helping or hindering us to carry them out. Every morning, his face, hands, and feet (and all body parts free from hair) are washed with room temperature water, and he obviously enjoys the procedure. When you pass your hand over his wet face, he often opens his mouth, catches water with his tongue and tries to get at least a drop, which reminds us of little kids, who usually do the same catching movements while being washed.

After washing, he is wiped dry with a towel, and he submissively offers you his face and limbs for this. When being combed, he follows the results with unabated interest. If you start examining the comb, apprehensively looking for insects, he literally stares at the comb, sometimes spotting and catching the parasite. Before you can say knife, he already has escorted the insect into his mouth, snapped his teeth, and eaten it. Joni cries when he is prevented from doing this.

Joni willingly presents himself for the search and apparently enjoys this "sybaritic" procedure. He lies, rolling from side to side, snoozing, holding his limbs up for the examination, and is ready to prolong this pleasure infinitely. It is widely known that, in Russian villages before the revolution, this "hair search" was a favorite pastime. You could often observe, especially during holidays, dozens of women engaged in a mutual "search." They did it not so much out of necessity as for the pleasure of falling into a sleepy nirvana of sorts invoked by the soft movements of fingers browsing in their hair.

The skin of the back of the chimpanzee's hand is usually very dry and begins to scale if not greased. Therefore, we often had to rub this skin with Vaseline or some other softening cream. The chimpanzee enjoys this rubbing, and sometimes he smears dense blots of the cream over his hand. With the same interest, the chimpanzee watches his nails being trimmed.

The chimpanzee also engages in his own long examination efforts. He can spend a lot of time looking at his hair [plate 33(3)], inspecting every tiny bit of fluff or splinter and removing them immediately. Joni assiduously browses in his hair, looking for insects. In following a running parasite, he gets nervous, breathes rapidly, and gives out a smacking sound with his lips; the movement of his lips seems to cooperate with the catching movements of his fingers. When he is lucky in this hunt, he kills the parasite with his teeth and swallows it. So, Joni obtains triple pleasure: He entertains himself with hunting, gets rid of the itch, and enjoys a pleasant gustatory sensation. No wonder he indulges in a self-search so energetically.

For hours, Joni can bite his fingernails or toenails with great seriousness and persistence in spite of any obstructing interference [plates 32(2), 32(5)]. Once, the chimpanzee started to bite the nail on his big toe. I tried for half an hour to stop this and to distract the animal, but he remained blind and deaf to all my appeals. When I applied force in an attempt to pull his toes from his mouth, he gave out a grunting angry sound, tried to bite me, and resumed his business with the same enthusiasm. Moreover, when he noticed that I did not leave him alone and continued to bother him, he moved as far from me as he could, turned his back to me, and kept biting his nails until the job was completed.

I have also observed this tendency to bite the nails in many lower apes kept in zoos. This applies only to healthy apes whose nails are in order and do not protrude over the fingertips; the nails in sick apes often are overgrown. Apparently, proper self-care is characteristic only of healthy species; the sick ones usually neglect themselves.

Almost after every meal, the chimpanzee, on his own initiative, thoroughly cleans food from between his teeth and sequentially examines them, using his nails as a toothpick [plate 32(6)].

My chimpanzee immediately notices even the slightest untidiness in his appearance and always tries to restore himself to an appearance becoming of a "decent" chimpanzee. He thoroughly wipes, with rags or simply with his hands, the spots stained with food. Every drop that happens to be on his hair is taken off by his hands or lips. He takes the strands of his hair that have been glued together with food residues into his mouth and sucks and cleans them. All the accumulations (profuse due to frequent colds) under the chimpanzee's nose are noticed immediately and taken off by the back of his hand. This wiping gesture was very widespread in earlier years among the Russian rural population, who usually managed to do without handkerchiefs. In spite of the fact that I had always used handkerchiefs during the chimpanzee's lengthy colds, he only very rarely used some kind of a rag for this purpose. The simpler and more direct wiping method (i.e. with the hand) became his habit after a long cold, and he started using it often and with no apparent need even when his nose was absolutely dry.

When his hands are busy, and the overflowing mucus reaches the edge of the lip, the chimpanzee finds another way of removing it: He sticks his tongue out and licks the fluid off, drying the edge of the lips from time to time. It is known that homeless children, who have not been brought up to use handkerchiefs, also use this method. But, compared to these children, the chimpanzee instinctively keeps himself clean. You can observe in these children the never-drying streaks under their noses, which apparently do not bother them. You never see such sloppiness in the chimpanzee, even during his worst colds.

However, the chimpanzee and children share one sloppy and unaesthetic habit: He used to clean his nostrils by picking the dried crusts out of his nose with his finger; he obviously enjoyed it. Characteristically, these crusts always went into his mouth and were swallowed. As much as I tried to prevent him from doing it, he did not drop this habit.

More amazing for me was the fact that a 4-year-old child from a very respectable family showed the same inclination to eat these dried crusts from his nose. The parents had to work hard to make the child get rid of this habit. It was even more surprising that this child was very fastidious; he was not willing to use his spoon if it had remnants of food from the previous course on it and demanded that all the dishes be changed after every course. Nevertheless, in this case, the child acted precisely like the chimpanzee.

If the chimpanzee happens to smear the edges of the lips, he extends them in a funny way in the form of a funnel. He examines them thoroughly, although with great difficulty, with his eyes aimed at one point and with using his index finger to wipe the lips [plate 32(3)].

It is interesting that the chimpanzee immediately notices pigmented spots when they appear on his hands and begins rubbing them persistently with his fingers and nails, scratching the skin almost to the point of bleeding. Moreover, even imaginary defects on his skin attract his keen attention [plates 34(5), 34(6)].

Once, I facetiously pointed my finger at the sole of the chimpanzee's foot and exclaimed, as if in amazement, "Look here, Joni, what is it?" And although there was absolutely nothing on the sole that could raise suspicions, the chimpanzee extended both lips forward, as he did for strenuous attention, bent his head, stared at one point, and began a long and obviously superfluous examination of the sole with his index finger [plates 34(5), 34(6)]. This unusual passion for self-examination, thorough self-care, and exceptional attention to appearance is based, I suppose, on his deep-rooted cleanness and tidiness instinct.

If the chimpanzee walks and sees a wet spot on the floor, he tries to step over it; if it is impossible to avoid it, he rests on his toes, raises his heels, and walks through the puddle. When walking in the yard, he carefully goes around any mud, but if he does smear himself, he thoroughly wipes the dirty spots on his body against the ground or against the grass or picks up a piece of paper or a rag and wipes these spots until everything is clean.

While in the cage, the chimpanzee never excretes at the place he usually sits; for this purpose, he climbs to the most elevated and distant places. In this connection, I recall an anecdotal episode associated with the chimpanzee's inclination to reach as high as possible for fulfilling his urgent needs.

One time, his small room was crowded with strange people; there was a group of student defectologists with Professor Rossolimo as head of the group. The little chimpanzee was to demonstrate his achievements in the framework of the "picking-the-sample" method; he sat outside the cage on a little table, surrounded by young people. Suddenly, in the middle of the test, he climbed to the ceiling of the cage and, according to his habit, let loose a fountain of urine, spraying everybody present with cascades of liquid.

I tried to train the chimpanzee to use a bed vessel by sitting him on the vessel when the defecation or urination started. He was about to get used to this habit, and after a week of training, he tried on his own to sit on the vessel when the need arose. However, he spent most of the time in the cage without supervision, in which case he fulfilled his needs where he pleased; this prevented this new habit from being established.

However, I know from the experience of V. L. Durov, the late trainer, that chimpanzees can be trained easily to use bed vessels, and that they retained this habit. I am absolutely sure that if I had been more persistent and strict, I would have received positive and firm results. In this case, the guardian was to blame, not the infant.

The instinct of the chimpanzee for cleanliness is associated with his distinctly expressed fastidiousness and revulsion, which are most evident during his adverse olfactory perceptions. For example, if you put a chimpanzee to bed and give him a blanket spoiled with his dried feces, he smells the blanket at different places with apprehension. When he finds the stinking stain, he emits angry barking, almost like that of a dog, throws the blanket away from him in disgust, and is not willing to take it any more. He also throws aside spoiled or smelly toys with the same anger and fastidiousness.

Once, I put sheets on Joni's bed, not noticing their foul odor. To my surprise, Joni did not want to go to bed that evening, although it was his usual hour for bed. In response to all my attempts to put him onto the sheets, he screamed and roared and was not willing even to sit on the bed. I examined the bed and the vicinity, looking for an object that might frighten Joni. Not finding anything, I attempted to put Joni to bed by force; he resisted violently, tried to break loose, screamed, stubbornly ran away from the bed, and almost bit me when I went to put him back on the bed. I slapped him a couple of times, hoping that he would give in, but he roared even more and persisted as before.

I turned the lights off, thinking that some invisible object in the room may have scared Joni; I wanted to put him to bed in the dark, but Joni kept crying and reeling back from the bed. Totally puzzled by this mysterious scene, I started looking through the sheets on which he slept and suddenly smelled the faint unpleasant odor of stale urine. I thought that it might be a good idea to get clean sheets and have the one on the bed washed. Immediately after I put clean sheets on his bed, the chimpanzee, not waiting for my invitation, lay on the sheets and began diligently to tuck it under the mattress, knead it, and lay it out, uttering a contented grunting sound. This was the proof that the chimpanzee's sense of smell is keener than that of humans, and that his disgust toward specific unpleasant odors is very strong.

I have to emphasize that the fastidiousness of the chimpanzee is "specific" because I often observed him, when he was bored in his cage with nothing to do, spreading his feces over the white walls of the cage.

Eating Instinct   1. Food and Water. Since the chimpanzee is so serious and particular about the ancillary hygienic procedures that upheld his physical well-being, it is only natural that such indispensable physiological needs as food, water, and sleep are objects of his extremely demanding care and attention.

First, the chimpanzee tries to satisfy his organic needs without any delay. When he wants to eat, drink, or sleep, he is afraid of losing his physical contact with me in my numerous duties of his morning toilet, feeding, and entertainment during the whole day until he goes to bed. Usually, he follows me relentlessly as I prepare his food. When I need to leave the room, he wants to accompany me, and if I do not permit him (fearing that he might catch cold due to the temperature differences between the rooms), he does not let me go, hangs on me, holds the door, and sticks his hands into the crack between the door and the door jamb, which prevents me from closing the door from the other side. He becomes deaf to my persuasions, roars vigorously, and makes an awful scene (similar to when he was in a depression), which shows the strength of his physiological needs and his energy toward their satisfaction.

If you are outside Joni's room and forget to feed him or put him to bed, he repeatedly gives out monotonous thundering cries, interrupted only by the clanging of the trapeze and the banging of the chimpanzee's fists on the wall. His crying is so hard and heartrending that you have to do everything in your power to satisfy his wishes.

No matter how hungry Joni is, he will not fall for just any food. He will smell even familiar food thoroughly before eating it. He is even more careful with any new food and usually is reluctant to try something new; he rejects it after he has smelled it for a while. Therefore, the first portion is given to him almost against his will. Such conservatism toward food resembles that of little children and of peasants of old Russia, who felt the same suspicion toward any new food.

Everybody knows an old adage: "Was der Bauer nicht kennt, das ist er nicht." ["What the farmer doesn't know, he won't eat."] But our chimpanzee, even when he enjoys the food, nevertheless smells it from time to time as he eats it. Certainly, the sense of smell governs his choice of food and is active during eating. For example, Joni immediately smells a small amount of butter in his food and categorically rejects such food; nobody can make him eat it.

Natural products in his food are carrot juice, water, milk, all kinds of berries, fruit (apples, pears, oranges, lemons, bananas, peaches, apricots, watermelon), vegetables (carrots, turnips, cucumbers, radishes), walnuts, sunflowers, chestnuts, and sometimes eggs. He eats these cooked foods: jelly, compote, boiled chicken breast, boiled cabbage, borscht, soup, hot cereal, candies, and cookies.

The chimpanzee also eats "inedible" things such as lime, plaster, chalk, graphite, and charcoal. He uses every opportunity to lick some ink. He willingly eats grass, various sprouts, petals, buds from trees, small sandstones, clay, wild strawberries, cobwebs, and small insects (apart from the already mentioned lice, there were also flies and spiders).

There are preferred things in every food group. For example, carrot juice is a treat for him; he always hails it with joyful hooting. You hardly come up to him with a cup of juice, he already extends his hand toward it and stares at it. If you are too slow in giving it to him, he cries and does not pay any attention to your attempts to engage him in play.

2. Chimpanzee's Way of Drinking.   When Joni eats semiliquid food (for example, jelly or carrot juice), he uses a special procedure to prolong the gustatory pleasure. He strains the juice between his closely pressed teeth.

He always meets his most usual food, milk, with warm welcome. The chimpanzee gets about 2 liters of milk a day three main meals[2]: 9 A.M., 2:30 P.M. and between 7 and 8 P.M. In the mornings, he eloquently notifies us about his awakening and his hunger. He makes noise with the trapezes, bangs with his fists or with various things on the cage walls, and does not calm down until the food is brought to him.

He hails milk with grunting and impatient movements, gets onto my lap, presses his body to mine, and greedily drinks the entire liter without stopping. He prefers warm milk (somewhat warmer than room temperature). If the milk is colder or warmer than that, he turns away from it after he has tried it or holds the milk in his mouth until it reaches the desired temperature and then swallows it. I consider Joni's craving for milk, especially for warm milk, as an indication that, in natural conditions, he apparently would have been fed with his mother's milk at his age.

It was not accidental that, in the early months of his life with us, he started sucking at my body (my neck or hands) so strongly when I held him that it was very difficult to detach him. Later, this "sucking" became a conditioned reflex for him: Every time he wanted to drink, he ran over to me and sucked my hand. I never mistook this as a signal of his being thirsty because my offer was never rejected, although in other circumstances, he would express his unwillingness to drink by the characteristic gestures of turning his head or his whole body away or shaking his head.

In general, the chimpanzee drinks a lot during the day, probably as a result of our relatively dry climate or as a function of the heating system in the building. Another reason may be the chimpanzee's extreme mobility.

The chimpanzee uses every opportunity to drink, even while he is washing. He willingly drinks slightly sweetened water. Often, he opens the faucet, takes some water into his mouth, and gurgles for a while before swallowing it.

Usually, I give him a drink from an enamel mug, saucer, or spoon, which I hold for him [plate 35(3)]. Joni tries to use a spoon by himself (for example, when he eats jelly), but he does it so awkwardly and unwillingly that he spills more food than he eats. When I am holding his mug, he affectionately touches my face with his hand.

Joni can use a mug or cup on his own, but he rarely holds it by the handle. Usually, he holds it by the edges [at the top and at the bottom; plate 35(2)] or places his fingers around the cup's bottom. He humps and drops his head so low that he immerses his entire upper lip almost up to the nose into the cup. It is very difficult for Joni to tip the cup and empty it, especially if the cup is fairly deep. Therefore, after he has finished drinking, Joni shifts his hand from the cup's sides to its bottom. Holding the bottom he straightens his body and throws his head back, moving the cup completely onto the nose and eyes, covering his whole face with it [plate 35(1)].

When the chimpanzee drinks from a mug, he uses both hands, but prefers resting the hand that holds the cup on his leg [plate 35(1)]. If this hand hangs in the air without support, he holds the cup less firmly and seems to be on the brink of dropping it at any moment.

The chimpanzee only very rarely takes a cup by the handle, but he does not hold it firmly enough. The cup is not steady, and sometimes he has to use additional support. He raises his foot to the level of his face and helps hold the cup with his toes under its bottom.

All these ways of chimpanzee's drinking from a deep bowl are obviously artificial for the chimpanzee. If, for example, he drinks from a shallow saucer or plate, he takes it in his hand unwillingly and awkwardly [plate 35(4)]. Most often, he bends over to the bowl, spreads his whole body on the ground, stands on all fours, and bends his elbows and knees [plates 35(5), 35(6)]. Then, he holds the edge of the saucer between his lips and sucks the liquid with his upper lip spread over the edge [plate 35(5)]. If there is only a small amount of liquid left on the plate, he shifts his lips to the center of the plate and drinks it from there [plate 35(6)]. He sometimes drinks in the same way even from a deep bowl when it is full; when there is not enough left to drink from the surface, he takes the bowl in his hand and tips the bowl to empty it. The chimpanzee living in the wild uses the same manner of drinking (standing on all fours) when he drinks from natural sources (i.e., rivers or creeks).

Sometimes, when eating dry food, Joni does not take it with his hands, but uses a primeval, savage mode. His mouth is on top of the food, and he totally ignores the advantages of his grasping hand.

3. Preferred Food.   Our chimpanzee's favorite natural food is sour cherries, grapes, bananas, oranges, and carrots. For cooked food, he also likes jelly and marmalade.

When the chimpanzee sees tasty food, even before touching it, he gives out a soft, joyous hooting. When he starts eating, this hooting transforms into a long, hollow sound that resembles a human cough. As he eats, it turns into a rapid, resounding grunting like that of a child. Once, I brought the little chimpanzee a box with 15 marmalade candies, and he began this contented grunting and barking. He took the first candy, smelled it all over, and raised it to his mouth. He bit off a tiny piece and apparently liked it because his grunting increased, but he was not in a hurry to finish it. Having tasted 2 more pieces, he turned to other candies. Then, he tried almost all the candies in the box, now eating them up, now hardly biting at them, now kneading them in his mouth and spitting out, now only touching them with his lips or tongue, now touching them with his fingers. Tasting a candy, he uttered a grunting sound. At the same time, he was tempted to try other candies, and only after all the candies had passed his brief inspection did he chose to stay with one of them, apparently the most enjoyable. From time to time, he took the candy out of his mouth, looked at it, and bit off another piece. The candy mellowed in his mouth, dripped onto his hair, and flowed over his fingers; he sucked the fingers, licking every drop of candy juice off his hair. Having finished with one candy, he started another one, of which he had already partaken, and ate it up using the same gourmet manners. Dealing with a banana, the chimpanzee quickly peeled it, eating the fruit itself, the tastiest part. After that, he scraped the inside of the peel with his teeth.

4. Eating Dry Food.   The chimpanzee usually bites dry food with his canines, gnaws it with his front teeth, and chews it with his molars. The strength of his canines can be illustrated by the fact that he gnaws the stone of a date with no difficulty, which even a human adult is not able to do.

The whole eating process is performed with extreme care, attention, and seriousness. The chimpanzee puts much energy into long preparation of certain fruit for eating; he willingly peels and cleans the fruit. For example, he separates, with his teeth and fingers, the peel from oranges, lemons,[3] bananas, radishes, or chestnuts and then eats the fruit. Joni takes an unpeeled apple in his mouth; chewing the apple, he meticulously separates the pulp from the rind and spits the rind out. He also throws out the thin rind of oranges and grapes and the seeds of plums, apples, and other fruit. Joni seems to like the yellow inside part of large carrots much better than their outer red part.

Sometimes, Joni eats his oranges in his own special way. He pierces the rind with his finger to the very core, takes the orange in both hands, and drinks the juice, squeezing it out through the hole until the orange is flat and dry.

Joni sometimes takes food (e.g., an apple) with his foot, carrying it all the way to his mouth; more often, he takes it from his foot to his hand. The chimpanzee sometimes bites at the fruit, holding it with his hand and his foot at the same time; more rarely, he holds it in both feet, bending over to it. However, he usually holds the food with his hands.

The overall eating procedure of the chimpanzee is marked by his search for the best ways of tasting and picking the food, by his starting from the tastiest pieces, by his haphazard manner of eating, and by his terrible wasting of the food. The more food is in the chimpanzee's possession, the more careless he is in using it, and the more he wastes and spoils it.

The chimpanzee is most wasteful at the initial phase of eating, when he bites and spits out more food than he actually eats, and when he hastily searches for the best pieces, discarding all the rest, which he afterward picks up and eats. As the food supply shrinks, Joni consumes the food by small pieces slowly and with great care, as though trying to extend the pleasure.

If the chimpanzee sees from far away that, for instance, sour cherries are coming, he begins to grunt, as if looking forward to the pleasure. As you come closer, the grunting increases and transforms into a ringing bark. Not waiting for our permission, the chimpanzee impatiently snatches a few cherries from our hands and quickly puts them into his mouth. If you give the chimpanzee all the cherries at once, he would not take the first cherry at random, but would pick it systematically and thoroughly. He attentively looks at them, fishes out the best and biggest ones, and eats them, grunting incessantly.[4] Even when the rest of the cherries look the same, Joni continues the selection, although there seems to be no point in doing this any more because eventually he eats them all. During the eating of the tastiest cherries, Joni's grunting turns into a hollow sound reminiscent of the expectoration or the hollow coughing of a human. The chimpanzee skillfully separates the cherry pulp from the cherry stones, but if they get accidentally into the plate with the pulp, he immediately throws them out, using his fingers.

5. Preferred Eating Conditions.   The satisfaction derived from eating is determined by the mental condition of the animal, which is dependent on the chimpanzee's impressions. First, the chimpanzee eats with greater appetite when surrounded by friends. He grunts expressively and rapidly, once in a while extending his hands to one of them, grunting even more loudly.

When the chimpanzee is put into the cage and, as compensation for his lost freedom, is given his favorite food (sour cherries, bananas, or candies), he remains indifferent to the food for a long time and seems not to notice it. When finally he starts eating, he does not express his gustatory pleasure by his usual grunting He eats silently. In contrast, when Joni is outside the cage, he devours even less attractive food (which he earlier rejected) greedily and with a smacking and grunting sound, explicitly showing the pleasure from pleasant gustatory sensations.

The chimpanzee eats with special appetite if eating alternates with play: Taking a piece, he runs away and eats it; then, playing, he runs back to you for another one. Or, you pretend to be catching up with him to take the food back from him.

The chimpanzee likes to eat looking out the window, watching the street traffic of people walking or riding horses. If Joni sits on my lap while he eats, he does not stay still. He fumbles at the creases of my dress, stares at my face and feels it with his fingers, or reaches out his leg to hitch and drag over a piece of furniture.

Watching the chimpanzee eat (if the food is not too tasty and he is not too hungry), one can see that he is bored with the eating process itself and tries to enliven it by any means. In this respect, he strongly resembles children of the same age, who are known to consider eating an unpleasant duty and are reluctant to abandon their play for a meal; they can sit at the table only after extensive training, and even then, they use various tricks to get some entertainment. Everyone knows that the children living in rural areas eat all the time while they play or run, stocking themselves up with enormous pieces of bread or anything else that is available.

The chimpanzee especially enjoys eating something he has found on his own or the food he has obtained surreptitiously, using special tricks.

6. Food That the Chimpanzee Finds by Himself.   In the wild (in the field or in the woods), the chimpanzee makes ingenious attempts at exploring the territory and finding suitable edible material: fresh sprouts of plants, buds or petals, strawberries. He hastily picks and eats them. Several times, Joni found his way into the food cabinets and ravaged edible products (i.e., sugar, dry fruit, candies, or cookies).

Joni persistently tries to gnaw off pieces of lime in his room; he climbs for this purpose to the ceiling, hangs for hours on the wall, and chews at the protruding parts of the ceiling. He uses every opportunity to taste lime, and because he is forbidden even to scratch it off, he resorts to some tricks. He does not try to touch it when we are around, but as soon as we leave his room, he uses various devices to open the cage and immediately climbs to the ceiling, immersed in this destructive activity until we come back. As soon as he hears one of us coming near his room, he quietly comes down and sits innocently, completely unaware of the fact that his continuous chewing and his lime-stained nose give him away.[5]

Once the chimpanzee takes a piece of lime into his mouth, no persuasion can make him give it up, although he easily dumped other inedible products from his mouth and gave them to us on our first request. This was an indication of how much Joni loved lime.

Needless to say, the chimpanzee swallowed very greedily the insects he caught (i.e., flies, spiders, or parasites [lice]).

Hunting instinct. We could often observe the chimpanzee running all over his room, getting on couches and chairs, following flies, catching them either with his hands or with his extended lips, and immediately eating them. With the same intense attention, the chimpanzee inspected his hair, looking for parasites; he skillfully caught them and snapped them with his teeth. I tried to break this unaesthetic habit of his by taking the parasite from him, but his bitter crying proved how much this prohibition deprived him of pleasure.

The chimpanzee never ate cockroaches (he only caught and killed them), probably because of his instinctive fastidiousness[6] toward certain insects.

Ownership Instinct   *Retaining the property*. The chimpanzee shows a strong ownership instinct over his food supplies. If I give the chimpanzee one of his favorite foods (for example, an orange) and ask him verbally and by gestures to share it with me, he either tries to eat it up hastily, turns his back to me for the rest of the eating, or eats with his hand covering his mouth. When I try to take a look at what and how he eats, he hides the food behind his side.

Joni is even angrier when he is denied his share, for instance, when I intentionally give a tasty piece of food to others instead of giving it to him. After the first moment of puzzling over why this piece "has missed" him, Joni immediately takes a swing at his lucky rival and tries to snatch the food from him.

The chimpanzee does not like when somebody (even one of his closest people) reaches his or her hands out to the food or to the bowl while he eats; he apparently considers it an infringement on his property. The animal quickly catches a hand stretched out to him, and it the person is strange to him, he even tries to bite it.

This ownership instinct of chimpanzee is not only over food, but toward all objects that are usually around him, that he has gotten used to, and that he has reasons to deem his own. Joni is most aggressive toward anyone claiming his property; the less familiar to him this "delinquent" person is, the stronger the chimpanzee's aggressiveness.

On Joni's very first day at our place, I wanted to take the case in which he was brought out of his room; he immediately bit me. Tampering with his sheets (even by his closest people), which was necessary for their daily airing, for a long period of time had been the reason for his angry protests and violent defiance. He became totally unresponsive to all requests, persuasions, yells, and even corporal punishment; he would not give up his rags for the world. He gripped them, holding them as fast as he could; he did not let them go even when he was dragged with them. He held them not only with his hands, but also with his teeth, and he pressed them to his body if they were about to slip away from him. At that time, the chimpanzee looked unusually detached: His eyes became dimmed and senseless, his mouth was open, his teeth were exposed. His entire body, his arms and legs, were endlessly moving, either hastily pushing the attacker away or tenaciously grasping the object of contention. This commotion was accompanied by the chimpanzee's incessant hoarse breathing. This scene usually had two different endings: Either the person had to give up his or her claims and pass the object into the fullest possession of chimpanzee, who finally brought it to the faraway spots of the cage and sat down on it; or, using a successful trick, the human managed to snatch the thing away from the chimpanzee, running the risk of embracing the angry animal that was ready at the moment to bite even his closest guardian.

This daily struggle for the chimpanzee's property was softened somewhat when the method of exchanging old sheets for new ones was implemented. Having joyfully received the new sheets, the chimpanzee was ready to part easily with the old ones. Once, Joni dragged somebody's lab coat into his cage, and when the owner of the lab coat was going to take his property back, the chimpanzee attacked him so fiercely, was so enraged, and yelled at him with such a sharp and violent hooting that the owner had to give up his rights.

The chimpanzee also passionately guards things not very useful to him, such as learning aids. In connection with this, I recall a demonstration session with several experts participating . I studied Joni using the selecting-a-sample method. Hastily passing a selected object to the experimenter, Joni accidentally dropped it to the floor. One of the honorable and venerable scholars, out of sheer courtesy, was quick to bend down and pick up the object. But, as he touched the thing, his "student" thanked him by hitting him on the head so hard that the professor understood right away that his gentlemanly deed had been misinterpreted; the professor had to return the object quickly to the animal so he would not become a suspect in the misappropriation of somebody else's property. Our learned friend got away relatively easily; another time, the mere touching of the chimpanzee's paraphernalia by a strange person caused such a violent reaction on the animal's part that it left tangible and lingering traces on the person's body.

The mere loss of the chimpanzee's "own property" enrages him. Joni struggles with the adversary to regain the object, sparing neither the man (no matter how familiar the adversary is to him) nor himself. His desire to repossess the object is so strong it is as if it were associated with his most urgent natural needs or his most favorite entertainment.

If you give the chimpanzee a most usual object he has totally ignored before and then you pretend to take it back, Joni turns into a real beast. Using his hands, feet, and teeth and roaring hoarsely, Joni pushes you with his chest and legs and bites you in response to the smallest hint of losing the object. As soon as you give this much desired object back to the chimpanzee, he immediately calms down; however, he does not express much interest in it after that. He puts the object aside, not paying any attention to it and virtually forgetting about it for a long time until the possibility of its loss appears again; then, his desire to regain it glows anew.

This passionate contention for his property is also aimed at people; it acquires the form of jealousy (see chapter 3, Social Service).

Accumulation of property. The chimpanzee not only guards his property, but also applies every effort to acquire it. Living in the wild, he picks up from the ground such objects as pieces of iron and stones, puts them into his mouth, and brings them home. Such manners of his resemble those of children, who are inclined to bring home worthless material from the street.

The chimpanzee is ready to take anything that has been shown to him, although he does not always use it. He cries and sometimes bites if, after showing him something new, we do not place it in his full possession. Once, I gave him a box with some granules that made noise when shaken [plates 36(1), 36(2)]. Joni expressed great interest in this box. He bent down to it, extended his lips forward, and kept staring at it. Fearing that he might open the box and eat the granules, I was going to take it back from him; he reached out for the box and began crying hard. Overcoming his resistance, I snatched it from his hands; he even started to bite my hands [plate 36(3)].

Preferences. For the entertainment, I often gave Joni piles of colored shreds, rags, ribbons, and laces to sort. This was one of his most favorite pursuits; he could sit for hours, concentrate on picking something from the pile, look at it attentively, and hang over his neck the ones he liked the most, things that were *bright, shiny,* and *long.* He would not give up these selected things for the world; he did not allow us to replace them into the pile. He brought them to his bed and put them there, guarding them from possible infringements. As a result of every one of these sortings, Joni came into possession of at least a couple of new things.

The chimpanzee was reluctant to give back to me the learning aids that were his favorite, the colored plates that were predominantly *blue* or *dark blue.* Every time Joni had the plates, he most often picked the blue circles, hid them in the safest place (his mouth), and played and ran with them. The theft was uncovered only when we had to put the plates back into the box.

Despite my requests, the chimpanzee did not want to return the plates to me and tried to bite me if I put my hand into his mouth to extract them. However, giving in to long persuasion, if he finally let them go, there always were a couple of circles left in his mouth. The procedure usually had several iterations; I often could not find the missing plate after thoroughly examining his mouth. Joni was so ingenious in his deception, he so masterfully hid the plates under his tongue or in the depth of his mouth between the gums, that it defied even the most complete search, which he did not always allow me to do.

Sometimes, if he took objects from the learning aids group that he was not allowed to take, Joni carried them out of my sight to the darkest corners of his room to play with them quietly and freely without running the risk of being caught red-handed. I have noticed that he enjoys stealing forbidden things very much.

As already mentioned, the chimpanzee loves to eat the food he has stolen. However, I observed several times that it was not only the food itself, but also the very process of stealing that attracted him. Once I caught Joni when he had an apple in his hand that he had just stolen from the desk drawer, but had not yet eaten. He generally was very excited when he took forbidden things (e.g., checkers) out of the drawer and ran away so that he would not get caught. He usually got away with all these thefts; they were left unpunished. Nevertheless, intentional stealing of the forbidden objects can be credited justly to the entire range of "chimpanzee's kin."

Of all three-dimensional shapes, Joni prefers spheres. The chimpanzee can often be observed running from one room to the next, surreptitiously snatching a ball and playing with it.

Nest-Building Instinct While sorting out piles of small things (e.g., rags or scraps), the chimpanzee finds a different application for them. Sitting on the floor, he lays them out around him, making a whole structure in the form of a nest. Sometimes, he tears the paper into small scraps, puts them under his body, and wallows in them, lying on his back and raising his legs into the air [figure 11(1)].

Joni willingly gets into a deep enclosed space (such as an empty case, carton, or basket), around which he places various rags and scraps, fencing it in on all sides. Often, Joni spreads his blanket on the floor, sits down in its center, and puts piles of small wooden blocks around him in the form of a fence.

As stated above, the chimpanzee will not go to sleep without getting his sheets. He spreads them and tucks them under his body in a businesslike manner; he shapes a part of the sheets in the form of a pillow and puts his head on it. But, the chimpanzee does not have a sense of topographical localization of his nest. He builds it in strange places and carries it around, conforming to the actual environment.

Usually, his sheets are on the small wooden platform in a corner of his cage. If the chimpanzee is worried for some reason, he brings the sheets down to the floor in a faraway corner of the cage, where he hides or considers himself the least visible. If it becomes cold, the chimpanzee, on his own initiative, brings his nest to the highest place in the room, a shelf under the ceiling.

One time, Joni built some kind of nest, and we watched him doing this. Apprehensively looking at me, he seized a long linen curtain, ripped it from the curtain rod, and dragged it along with him to the top of the cage. Holding on to the wire screen of the cage with his hand and both feet and carrying a curtain 3 meters long in his other hand, Joni climbed higher and higher. As Joni moved up, hanging only on his hands and pressing the curtain between his legs, he gradually lifted it up to the ceiling. The curtain accidentally hitched at the wooden boards or at the nails; the chimpanzee disentangled the curtain with great difficulty and finally spread it along the screen in a wide semicircle. He sat down on the shelf in the center of this semicircle, creating some form of closed retreat.

If the chimpanzee wants to lie down, he unceremoniously spreads the rags in all directions, tucks them under his body, and lies comfortably on his back or on his stomach for hours. If the chimpanzee is outside the cage and lies at some distance from the floor, he takes some measures not to fall. Once, I put him on a regular iron bed with legs. He slept all night, holding on with one hand to the bed poles, and every time I unclenched his fist and pushed his hand off the pole, he quickly reached out for support again.

While sleeping on a very wide and horizontal surface under the ceiling of the cage and fearing he would roll down, he held on to a belt hanging from the ceiling.

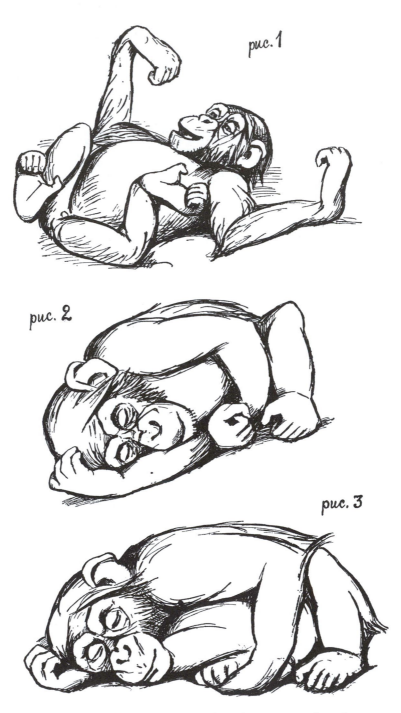

puc. 1

puc. 2

puc. 3

Figure 11. The lying postures of the chimpanzee awake and asleep: (1) merrily rolling about on his back; (2) and (3) asleep.

Sexual Instinct   Getting into his nest and lining it with soft cloth, Joni often rolls on his back or on his stomach. Sometimes, he holds a soft ball with both hands and both feet, kneading it, letting it go, and then catching it again and pressing it to his body. When during these manipulations the ball turns out at his groin, Joni presses his penis to it, gets excited, and makes some kind of coital movements. He breathes rapidly, his eyes become dimmed, his stare is blank, and his erection is visible. Later, I often noticed that Joni, while throwing a ball, ran after it and, catching it, held it with both his hands and pressed his tense penis to it, moving his body back and forth and angrily resisting my attempts to take the ball from him.

Similar sexual tendencies, accompanied by intense breathing, extension of the lower lip, and hunching of the upper lip, were shown toward Joni's bedroom vessel.

At other times, the chimpanzee's emotion of general excitability and any strong emotion, whether joyless, sad, or especially angry, always called forth his erection (see the text and the footnote on Joy in chapter 2; see also plates 7(4), 23(1)–23(6), 28(2), 28(4), 28(5)].

If you try to leave Joni alone in the cage, he protests; if you give him a sheet as consolation, Joni is not satisfied with the substitute and begins to bite the sheet violently. Then, pressing his body to it, he shows obvious signs of sexual excitement, if not sexual violence, over the inanimate object that has caused his angry feelings. When we left Joni alone in his cage, he roared angrily and hung on the cage screen. Eventually, his penis became tense, and he made penetrating movements into the holes of the screen.

Sometimes, during Joni's affective behavior, I observed that if the chimpanzee did not have suitable objects against which he could rub his erect penis, he grasped it with his toes and began rubbing. Our photographs [plates 29(1), 29(3), 30(1), 30(2), 37(3), 37(4)] show Joni's erection in association with the mental excitation of the animal, a condition that had nothing to do with his sexual behavior per se.

Chimpanzee's Sleep   The chimpanzee sleeps most willingly under the protection of his human guardian [plate 38(1)]. By 8 or 9 P.M., when the chimpanzee becomes tired, he calms down, does not leave me even for a moment, tries to get onto my lap, presses his body to me, and whistles and whines softly. I watch him go to sleep.

He slowly opens his huge mouth, uncovering his wide gullet; then, there comes a long, sweet, and contagious yawn. With every second, his eyes become more dimmed and senseless. He still tries to look, but the whites of his eyes become so visible that he seems not to see anything. Trying to resist an unsurmountable force, he rapidly blinks, but sleep comes closer, bringing his eyelids down. His eyes close, although his hands still move weakly, diddling with objects that are within his reach. Here comes a slight noise, and in no time, the chimpanzee's eyelids open widely and fearfully; his dim glance surveys the surroundings, but in the next second, powerful sleep puts its heavy seal on his eyes. His eyebrows shift down, the bridge of the nose wrinkles, and his eyelids close so tightly that only the tips of his dark eyelashes can be seen. The chimpanzee finally is asleep; his facial expression is tense and worried. As time goes by, this tension eases, but it does not disappear completely. His sleep is light and nervous; his whole body often shakes, and his arms, legs, and head twitch.

The poses of a reposing chimpanzee are very diverse [figures 11(2), 11(3)]. He usually sleeps on his side; he turns away from light, rolls himself into a ball, and presses all his extremities to his trunk. His hands serve as a pillow; most often, he comfortably puts his head on his right or left hand, depending on the side of his body that is close to the ground, and rests his temple or the side of his face on his palm or on the inner part of his forearm [figures 11(2), 11(3)]. The other arm, which is not participating in making a pillow, often embraces his legs. They are raised to his chin, so he thus fastens their divergence from the trunk and creates a compact position conducive to his staying warm.

It has been noticed that this pose of chimpanzee changes depending on the room temperature. When it is cold, the chimpanzee either lies on his stomach with his face down, tucking his hands and feet under his body, or lies on his back, embracing himself with his arms and legs, drawing his legs up to his stomach, and completely covering his trunk with them. On hot summer days, the chimpanzee sleeps on his back, throwing his extremities to the sides, his legs upward at a right angle to his trunk, and sometimes holding his feet in his hands.

I often noticed that, if given some hay to lie on, the chimpanzee tried to make a bed out of it, whipping the scattered stems into a mound that resembled a pillow.

The chimpanzee prefers soft sheets, which he uses as a mattress when it is cold, but he also tries to make himself a blanket. For this purpose, he uses rags, with which he covers his legs, sometimes up to the middle of his trunk. In our conditions, for which the temperature was never below 15–17°C, I never saw the chimpanzee covering himself up to (or over) his head. Moreover, fearing that the little animal might catch cold, if with good intentions I covered him up to his head with a blanket, he took the cover away from the upper part of his body, carefully tucking it between his legs.

I often heard the chimpanzee snore the way humans do.

Love of and Struggle for Freedom

Sometimes, you could observe the chimpanzee huddle from the cold; nevertheless, he would not allow you to cover him. In the beginning, I did not understand the reason for this rejection. But, having analyzed a series of similar events, I concluded that it had something to do with his self-preservation instinct, which forbade us to impose even small limitations on the freedom of his actions.

In fact, the chimpanzee vigorously objects to any restrictions of his movements and does not tolerate any kind of tying him up. As already stated, he never allowed us to bandage his fingers, not to mention apply compresses to his trunk.

When it was cold, the chimpanzee prevented putting on any kind of apparel (for example, a soft woolen coat). If, however, you did succeed in putting it on him, at the first opportunity he would take it off, using his hands and his teeth until he completely freed himself of it.

One time, it was necessary to wrap him in a shawl because I needed to transport him by car and then by train. As long as he sat on my lap in a closed car, he did not protest. But as soon as we entered the train station and he found himself in a crowd of strange people, he became edgy, crawled out of the shawl, tried to break loose, bit me a few times, and calmed down only when I freed his hands and left only his legs covered.

If you try facetiously to hold the chimpanzee by his arm or leg or by his fingers or toes, he becomes violent, vigorously beating you off with his free extremities. He breaks from you, risks dislocation of his joints, wriggles his whole body, dashes around the room, fidgets, and tries to take his lost freedom back. When he does not succeed, he clutches at your dress, threatens you with his open mouth, holds your restricting hand with his teeth, and attempts to bite the hand and increases the pressure of the teeth depending on the force and duration of your resistance. If you persist and do not let him go, he becomes violent, utters hoarse sounds, almost chokes, and almost closes his eyes. Following you with a dim, senseless glance, he tries to scare you with his open mouth, as if ready to attack you and to tear you to pieces. If you let him go, he calms down immediately and amicably plays and romps with you as though nothing has happened.

All these examples of the chimpanzee's freedom-loving tendencies reveal that their origin is the self-preservation instinct of the animal. Any restriction of the chimpanzee's freedom, especially of the movements of his most effective defensive tools, his arms, immediately deprives the animal of his confidence in his ability to defend himself. He by all means tries to come out of this situation he considers dangerous to his well-being.

No wonder that, at the train station in the crowd of strange people, our attempt to cover his hands called forth his vigorous resistance, while in familiar surroundings he could tolerate such covering.

When separated from open spaces and limitless, powerful, virgin, moist, and warm African jungle and then imprisoned in a cold country in the stony bowels of a multistory building in a small room and in a cramped cage, the sanguine nature of a chimpanzee ceaselessly longs not only for the freedom of actions, but also for the freedom of movement. And he uses every opportunity to broaden the sphere of his activity.

If Joni is closed in a cage, all his designs and intentions concentrate on getting out of the cage and into the room. If he is in a small room, he cannot wait to go out into the corridor or into the next rooms.

While in the countryside, Joni does not settle for staying in the house, but comes out on the porch. From there, he climbs to the roof and moves along the fences to the neighboring buildings, getting to the very top of them [plate 15(1)].

There is nothing more pleasurable for the chimpanzee than to accompany us on our excursions into the fields or woods, though he stays close, not brave

enough to undertake his own reconnaissance. He apparently fears possible encounters with animals—horses and cows, of which he is extremely afraid.

You have to experience the awkward and painful procedure of putting the chimpanzee into the cage to understand and to feel how hard this confinement is for
him and how he tries to avoid it. You open the door of the cage and lead him
inside, but he does not want to be led. He resists you, cries, roars, and clutches at
your dress, at the door jamb, or at the cage screen. You push him through and
close the door, and he starts an ear-splitting roar that subsides only after a while.

If the cage is closed with a latch and you leave, he immediately sticks his
index finger through a screen hole, pushes the nearest end of the latch, and
triumphantly gets out. To prevent this, I tied up the latch with a string, but the
chimpanzee reached the string, drew it to himself, ripped it off the latch, and
then opened the latch as earlier, by pushing it. I used a stick instead of a string,
but the chimpanzee also found a way of removing it and got out again. I resorted
to my last, and seemingly very reliable, method: I padlocked the latch. What did
I find? If I left the key in the padlock, the chimpanzee, using ingenious tricks,
reached out, drew the padlock toward himself, turned the key, opened the latch,
and freed himself again.

With various degrees of success, we used the following restrictive stimuli: a
sharp cry, the swishing sound of a whip, knocking on the floor with a stick, and
if all of that failed to bring any results, whipping Joni or showing him a picture
of a chimpanzee, a mask of a human face, a stuffed head of a wolf, or a cormorant.
We have to admit that the least effective method was whipping the animal because at moments of nervousness and anxiety he was almost insensitive to pain
and would not budge.

Among more peaceful methods of putting the chimpanzee into the cage, the
most effective was presenting him with a new object that riveted his attention
and curiosity. For a few seconds, he stopped his resistance to being put in the
cage; this short time was enough for me to latch the door and padlock the cage.
Later, having experienced my treachery, the chimpanzee either sat in the door-
way or tried to satisfy his curiosity as quickly as possible and tried to prevent me
from closing the door. But, I became exceptionally ingenious, preparing for him
intricate surprises wrapped in paper, put in a series of boxes, or put in smaller
boxes enclosed in bigger ones. Unpacking and opening these bundles and boxes
distracted Joni for time sufficient for me to close the cage. Subsequently, Joni
was afraid to lose his freedom again; he always sensed when this unhappy moment was coming.

In the mornings, after the first meal, the chimpanzee is usually put into the
cage; soon, he became more cautious due to this experience. As soon as he has
finished drinking his milk, he impatiently looks up and escapes from my hands.
He rushes onto the roof of his cage, near the ceiling; he can sit for a long time
totally out of reach.

Not being able to get him, I appeal to Joni verbally and by gestures, urging
him to come down; it is all in vain. He is unswayable and completely indifferent
to my entreaties because he knows from experience what is going to follow.

Therefore, he prefers to wallow on the roof for hours, doing nothing but remaining free, rather than have access to various entertainments in the confinement of the cage. The chimpanzee will not trade his freedom for tasty food or toys; he will not give in to caresses or threats.

I get impatient and bring a ladder to get onto the cage's roof; Joni immediately comes down, crawls under the cage, dashes around the room, and dodges the pursuit so adroitly that it makes catching him virtually impossible. Sometimes I toss up scary objects (for example, a mask of a human face); Joni gets angry, curves his mouth, and dodges, but does not shift a bit.

I could make Joni come down from the cage by showing him something new or by staging an assault on me. In the first case, Joni is ready to sacrifice his freedom out of curiosity; in the second, he feels sorry for me and is going to rush to my rescue any minute. However, he is ready to retreat, correlating his closeness with the intensity of the assault. That is, when the assaulter is more aggressive, the chimpanzee may come very close to him and even bite him; when the assaulter is less aggressive, the chimpanzee may come down only halfway from the roof and glance threateningly down.

To keep the chimpanzee in my power, I held him by the hand or by the foot when he was finishing his meal to prevent him from running away, but he found a way to outwit me. Now, he did not want to empty his mug of milk and left a few gulps in the mug. I spent a long time trying to persuade him to finish the milk; this put off the moment of placing him into the cage. However, sometimes during such persuasion I eased my attention and forgot that I had to hold him; he took advantage of this moment and rushed onto the cage roof and went up to the ceiling or got under the furniture into the faraway corners, from where he could be extracted only by force.

Sometimes, I would only reach out my hand to Joni and say, "Come with me," meaning to go to other rooms. He would immediately come down from the cage and allow me to take him on my lap, often getting into a trap, because after a short sojourn in that other room, he would be brought back into his room and into the cage. In wintertime, he was wrapped with a blanket over his head when taken to other rooms to protect him from catching cold. After a series of such wrapping procedures, Joni started to construe them as a conditioned signal to go out to other rooms. He would put any rag on his head, expressively extend his arm toward me, and sit in such a position for a while, glancing at me from under the rag. If I attempted to take the rag off his head, he grunted belligerently, snatched the rag from me, and put the rag on his head again; hugging me by the neck, he hung onto me and was absolutely ready to follow me anywhere. If I objected, he attacked me, pounded me angrily, and tore the rag with his teeth, as though taking it out on the rag for thwarting his passionate desire to leave the room. He did not calm down until we took him out.

The chimpanzee prefers to sit or lie motionlessly outside the cage beside me for hours rather than be in the cage filled with toys and entertainment.

All these examples prove the chimpanzee's persistent intention to gain and preserve his freedom and the ingenuity he employs for this purpose.

Self-Preservation   Having obtained freedom, and left to his own de-
Instinct   vices, the chimpanzee bears himself apprehensively,
distrustfully, and suspiciously. When you observe the chimpanzee in nature, you
can see clearly how defenseless and helpless a creature this little "half-human"
is, deprived of his homeland, of his kin, and of his family. No wonder that,
finding himself in such a strange and artificial atmosphere, the little chimpanzee
worries, most of all, for his own well-being. He is frightened by everything unexpected.

1. Fear Forms and Stimuli That Cause It.   A sudden sound, cry, or noise scares
Joni. He immediately rushes under human protection; his face gets pale, his heart
beats rapidly, and tiny drops of sweat appear on his face. Opening his eyes wide,
he nervously looks around and does not calm down for a long time.

Auditory frightening stimuli. Once, the chimpanzee was extremely awed by the
sound of a bursting paper bag; he took his head in his hands, pressed his face to
the ground, and shielded his head, instinctively defending first the most valuable
part of his body. When, after a while, Joni mustered enough courage to raise his
head and look around, his eyes were open wide and expressed disbelief, as though
asking, "What is it all about?"

The same panic is caused even by a faraway sound of gun fire, by the smacking
sound of lips in imitation of gunfire, or by the sound of a whip hitting the floor.
I suppose that Joni had some fearful experience associated with a gun, and this
was the reason for his panic. You might take a stick or a cardboard tube and aim
it at the chimpanzee, smacking your lips at the same time, and he would fall onto
the ground in horror. I would knock suddenly from under the table on which he
was sitting; not knowing who caused the sound, he immediately would press his
trunk to the table, look apprehensively around, and search for the violator.

When you loudly call for somebody in Joni's presence, he perceives this as a
signal of danger. He comes closer to you and gets fluffed up and worried. At that
time, he most vigorously resists putting him back into the cage or your leaving
the room.

Often, in the process of a long and hard struggle with the chimpanzee, when
we put him into the cage or drove him off the cage roof, it was sufficient to knock
with a stick or smack with a whip on the floor, and he obediently rushed into
the cage. We are absolutely sure that it was the sound and not the sight of the
hitting objects that scared him because, first, he had never been beaten by a stick;
second, he had never been afraid of a whip due to his insensitivity to physical
pain.

The more unusual the sound, the more threatening its perception. Once, when
the chimpanzee was sitting quietly on a bench near the house, he suddenly heard
the sound of a horn. He threw his arms into the air, fluffed up his side-whiskers,
pressed his toes into fists, and began to stare into the distance, extending his
tightly closed lips forward. He sat motionlessly in this pose until this threatening
stimulus was over [plate 19(2)].

A sudden cry is even more frightening for the chimpanzee; it alerts him instantly. Even when the chimpanzee's behavior is aimed toward achieving a certain goal (e.g., not giving you back an object) and it seems that no external force can distract him from doing this, it takes only a strange whining cry, and he becomes frightened and stares at you with his eyes wide open and full of fear. He stops his defiance immediately.

The chimpanzee cannot get used for a long period of time to the clicking of a photo camera; he is startled every time he hears this sound. During the film shootings, he tries to sit closer to me and asks to get onto my lap. If I refuse to take him or I wave him off, he closes his eyes in utter despair, bends his head, and roars until I come up with some entertainment to console him.

The hitting of a piano key, especially in low tones, and loud and forceful sounds are extremely scary for the chimpanzee. First, he nervously dashes around the room as if looking for the source of the sound, then angrily attacks and bites suspicious objects. Then, having related the sound to the object (e.g., the piano key), he curiously looks at the keyboard and watches me hitting the keys, but he cannot make up his mind whether to touch them himself or not because his sense of fear subdues his sense of curiosity.

It is natural, that such sudden, unlocalized, powerful sounds as thunderbolts call forth extreme fear in the chimpanzee. At the sound of a thunderbolt, he fluffs up, presses his body to a human, and fearfully looks at the sky, searching for the source of the sound.

Luminary frightening stimuli. Suddenly appearing light scared the chimpanzee no less than a sudden sound. When photographed in the evenings, he instantly left the table on which he was sitting; he was consumed by fear, roared violently, and refused to come back for the next shooting. Even after many snapshots taken by flashbulb, so he had every opportunity to be convinced of the total harmlessness of the flashbulb, he never got used to it and startled every time it was used. The same reaction of fear in the chimpanzee could be observed during thunderstorms and flashes of lightning.

Tactile threatening stimuli. The chimpanzee becomes extremely frightened at sudden tactile perceptions. If you jokingly throw a paper ball at Joni, his whole body starts shaking. He throws his arms into the air, closes his eyes with his hands, rolls himself into a ball, and convulsively touches the spot where he was hit.

The chimpanzee is frightened more easily in response to an unforeseen touch. If he sits on a chair with his face turned away from you and you touch the other side of his face with your finger, he will fall immediately to the floor, run away, be afraid to turn around, press his body and his face to some object, and roar. He will not look up and go away until he is taken by one of us.

The chimpanzee becomes even more frightened at sudden and more tangible shocks. One time, Joni fell from a fairly low trapeze. He started to cry and rushed into my arms. I am convinced that it was the mental shock, not the physical pain, that made him cry since usually he did not cry when he hurt himself badly in climbing the trapezes or swinging on the ropes. Once, a chair fell on Joni, and he immediately turned pale, rushed to me, and pressed his body to me; I could hear his heart pounding.

The chimpanzee's spells of anxiety and fear often are accompanied by the involuntary evacuation of his bowels, but he does not always fluff up. On the contrary, during sudden and strong fear, the chimpanzee's hair is entirely straight and close to his body.

New threatening stimuli. The chimpanzee is less afraid of new impressions than of unexpectedness. He is apprehensive toward every new thing in his surroundings. He is suspicious toward every new person and tense in any new circumstances. When you show a new object to the chimpanzee (e.g., a simple glass box), he looks at it very intently, extending his lips. He fearfully reaches out, trying to touch it with the tip of his index finger and ready to withdraw it at any moment. Then, he raises his exploring finger to his nose, smells it, and only after that decides to engage in closer contact with the object.

We have already mentioned the disgust with which the chimpanzee treats unpleasant olfactory stimuli, but some smells (e.g., the smell of raw meat) cause not only his strange anxiety, but also quite definite fear. We discovered this by accident. Once, the chimpanzee was running from one room to another, emitting the usual short hooting sounds as he often did after spotting something strange. I followed him to the kitchen, where he repeated this sound a few times; his hair was fluffed up. I ascribed his anxiety to the abundance of new things in the kitchen; because his attention was not directed to anything in particular, I simply took him out of the kitchen. After a while, I heard the same sound coming from the kitchen again; I went there and saw Joni bending over a piece of raw meat. He smelled it, then ran away from it and rushed to me. I took the meat and carried it over to the chimpanzee. Joni withdrew, ran aside, and looked at it as though mesmerized. I came closer to the animal with the meat in my hand. He ran away from me again to the opposite side of the room, hurriedly turned around, and got into the hidden corners or under the tables or chairs. When I can closer to him, he started dashing around the room looking for a refuge. When I touched him with the meat, he shivered and, smelling the spot I touched, wiped it off with his hand as if trying to remove the smell.

Once, when the stove was burning and the room was filling with smoke that irritated the eyes, Joni got frightened and hurried to me for protection. Joni once tried to locate an invisible enemy (turpentine, with which he had been rubbed).

Other new olfactory stimuli also generate fear in the chimpanzee. One night, Joni was given new washed and air-dried sheets. For 2 hours, he stubbornly resisted lying down on them: Every time I put him to bed, he fled, roaring. Not understanding what was going on, I gave Joni something to eat and to drink, but this did not help. Joni continued his crying and his defiance. To check whether he was well, I started playing with him, and he immediately cheered up and stopped crying. When I put him to bed again, he resumed crying. I sat him on my knees for some time, and he went to sleep at once. But, as soon as I tried to put him to bed, he woke up, and with the same fear, recoiled from the bed, smelling the sheets. Then, it dawned on me that Joni was afraid of the smell of freshly washed sheets. I put my apron on the bed; after smelling it, Joni tucked it under himself and closer to his face; he fell asleep at once and slept like a log.

The chimpanzee is also afraid of empty dark hollows, heights, and various stimuli previously unknown or incomprehensible for him. For instance, Joni accidentally opened the lid of a big basket and, seeing the void, got nervous; he looked intently into the basket, then ran away from it, fearing for his well-being.

Concerning the chimpanzee's fear of heights, every time I observed him looking from the balcony of a two-story house or taking a look over the banister of a stairway into the depth of the three-story staircase, I noticed his extreme tenseness and desire to stay away from these dangerous places. However, when the chimpanzee was allowed out of the cage, he willingly mounted the high roofs of two-story buildings and could walk the ridges of the roofs for a long time, apparently not afraid of falling down.

The same tense distrustfulness was observed in the chimpanzee toward newcomers, whom he typically subjected to a preliminary examination of timid probing, tentative touching, and attentive smelling, which always preceded his entering new communication.

When we were going to take him from his previous owners, my two colleagues and I once were the objects of his apprehensive attitude toward people strange to him. In spite of the fact that we had not applied any repressive measures to the chimpanzee, he objected when any of us tried to take him in our arms. He beat us off violently and dashed around the room, fleeing us like the most savage beasts. He cried, yelled, bit us, and was so enraged that he could be taken only by force. With the help of his owners, we caught the animal and put him into a temporary transportable cage.

Fear of a crowd. The chimpanzee becomes frightened in a big crowed of people strange to him. Once, there was a group of 20 students that came to his room. As soon as Joni saw this crowd, he pressed his body to the ground, shaking and roaring desperately. He began to calm down only when he heard my voice and saw me, but he could not recover totally from fear and continued shaking for some time.

Any new surroundings that contain many scary surprises provoke fear in the chimpanzee. When we brought him to our flat, his first reaction to his new room was fear and the resulting state of deep depression. In the beginning, he sits still, afraid to make a free movement, and bears closer to the people, surveying the room attentively. His eyes travel from one object to another. Most of all, he is afraid to be left alone, and he uses everything at his disposal to keep his human protector around.

Fear of loneliness. Left alone against his will, the chimpanzee shrills; involuntary evacuation of his bowels often results from fear. Then, the chimpanzee gets into the darkest corner, sits there gloomy and motionless, and does not want to do anything. If somebody comes in, he rises from the floor, starts playing, and gets onto the trapezes and uses them like a skillful gymnast. Should the person leave, the chimpanzee gloomily comes down and goes back to his corner.

If someone appears beyond the glass door of his room or if he sees me from far away through a partially open door, he immediately gets excited and starts playing. As soon as he loses this illusory communication and protection, he becomes depressed, indifferent, and passive again.

Feeling sorry for the little chimpanzee, I put on a lab coat, covered my head with a kerchief and lay on a couch in Joni's room, turning my back to him. It was enough for the chimpanzee to come out of his corner, stop worrying, and begin swinging on the trapezes. In a while, I left the room, made a big doll out of a cushion, put the same lab coat on it, covered its "head" with the same kerchief, and asked someone to bring the doll to the chimpanzee's room. The doll, with the half-covered face and with its back to the chimpanzee, was placed on the same couch, in his full view.

In the beginning, as a result of even this passive existence of the doll in the room, the chimpanzee became less nervous and more active. He started to cry and moan if somebody was going to take the doll away, became nervous, and grunted when clapped on the back. Later, when he got used to his new conditions somewhat, the dead motionlessness of his companion did not satisfy him anymore, and this substitution lost its purpose.

The chimpanzee also becomes depressed in a new situation, when he is in a closed car. During an hour-long trip he sits totally baffled, pressing his body to me. When the chimpanzee was transported in a cab on the way from his previous owner's home to our house, he was placed into a transportable cage hastily made out of wooden poles. Finding himself in these cramped and unusual surroundings, Joni started to cry and shout, discharging earsplitting weeping sounds. But, as soon as I pushed my hand into the cage and took the chimpanzee by the arm, he calmed down at once and sat timidly during the whole trip.

Once, the chimpanzee was overwhelmed by fear at a railway station, where he was surrounded by a crowd of strange people swiftly walking back and forth. Sitting on my lap, the chimpanzee tossed from side to side, worried, and was ready to attack any inquisitive onlooker who would be brave enough to come too close to us. He was especially outraged when, trying to curb his aggressiveness, I was going to wrap his hands with a kerchief, which made him feel even more helpless.

The chimpanzee showed the same fear when sitting in the compartment of a train. Despite being among his closest people, he was very tense and suspicious. He did not take his eyes off me and did not leave me for a second, and as soon as I came up to the compartment door, he immediately caught me, hung on me, and would not let me go, yelling and crying if I tried to break free.

The presence of many new unexpected and frightening stimuli exacerbates the chimpanzee's fear. Once, to entertain the chimpanzee, I brought him into a room where a Christmas tree stood illuminated. There were many people in the room; they talked, laughed, and made noises. Contrary to my intention, instead of pleasure, I brought only trouble to the chimpanzee. He shook all over from the fear of being in such unusual circumstances, incessantly roared, and could not calm down for a long time.

Joni was also very frightened when he was brought to the shooting of a film. Seeing the bright electric light, the cumbersome filming equipment, and many new people, Joni got fluffed up, kept rising and sitting down, apprehensively looked around, stared at the sharp outlines of illuminated objects and their shadows, shuddered at a slight noise or rattling of a camera, and rushed to my arms, looking for protection from imaginary danger.

Another time, I decided to wash the chimpanzee in a bathtub. I never suspected that this procedure would provoke such strong resistance on his part because putting him into an empty tub had not caused any problems before. Water per se had always been his favorite object of play. However, immersing the animal in the tub filled with water turned out to be such a scary procedure for him that I had to struggle with him fiercely before I managed, with the help of others, to wash him. He tried to break free, bit me if I was too persistent, wriggled his whole body, got out of the tub, and did not allow me even to splash him with water. He violently roared and was so enraged that we had to limit ourselves with washing only certain parts of his body.

During our walks with the chimpanzee, he is very tense and fearful. Coming out of the house or coming down the stairs, he is already fluffed up; he follows me without falling even a single step behind. At that time, his lips are pressed tightly to each other, and his upper lip is hump shape [plate 20(1)]. If Joni encounters an unusual object (e.g., a dead bird) on his way, he gets frightened, his hair bristles yet more, and his whole head seems to have a crown of bristling hair [plate 20(2)]. As a result, he bears as close to me as possible

But, even being that close to me, Joni is still far from calm. Since he runs on all fours, his field of vision is quite narrow; every 10 steps, he stops, stands up vertically, looks around [plates 7(1), 7(2)], listens attentively, and if he does not see anything that would rouse his fear and he is sure of his safety, he assumes his previous horizontal position and continues on his way. In the forest, he is most afraid of noises; he peers into the distance, climbs the trees at the first sign of danger, shivers, and comes down from the trees at the slightest crack of a broken twig. In open spaces, he is most apprehensive toward possible encounters with animals, with the intensity of his fear proportional to the size of the animal.

Fear of animals and moving objects. Small animals, such as cats, hens, or ducks, did not fill him with fear, moreover, he never failed to frighten them himself. Big animals, such as pigs or sheep, aroused his suspicions. Huge animals, such as horses or cows, imbued him with awe.

During my walks with the chimpanzee, when he saw a cow, he immediately started to moan; he reached up to me, asking me to take him in my arms. If I lingered for a moment, he clutched at my dress, threw his arms in the air, beat himself on the head, and opened his mouth wide, roaring at the top of his voice. He crossed his hands on the crown of his head in despair [plates 30(1)–30(3)], closed his eyes, took his hands off the crown of his head every few seconds, tensely followed my every gesture, shivered, and waited for my response movement. If Joni saw that I was adamant, he became desperate and cried until he lost his voice, but as soon as I extended my arms to him, he rushed to me and hung on my neck, still trembling from fear and all sweaty. He held me firmly and was ready to follow me anywhere.

During our country trips in a carriage, the chimpanzee, even while under my protection and sitting on my lap, looks around all the time. When he sees cattle or horses headed in our direction, he recoils from them in fear and strongly presses his body to me. When our coachman whips the horses, Joni worriedly hoots in time with the thrusts of the whip.

The chimpanzee is also afraid of big, burly people. The first days of his life in our house, Joni feared only men and obeyed their orders instantly, while women had to struggle with him for hours to make him perform the same action. It was obvious from observation of our day-to-day life the Joni avoided men and preferred to make friends with women.

However, Joni experienced fear toward two miniature reptiles: a grass snake and a tiny, slumbering Middle-Asian turtle. Even when shown the turtle from a distance, Joni covered his face as if shielding it with his hand, turned his back to the turtle, glanced at it hesitantly, and at the smallest movements of its head, took off, hiding in the remote corners of the room. This could be an example of the well-known consternation toward snakes on the part of apes; the turtle partially resembled a snake in the form of its head and neck.

Also causing Joni fear were stuffed animals, even animal furs, and pictures of animals and inanimate objects. This fear subsided only after Joni got to know these frightening objects very well.

Totally unexpected for me was Joni's reaction to a large picture (a drawing by Shpekht) of a chimpanzee. When I showed him the picture for the first time from a distance, the chimpanzee, staring at his own image, started to utter low, barking sounds and evidently was worried. At the second demonstration, he could not take his eyes off the picture, but began to bark only when I tapped on the back of it. As soon as I brought the picture closer to the animal, he immediately retreated to the corner and hid there behind hanging clothes.

Later, I used this picture as a threatening stimulus, which helped me to overcome the chimpanzee's long resistance to placing him into the cage. I had only to show the picture to Joni for a moment, and he instantly ran away from me, which gave me the opportunity to close the cage door. If I took the picture away from the cage, he would not calm down for some time and continued to look at the spot where the picture had been. If, however, he saw that the picture remained in the next room, he sat in his cage subdued and played passively and unwillingly. If I happened to leave the picture in the cage, he violently cried, attacked it, and showed all signs of extreme fear.

Once, in the heat of Joni's play in the middle of a big crowd, somebody brought in a picture of a chimpanzee. Joni's face instantly became dull and pale; he quieted down, as though mentally traumatized, and would not play any more.

Characteristically, a picture of an orangutan shown to Joni at the same time and drawn by a different artist did not leave the same strong impression on the chimpanzee. Moreover, when I put this second picture on the cage floor and left the chimpanzee alone with it, he fearlessly stepped on the picture and walked over it. The chimpanzee was afraid of a third picture (by Kunert) of a gorilla, also shown on the same day, but significantly less than of the picture of a chimpanzee.

The chimpanzee experienced the same fear at the sight of a big (twice life size) painted mask of a human face. Joni was afraid of the mask not only when it was put on people's faces, but also when it was in somebody's hands or when only the nose of the mask was sticking out through the crack of a partially open door. Seeing the mask, the chimpanzee became fluffed up and stayed motionless, emitting a long, dull, abrupt sound (as observed for fear). He did not take his eyes off the mask; at its slightest movement, he ran in the opposite direction, looking around his shoulder as if checking whether he was being followed. If he saw it coming closer, he got under the furniture or hid anywhere he could find a hiding place.

One time, Joni did not want to come back to his cage. Trying to put him in as quickly as possible, I threw the mask in his direction. He shrilled, his whole body started shaking, and he bolted into the cage. Showing him the mask was accompanied by the knock on the door; later, the knock alone alerted the animal and was itself a frightening stimulus that successfully substituted for the mask and that served as a tool for curbing the defiant animal.

Joni feared all kinds of furs and hides of animals; the degree of fear might be different in different cases. He was very suspicious toward fur hats or muffs; he did not even want to touch them. He was afraid of a wolf skin spread on the floor and was panic-stricken when shown the hide of leopard.

When Joni saw the wolf skin for the first time, he gave out a hollow long sound "o-o" (which he often did when he saw an unusual frightening thing). Resting his arms on the floor and bending down to the ground as low as he could, Joni stared at the hide, reluctant to come closer, and then walked away from it completely. As long as the hide was not moving, Joni did not pay attention to it. But, as soon as I put my hand under the hide and shifted it a little, he became alert, extended his lips forward, and tried to find a place to hide. When I raised the hide and came closer to Joni, he dashed around the room, totally beside himself. He cried, groaned, and climbed up and down the cage, looking for proper refuge. He roared from fear if he failed to find one and calmed down only when I, having dropped the hide, took him in my arms.

The leopard hide was shown to Joni in a window across from the balcony where he was sitting (from a distance of 5 meters). When he noticed the hide, the chimpanzee became "glued" to the table, and his whole body expressed sheer terror [plate 19(1)]. Joni stared at the object with his eyes open wide. He squatted, bent his trunk forward, and rested on his arms as though getting ready to run away. He opened his mouth, tightly stretching his lips to the sides and uncovering his gums. He seemed to be hypnotized by the hide, but when I started to move the hide around, he rushed, roaring, for the protection of a human, trying to disappear or flee as far as possible from danger.

The chimpanzee was also overwhelmed by panic when he was shown the stuffed head of a wolf with bared teeth. Seeing such a head, the chimpanzee dashed away, turning his head around every moment. After reaching a safe place, he stared at it, giving out an earsplitting, ringing, or hollow barking like that of a dog, which sometimes ended with a long whining sound. This wolf head served for some time as our tool for putting the animal back into the cage. We used it in combination with the words "wolf is coming," after which the chimpanzee flinched and hurried inside.

Joni also feared a stuffed rampant bear. Even if Joni was in my arms when I was passing the dummy, the chimpanzee bristled and shivered. If I happened to stop near the bear, Joni immediately dashed away as fast as he could.

Characteristically, the first demonstration of stuffed birds (e.g., duck or cormorant) also scared the chimpanzee, only not so severely. He got used to them considerably faster; eventually, they stopped causing him fear at all.

Moving objects were also a short-time source of fear for the chimpanzee. Once, I gave him several flat black wooden shapes. The chimpanzee fearlessly took them, but as soon as one of the shapes fell, he immediately recoiled from it and was reluctant to touch it after. Another time, to entertain the chimpanzee, I gave him a trash bin made of wire. Joni began to play with it, but as soon as the bin fell on its side and rolled over the floor, he hooted, fluffed up, and quickly retreated, looking back apprehensively. Such behavior was reminiscent of that of little children, who often are afraid of mechanical toys.

The chimpanzee has a certain tendency to "refresh his fear" by familiarizing himself repeatedly and more closely with the scary object. Perhaps, his desire to dispel the fear is the basis of this self-intimidation. "In order to defeat the enemy you have to know him better," an old proverb says. The chimpanzee unknowingly follows this principle in dealing with the real or imaginary enemies: He overcomes the fear either by familiarizing himself with the frightening objects or by taking a series of precautions that alleviate or eliminate the possible harm they may cause.

For instance, although Joni was afraid of the mask, it was evident that he tried to look at it more closely when he was not in danger, was far enough away, or was surrounded by people he knew. Although the wolf's head scared Joni and he ran away from the curtain from where the head had been shown, he nevertheless lifted the curtain and examined it attentively, smelling and inspecting the floor behind it. The fear of the stuffed bear did not deter him from running up to the door, opening it a crack, and peeking inside to look at the frightful object and then again run back and forth.

The demonstration of the frightful mask of the cormorant to the chimpanzee often was accompanied by rhythmic knocking on the door; later, the knocking itself was sufficient to cause fear in Joni. It is interesting that, when under my protection and out of danger, Joni himself tried to reproduce this rhythmical knock, intensely watching for the results. In this case, the chimpanzee's face assumed the same concentrated and frightened expression as when he reacted to scary objects. His lips were closed tightly and bulged somewhat forward; the upper lip hung over the lower one [plates 20(1), 20(2)]. His pose was tense; his movements were nervous, vigorous, and hasty; and his arms and legs were ready to carry him away any minute. Joni, in fact, fearfully fled at the appearance of any frightening object, which did not prevent him from new attempts to challenge the dummy.

It is difficult to say exactly what the stimulus for this challenge was. As mentioned, it was his desire of the repeated familiarization with the frightening object to alleviate his fear, his curiosity overcoming the fear, the joy of playing with dangerous things, his plain imitation of those around him, or the intimidation of other people.

It is not so easy to determine the reason for these actions on the basis of the few facts we have stated here.

Methods of self-defense. The chimpanzee's reaction of fear invariably includes the self-defense reaction. In case of danger, he tries to run away as far as possible, hide safely, be less conspicuous, and use the protection of a human. Even if someone raises his or her hand against the animal or facetiously aims a stick at him, he throws himself on the ground face down, presses his arms and legs to his body, and leaves only the least sensitive and valuable body part, his back, open. As we mentioned briefly, the chimpanzee often uses rags and scraps for his self-defense.

Fleeing from real or imaginary animals, Joni hides his face in the creases of curtains, holds them fast, and then calms down at once, apparently feeling under their protection and consequently out of danger. Sometimes, when the chimpanzee roared; did not want to come back to his cage; was not obeying orders, threats, or lashing by the whip; and was getting under the furniture and catching me with his hands, it was sufficient to give him a small rag. He grasped it greedily and, without any further delay, went into his cage, where he sat down on the rag, tucking it under his body, or he covered himself over his head and sat still, as if convinced of the safety of his position.

Sometimes, when frightened of an object, Joni quickly runs away from it. He tries on his way to pick up a rag and drags the rag with his foot or slings it over his shoulder, as though using it as an aid. He becomes very distressed and protests fiercely when somebody tries to take it from him. Joni always holds these precious protectors under his surveillance. If he does not sit on the rags, he keeps them nearby, glancing at them from time to time. He even takes them with him when he climbs the trapezes or the cage or when he runs around the room. The rags will trail some 1 to 2 meters behind him, and he is not discouraged when they hitch on or stick under various objects, which obviously hampers his movements.

Closed in his cage after vigorous scenes of struggle and resistance, Joni quiets only when he gets hold of some sheets and sits on them, arranging them around him. Moreover, Joni's propensity for rags and scraps is so compelling that he uses every opportunity to pull out clothes or scraps of cloth from wardrobes and drawers. He hastily brings these things behind the partition to his place. It is absolutely impossible to take these things back from him; he holds them so firmly and struggles with you so energetically that you are more likely to rip the things apart than take them back.

As mentioned, the only way to get something back from Joni is to trade something else for it. Joni will not give the thing back to you before he gets an object in exchange, which he will subsequently use in the same manner.

If the frightening object does not cause extreme fear in the chimpanzee, then the self-defense reaction, including the emotional element of fear, will evolve into the offensive reaction, with the rage and anger emotions predominant.

2. Anger Stimuli and Forms of Manifestation. The chimpanzee not only defends himself, he also tries to act toward intimidation or elimination of the unpleasant or hostile object. We pointed out the angry revulsion that Joni feels toward sheets that smell bad.

Inanimate stimulating objects. Once, I intentionally covered Joni with the sheets he had rejected previously. He quickly took them and started ripping them fiercely with his hands and teeth. He threw the sheets on the floor and trampled them, his whole body trembling.

Sometimes, we left a dead bird (e.g., a hazel hen or grey hen) on the lawn where the chimpanzee usually was. Noticing the bird, the chimpanzee became nervous, fluffed up. He touched the bird only with his index finger [plates 20(1)–20(3)], then smelled the finger and subsequently stayed far away from that place. If we threw the bird right in front of him for the second time, he pulled a grimace of disgust and shoved the bird aside with a swift gesture.

Joni experienced the same fear toward stuffed birds (e.g., a duck), but as he got used to them and was convinced of their harmlessness, he was likely to attack them. Once, I gave Joni a stuffed duck that had scared him previously [plates 22(2), 22(3)]. Shielding himself from the dummy with one hand and baring his teeth, the chimpanzee took the duck in the other hand, pulled it closer to him, and then pressed its neck between his feet, extending his lips. With a gloating expression on his face, he started plucking the bird vigorously and hastily so that the feathers were flying in all directions. Joni did not stop until the bird was plucked almost naked.

The chimpanzee used defensive gestures evolving into aggressive ones, when he was shown a dead magpie. Sitting on the table and spotting the bird from far away, the chimpanzee became fluffed up somewhat and first got ready for defense. He pulled one leg up almost to the chin, its toes aimed forward. He raised both hands and covered with them his forehead. The above-the-eye arcs formed a visor above his intent eyes; the visor was ready to be pulled over his eyes when needed and to protect them from possible hazard [plate 22(4)]. In a moment, seeing the bird getting closer, the chimpanzee turned his side to it, bared his teeth, covered his face with one hand, and made a hesitant waving gesture with the other. As the imaginary aggressiveness of the bird increased, the chimpanzee's mouth curved, and his teeth became more exposed and the gestures more rapid, frequent, and compelling [plate 22(1)].

The chimpanzee often clenches his fist,[7] aiming it at the stimulating object and trying to hit it as strongly as he can [plate 22(3)]. Sometimes, shielding himself from the object with one hand, Joni threatens it with the other one.

Joni reacts to showing him a live hen in the same manner.

Phase 1. The elements of self-defense, keeping the frightening object at a distance: defensive lifting and setting his foot forward, with the pushing-aside gesture of his hand usually accompanied by wrinkling of the upper part of the face and baring of the teeth [plate 20(4)].

Phase 2. The threatening gesture of the hand toward the stimulating object and touching it accompanied by the rippling cramp of the upper lip and raising of the upper edge of the lip to the level of the canine teeth [plate 22(6)].

Phase 3. Complete baring of the teeth, throwing the clenched fist forward [plate 22(3)].

Phase 4. Grasping and pulling of the unpleasant object.

The chimpanzee's reactions to real and stuffed animals is much more aggressive than to birds. The movements that accompany these reactions are extremely expressive.

If I put the fur of a squirrel on the floor, Joni gets agitated: His hair bristles, he does not take his eyes off the fur, and he assumes a threatening pose. Positioning his body at an angle to the ground and resting on extended arms, Joni quickly shakes his head from left to right, bends his body forward a few times, pulls back as though getting ready for a leap, and threateningly knocks with his folded fingers and toes on the floor, but he cannot make up his mind to touch the fur. Seeing no change in the position of the fur, Joni becomes bolder and resorts to more tangible ways of treating the hateful object. He runs around the room, brings a chair, hastily dumps the fur onto the chair, takes the leather case of a watch, puts it on top of the fur, and then apprehensively takes the case back. He picks up a handkerchief and, holding it up in the air and waving it, runs up to the fur on three extremities and lashes the fur with the handkerchief a few times.

Not feeling any resistance on the part of the hostile object, Joni gets bolder and begins to beat the fur with his fist; he catches the edge of the cloth lining with his teeth and turns the fur inside out, looking closely at the cloth. Although he has been in close contact with the fur, he is still afraid of it; he does not touch the fur itself, but the cloth, jumping aside at the slightest movement of the fur. I roll the fur into a tube. The chimpanzee bristles again; with a visible effort, he pulls a big chair closer to the fur. He topples the chair right onto the fur, stands on the toppled chair, and quickly comes down to the floor. He brings a trash bin and puts it on the fur or throws a little box, which he has just picked up from the floor, at the fur.

Then, Joni applies new offensive tactics. He starts rolling the chair over the fur. He carefully pulls the fur with his hand, knocks on it with his fist, jerks and bites it, and tries to hitch it by the cloth lining, turning it over and over again. Finding a hole, Joni puts his finger into it and pulls the fur closer to him, tentatively at first and then more confidently. After a while, he feels comfortable enough to assert himself as the master of the situation: He plucks the fur with his teeth and spits it out, knocks on the fur with both hands, steps or wallows on it, and rises and presses it to the floor with all his weight again.

When I make a semblance of a live creature out of the fur and make this creature fiercely hit the piano keys, Joni fearfully picks up any object and tries to throw it at the fur or hastily runs around the room, dumping everything he can get his hands on to the floor.

The chimpanzee's reaction to the stuffed cormorant is also very interesting. At the first demonstration of the bird sitting on a chair, the chimpanzee shows all signs of great fear and mental confusion. He runs away from the bird, hides under the table, gets out of there, climbs the table, quickly runs over it, jumps down to a chair from there, runs over the chairs, comes down to the floor, and fidgets, seemingly unable to find the right place for himself. His hair bristles all the time. He does not take his eyes off the dummy, and his lips are tightly closed. They extend forward in the form of a promontory, and his upper lip hangs over the lower one [plates 20(1), 20(3)]. From time to time, Joni grunts abruptly, stands vertically, rests on one arm, crouches and rises a few times, bends and then straightens his knees, and extends his arm toward the stimulating object.

After the chimpanzee had familiarized himself with the dummy and found it absolutely harmless, he did not experience such strong fear any more; he did not assume the extravagant poses or demonstrate the threatening and offensive gestures. Nevertheless, Joni showed certain anxiety and aversion toward the bird for a fairly long time, and if the opportunity arose, he tried by all means to get rid of the loathsome creature or inflict fear or some kind of trouble on it.

If I put the dummy on a chair in the room where Joni is playing, the chimpanzee becomes worried and makes various attempts to take the cormorant out of the room. He holds a chair by its back or leg; not seeing the bird from there, he probably feels safer. He begins to drag the chair around the room, with a very concentrated expression on his face and with his mouth tightly closed. From time to time, Joni leaves the chair, runs away from it, and looks at it from where the dummy is visible. Then, after seeing the bird sitting unperturbed at the same place, he resumes the chair dragging, watching for results.

If all this is useless, the chimpanzee begins to jerk and shake the chair, trying to dump the worrisome creature to the floor. He recoils at its slightest movement and runs away to a remote corner of the room. In a moment, the cycle resumes: The chimpanzee is again after the bird, throwing things and waving rags at it.

When the stuffed cormorant is placed where the chimpanzee can see it but cannot reach it (for example, on a high wardrobe), he becomes very angry. He assumes expressive threatening poses, runs around the room, or fiercely knocks on the wardrobe doors with his knuckles; or overcoming his fear, he attacks the wardrobe, trying to get on it. Failing that, Joni dashes back and forth, dumps to the floor everything that happens to be in his way and may bite his closest human friends, even me.

The picture of a chimpanzee (by Shpekht) and Joni's own image on photographs or in a mirror call forth an angry reaction in the chimpanzee. As mentioned, Joni was very much afraid of the picture in the beginning, but later, he started to attack it. He barked at it and even attempted to tear it apart.

Joni was very aggressive toward his own photographs. He stared at them intently, touched them with his teeth, snatched them from our hands, threw then to the floor, and scratched them with his nails, particularly in the face area. He did not stop until the picture was torn to pieces.

Showing photographs of people to Joni did not cause affective reactions in him, although he looked at them very attentively. When an album with photographs of various animals[8] was demonstrated to Joni, his reaction was differentiated. For example, he quietly looked at the pictures of birds, but he was belligerent toward the photographs of animals with expressive faces and with protruding and flashing eyes (such as those of apes or tigers). He banged with his fist on them and tried to destroy them by any means.

Even the flat wooden figure of an orangutan made the chimpanzee aggressive. Seeing it for the first time, he made a typical angry grimace, wrinkled the upper part of his face, and partially closed his eyes. Then, taking the orang's chin in his fingers, he began pulling the figure to himself with abrupt movements, trying to destroy it.

Most certainly, the chimpanzee is worried when he sees his image reflected in a mirror [plates 39(1)–39(4)] or in plain glass. In the beginning, he stares at the image and opens his mouth in surprise [plate 39(1)]. Discovering the face of an ape, he reaches behind the mirror and makes catching movements with his fingers [plate 39(2)]. Then, seeing that all his attempts are futile, he clenches his fingers in a fist and ferociously and abruptly hits the image in the mirror [plate 39(3)]. As the loudness and frequency of this beating increase, he falls into a rage, and his face becomes extremely expressive. He tosses his head back, partially closes his eyes, and partially opens his mouth; his upper lip jerks in a rippling cramp, his right or left canine uncovers for a short time, and he emits a hollow, coughing sound in time with the hit. With every new hit the vigorousness of his offensive increases. He slips into a violent frenzy; his eyes almost close. He continues banging as if in sheer madness [plate 39(3)].

Another time, the demonstration of a mirror to Joni (and even hanging it in his cage) does not cause such strong curiosity and such extreme aggressiveness, although his reaction is far from peaceful. Seeing his image, Joni opens his mouth quickly and widely and gives out a short guttural sound, as though choking. His lips are pulled out to the sides, and both rows of his teeth are bared. Then, not taking his eyes off the mirror, he either bangs on it with his hand, clanging with his teeth loudly, or starts to open and shut his jaws rapidly with a cracking and threatening sound. This repeated opening and closing of the mouth makes him yawn three or four times. Sometimes, looking into the mirror, Joni begins rapidly shaking his head from right to left, or passively sticking out his lower lip, he shakes his head up and down.

Seeing that his duplicate has an aggressive mien and becoming excited, Joni perhaps does not understand or forgets that he is the cause of it. Therefore, he becomes even more angry. He strongly slaps with his spread palm on the glass and runs aside, all fluffed up. He beats the cage wall with his fist, attacks the mirror again, and grasps the rope holding the mirror and fiercely pulls at it. He brings his face close to the mirror and, staring at his image, tries to catch this elusive duplication.

Suddenly, Joni starts biting himself while still looking at the mirror. He rips apart a rope that lies nearby and then again shakes his head, bares his teeth, and opens his mouth. Standing on all fours in front of the mirror, he jumps from his feet to his hands and back, his mouth open; his jumps become so rapid that the outlines of his body become blurred.

Sometimes, Joni gets carried away when, standing in front of a mirror, he begins to spin around. Then, he runs away from the mirror and vigorously shakes the cage as though trying to break it. Seeing that his imaginary adversary remains whole and intact, Joni may apply a different method of its elimination. He takes water, or sometimes even his urine, in his mouth and sprays the mirror while lying on the floor with it, thereby blurring the image and freeing himself from the irritating company of his twin. Later, when he got used to his reflection, Joni entertained himself by sitting in front of it and making cracking sounds with his lips [plate 39(4)].

Finding his reflection even on the miniature metal balls on the back of a bed, Joni gets on the bed, knocks on the balls with his teeth or with his hands, and tries to loosen them. If the balls do not yield, he starts biting them.

Seeing the outline of his head on the shiny surface of a stapler, Joni, his mouth open, passes the stapler in front of his eyes, following intently the movements of the outline, tossing his head back, now bringing the reflection closer to his eyes, now pushing it away from him. He rattles with his extended lips and reaches out to the reflection with his hand as though challenging it.

Joni is fascinated with a metal shiny bell and its reflecting surface. The chimpanzee smells the bell, knocks on it with his knuckles, and assumes ostentatious bowing poses or poses meaning an invitation to play. Apparently, this reflection makes Joni perceive the bell as animated object capable of a response reaction.

Once, being with Joni outside our laboratory, to my surprise I noticed that Joni was making various funny grimaces and waving his hand. Turning around, I saw that he had found his reflection in a big cheval glass, which was far away from me behind my back, but in front of him. I kept watching. Of course, Joni did not limit himself to this passive role. He rushed to the mirror on all fours; on his way to the mirror, he stopped a few times and tapped on the floor with his feet. Coming up to the mirror, he began to knock on the mirror stand with his knuckles and to beat on the mirror glass with his open palm. Joni retreated several times, staring at his image. Then, opening his mouth wide and baring his teeth beyond his gums, he hopped toward the mirror and forcefully knocked on the glass with his teeth, increasing the strength and frequency with every knock [figure 12(1)].

Fearing that he might shatter the mirror to pieces, I pushed Joni aside, but he wriggled himself out and resumed beating on the glass. After Joni's forceful removal from the mirror, he did not quiet down and tried to intimidate, tease, and challenge his mirror brother from a distance.

The chimpanzee assumed the vertical position, waved his hand threateningly at the mirror [figure 12(2)], or crouched; the frequency of these intermittent movements increased. At last, he stood upright, raised his hands, bared his teeth, moved over to the mirror, and began banging on the glass with his knuckles profusely awarding his reflection with punches [figure 12(3)]. Afterward, it took much effort on my part to prevent the chimpanzee from breaking the cheval glass because every time Joni saw his reflection, he was filled with a strong desire to engage in a fight with it.

Joni also stared at his reflections in the window or door panes and barked with a hollow sound almost like a dog. Sometimes, he waved at the mirror with a rag, got fluffed up, and dashed toward the mirror, with his mouth open, his teeth bared, and his feet tramping.

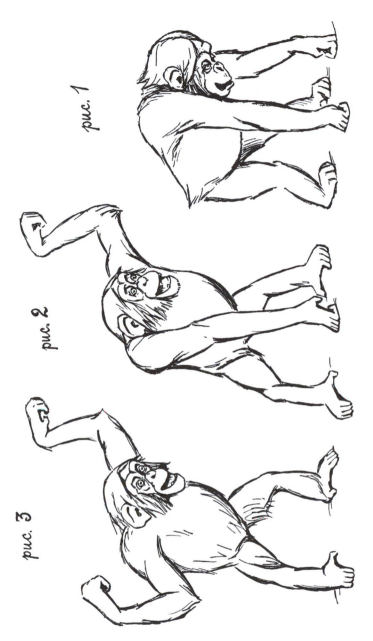

puc. 1

puc. 2

puc. 3

Figure 12. The postures of the aggressively disposed chimpanzee (in reaction to mirror): (1) threatening gestures; (2) lifting arm, ready to hit image; (3) on his way to attack.

A small stuffed chimpanzee (4 months old) impresses him even more than the mirror reflection. He gets excited and stands upright; baring his teeth, he looks at the dummy [plate 18(1)]. Then, resting on all four extremities, he extends his arms forward, repeatedly bangs on the floor with one foot, opens his jaws wide, bulges his lips, and follows the stimulating object with his eyes [plate 18(2)]. Then Joni rises again to his feet and frees his hands; straightening his body, he tosses his upper lip up and forward, baring the entire upper gum. He fiercely attacks the stuffed chimpanzee, dashing toward it with his whole body.

Both live and dead animals call forth extremely vigorous response reactions, predominantly aggressiveness. The magnitude of this aggressiveness is dependent on the kind of animals and their behavior.

Once, I showed the chimpanzee a recently killed hare. The chimpanzee crouched from unexpectedness and fear, pulled an angry grimace, and showed his teeth [plate 21(1)]. Soon, he recovered, rose to his feet, clenched his fingers into a fist, extended his arm, and with the same grin, tried to punch the hare, but unsuccessfully, because I moved the hare away from him just in time [plate 21(2)].

Joni even more got angry. He wrinkled the upper part of his face, curved to the side his tightly closed lips, and boldly extended his arm toward the hare. I outmaneuvered the chimpanzee again, and his threatening hand hung in the air [plate 21(3)]. Then, Joni stood on all fours with his body at an angle to the ground and extended his arms forward. Lowering his head somewhat, as though getting ready for ramming, he opened his mouth and started rocking and jumping quickly. He jumped from hands to feet and back, every time increasing the pace of these movements [plate 21(4)]. Suddenly, he freed his hand, took it off the ground, and made repeated waving movements with it. He wrinkled the upper part of his face, opened his mouth wide, and, lowering the upper lip, began shaking his head from right to left so rapidly that the lip also shook and trembled [plate 21(5)]. At last, the sense of anger apparently overpowered the sense of fear to such an extent that the chimpanzee, ignoring any danger, rushed toward the hostile object, gripped it, and began pulling at its hair [plate 21(6)].

When shown to the chimpanzee the next day for the second time, the same dead hare does not cause in him any fear, only anger. The chimpanzee stands up at once, wrinkles his lips, and tries to reach the hare with his hand. He threateningly waves at it and punches it with his fist. Given the opportunity, he plucks its hair and pulls at its paws.

If I did not leave the hare to Joni's mercies, he would spill out his anger in a different way. He would bang on the wall with a flat hand or knock on the table with his knuckles with ever-increasing strength. Having finished knocking, he would dash toward the irritating object.

If I tried to impose the hare on the chimpanzee, he would turn his back or his side to me and turn his head and eyes away from me, trying to shield himself from the dummy with the back of his hand or even with his elbow.

Live stimulants (live resisting animals). This anger can be greater when it is aimed at animals that put up resistance to the chimpanzee, for example, a cat. Joni cannot remain indifferent even when he sees a cat sitting calmly; the fight is inevitable at every encounter. First, the chimpanzee stares at the cat, assuming his usual threatening poses [plates 21(4), 21(5)]: He rises to his feet, leans forward somewhat, shakes his head, bares his teeth, thumps with his feet or waves with his arms right in front of the cat's face as if teasing it, and tries to pinch the animal. The cat, defending itself, shows its claws; this brings about additional anger in the chimpanzee. He gets a scratch and becomes relentless in pursuit. He is violent and cruel in attacking the cat. He runs and climbs tirelessly after the cat, catches up with it, and runs it to death. He finds it in the most hidden corners, pulls it to himself, bares his teeth, and performs on it a literal vivisection. Joni pinches and squeezes the cat and pulls at its hair. Holding it with one hand, Joni waves his other hand in front of its face and beats the cat on the head, aiming at the eyes and at the forehead; the punches become more frequent and strong, and finally he pushes the cat away from him.

Mounting such an offensive, Joni does not forget about the defense. When the victim happens to wriggle out from him, he immediately turns his side or back to it. He usually holds one hand in front of his eyes, turns his face away from the cat, pulls up and curves his upper lip, and uncovers his canine (as in anger). A light cramp ripples over the chimpanzee's lip from right to left, uncovering his left canine for a moment. Then, Joni begins beating the cat at random with the back of his hand. Sometimes, taking the cat by the tail, Joni drags it behind him with a contented smile, like a victor with a trophy, not paying any attention to the desperate cries of the victim.

If the cat breaks free, it can find a refuge only behind the heavy furniture, such as the piano, which stands close to the wall and where Joni simply cannot fit. In this case, he patiently stalks the cat near the place where it vanished, making threatening gestures with his hand. If the chimpanzee sees the cat somewhere between the furniture, he goes around that area, approaching the cat from different sides. If there are two exits from the refuge, he runs from one to the other, afraid to miss the moment of the cat's appearance. The chimpanzee is so persistent in stalking the cat that it is very difficult to distract him from this pursuit. When the cat is out of Joni's reach (for example, on the top of a wardrobe), he spills out his anger in a different way. He runs around the room; bangs with his fists on the furniture, on the wardrobe where the cat sits, on the walls of the room; shakes the door and bangs it open and shut; and generally behaves like a madman.

The chimpanzee shows such anger not only toward animals that sometimes are aggressive to him, but also with respect to some inert and helpless creatures. Moreover, the smaller and more helpless the animal he is attacking, the stronger Joni's destructive instincts and his propensity to violence. There is no other activity than these tortures in which he can be involved so willingly and for such a long time; he never misses the opportunity to hassle his smaller brothers.

Right at the first moment of Joni's encounter with a live smaller creature or of our demonstration of a miniature animal (for example, a small rabbit) to him, Joni bares his teeth up to the gums, waves his hand with folded fingers, reaches out for the animal, and kneads, pinches, and batters the animal, not paying any attention to its desperate cries. This ability of pursuing and attacking the small and the helpless makes Joni dangerous in the wild. He also runs after small children without restraint, attacking and biting them, and after all medium-size domestic animals, such as hens, ducks, pigeons, baby pigs, or dogs.

The second incident is no less characteristic. Accompanied by three people (two men and a petite 13-year-old girl), I came for the first time to the store where Joni was being sold. After our curiosity had been satisfied, we started talking business with the owners and left the animal to his own devices. He pranced around us and played with a small dog; he squeezed and pinched it, awarded it with punches, and drove it from place to place. Suddenly, he hopped up to the girl and bit both her legs so hard that they started to bleed, and we had to go to the doctor. Later, while in our house, Joni remained dangerous to that girl; he tried to bite her at every opportunity, although he had never been an object of her provocations.

I have also noticed that Joni picks the younger people (for example, 15 or 16 year olds) from a big crowd and tries to pinch or even bite them. A 14-year-old boy, who had come with adults and calmly sat on his chair, literally was suffering because the chimpanzee constantly ran over to him and bit his legs.

Joni's unrestrained tendency to attack "the small ones of this world" is especially visible in his relationship with tiny animals and insects.

Revulsion. The little chimpanzee immediately starts beating tritons and frogs when we show them to him. He cannot stay quiet even when he sees frogs sitting calmly; he feels certain revulsion toward them. At first, the chimpanzee attentively looks at the frog, extending his tightly closed lips toward it [plate 19(3)]. Then, he makes the characteristic grimace of revulsion and waves at the frog with his fist. If Joni's hand touches the frog's skin, he smells the hand and thoroughly wipes it with something. If the frog is still there before his eyes, he stands on all fours facing it, staring at the creature. The chimpanzee pushes my hand with the frog away from him, trying not to touch the frog. His facial expression is like the one he had during his attacks against the hare or the stuffed cormorant: His lips are closed tightly and extended slightly forward; his upper lip swells in the form of a hump and sharply protrudes over the lower lip [plates 20, 21(1), 21(2)].

The chimpanzee pursues every fly in his sight and tries to catch, kill, or sometimes eat it. Although bugs, cockroaches, and caterpillars bring out a feeling of distaste in the chimpanzee, they get brutally thrashed by him anyway. As soon as Joni spies a crawling cockroach, he catches up with the insect against all odds, does not take his eyes off it, and gets fluffed up. Then, Joni slaps it with his hand or crushes it with his knuckles. He stares at its remains and spreads them over the floor until there is only a wet spot left.

This "bloody business" finished, Joni smells his fingers with a grimace of revulsion on his lips and then assiduously wipes the fingers dry (on the wall or on the floor). Even after his hands are completely dry, he smells them many times and continues to wipe them to eliminate the last traces of the odor.

The chimpanzee was very enthusiastic about the cockroach hunt. He stalked them near the cracks in his cage and searched for them in the cabinets. To speed up the hunt, he often took a stick or a straw and stuck them into the cracks, driving the cockroaches out. Joni fiercely attacked the cockroaches and killed them as soon as they emerged.

In the same manner, Joni beats a crawling bug with his hand and then smells the hand. If he sees that the bug is still crawling, he covers it with a rag and beats it through the rag, apparently not willing to touch the insect because he feels disgust toward it, which is incomprehensible to us. The fear of the bug may be the reason for that. Covering the bug with a rag prevents Joni from making an accurate hit; as a result, the bug usually survives and crawls out from under the rag. In such a case, Joni repeats this procedure until the bug is killed.

If the chimpanzee does not have a rag at hand, he hits the bug with the back of his hand, particularly with the hairy part of the hand; sometimes, he uses his arm or elbow for this purpose. If the bug remains motionless, Joni touches it with his lips, but as soon as it starts moving its legs, Joni recoils and then punches away at it with special strength and anger.

The overwhelming majority of cases in which the chimpanzee reacts aggressively proves that the anger emotion appears in Joni as the counterbalance of the fear emotion. The anger emotion relates to the self-preservation instinct and represents a reaction of self-defense.

People who irritate the chimpanzee. The same anger appears in the chimpanzee not only with respect to unfamiliar or frightening animals, but also toward strange people and unfamiliar objects without past benefits to him. For example, at our first visit to the zoological store where the chimpanzee was being sold, he resisted our every attempt to touch him or take him in our arms and tried to bite us, even after we stopped paying attention to him and were engaged in conversation with the owners.

During his first days in our house, Joni bit us at every contact with him. Women, the least resisting to his wild mischief, were affected the most. My little sister was particularly vulnerable; he constantly tried to attack and bite her. As the chimpanzee was getting used to us, this angry biting stopped, but his belligerence toward unfamiliar people remained.

Joni was extremely unfriendly to our few guests. He tried to bite them, or if we isolated him from the guests, he threatened them with a stick (the one that caused fear in the past), threw the stick at them, and took a rag and rotated it in the air, trying to reach the strangers. Sometimes, Joni, fearing the strangers, covered his face and his head with a rag and then attacked the newcomers more persistently and bravely.

Even toward people to whom he is attached he shows angry feelings and can become dangerous. This anger stems from the same source of self-preservation and usually is associated with improper satisfaction of his physiological needs due to illness, tiredness, or corporal punishment. Sometimes, Joni becomes angry as a result of our noncompliance with his desires or our resistance to his behavior.

For example, when Joni wants to eat, drink, or sleep, he becomes especially excited and angry at the slightest departure from the orderly satisfaction of his needs or at our failure to understand his demands. Joni sometimes is hungry in the evening when the food is not ready yet. To preclude his whimpering, I try to entertain him with some toy; he impulsively snatches this gift irrelevant to his needs from my hands. He fiercely gnaws it and testily throws it away from him.

In the evenings, when he is ready to go to sleep, he would not leave me alone for a minute; but when I must leave him eventually, I can hear him throwing things all over the room and crying if I do not come back.

Sometimes, when I am late coming to his room during the day, Joni expresses his anger very conspicuously. He pushes the trapezes, knocks on the walls or doors with his knuckles or with his flat palm, makes noises by all means available to him, overturns the furniture, throws sticks and wooden toys, utters a typical hooting heightening sound that ends with abrupt screeching thrice-repeated barking, and punches with two fists on the cage wall. If I give him a rag as a means of consolation when leaving, he immediately throws it away from him. When I insist and try to impose the rag on him for the second or third time, he tears it to pieces with his teeth, brazenly looking at me. Sometimes, when there is no object at hand on which Joni can take out his anger, he starts biting his hands and feet.

By evening, when the chimpanzee is already tired and wants to sleep, he allows only me to stay around. He loudly protests if any other person (even someone from our group) comes near. He either attacks and bites the person, grunting angrily, or abruptly hoots or threatens the person with his hand an pushes the person away.

Joni is especially distrustful and vicious when sick. He does not let strangers enter his room and attempts to bite them; he is very apprehensive and aggressive toward our household and always prefers to be left with those he trusts most and to whom he is most attached (as though striving for more safety). In trying to make this person to stay with him, Joni not so much asks for it, as he angrily demands it, getting furious if we fail to satisfy his wishes completely and even biting the people closest to him.

When Joni eats, he never allows any one except me to touch him. If someone from our household pretends to be reaching for his food, he grins, uttering a short coughing sound "ah-ah," and fiercely grips the "challenger's" hand.

Irritating conditions of everyday life. We noted above how widely the chimpanzee reacts to any infringement on his things, how fiercely he struggles for his possessions, and how upset he is if the desired objects are taken away from him. Any attempt to deprive the chimpanzee of his dear freedom calls forth angry behavior. When I catch him and forcefully put him into the cage, even I become an object of his violent attacks. He roars, shrills, fidgets, and tries to bite me.

Joni immediately makes an angry grimace and defends his rights using his teeth in response to simple holding him in one place against his will, gripping his arm or leg, or squeezing his fingers (see the section, Love of and Struggle for Freedom above). Joni is yet more enraged when, to restrict his dashing around the room (for example, during the clean-ups), I tie him to a certain spot. He perceives any attempt to put clothes on him as an infringement on his freedom; therefore, it produces his angry reaction.

Needless to say, Joni becomes angry when we inflict pain on him (during punishment), especially when the punishment is executed by a person not very close to him. The chimpanzee reacts aggressively and becomes so enraged that he may become dangerous.

Vindictiveness. In some cases, we certainly may be dealing with the chimpanzee's vindictiveness. Once, Joni burned his hand on a hot iron stove. After a while, he covered himself with a rag and began to attack the stove, slapping it with his palm.

In exceptionally rare cases of punishing Joni with the whip (for example, when he bit a little girl), we have noticed that if the punishment is imposed by his masters, he endures it submissively. He sits still, makes a typical angry grimace at every hit, bares his upper teeth, curves his lips as though having a cramp, and wrinkles the upper part of his face, but he does not mount an offense. At the most, he takes out his anger only on the punishment tool: He snatches the whip from the offender, bites it, and furiously throws it away from him. But, if a stranger dares to push him slightly in response to his bite, he becomes enraged beyond himself and attacks the person with enhanced strength and anger.

Sometimes, however, Joni also tries to take vengeance on the people closest to him, but he never is carried away to the extent as toward strangers. For example, he is at fault and I threaten him with my hand and then spank him one time. He obviously becomes angry, bares his teeth, curves his mouth, punches me in turn, and runs away. When he experiences pain from an accidental encounter, he starts breathing rapidly as if offended, as in cases when he is caught in the course of a game of pursuit. If the chimpanzee is in his cage (not free as in his room), he apparently feels more confident, is more revengeful and aggressive in response to punishment, and even may often bite a person from our household.

I have to mention again the case of restrained anger (see the section, Treatment of Chimpanzee, above), which indicates that the chimpanzee differentiates his aggressiveness depending on the person causing the anger and on the place of occurrence. He may tolerate something only under certain conditions; what one person is allowed to do with him may not be the case for others. The proverb "Quod licet Iovi, non licet Iovi" ["What is allowed to the farmer may not be allowed to the ox."] also applies to the chimpanzee's etiquette.

We have mentioned the unusual inclination of the chimpanzee to imitate the mood and feelings of the people closest to him; the anger feeling is no exception.[9] Moreover, we are certain that Joni associates himself more easily with humans who express anger.

| Communication Instinct | Concerning the chimpanzee's social attitudes, we have to emphasize that, like any child, the chimpanzee strives to find in a human a protector, guardian, and nurse to substitute for |

zee strives to find in a human a protector, guardian, and nurse to substitute for his lost mother. If he lived in the wild, he would have retained direct and close contact with her at that time.

Expression of affection and attachment. Not surprisingly, the chimpanzee is more inclined to communicate with women than with men because women, with their maternal instinct, better understand and fulfill his infantile demands. He immediately distinguishes the woman who feeds him and runs only to her for protection from real and imaginary fears. He is completely calm only when she is present, trusts only her when he is sick, and especially needs to be closer to her. He is extremely upset during the temporary absence of his protector and makes persistent attempts not to let her go. He is very glad to see her come back and demonstrates to her his affectionate compassion; he quickly conforms to her mood, never gets angry with her, even restrains himself from involuntary, reflex-based angry feelings that appear toward her. He is extremely sensitive to any reprimand and punishment on the part of his protector, immediately runs to her when he sees that she is upset, and attacks her imaginary offenders, even if they do not present serious danger to him. The chimpanzee's attachment is character-ized by one more feature: instability or inconsistency. If for some reason he loses one protector, he immediately tries to find another one, as though fearing to remain without custody; he momentarily abandons his past affections for the new ones that he currently needs.

Living at a zoological firm for a week, the chimpanzee became closely attached to his temporary master, the company's female owner. He had never bitten her despite all her manipulations of him; he immediately rushed to her (not to the male owner) when he saw us. When Joni was about to be placed into a transport-able cage, he ran to the protection of the woman and embraced her by the neck with such strength that two sturdy men could not tear him from her. He beat his arms and legs, roared desperately, and bit everyone except his female master; when he did not find her support and saw that she had become an accomplice in the business of putting him into the cage, he dashed around the room and went to its remote corners. When he was dragged out of there by force, his despair had no limits: With earsplitting screams, he rushed to his treacherous protector and resisted everybody so strenuously that it took five people (three men and two women) and 2 hours to catch him and put him into the cage. When Joni was finally in the cage and the cage was put in the cab, he kept roaring desperately, but as soon as I stuck my hand between the bars in the cage and took him by the hand, he immediately calmed down and remained quiet during the rest of this long trip.

I expected that, after he entered our house, which was totally strange to him, and was released from the cage, he would demonstrate his despair and his anger with the same strength. I was amazed when Joni, right after we opened his cage, rushed to me and hung on my neck in the same manner as he did with his previous owner. He did not want to leave me for a second and clung to me so strongly that you could hardly believe that he had responded with angry and persistent biting to all my attempts to approach him with affection just 1 hour ago.

Although he was peacefully disposed toward all adult members of our family, he definitely "favored" women. He would go on a woman's lap more trustfully and would play with women more willingly and for a longer time.

In connection with Joni's desire to provide himself with protection, I recall an episode that took place much later. Once, in spite of all our persistent efforts to prevent Joni from biting a girl, he found a moment and a place and bit her finger hard. My husband immediately grabbed the chimpanzee by the neck, bent him to the ground, and began spanking his back soundly. Joni gave out a loud roar and rushed for my protection, continuing to cry desperately. When I awarded him with a spank, too, he cried even more loudly and ran to another woman from our household and sat down beside her submissively and quietly, as if looking for her protection.

Joni's attachment to a certain person is determined by how long he has known this person. During his first days at our house, he got used to me very quickly and preferred me to all others, but if someone else happened to spend a day with him, he tended to express more sympathy to that other person than to me. Joni was more attracted to that person, played with him or her more willingly than with me, and cried when his new protector was leaving him to my care.

However, I regained the chimpanzee's attention and sympathy after I spent a day or two with him without anyone's interference, and then I conquered his affection so successfully that I seemed to have turned into a willing slave to this little despot.

My every departure caused Joni despair and usually was accompanied by forbidding scenes of his struggle and resistance. He fiercely screamed, cried with all his strength, and slyly used all means in his possession to keep me near him (see the section, Sadness, below). Later, Joni let me leave him more freely, but every time I left, I turned around and saw his stare following me.

Each of my visits to Joni's room was a source of joy for him. He got excited, fluffed up, rose to his feet, and met me with the ringing, shrilling, hooting sound that evolved into high-pitched barking like that of a dog. He reached his hands out to me, rushed into my arms, pressed his body to me, and touched my neck with his half-open mouth and breathed rapidly.

After 10 months in our house, Joni became attached to both my husband and me so strongly that every time he saw us, through the glass in his room, leaving for a ride on horses or for a trip in the forest, he gave out a thundering scream, which could be heard at a great distance. He would not stop until he lost sight of us. Through a double window pane, Joni would spot our appearance at the fringe of the forest and would begin hooting in excitement. He often recognized my voice through the door of his room and responded to it with a joyful hooting sound. As noted above, Joni associated the sound of the elevator stopping in front of our apartment with our arrival, and we heard how vigorously and loudly our little prisoner welcomed our return.

His preference of me over my husband was evident and can be illustrated by several examples. When we were with Joni at the railroad station to leave for another city, my husband wanted to help me carry Joni and started to take him in his arms. The animal resisted violently and would not let him do it.

When three of us (including Joni) occupied a train compartment, Joni cried, held on to me, and did not let me go out for even a second because he was unwilling to remain with my companion (which he easily did in our day-to-day life). Moreover, when my husband was going to take the chimpanzee in his arms by force, Joni even tried to bite him, which he also never did at home.

For another example, three of us were walking in the forest, looking for mush-rooms or berries. I stopped at some place, my husband walked on; Joni followed my husband for some time, but then returned and stayed with me.

In the third example, my husband went in the direction of our house, and I remained by the creek. Joni again followed him for some time, but as soon as he saw that I was left behind, he abandoned his companion and returned to me. If we parted in the forest, Joni followed me, not my husband. When Joni sat with my husband on the open porch of our house and saw me coming to the house from far away, he took off and ran to meet me, his entire body expressing delight.

After 2 weeks in our house, Joni did not let anybody except me feed him, stubbornly turning his head away from the food at the usual meal time. He flatly rejected the most persistent and endearing offers and did not eat until I came and fed him.

Once, I fed him at 9 A.M.; at noon, his usual breakfast time, he refused to take food from others. He closed his lips when offered a mug of milk, turned his head away from it, and remained hungry until 4 P.M.[10] When I came back and handed him the mug, he greedily grabbed it and drank the milk without pause until it was dry. He was obviously hungry, but preferred not to take anything from someone else's hands.

Sometimes, he is ready to overeat or will eat unwillingly because he wants to spend a few more minutes with me. Joni expresses certain affection toward me most often during his meals. When he drinks his milk or eats the delicacies I have brought him, he will touch my head carefully with his fingers or grab my chin with both his hands [plate 35(3)], press his open mouth to me, and utter a ringing grunting sound. I am inclined to regard these gestures and movements of the chimpanzee as an external expression of higher exhilaration, a form of gratitude, that accompanies his friendly, affectionate feelings.

As noted, when Joni wants to eat, drink, or sleep, he literally does not take his eyes off me; he follows me everywhere, staring at my face all the time. He has developed a special conditioned reflex for the expression of his desire to drink: He runs up to me and sucks at different open parts of my body (at the hands, neck, or face), and every time he greedily drinks the water that I offer him in response to his expressive pose.

Joni allows only me to put him to bed and does not want to go to sleep in someone else's presence. He willingly falls asleep on my lap and can sleep a long time in such a position [plate 38(1)]. He would be even more willing to sleep in a bed with me and protests when I do not permit this.[11]

On the first days at our place, out of necessity I had to make a bed for him in a case on the floor, in the corner of my bedroom. He kept getting out of the case and onto my bed; he wanted to sleep beside me to be closer to me. His previous master, the owner of a zoological firm, told me that the chimpanzee showed a similar tendency when living in her house; he always preferred to sleep on the same bed with his master or with the maid, but not alone.

As mentioned, when Joni is sick and feels absolutely helpless, he is drawn only to me and does not want to leave me even for a minute. For example, on the next day after filming the movie, Joni's eyes ached after he had been blinded by the bright light; he closed his eyes and could not look at the light at all. Because of that, he felt more helpless than he usually did; he either sat in a corner, where he felt safer, or clung to me and firmly protested against my attempts to leave him even for a short time. At such a time, he wants to get onto my lap; he presses his body to me; affectionately looks at me; sits motionlessly for hours, which he does not do normally; and examines my face and touches my hair [plate 41(2)].

During sickness, he is especially sensitive to my attitude toward him, which manifests itself in the tone of my voice, to which he usually does not pay much attention. If he hears sharp notes of disapproval, he extends his lips in the form of a funnel as if he were about to start crying, or he begins breathing rapidly. He grabs my chin, and as if wishing to make me mellow or to ingratiate himself with me, he takes my finger into his mouth and sucks it. If he does not obey me at this time and I have to resort to loud, strict orders, he starts crying and will not calm down until I caress and embrace him. When he is sick, he affectionately responds to my loving care more often than usual.

If I put the chimpanzee in my bed to quiet him down during his illness, he is extremely glad. He breathes rapidly, grabs my hands, presses his half-open mouth or tightly closed lips to me, slightly squeezes my cheek or sucks it, and his whole body shakes. The stronger his lips squeeze, the faster the rate of his breathing.

I interpret this touching with the chimpanzee's open mouth or with his tightly closed lips when he is joyful as an antecedent of a kiss born of his desire for tactile contact with the living creature who presents him with that joy. However, Joni's previous owner demonstrated to me the chimpanzee's ability to kiss in an entirely human manner. In fact, in response to her extended lips, Joni extended his lips. There was a smacking sound, and the kiss was created. Because of my fastidiousness and also for hygienic reasons, I was not inclined to induce his kisses, so he dropped this habit of kissing with extended lips. Under appropriate circumstances, he used a more natural way of kissing for the expression of joy and tenderness: He touched a human's skin with his open mouth and lightly pinched the skin with his lips.

I observed such touching by Joni only toward the people closest to him and others who were especially well disposed to him, for example, me and my husband. I never noticed this with respect to strangers; Joni's touching of strangers was limited to the use of his hands (the antecedent a handshake).

The degree of Joni's sensitivity to any reprimand on my part can be judged by the fact that it is absolutely impossible to use strictness during my lessons with him. Even when I simply wave him off if he is taking the wrong object and foisting it on me, let alone my abrupt cries of "wrong," Joni becomes worried and confused, whimpers, and cries so hard that he ceases to understand anything. If I am slow to comfort him, he extends his arms and begs to be taken onto my lap; if I refuse, he cries violently and totally forsakes his work. Even after I have comforted and caressed him, he recovers only after a while and cannot resume his exercises at once. During the first months in our house, he refused totally to conform to my requirements.

Without someone else's help, I cannot leave his room. He entirely ignores my orders to return to his cage, my angry tone and loud threats notwithstanding. At the same time, he is more willing to obey other people, for example, my husband. Moreover, in my presence, Joni refuses to obey orders of people to whom he usually has complied; yelling and crying, he rushes to my protection and stubbornly asserts his "rights" until I leave. Only when he sees that I am not there and that there is no one to count on does he quickly obey the order.

If Joni does not obey me, I only have to threaten him with leaving the room, and he immediately surrenders. If he gets carried away and still persists and I execute my threat, leaving him with another person and waving at him as if in rejection, he falls in despair, shakes, runs to the door that closed after me, and fiercely gnaws at it.

He is beyond comfort until I make peace with him. If I sit down at some distance from him when I enter the room, he extends both arms to me. If I do not take him, he puts his arms over his head, covers his eyes with them, cries, and then looks at me; staying in the sitting position, he furtively moves his legs, slowly getting closer to me. If I push him away somewhat, he roars fiercely; if I stay intentionally indifferent, close my arms, and do not encourage him with any single movement, he forcefully pulls my arms apart and tries to get onto my lap through the small hole he has just created. He carefully sticks his arm, then his head, into the hole; seeing no resistance on my part, he slides his trunk in, takes a firm seat on my lap, and as if asserting that he is in a safe, pleasant place and is restoring a friendly relationship, he calms down completely. During this procedure, if I say softly, "Well, come right over," he immediately takes off, rushes onto my lap, presses his body to me and only then seems to be entirely happy and at peace.

Only much later (half a year after his arrival), Joni began to obey me, complying with almost all to my requirements. For 2½ years during his life in our house, there were only two accidents when he severely bit me. The first accident was on the very first day of his life at our place. Joni suddenly and strongly bit my finger when I was going to take away the cage in which he had been brought from the zoological firm. The second accident happened 6 months after the first one and was associated with the painful procedure, for the chimpanzee, of putting him into the cage. Alone in the house and unable to cope with Joni, I wanted to put him into the cage for the night. I had to resort to the broom, of which he was usually very much afraid. In fact, seeing the scary object, Joni immediately got into the cage, but did not let me close the door behind him. Trying to push him away from the door, I stuck my hand with the broom into the cage; at that moment, Joni bit my hand with such a force that it bled.

Sometimes, I noticed that Joni even held back the external manifestations of his angry feelings toward me. I have described already the incident when I inadvertently hurt the chimpanzee while applying an ointment to his nose. He prepared to bite my hand fiercely, but at the last moment, thought better of it and only lightly pressed the hand with his teeth, not leaving any trace of a bite on it. On the first days at our place, when Joni was playing, he tried to bite my hands somewhat, but as soon as I pretended to be crying, he stopped biting at once and looked at me with a question in his eyes.

When he is being punished with a whip or waved at with a rag, as in the case when he bit a country girl, he becomes angry and grabs the punishment tool, leaving intact the primary perpetrators, my hands, obviously trying to avoid hurting them with his teeth. After I scared him with the mask, fur, or broom, Joni, gained access to these things and fiercely destroyed them, at the same time leaving me intact. I am certain that sometimes Joni expressed his sympathy to me.

Expression of sympathy and intercession: Vindictiveness. If I pretend to be crying, close my eyes and weep, Joni immediately stops his play or any other activities, quickly runs over to me, all excited and shagged, from the most remote places in the house, such as the roof or the ceiling of his cage, from where I could not drive him down despite my persistent calls and entreaties. He hastily runs around me as if looking for the offender; looking at my face, he tenderly takes my chin in his palm, lightly touches my face with his finger, as though trying to understand what is happening, and turns around, clenching his toes into firm fists [plate 42(1)].

The more sorrowful and disconsolate my crying, the warmer his sympathy: He carefully places his hand on my head, extends his lips toward my face and looks into my eyes compassionately and attentively [plate 42(2)]. Then, he stands up in the vertical position, touches my face [plate 42(3)] or my hands with his extended lips, and slightly pinches my skin (as if kissing); sometimes, he touches me with his open mouth or his tongue [plate 42(4)].

If I keep crying and cover my face with both hands, he tries to take my hands off my face. He looks under them and becomes more worried and fluffed up. He looks around, folding his lips in the shape of a promontory, and slightly groans and whimpers, as though getting ready to start weeping [plate 37(1)]. The more I cry, the stronger his confusion. All his hair bristles; his side whiskers stick out to the sides. Standing up in the vertical position, he extends his arm and stares into the distance, as if looking for the object that bothers me. He rises and crouches, immersed in a long thundering hooting.

During my fictitious crying, although Joni sees nothing suspicious around, he nevertheless extends his lips in the form of a funnel and hoots sharply and angrily. His toes clench into firm fists, and his penis becomes tense. Then, he performs a series of gestures and body movements that reflect his anxiety, which is tinged with angry feelings [plate 37(3)].

He would either take a rope and, with his eyes closed, vigorously whip himself or he would start biting his hands and feet, attack his sheets and rip them apart, or bang with his hand on the wall. If I cover myself with sheets and cry, Joni viciously pulls the sheets off me, waves them in the air, and impetuously throws them aside.

Sometimes, in response to my simulated crying and with no culprit around, Joni runs up to me, throws the books that lie near me on the floor and ferociously drives his teeth into them. He pulls the tablecloth from the table at which I sit, tries to take some objects away from me (for example, he twists the piano candle-stick off the piano), knocks abruptly on the lid of the piano, snatches the notebook from under my hands, carefully touches my face with his hand, extends his lips, and again knocks on the piano with his knuckles.

If I increase my crying and lie on the bed, Joni worriedly runs around me, gets onto the bed, deeply scratches my chin, knocks on my head with his fist (as he usually does when he invites me to play with him), strongly pulls at my hand, pinches my finger or the skin of my hand, and increases the pressure as my crying increases. Then, he dashes around the room again, tumbles the chairs, drags a chair onto my bed, throws everything he can lift on the floor, jumps, brushes various objects on his way, and produces indescribable noises and commotion.

Once, while in such boisterous disposition, he turned around and saw his reflection in a big mirror. He immediately came up to the mirror and began banging fiercely on it with his fist, punching his image as if it were his most vicious offender.

At the time of this violence, if some stranger is unlucky enough to enter our room and appear before Joni's eyes, Joni receives the stranger as his most embittered enemy. He spills out all his rage on the stranger, savagely attacking and biting.

If I do not stop crying for a long time and the danger threatening me has never transpired, Joni begins to get bored. He tries to invite me to play with him with his usual urging technique. Looking straight at me, he batters my forehead with his knuckles, quickly increasing the force and the pace of the knocking; he will not stop until I return to normal [plate 37(4)].

If Joni sees an attack staged on me by some stranger, he takes active measures to help me out of this trouble, and he takes revenge on the offender. The magnitude of this revenge corresponds to the magnitude of his disposition toward both the offended and the offender.

His intercessions for me are usually more fierce and aggressive than for anyone else, but if I am offended by someone of us, his revenge is somewhat more restrained. In such cases, his special disposition and sympathy to me is most evident.

In the inverse situation, his behavior is based on the same principal: If I attack anybody from our household, Joni becomes worried. He tries to stop me, applying only lenient measures against me. If I, or someone else from our household, play the role of offender of "strange" people, Joni immediately associates himself with us and joins our attack very aggressively. If strangers are brave enough to assume the role of oppressors against members of our household, Joni takes his revenge on them commensurate with the extent of his sympathy toward the offended.

If my husband pretends to be beating me, threateningly waving at me with his hands, and I pretend to be crying, screaming, and groaning, Joni gets worried. He appears from under the chair, assumes the vertical position, extends his arm majestically toward the offender, and utters a long sound, "oo-oo-oo." If the offender does not pay any attention, does not stop beating, and my groans continue, Joni hits the offender with his hand; if that does not help, the chimpanzee throws himself at the offender and tries to bite. If someone familiar to Joni takes up the role of oppressor, Joni, without any warning (without so much as extending his hand), takes off and tries to hit the person.

If we exchange our roles and I pretend to be beating my husband, Joni, as above, rises to his feet, runs over to me, attacks me, attentively looks into my eyes, and carefully follows the scuffle. If Joni does not see it coming to an end, he runs up to me, slightly hits me with his hands, carefully pinches my hand with his teeth, and increases the pressure as the plaintive cries of the offended become louder.

During the second staged scene, when I attack a stranger, Joni immediately associates himself with me, hoots, utters shrieky and abrupt barks, joins me in hitting the person, and angrily seizes the person with his teeth. If the offended person comes from a group of familiar people that has not participated in previous attacks against us, yet nevertheless is being beaten and is crying in Joni's presence, Joni shows certain sympathy toward the person. In that case, Joni does not take part in the scuffle, but tenderly touches the person with his hand or with his lips or lightly taps the person with his knuckles, inviting play.

Joni behaved in the same manner while living with his former owners. If his master pretended to attack his wife, Joni immediately bit him. In the inverse case, when the wife was "beating" her husband, Joni only hooted and looked worried.

Joni takes up the role of the defender only if it does not mean any danger for him. Otherwise, he cowardly retires and leaves his ward to the mercy of fate. At the appearance of a frightening object (for example, a mask), the animal runs away from the object as far as possible, preoccupied only with hiding himself.

The chimpanzee's sense of affection and love bears outright egocentric features.

Jealousy. In the section that discusses ownership instinct, we described how persistently Joni protects his things against even the people of our household and how unwilling he is to share tasty food even with those to whom he is close and strongly attached. As also noted, the chimpanzee holds an everyday despotic monopoly over his guardians, and he treats them with apprehensive jealousy. He considers his last two masters, me and his previous owner, literally "his own." He not only defends us ardently against any staged attacks, but also will not let anybody touch us and will not let us pet any other living creature.

If one our people comes up to me in Joni's presence and kisses and hugs me, Joni fluffs up, hoots, extends his lips, runs over to that person, makes a warning gesture with his hand, and does not leave me until the other person retires. When I overtly express my sympathy to someone and stroke the person's head or kiss them, Joni develops the same response reaction of anxiety and threatening dissatisfaction with respect to the person.

Joni cannot stand if I take a cat on my lap and cuddle it tenderly. He rushes to me, tries by all means to spoil our entente and to disturb the cat somehow. He pulls at the miserable creature's fur, beats it on the head, tries to drag the cat off my knees, and does not stop until the cat is away and out of sight.

In fact, "human" jealousy also appears in the sense that there is the egotistic sense of love that overwhelms the person who strives monopolistically and greedily to take into possession everything that belongs to the object of love (from a physical relationship to the most private thoughts). Such love craves for unlimited, enslaving control; such love is jealous and intolerably stingy toward any form of sharing of the coveted object with other persons.[12]

I happened to notice the distinct sense of jealousy in little children and in many animals, for instance, in dogs.[13] It is known widely how sensitive children are to any instances of inequality of treatment within their groups by adults, how offended children are by the slightest symptoms of preferences by their superiors.

In my opinion about Joni's jealousy toward his guardians, I am not inclined to consider these facts as the chimpanzee's simple intercession for people with whom he sympathizes due to his not taking into account or misinterpreting the emotional relationships within the group of people around him. The following concrete reasons prove my point:

1. Joni properly uses offensive and protective gestures himself; therefore, he knows from his own experience what they mean.
2. Joni excellently understands, as in the case of a staged attack, even in the process of closest contact among participating actors, who is attacking and who is being attacked and reacts aggressively toward the former and favorably toward the latter.
3. Joni follows strictly the role changes among the same participants and never makes a mistake when taking on himself the role of a defender of one person and the role of an executioner of another.
4. Joni never saw any animal express aggressiveness toward me; therefore, he could not assume the role of my savior against the attacks of a cat.

All these reasons can be the reliable base for filing the above-mentioned emotional reactions of the chimpanzee under *jealousy* stemming, in its very deep origins, from the egotistic sense of ownership.

But Joni sometimes tried to do me favors.

Social service. Once, Joni, scrutinizing my hand, noticed a small dried pimple [plate 41(5)]. He became excited at once, grabbed my hand with his feet, and carefully touched the pimple with his fingers, breathing rapidly, extending his lips forward or pulling them to the sides, and half-opening his mouth. Every second, he took his eyes off the object under examination, raised his head, and wrinkling his forehead, looked at my face. If my face was totally calm, he started picking at the pimple with the same curiosity and apprehensive attention. However, if I simulated an exaggerated grimace of pain, he removed his fingers from my hand and only stared now at the pimple, now at me, not making up his mind to touch me as if he were not willing to inflict pain on me.

He vacillated for some time, shifting his eyes from my face to the pimple, then suddenly made his final decision. With rapid, abrupt movements of his fingers and nails, he tried to tear the pimple off, opening and closing his mouth and extending and rotating his tongue in time with his picking movements. He stopped this "surgical operation" only when the pimple was gone.

The chimpanzee kept these operations in his memory for a long time and tried to resume them. In a few days, amid his other pursuits, Joni suddenly grabbed my hand, turned it, pulled up the sleeve, and looked for the pimple. If I resisted and did not allow him to pull up my sleeve and covered the hand, he whimpered in discontent, struggled with my hand, and roared.

The same curiosity in Joni is generated by a blue anchor boldly embroidered on the white surface of my sailor's jacket. Seeing the anchor, Joni suddenly grabs my arm, turning the sleeve close to him. He stares at the anchor and, opening and closing his mouth in a funny way, traces the anchor with his index finger, feeling its every convexity [figure 13(1)].

As the chimpanzee immerses himself in this examination, his mouth opens wider and wider and eventually remains motionlessly open for a long time [figure 13(3)] Joni becomes totally blind and deaf to everything around him. He does not respond when called by name, does not look at what I show him, breaks his hand loose when I intend to take him away, moves closer to the object of his interest as if striving to get rid of interference, and continues with his business with unabated fervor.

Sometimes Joni uses both index fingers. He traces, strokes, and feels the convexities simultaneously in different parts of the pattern [figure 13(2)]. Touching the object with his fingers is sometimes interrupted by touching it with his extended lips or by his feeble attempts to peel it off with his teeth. Often, you can see how Joni, in time with the crawling movements of his fingers over the embossed object, performs catching movements with his jaws; he clacks his teeth or smacks his lips, as when he is looking for parasites in his fur. Professor Yerkes thinks that such chimpanzee's behavior can be considered a precursor of the intention of providing medical assistance.

Joni shows similar behavior when he accidently finds birthmarks, pigmented marks, or scratches on his skin or on the surface of a human body.

Imitation (chimpanzee's emotional solidarity with humans). Describing the imitating ability of the chimpanzee, we have to mention that the chimpanzee becomes infected with the mood of a human rapidly and easily. If I pretend to be crying, Joni immediately stops his most lively play, runs over to me, puts his hand on my head, extends his closed lips toward my face and compassionately looks into my eyes [plates 37(1), 42(2), 42(3)]. He reveals the same silent compassion when I, taking an animal (for example, a small lamb or cat) in my arms, begin to groan slightly as if complaining [plate 43(1)]. If I show certain aggressiveness toward that animal, waving or yelling at it angrily, Joni becomes belligerently excited; his side-whiskers bristle, and extending his lips in the form of a funnel, he emits a sharp, abrupt hooting, staring at the stimulating object [plate 43(2)].

The chimpanzee most easily becomes inflicted with fear. Even if Joni is near me and there are no suspicious frightening or dangerous objects around, as soon as I start groaning plaintively and restlessly, Joni gets scared and worried. He runs around me, as if looking for the violator of peace, stares at the distance, extends his lips in the form of a funnel and emits a hooting sound. He does not

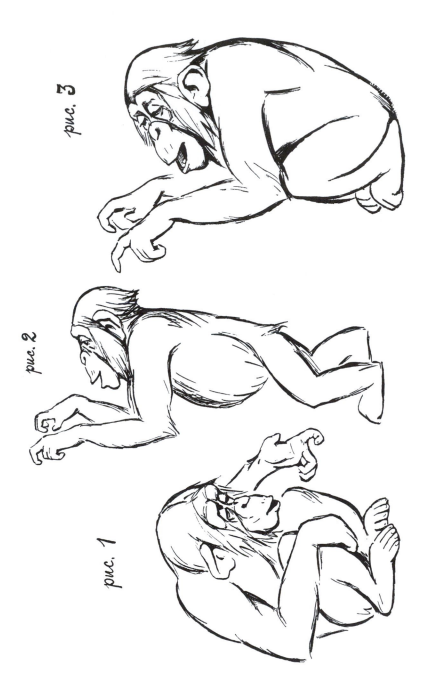

Figure 13. Manual exploration of intriguing objects: (1)–(3) exploration of a tissue sown in relief.

puc. 1

puc. 2

puc. 3

calm down until I stop groaning [plate 43(4)]. The merry, lively mood of the people around chimpanzee infects him with a joyful feeling; this immediately shows on his face in his smile [plate 43(3)].

For example, if the chimpanzee is in his cage and I begin fearfully screaming and knocking on the door, he gets worried. He nervously looks around and startles every second. He groans softly, whimpers, turns his whole body to one side or another, and looks for the invisible enemy as if getting ready to face the danger armed to the teeth. The more I scream, the more worried he becomes. He dashes around the cage and sticks his finger through the mesh of the screen, willing himself closer to me. At the slightest noise, he gives out a violent, monotonous roar, spins as if he were stung, extends his arms entreatingly to me, falls onto the ground face down in fear, and quiets down only when I take him under my protection.

When, for instance, I abruptly throw some inanimate object away from me, Joni immediately fluffs up, bares his teeth, hoots or barks, catches up with this object, grabs it with a hollow grunting sound, and tears and bites it, trying by any means to destroy it. Sometimes, in time with my throwing some toy, he emits angry barks like those of a dog.

Once I threw a wooden stick away from me. Joni immediately rushed after it, caught it, and tried to break it with his hands. Failing to do that, he started to break it by resting its middle on the sole of his foot, using an entirely human way of doing this. After breaking the stick with difficulty, Joni threw both halves aside and unsatisfied with the result, picked them up and threw them farther away and out of sight.

Sometimes, in Joni's presence, I make the dangerous experiment of attacking a man. Joni becomes violent. He joins me at once and is ready to rip the victim apart, so that I have to "have mercy" on the man to avoid an unhappy end to the experiment.

In addition, I want to mention the extreme hatred with which Joni treated the telephone receiver. Every time he got access to it, he took it in his teeth and fiercely gnawed at it. I suppose that, seeing and hearing telephone conversations through the receiver, Joni considered it a peculiar creature. Since the accentuated tone of telephone conversations usually was sharper than the tone of real conversation, it is possible that Joni found our attitude toward that creature to be aggressive and, in solidarity with his masters, was ready to demonstrate aversion toward this object.

Even my sharp cry and premeditated aggressive gesture toward both animated and inanimate objects bring about abrupt hooting, angry barking, and threatening offensive gestures from the chimpanzee. Sometimes, I happen to spank Joni facetiously for some disobedience or minor misdemeanor. He immediately waves his hand at me in a threat, bares his teeth somewhat, and tries to hit me, although tentatively and softly.

If I punch a big ball, Joni also grabs the ball, bites it, and unable to destroy it, runs around the room in frustration, all fluffed up. He tumbles chairs and, running up to the ball from time to time, hits it with an open hand or with his knuckles.

Joni also joins humans in their joy. In a noisy, cheerful crowd, Joni becomes so excited that it becomes difficult to quiet him. When a nice guest familiar to Joni enters the room and I joyously greet the person verbally and with a smile, Joni seems to join me. Taking me by the hand [plate 43(3)], he hoots in joyful excitement, barks, and with a smile on his face, starts playing with the newcomer.

With calm, quiet people, Joni bears himself delicately and carefully. With gloomy and suspicious ones, he is apprehensive. With the abrupt and impetuous ones, he is aggressive. With merry ones, he is vivacious. Like a sensitive resonator that picks up and amplifies a certain note, Joni's reaction is in harmony with the emotional timbre of the human.

This takes place particularly when a human imitates sounds emitted by a chimpanzee. For example, if Joni begins his long hooting and you start to repeat after him, he continues to hoot with even stronger inspiration. He increases the volume and duration of the sound. He finishes one sound phrase, and according to the ritual, on hearing your resumed hooting, picks it up again and enthusiastically continues it until you stop.

Even the repetition by a human of a grunting, guttural sound of the chimpanzee that accompanies the animal's pleasant gustatory sensations stimulates Joni to grunt more sonorously and at a greater pace. If the human does not fall behind in the repetition of the sound, Joni tries to outdo him; the grunting ceases to be intermittent, and the sound evolves into a groan that becomes longer and longer. Joni turns himself on with this joyful sound to such a degree that he can no longer sit still. He gets near the human, touches the human's face with his own, embraces the human, pinches the person's cheek with his lips, and breathes rapidly, his whole body shaking.

Joni tries to reproduce not only the inarticulate sounds (similar to his own sounds) of the human voice, but also the knocking sounds emitted by humans. During my rhythmic knocking on some hard surface with my fist, Joni is burned with the desire to knock. He batters his chest with his fist, beats on the table with his knuckles or on the wall with his open hand, and tries to pick up the rhythm, often succeeding. He gets excited emotionally, fluffs up, opens his mouth, and mounts a challenging attack against me, as if inviting me to a fight.

I have watched many times how Joni, hearing a dog barking, starts to imitate it, sometimes in time with it, sometimes lagging behind, sometimes as if producing an echo. If he can at the same time watch the dogs as they chase an animal, he follows their movements with tense attention and barks with enthusiasm as if participating in the chase from a distance.

You can provoke chimpanzee's grunting any minute. He cannot help but join you and repeat this sound. At the same time, he fluffs up and shows some signs of anxiety.

Joni's hand clapping represents a gesture borrowed from man. This clapping usually coincides with his elevated, cheerful mood, for example, when we enter his room, when he is surrounded by a lively crowd, or when he is being entertained.

Tendency for communication. Concerning the chimpanzee's social senses of a different nature, we must note his strong propensity for communication. It has been known widely that "man is a social animal." It must also be said about chimpanzee that his life is unthinkable without communication because only communication makes possible the full manifestation of the chimpanzee's innermost desires.

In describing the chimpanzee's sadness, we clearly showed the despair or even complete physical frustration caused by his loneliness and how ingenious he was in his attempts to keep a human beside him. In the section dedicated to fear and love of freedom, we cited different threatening or peaceful ways of putting the animal into the cage, the heartbreaking scenes of the chimpanzee's resistance, and the chimpanzee's struggle for freedom and for communication with man.

This resistance to being left alone was observed not only during the first days of Joni's life in a new environment, which may have instilled fear in the animal, but also much later, after he became accommodated to his new surroundings. It was obvious that all his intentions and efforts were aimed at involving some living creature, first of all a human, into communications with him; no object was more interesting to him than a human. After a few weeks of living in a new place, Joni stopped playing "a miserable ape" when left alone (he would not retreat into a dark corner and hide there); however, you could see clearly that he was far from happy.

Left alone, Joni obviously becomes bored and does not want to do anything on his own. He either lies motionlessly, rocks the trapezes slightly, or languidly gnaws at various objects. As soon as a human comes into his cage, Joni changes. He rises to his feet, gets excited, dashes around the cage, plays, finds entertainment everywhere, and climbs the trapezes, using them as if he is the most skillful gymnast (plate 40).

You expand this communication, surround him with people, and you will see what an inexhaustible source of energy, liveliness, and mobility this small, frail creature will reveal, the same animal that only a minute ago seemed inert and phlegmatic. Having been allowed a couple of times into the next room and having experienced the "joy" of communication with people, Joni, hardly awake in the early morning hours and hearing that the house also is awakening, begins to fidget in his corner. He restlessly grunts, plaintively groans, endlessly whines, or pushes the trapezes against the cage walls with all his strength, blatantly calling us in.

If you watch him through a peephole, you will notice with sadness that the little prisoner is engaged in the life outside the walls of his prison. He will listen carefully to every noise behind the door and scream. He will hear steps, quiet down, and sit still; the steps will get nearer, and he will become tense and alert and stare at the door; the steps will retreat and he will give out a violent roar. If you click the lock of his cage open, you will hear how impatiently he rushes to the cage screen and hangs on it, not taking his eyes off the door. If you enter the room where his cage is, he will follow your every movement, plaintively groaning all the time and shifting his eyes from the cage door to the room entrance and back. If you linger at opening the latch for a second or two, his anxiety will increase: He will aimlessly run around the cage as if failing to find the right place, climb the screen, jump up onto the shelf, and jump down to the floor. Or, he will sit down suddenly on his sheets in his corner, depressed and with a strikingly gloomy expression on his face, he will tuck the sheets under his body. Or, he will dash to the screen and look at you, fear and hope in his widely open eyes, which seem to be ready to pop out of their orbits. He apprehensively looks around, following you with his eyes and plaintively, endlessly groaning, his mouth open wide, his gums pulled up, and all his teeth bared. With each second of delay in his release, his quiet, intermittent groans become longer and sharper; he puts his hands or only his fingers through the mesh of the screen, desperately crying all the time. His mouth opens so wide that the upper part of his face wrinkles up, his eyes narrow, and his lips are stretched to the sides so widely that it seems they are about to burst from tension.

You hardly open the door of his cage, and he jumps into your arms, embracing you with all four extremities. He immediately calms down because he knows that you are in his power. He will not give up his freedom and your company easily.

If you decide otherwise and intend to leave the room for some reason without letting him out of the cage, and you say to him, "Play, Joni," and point your finger at the trapezes, he fiercely pushes, bites, rips, and destroys them and bangs on the walls. If you say to him, "Joni, you are not allowed to come out," he turns his back to you as if offended and sits whimpering. He turns his face to you from time to time as if checking out your decision, then he turns away and for a long time does not want even to look at you. If, to reinforce your point, you wave your hand at him and sharply say to him, "Joni, you must stay here," he rapidly shakes his head, reaches out to you with his hands, and greedily catches your finger in his mouth, as if sucking at it, which has always been his way of affectionate pleading. In this case, if you cannot fulfill his persistent desire to communicate with you and you leave the room, he falls from the screen straight onto the floor, fiercely rolls over the floor, emits volleys of roar, tumbles over his head, and shows complete despair, often accompanied by evacuation of his bowels. An assessment of the power and duration of the sadness affect can be made on the basis of the fact that once, after being in such situation, Joni refused to calm down and cried for 3 hours.

Joni happens not only to feel offended by people, but also may hold a grudge against them. He will not take even the tastiest food from the hands of a person who has driven him into the cage, and when this person tries to resume friendly communication with him, Joni hides (for instance, behind hanging clothes), turns his back to the person as if unwilling even to look at the person, or gloomily rejects the signs of affection. But this mood is fleeting, and it takes some sincere patience to restore peace and his amiable disposition toward the offender.

Even if Joni has been allowed out of the cage into the room and a human keeps his company, the animal nevertheless has a desire to go further, to the next rooms. In the surroundings familiar to him, Joni eats and plays much less willingly than in new surroundings. Brought into another room, put into the company of people, Joni changes completely. What unlimited and lively joy; this creature shows his buoyancy, mobility, noisiness, and cheerful spirit, which appear in larger quantities when he is surrounded by a larger group of people.

Forms of friendly communication. Running into a new room and seeing many people, the chimpanzee worriedly fluffs up at first, then hoots; the thundering modulated hooting sound ends with abrupt, high-pitched barking. With this loud barking, the chimpanzee runs over to every person present, smells them, touches them only with his finger (if these are strange people), officiously slaps the people with his hand, or slightly punches the people on the face with his fist. He runs from one person to the next, makes a series of aimless movements, hitches or hits things, bestows his attention on everybody, and cheerfully invites everybody for collective play. His mouth is stretched into a smile all the time; his eyes glisten. His arms and legs move incessantly, and his frail body dashes around the room as if picked up by a squall. He strives to involve all the people around him in playful communication. He is ready to play and frolic with no end, and he is upset when he does not see a response to his insistent invitations. However, in reaction to even a slight response, his impulses become unbridled and wild. He is carried away to such an extent that he interferes with everybody. He throws himself under people's feet, begins biting seriously, does not mind threats or yells, and calms down only when threatened to be put back into the cage.

We must emphasize that the chimpanzee treats different people differently. First, he differentiates between his own people and strangers with total trust toward the former and apprehension and fearful curiosity toward the latter. In the case of the stranger, Joni carefully touches the newcomer with his hand and then smells his hand. If the stranger is totally passive, Joni begins close communication and thoroughly examines the person. Joni traces the stranger's face apprehensively, shuffles the person's hair, and tensely stares at every facial feature and touches it. From time to time, Joni brings his finger to his nose and smells it again. Sometimes, the chimpanzee brings his face close to the stranger's face and smells at its different parts or tenderly touches it with his lips. He often thoroughly examines the clothes. He will touch the dress with his finger and explore every button and every crease with unabated interest. Or, he will sit down on the floor and raise a woman's skirts, look under them, touch the shoes (he pays special attention to high boots as most unusual objects), put his hands into the pockets, and take everything from the pockets and inspect it.

If the newcomer makes an abrupt accidental gesture during this procedure, Joni gets scared, retreats, assumes a defensive pose, and is ready to use his teeth. If the examination is running smoothly, Joni gets more daring and officious. He sits beside the stranger, slaps on the floor around him with his hands and feet, jumps, throws himself at the stranger's shoulders or back, slaps on the person's head or back with his open palm, or batters with his knuckles. Often, the chimpanzee grabs the guest by the dress, which he pulls with his hands or teeth, trying to hitch it on his canine to make the ripping and holding easier. As time goes by, the chimpanzee becomes more and more audacious. He begins a wild play of attack and defense against the visitor, using his teeth; he gets so excited in this role that we have to rescue the guest and detach the person from the boisterous animal that has seized the victim with his hands and feet and keeps holding so fast that it usually takes two men to disengage the animal.

The more fearless, trusting, and passive the newcomer during this procedure of examining, smelling and probing, the faster Joni's accommodation to the person, and the easier he makes friends and plays with the person, biting lightly at the same time. Vice versa, he is especially vicious and filled with apprehension and disgust toward people from whom he is detached out of fear that he may bite them or who make abrupt defensive gestures at the time Joni's anger and creativity are aimed at somehow pulling off a bite.

We have to emphasize Joni's differentiated treatment of people according to their age and sex. We have already mentioned Joni's despotic and hateful attitude toward small children (as well as toward small animals). He attacks them like a hawk attacks a chicken and tries to bite them.

Toward teenagers, Joni shows cheerful playfulness tinged with visible anger. The mere appearance of a teenager apparently incites heated feelings in Joni. He immediately takes off, stands on all fours or assumes a semivertical position, and resting on his arms, stares straight at his partner as though teasing. Or, he rapidly shakes his head up and down, right and left; he opens his mouth, baring his teeth and shaking his freely hanging lower lip. Or, straightening and bending his trunk, he slaps on the floor with his open palm or knocks on the floor or on the walls with his knuckles. Sometimes, he stands face to face with his imaginary enemy, stretches himself into the vertical position, leans to the wall, taps a couple of times on the floor with his feet, shakes his head, and then suddenly seems to charge straight at his competitor. However, at the last moment, he manages to pass the person, but so close that, in passing, he brushes the person's legs and clothes with his hands and hits the teen. Then, running aside, the chimpanzee stops and turns around as if checking the effectiveness of his challenge.

Such intimidation and teasing of small children is Joni's favorite pursuit. If Joni sits beside me and sees boys curiously staring at him, he immediately cheers up; holding to my hand as if fearing that I am going to leave him, Joni bares his teeth and smiles vividly, putting his lower lip forward. His eyes glisten, his side-whiskers rise, and his hands make teasing, waving movements in the air. If the boys retreat somewhat, Joni calms down and keeps silence; if the boys get bolder and come nearer, Joni seems to allow such liberty. He stares straight at them, with his mouth tensely open, but as soon as they come within his reach, the chimpanzee, still cautiously holding my hand, suddenly throws his whole body at them and, with a challenging grimace on his face, tries to yank, pinch, or hitch them. He joyfully smiles to see how they stampede and scatter, only to resume in a minute the mutual teasing, assaults, and retreat.

Often, Joni puts a rag on his head and charges at an imaginary enemy, biting the enemy from under the rag. Sometimes, covering himself in this way and seeing moving figures through the cloth, Joni runs around the room as in a game of blind man's bluff, scaring now one, now another teenager, who scatter from him with shrieks, to the obvious pleasure of the assailant. A similar reaction appears in Joni toward inanimate objects, his reflection in a mirror, the pictures of apes, and especially things he has not seen before.

Joni pesters all the teenagers who show up in our house; he grabs them by the legs and fumbles at their clothes. When they break free in dismay and beat him off with their hands, Joni painfully bites them in the calves and fingers. He behaves similarly with live animals; he pulls their tails and paws and awards them with punches in the face if they do not want to play with him. The same holds for stuffed animals and even for inanimate objects. Joni treats young men the same way; they allow him to play every prank on them and therefore are his favorite company. In the company of elderly people with long beards, Joni usually restrains himself somewhat and shows more respect and obedience.

Joni is more delicate and careful with women than with men, even during play. He often shows tenderness and affection toward women and makes quiet, timid, soft movements. For example, our maid, a very quiet and passive woman (with whom he never had any clashes), won even greater sympathy with the chimpanzee than I did. Every time she came, he slid from my lap, reached out for her, got on her lap, pressed his whole body to her, touched her face with his extended lips, carefully traced her features with his fingers, shuffled her hair, and examined her hands. He looked at every little stain, every little scratch, and pinched it slightly, looking at her face from time to time and immediately stopping the pinching at her first plaintive sound. Sometimes, Joni would sit on her lap for a long time, pressing his body to her, and would not come down despite all my appeals to him.

Thus, it served everybody right: Cowardly behavior must be punished, and Joni is willing to torture a timid person, to scare and bite that person. Bravery must be encouraged, and Joni bears himself with the brave as if they were his peers, not crossing the boundaries of reasonable behavior. Affection and tenderness brings out a tender and affectionate response in the chimpanzee.

It is definite that, when playing with people, Joni wants the initiative for the communication always to be in his hands; as soon as the human dares to be too careless and during his play with the chimpanzee takes on the leading role, Joni becomes tense and aggressive. Finding himself up against long odds and in an inferior position, Joni gets angry. He grunts fiercely, attacks his partners, violently bites, tears everything apart, and gets so ferocious that he looks almost insane. This dangerous play must be stopped forcefully.

Chapter 4

Chimpanzee's Play

Active Play   1. Play with Live Creatures.   Active play with live creatures represents an essential need for the chimpanzee, sanguine by temperament, a child by age, a prisoner by the conditions of life in captivity (in a cage), and a member of a herd by habits in the wild. That is why movement for the sake of movement is his unalterable, unquenchable desire; active play that becomes more lively with human participation is his preferred favorite pastime. He can engage in it for hours, from dawn to sunset, day in and day out.

Left alone, Joni motionlessly lies on the cage floor, his belly up. He idly rolls from side to side and entertains himself with the movements of his hands and feet [figure 11(1)]. But, when one of us enters the room and says, "I got you!" he jumps to his feet as though stung and runs for his life, fleeing us as if we were his most fierce enemies.

When Joni is outside the cage and you appear suddenly before his eyes, he immediately takes off, meets you halfway, fumbles at your apparel, grabs your arms, runs off, and looks defiantly at you. At your slightest attempt to catch him, he takes to his heels. He gets onto couches, armchairs, and wardrobes. With a big rumble, he moves from one place to another, up and down, down and up; he leaps, jumps, climbs, and runs. He gets on the trapeze, on the cage screen, or under the furniture. He gets stuck under the planks of the tables and chairs, sometimes badly hurting himself, but he does not pay attention to it and continues his ceaseless dashing around as though escaping a deadly threat.

If you keep catching up with Joni, he feverishly runs aside, now withdrawing, now getting closer, as though teasing you by seemingly making the capture easier. But, just as you grab him by the arm or leg and for a moment hold him there, he becomes angry, breathes hastily, frequently gasps, and wheezes; his eyes become cloudy, vacant, and dull. He applies every effort to break free from you. If you put up special resistance and hold him by force, he becomes really violent, using his hands and feet and wriggling with his entire body. He tries to beat you off, snaps at you, pushes, and becomes more and more breathless. His voice becomes totally hoarse, his face slightly reddens, his eyes narrow, his mouth opens widely, and two rows of his shiny teeth fiercely click shut and open every second, getting at you here and there, while he vigorously makes your hands let him go. You let him go; he calms down for a second and, having come to his senses, starts playing with you with the same enthusiasm and with his usual alluring tricks.

Commanding a vantage point in the room, he will try to hook you, to yank you with the toes of his dangling foot. Or, after he has come back down to the floor, he will throw himself under your feet like a ball; then, in a blink of an eye, he will run away from you. Or, he will stand at a distance and, bringing down his foot, shake his head, rattle his lips, pound the wall with his fist, bend down or straighten up, strike the floor with his outstretched hand, and not taking his eyes off you, he will attack you energetically, wave at you, come up very close to you, and either bravely knock on your forehead with his knuckles or tug at your dress (if you are a woman) or at your beard (if you are a man). Then, he will run aside quickly, provocatively looking at you over his shoulder.

Even if he does not see you following him, he still will not stop; standing on all fours, he jumps from feet to hands as if charging at you from a distance to intimidate you. If you do not give in, he runs around the room doing things he usually is forbidden to do: hitching and throwing things to the floor, slapping with his hand everything that comes his way, seizing and ripping clothes off the hangers, opening the wardrobe doors, and knocking on the windowpanes, almost breaking them.

Now, against our will, we have to engage in the business of taming him to put to an end this tantrum, which in 5 minutes has turned a tidy room into ashes, into a dump of chaotically scattered, broken, and torn things from the hands and teeth of our unbridled, wild pupil.

You catch up with him and prevent him from taking and tearing things apart, and that is exactly what he needs. Filled with impulses to move and inspired by the sense of resistance, the chimpanzee continues his uncurbed activity with increased energy.

Observation of a playing chimpanzee reveals that he prefers running from you to running after you. If I intentionally switched roles and start running from him, he either did not run after me altogether or did it only unwillingly, passively, for a short time, not being inspired by the role of a catcher.[1]

Hide-and-seek game. Sometimes Joni hides from me and gets under the cage into a remote corner where he is unreachable. He feels safe there as long as he sees only my feet near the cage. As soon as I bend down and our glances meet, he understands that he has been discovered and begins to breathe with aggravation as if I have already caught him, although I am still unable to come closer to him.

I observed in human ontogenesis that the intention to run away from somebody appears in small children much earlier and manifests itself more powerfully than the intention to run after somebody. With respect to the chimpanzee, two corrections must be made. The chimpanzee surrenders to the role of a chaser only when he follows inferior animals (such as cats, small dogs, or baby pigs). The chimpanzee becomes inspired when chasing is combined with seizing something from the one that is being caught.

The following example can illustrate the first case. Seeing a small dog, Joni immediately starts chasing it; he runs around the room as if insane, catches up with the dog in the most hidden corners, squeezes it, or pulls at its tail and paws. If the poor victim shrills, barks, and tries to break free, Joni lets it go, but at the next moment, he continues to chase it with the same energy until the animal gets completely exhausted. Joni does not look tired or feel sorry for the animal while playing the cat-and-mouse game. If I were permitted to rephrase the old adage, it would be "Toys—for the chimpanzee; tears—for the cat."

The situation of the poor dog, indeed, is not much better than the situation of a mouse caught by a cat. If the dog tries to bite Joni, Joni bites it back much more severely. If the dog, exhausted by this relentless chase, stops running and sits calmly in the corner, Joni awards it with punches in the face, driving the dog away from its place. If the dog, becoming used to the torture, does not respond even to this, he pulls it out of the corner by the paw or by the tail and throws it away from him so forcefully that the dog is almost smashed against the wall. Then, in a last effort, the dog flees from its despot, providing a few more pleasant seconds of the chasing game. With the same vigor, Joni chases baby pigs in the village, also driving them to exhaustion.

The chimpanzee heatedly chases a human if the ape wants to take away from the person the things that he has been denied having. He runs with enhanced excitement if he steals from you, against your will, some important object; he considers it his prize, fervently and persistently protects it from your encroachments, and resists you with all his strength.

Play of expropriation. Play aimed at taking possession of things is no less desirable for the chimpanzee than play that involves chasing and catching. He immerses in them so passionately that, more often than not, he crosses the limits of play and takes it seriously. In the beginning, Joni's facial expression is friendly and lively, and his methods of expropriation are light and playful. After many frustrated efforts to get what he wants, Joni becomes so excited that he does not spare the desired object, his rival, or even himself, and he acts as if in a delirium. While in other pursuits Joni shows great unsteadiness and his interests shift, in the play of expropriation the chimpanzee demonstrates extreme persistence in getting the desired object, although he often remains fairly indifferent to the object itself. For example, if I take a cork, for which the chimpanzee has shown no interest, and pass it in before Joni's eyes, he immediately is overwhelmed by the desire to take it from me. He catches my hand, but I dodge, and he fails to grab it.

The chimpanzee gradually increases the pace of his movements; with one hand, he tries to hamper my manipulations and to stop my hand; with his other hand, he snatches the cork from me. When he is close to success, I suddenly press the cork in my fist at the last moment. Joni gets nervous: He immediately takes my hand to his teeth and tries to unclench my tightly closed fingers and to take the object away from me. For me, the play is not worth the sacrifice; to save my hands, I take them off the chimpanzee and raise one hand, holding the cork in two fingers. Joni quickly rises to his feet, stretches himself into the vertical position, and raising his hands, tries to reach my hand that has the cork, but fails to do that.

Then, Joni applies another trick. He entwines his body around my straightened arm, as he would around a stick or liana, and climbs it as though it were a tree, resting his feet on my body. He quickly reaches the cork, but it escapes him again: I take it in my other hand. The chimpanzee stretches his body toward my other arm, hangs on it, and pulls the hand with the cork to him, but I suddenly drop the cork on the floor.

Joni at once dashes down, trying to grab the cork before me, and it is here that the struggle for possessing the object starts. Grunting angrily, Joni pulls my arms aside, drives his teeth into my hands, which hide the disputed object. When he sees parts of the cork sticking out of my hand, he applies his entire weight to my arms to isolate my hands from me and to get them under his unlimited rule. When he is spry enough to snatch the cork from me, he promptly puts it into the safest place, his mouth. When I try to get it from there, defying the danger for my fingers, Joni writhes his whole body in rage, tumbles over the floor, and breathes hastily and with grunts. He pushes me away from him with his hands and feet and bites me so angrily that he appears to be protecting his "daily bread," which is needed to satisfy his unbearable hunger.

The following moment is even more interesting. I stop fighting for the cork, and Joni spits it out in a minute. The cork lies on the cage floor completely unattended while he plays with other things. This reminds us of a phrase, "Happiness is not in being happy, but in the process of achieving it."

The second example is similar to the first one. I show Joni a thimble on my finger. He immediately is engulfed by the desire to pull it off my finger, but I shift the finger aside, and the thimble "passes" in front of his nose. I knock with the thimble on a window sill and then take the hand away. Joni thinks that the thimble remains on the sill. He stands on his toes, reaches the sill, and sees that the thimble is not there.

At that moment, I put the thimble on another sill where Joni cannot see it. By the time Joni gets on the other sill, I move the thimble onto a nearby chair. Joni climbs down to the floor, comes up to the chair, and staring at the thimble, tries to grab it quickly, but I make it before him and put the thimble on the sill again. This time, Joni, to climb on the sill faster, gets on a chair, not taking his eyes off the thimble and getting his hand ready for a quick grab. But, I outdo him again and transfer the thimble to its previous place. Then Joni, to avoid the long procedure of coming down from one window and getting onto another, manages to jump from one sill to the other; he seizes the thimble, puts it into his mouth, and rushes from me, but accidentally drops it on the floor. I take the thimble, put it again on my finger, and raise my hand. Joni gets on me, stretches his body, hangs on my arm, reaches my hand, and grips my fingers. My arm gives in, but I clench my fist, and the thimble disappears from the chimpanzee's sight again. Joni becomes enraged; he fiercely attacks my arm, drives his teeth into my hand, tries to unclench the fist, breathes rapidly, grunts, and bites me so mercilessly that I have to give up the desired object.

It may seem that this thing, contested so passionately and acquired with such difficulty, will be the object of his zealous and loving care. But, you will be amazed when you see that the chimpanzee, having received the desired object, hardly looks at it and puts it aside as uninteresting. You would hardly believe that he fought for this object so persistently only a minute ago.

Often, Joni himself invites me for similar play. For example, he throws a little ball at me; I catch it and throw it back to him. We continue this throwing for a while, and then the ball falls short of me. I try to pick it up; Joni takes off from his place and snatches the ball from my hands. I resist, and the scuffle for possession of the ball begins. Sometimes, Joni gives me an object, for example, a belt[2] or a rag, and when I obediently take the object, he immediately snatches it from my hands, attacks me, and desperately pulls at the object if I do not give it up. Losing the struggle, he becomes even more excited and will not calm down until he gets the ball back; rather, he is ready to tear the thing apart than give it up. He contests the object with such an overwhelming fervor that, at that moment, he seems to be blind, deaf, and senseless to everything that surrounds him. He does not respond to yells, threats, or even spanks and does not for a second lose sight of his final objective or lessen his attention and his energy aimed at its achievement.

Any *Homo sapiens* in any circumstances could learn from the little chimpanzee how to achieve a desired goal. Here is the secret of success: You should be able not only to long passionately for the goal, but also be ready to sacrifice your well-being, muster all your physical and moral strength; you have to be able to suffer, struggle under the utmost tension of your willpower, and never give up, not for a moment weakening your alertness or taking the goal out of the center of your attention.

Play involving struggle—the contest in agility. Any play involving elements of struggle, resistance, or competition is a highly desired activity for the chimpanzee; he willingly engages not only in the agility-of-catching contest, but also in the agility-of-running-away competition. For example, one of his favorite things to play is the dangerous and risky game that imitates the struggle for freedom.

The chimpanzee enters the cage and stands near the entrance. If you pretend to start closing the door behind him, he immediately runs out of the cage; he runs back into the cage in a second, looks defiantly at you, and knocks with his hand on the floor. If you do not move, he retires deep into the cage; as soon as you are going to close the door again, he dashes outside.

In time, the ingenuity of both players in this agility contest becomes more and more subtle and the manipulations crucial for the victory more refined. While in the cage, Joni becomes even more audacious. He runs around the cage, slaps on the walls, and defiantly allures his rival to take advantage of his seeming indifference to freedom and to close the door. This continues as long as you remain passive; your slightest movement is enough for him to become alert and to glance at the door and at you. He starts to relate his position to the distance between him and the door; when you, trying to outplay him and after standing motionlessly for a while, suddenly slam the door shut, he manages to sneak outside at the last moment.

But, when you are faster and close the door before he is able to sneak out, the defeated little prisoner gives out such an earsplitting and desperate cry that you hastily console him and return to him the freedom that he carelessly lost in the game. This bitter cry is more eloquent than words: It tells you how high are his stakes in this game.

Nevertheless he runs that risk, and even after his defeat and loss of freedom, he strives to resume this intriguing game. He is like a gambler, for whom the pleasure of playing increases with the stake values, and the desire to win losses back stimulates staying in the game, sometimes more powerfully than the prize itself.

Another kind of the fight-for-freedom game, the contest in the agility of catching Joni himself, is not so risky, but no less lively. The chimpanzee lies on his back and bluntly extends his foot toward you, but as soon as you are ready to grab the foot, he immediately pulls it back. You pretend to renounce your claims to the leg; he teases you again, almost reaching your body with his foot, but as soon as you make a movement, the foot immediately springs back to its owner. Sometimes, you manage to catch the foot and hold it; if it does not last long, Joni does not protest and is ready to continue playing with the same energy. But, if you linger for a little longer in setting the foot free, Joni's mood changes drastically; he begins jerking his foot from you, risking dislocation. He takes the play seriously, gets angry, breathes hastily, spins, tries to break the foot loose, and even bites you if he does not manage to succeed right away.

Similar play is conducted with his fingers. Joni sticks one of his fingers (usually the index finger) through the mesh of the cage screen. Standing outside the cage, I try to grab his finger; he quickly withdraws it from one hole and sticks it through another, away from me. By the time I reach this new place, Joni already has moved to the opposite corner of the cage and now teases me with his protruding finger from there. He endlessly changes places, the duration shown and the length of the protruding part of the finger, swiftly dodges my pursuit, and promptly jerks the finger from me if I happen to catch it. Sometimes, if I capture his finger and hold it, Joni finds another way out of this unpleasant situation: He sticks another finger out as new bait. Hunting in two places at once hampers my chase and gives the playing animal more chances to continue safely his play of teasing and dodging.

The chimpanzee's most lively play is that with human participation. It is the person who romps with him, chases, catches, squeezes, swings, tumbles, tickles, and rolls him that brings him special pleasure.

Tickling. In response to playful touching of the chimpanzee's armpits and the lower part of his belly, he smiles, opens his mouth wide, pulls and curves his lip at the corresponding side, and catches your hand and pushes it aside. If you pester him again, slightly pinching or suddenly touching him in different places, he desperately beats you off with his legs and arms, turns his head, and tries to catch, with his fingers and teeth, the hands that are bothering him, and he bites them slightly. If you do not stop and hold him faster and longer, he becomes more and more excited. He bares his teeth, closes his eyes, breathes rapidly, falls on his back, fidgets, rolls from one side to the other, rotates his body, lies on his back, and sticks his feet into the air [plates 30(4)–30(5)]. From time to time, he opens his eyes a little and follows you with his dim, senseless glance. Then, as the tickling continues, he closes his eyes tightly again and chokes with hoarse breathing. Now, he gropes for you with his eyes closed and catches you angrily, violently, and mercilessly; he madly attacks your hands and tries to hold them more firmly and bite them more solidly and tangibly.

If you retreat and leave him alone, he calms down at once, but soon he begins badgering you. He waves at you, pinches you, pulls and pushes you, catches and lets you go, and invites you to a new scuffle; he is ready to romp with no end.

2. Objects to Ride and to Drag.   Speaking about play with human participation, I want to mention riding, which Joni enjoys so much. His riding patterns are very diverse.

Joni most willingly sits on his friends' shoulders and loves to ride on their backs. He can ride endlessly and will not let his victim go despite energetic resistance. He drives his fingers and teeth so strongly into their clothes that you hardly can tear him off.

Joni also willingly rides a flat, wooden cart that I pull around the room. Sometimes, he sits on one end of a piece of cloth spread on the floor, and I pull him by holding the cloth by its other end. In order not to fall down, he holds with his hands or, at a higher speed, with his teeth, the cloth edges, or my extended hands, while I face him and move backward. But the most enjoyable for Joni is riding an overturned chair. Knocking over a chair on the floor so that its back lies flat, I will sit Joni on this back and drag the chair by its legs from one end of the room to the other.

In all these cases, Joni shows certain impatience and often abandons the passive role of a rider. At every stop or turn of his "cab," Joni suddenly jumps down and makes a few rounds about the room, only to resume his ride in a moment. At a lower speed of the cab, he jumps down to run around a little.

Sometimes, tired of waiting for my straying passenger to come back, I put a more "predictable" object (for instance, a shoe) on the back of the chair. Joni cannot stand it: He immediately pulls his substitute down even though he is not always going to take its place. Once he has run around to his satisfaction, he is willing to resume riding.

There is no doubt that Joni loved riding, but the ways of indirect movement that I offered him apparently were too slow for him compared with his own running, and the excess of potential energy (*Kraftuberschuss*[3]) accumulated in the chimpanzee during passive sitting was so great that he had a powerful urge to take off from time to time to discharge it in the fastest and most effective way.

Any round object gives rise to Joni's temptation of rolling it. Whether a small wooden ball, rubber ball, egg, or orange, he immediately starts rolling it over the floor; he pushes it, runs after it, grabs and throws it again, catches it, plays with it as if it were an animated object, and runs from it when it happens to bounce and come back to him. Joni gives out a sad breathing sound when a rolling object touches him or catches up with him.

Joni has many ways of moving a round basket. He either rolls it from side to side or, taking it in its teeth, drags it over the floor, pushing it in front of him; or, he sticks his head into the basket and pushes it forward.

Overturning a chair, Joni harnesses himself into its legs as if they were shafts.[4] He drags the chair over the floor, swings it, and then abruptly pushes it forward, putting the chair into a fast slide.

It is very interesting that he tries to use every object that he rolls as a carriage for independent riding. He obviously is puzzled when, sitting in or on the object, he does not find himself moving. For example, Joni pushes a cart or an abacas; it has trundles over the floor and stops. The next moment, Joni sits in it and waits for a few seconds. The "carriage" does not move; Joni rises and pushes the cart again, and it rolls. Joni sits on it again and waits for it to start moving.

After several failed attempts to implement this method, Joni thinks of something else. Pushing the cart and running after it, he tries to jump in it while it is moving, but as soon as he loads his body into the carriage, it, quite naturally, stops.

Completely dissatisfied with this motionless sitting, Joni applies another method of action. He straddles the cart, rests his feet on the floor, and begins to participate in the movement of the carriage, pushing off against the floor. It works; the chimpanzee is moving, although slowly and with difficulty, and a happy smile lights his face.

Having used this method of independent riding once, Joni applies it with irregular success to many different carriages. He sits on a cushion, rug, piece of cardboard, slippers, or overturned bed vessel and, in the same manner and resting his feet on the floor, moves around the room, crossing it back and forth.

Joni tries to employ the same technique with a small wooden ball or an orange, but he suffers defeat and falls on his face. Is there anything that Joni does not use for riding? When he sees sanded wooden panels leaning on the wall, he gets onto them and slides down in the sitting position. If he notices an open door, he at once hangs on it, holding the door knob, and rides on the door back and forth, from time to time pushing off against the floor with his feet or against the wall with his hands.

3. Running (Free and with Handicaps).   Only those who have seen even once the fervor and enthusiasm with which Joni devotes himself to running can understand how much the chimpanzee enjoys this overwhelming movement. He dashes around like a colt set free from a dark stable into the vastness of the fields, like a hound roaming around in search of prey, or like a swallow playing in the air.

The wider the arena for the chimpanzee's movements, the more precipitous and unhampered his running. I recall that Joni was placed in an apartment with a number of joined rooms and later moved to a summer home with a terrace around the whole house. This gave him the possibility of touring around, and he was engaged in running for hours, enjoying the very process of movement and not looking for any other entertainment.

Confined within one room crowded with the furniture, Joni certainly runs less enthusiastically; therefore, he makes up various tricks along the way. Running on all fours, he suddenly falls on the floor, resting on his forearms, and moves along only using his arms as if he were skiing. He squeezes an object in his groin, between his thighs and belly, or between his head and shoulder, and he tries to run and not drop this object. When it slips from him and falls down anyway, Joni breathes sadly and hastily, like a pickpocket; picks up the fallen object; puts it back into the same place on his body; and runs again, this time faster, as though fleeing from pursuit.

Sometimes, Joni runs, holding some of his wooden toys (most often a small ball) in his mouth. Or, gripping a long rag in his foot, he dashes around the room; the rag hitches on the furniture and hampers his run. He has to free it and himself quickly using his hands and teeth, and this complication adds pleasure to his running.

Often, Joni holds a rag in his foot and engages in a quick circular motion around a table or chair leg. He becomes excited and breathes rapidly when the rag gets wound around the leg and obstructs his further movements.

The bigger the obstacle that hampers the chimpanzee's movements, the faster he runs. For example, he takes a belt with a long rod in the form of a hook tied to it; then, he puts a stick into his mouth to prop it open and runs. He gets stuck often because the rod hitches on various objects.

Suddenly, Joni takes shoes and drags them by the shoelaces, going in narrow passages or under the tables or chairs. During this rapid movement, the laces inevitably get wound over the furniture legs, which impedes his running. Joni breathes rapidly and makes hasty, energetic efforts to extricate himself; he runs away again, but still prefers the thorny path to the beaten track.

During Joni's fast running or when an entangled object suddenly breaks loose and bounces back so strongly that it hits Joni, he often gets excited, turns around, and tries to break off. The object hits him yet more severely, and this seems to pour oil on the flame and makes the play more heated.

Sometimes, Joni grabs a very long rope and dashes back and forth, up and down, gets onto the shelves, and hangs on the trapezes. The rope trails behind him in his aerial evolutions and lashes surrounding objects and Joni himself. Joni gives out a sad, aspirate sound, but does not drop the rope. If the rope is caught fast and holds Joni, he gets stuck along with it, still not willing to give it up. He persistently tries to untie it, and disentangling himself, he resumes his activity until he drops in exhaustion.

The more knocking, rattling, and thundering the chimpanzee produces during dashing around, the more enjoyment the play brings. How joyously he runs, for instance, around the metal leg of a bed, holding a metal chain in his foot. The chain hitches on things and thunders at every turn; he winds it around the legs clockwise or counterclockwise. He finds himself tied to the bed, grunts, breathes sadly, and is unable to rip the chain off and break free. He jerks the chain ceaselessly and strongly, but to no avail. He falls into a rage as he feels the resistance, but nevertheless is not willing simply to let it go.

We can state with certainty that the resistance is a pleasant and exciting stimulus for Joni. He looks for any opportunity to create for himself the semblance of a struggle and competition, which gives rise to his counteractions, sometimes bringing him to the point of boisterous frenzy or even to insanity (as in chasing baby pigs or dogs).

4. Getting Through.   Joni tries, with enthusiasm, to overcome the resistance of a material. For example, he takes a rag and sees a small hole in it. He thrusts his finger into the hole at first, then his hand, then his arm, stretching the rag with effort and ripping the cloth apart. Then, he tries to put his head, and finally his whole body, into the hole. If the cloth is firm, unyielding, and resists his efforts to get through, Joni grunts sadly, employs his teeth, pulls the cloth apart with his hands and feet, and clears the passage, getting into it with all his strength. Once through and free again, Joni sits with a contented smile, apparently experiencing gratification after succeeding in solving a difficult problem [plate 44(3)].

Once, Joni received in his possession a T-shirt made of mesh cloth; he accepted this shirt as an object suitable for play. Seeing through the holes, Joni tried to thrust his head and then both hands into the sleeves; the cloth stretched but did not tear. Joni got angry and tried to pull it off, but only tightened it and stifled himself with it more and more. His angry excitement increased, and his breathing was rapid and uneven. He opened his mouth, bared his teeth, and partially closed his dim, blank eyes as if they were covered with a thin film. Totally enraged, he started fidgeting and somersaulting; he applied all efforts to break free from this prison. At last, he freed himself with great difficulty, but as soon as he got his wind back and had recovered from fear, he resumed his attempts to get through the cloth again and again. Later, when Joni was busy with other play, I quietly stole the T-shirt and threw it into a remote corner. As soon as Joni noticed my thievery, he roamed around, found the shirt, and tried to put it on again. Difficult as this repossession was for the chimpanzee, his desire to become entangled again did not abate.

Sometimes, Joni is ingenious enough to create various loops to use for entertaining tricks. Taking a long piece of cloth, Joni throws it on his back or neck; with his arms crossed, he tightens this cloth around his neck as if strangling himself. He starts grunting when it hurts him more than he can endure; then, Joni loosens the pressure a little, as if resting and recovering his strength, only to squeeze his neck again with refreshed energy.

Once, Joni pulled a rope that held the seat of the swing from below so that a loop was formed. He immediately appreciated its merits: He put his head and hand into it, although his head made it with much difficulty. Not in the least confused and breathing hoarsely, Joni nevertheless tried to fit his entire body into it. It took a lot of effort, but he succeeded this time, too. However, as a result, his body was so pressed to the seat that he could not move. He did not like this and began fiercely to extricate himself. When he failed to do so, he became rampant. Raging and using unbelievable gimmicks and all his strength, he freed himself at last.

It seems that no temptation would make Joni repeat his exploits, but this is deceiving. In a minute, having hardly recovered from this mishap, Joni started the same procedure all over again.

One time Joni found a hard rubber ring and used it for his voluntary detainment [plate 45(2)]. He passed his hand into the ring and pushed it higher and higher up; as his arm became thicker, the ring squeezed it more and more. At first, Joni was afraid of the pressure and quickly pulled the ring off. But, having done that and becoming convinced of its safety, he dared to put it on again, either on his arm up to the shoulder or on his leg or head, holding the ring with his mouth and pushing it down at the back of his head. When the ring slid down onto his neck, Joni carefully held it with his hand at the front of his neck, as if fearing strangulation, and then bravely brought it down, being able to breathe in this resilient, but firm, vice.

Once, as a joke, I put an iron ring on Joni's finger. Although the ring did not squeeze the finger, the chimpanzee nevertheless wanted to get rid of it immediately. It was not a very easy thing to do because the knotty swellings at the second joint of the finger obstructed sliding the ring in the opposite direction.

After a few failed attempts to jerk the ring off, Joni became irritated, breathed rapidly, bared his teeth in anger, and used his usual (meaningless in this case) methods of freeing his body from the loops: He writhed his whole body, stood on his head, somersaulted, and fidgeted completely unaware of the fact that these tricks were totally out of place and that he was getting nowhere. He eventually succeeded only with my help.

Joni really enjoys long shoe-laces tied together at the ends; they constitute a suitable material for his manipulations, first of all for entangling and getting through [plates 46(1)–46(4)].

Joni stretches a shoelace with his hands, feet, and teeth [plates 40(1), 40(2)]; he turns it, winds it around himself, makes new smaller loops, stretches these loops with this hands and feet, and puts his head [plate 46(3)], foot, or hand into these loops. He gets stuck in one place and frees himself in another. Sometimes, if the lace withstands his attempts to tear it, he is entangled and tightened as if caught in a net [plate 46(4)]. Then, he grunts, moans, and makes incredible efforts to get out; he cannot be freed even with outside help, not until a radical measure (i.e., cutting the lace) is applied.

Sometimes, Joni creates a vice of an unexpected kind. Finding a metal tube bent in the shape of an arc, he puts it on his neck with great difficulty and does not want to give it back to us. The narrow, stiff arc squeezes his throat. Joni grunts, becomes angry, and tries to free himself from this self-inflicted yoke. With his freedom back, he tries, in a minute, to clamp not only his neck, but also his trunk with the tube.

A firm, inflexible vice scares Joni. Once, I squeezed his finger by a paper clip; although the pressure was not too strong, Joni nevertheless was quick to take it off, hooting worriedly. He never touched it, let alone clamped his fingers in it, after that.

Once, Joni saw a small round wicker basket with a hole in its bottom; he immediately began putting his head through this hole. The stiff loop almost strangled him; sharp edges of the broken wicker pricked his neck. Joni tried to turn the basket back, fidgeted, and pulled the basket off with his teeth, hands, and feet. With extraordinary effort, he freed himself, but resumed the whole procedure because he liked it so much. Suddenly, he came up with another entertainment. He made a new hole in the basket, put his head through it with great difficulty, took the edge of the basket in his teeth, and walked around the room with this contraption on his head, grunting rapidly and loudly and suffering from apparent discomfort.

Soon, even this fun did not satisfy him, and he ventured on new ideas. As soon as his head had barely made it through the hole in the basket, continuously grunting, he also tried to put his arms and his trunk through the hole. In getting through, he became so squeezed by the wicker that you would think he could suffocate, but he stoically pursued his objective. With incredible effort, he put his trunk through the hole and was girdled by the basket around his stomach. You could see clearly how uncomfortable, painful, and unusual it was for him: In walking around the room, he could not coordinate his movements and fell on his side, brushed surrounding objects. He would not mind getting rid of the basket, but this was not easy now. He rolled over the floor to take it off, but this maneuver only exacerbated his pain and increased the clamping pressure. Grunting angrily, he turned left and right, grabbed the closest wicker with his teeth, gnawed at it fiercely in the hope of freeing himself, but this was in vain. He raved, became furious, closed his eyes, breathed hoarsely and in gusts, gnawed at everything he could lay his teeth on, fidgeted, writhed, pulled the basket with his hands and feet using all his strength, and freed himself of it at last, completely exhausted.

Sometimes, Joni would take that basket by the protruding wicker and try to drag it through narrow passages between the furniture. The more delays there were on his way, the more agitated he became. These delays gave rise to anger and excitement. His movements became impetuous and uneasy, his teeth were bared all the time, and you could hear him breathing sadly.

If nothing was at hand suitable for creating obstacles, Joni invented his own methods of self-clamping. He came up to people quietly standing or sitting, grabbed them by their feet, pushed their legs closer to each other, and tried to get through the narrow hole formed, grunting and groaning if he could not do the trick.

The same principle of struggle, with the compelling intent to exercise his ability for resistance, is the basis of the chimpanzee's play of another kind. For example, the chimpanzee lies on the floor near a chair or armchair, lifts this piece of furniture by its leg, and puts this leg on his hand, foot, or even neck or stomach, thus pressing and clamping himself to the floor. Depending on the weight of the pressing object, he emerges from under it using more or less energy, groaning, grunting, and getting angry, as if fighting the intrigues of a fierce enemy.

A few times, I caught Joni when he was trying to put his body under the open door of a sideboard to become trapped between the door and the floor, creating conditions under which it would be hard to get out. When Joni could not create such conditions, he made other inconveniences for himself. For example, lying on his back, he tucked under his head or trunk certain coarse objects, wooden planks, or balls. He rolled them under his body and fidgeted over them, breathing sadly, as though trying to push them out from under his body and slide off them on a smoother place. Having done that, he got back on these objects and continued to fake the sounds of frustration and the actions of protest.

We see profound meaning in this peculiar construction of traps, snares, strangling loops, yokes, collars, entangling nets, and other devices that hold the chimpanzee in detainment and obstruct his release: The meaning is defined in the play theory of Groos.[5] During such play, a young animal instinctively learns to interact with inanimate objects, which can be useful in its future struggle for existence. The animal persistently trains its diverse motion habits, learns to endure pain or inconvenience, and by all means, cultivates its motion ingenuity in the process of overcoming sudden mechanical difficulties.

You can only wonder how, in captivity, with the chimpanzee supplied with everything he needs and a comfortable, almost parasitic, life, the powerful manifestations of the self-preservation instinct still remain strong. This compelling call urges the little chimpanzee to use, in his self-development play, the most unusual things and the most artificial situations.[6]

Among play associated with the chimpanzee's independent movements, we also have to mention swinging, spinning, hanging, jumping, and climbing.

5. Swinging, Climbing, Hanging.   Joni highly enjoyed rocking on the swing. At first, he apprehensively climbed onto the swing and, sitting on the rocking board, was reluctant to entrust his body to it. He kept holding the nearby trapeze, which he knew was reliable. Soon, he became used to the swing, and it became one of his favorite entertainments. When Joni was rocked on the swing, he radiated vigor and joy: A broad smile did not leave his face, his eyes glistened, and his hands made playful, waving, hitching movements every time the swing came close to the surrounding people.

Joni could rock for hours after easily mastering the way of rocking by pushing off the wall with his extended feet or hands. The first weeks after the swing was installed in his cage, Joni was on it all the time; even if he was not rocking, he would rather sit on it than on the motionless shelves or benches [plates 40(1), 40(3)].

All kinds of trapezes, in the form of rope ladders with wooden rungs, cords with knots [plates 40(1)–40(4)], on bars hanging on ropes, were no less entertaining for Joni. Joni performed many gymnastic tricks in his aerial evolutions. It seemed that Joni felt himself in his own elements there, like a fish in water, a bird in the air, a wild horse in open spaces, a squirrel on tree trunks, or a lizard on the ground. The most skillful gymnast could not compete with the chimpanzee in swiftness, agility, resilience, and diversity of movement.

With the spryness of a big, strong, wild cat, he climbed the stairs almost to the ceiling; easily sitting on the board and holding the rope with one hand, he freed the other hand for playful movements or for hitching an adjacent trapeze [plates 40(2), 40(3)]. He would take his foot off the board, hang only on his hands, and rock; or, he would lower one hand and swing back and forth, hanging on one hand. He also would attach himself to the swing with his foot and hand and hang sideways, lowering his other hand and foot; or, he would detach himself completely from the swing and jump down.

There was a thin string hanging temptingly from the ceiling. Joni grabbed the string and climbed it; he dashed down to a rope hanging below, caught it with one hand, and hung on it. He swang back and forth a few times, hanging on this rope; every time he got close to a wall, he pushed off against it with his feet or his free hand, thus restoring the abating amplitude of the rocking. In a second or two, Joni's gymnastic pattern changed. He attached both hands to two parallel ropes. He dashed in the air, let go of one rope, jumped to the other, then jumped back to the first rope, and so forth.

Spinning. This swinging and jumping also became boring, and his gymnastic tricks got more complicated. The chimpanzee noticed a loop at the end of one of the hanging ropes and immediately used it for new pranks. Pulling his body through the loop with great difficulty, Joni rocked in the air for some time. Crouched and squeezed in the loop, he pushed against the walls and hitched objects with his free hands and feet.

This exercise also was abandoned soon, and another one appeared. Joni thrust one leg up to his thigh into the loop and, taking the rope in his hands and feet, rocked in this loop as if on a giant swing. Pushing off against the walls and other hard objects, he twisted the rope, pulled his leg up, held the rope in his teeth, and began spinning in the reverse direction. Joni liked this very much, and his performance got longer each time. You cannot help laughing when you watch the black, scrawny figure of the chimpanzee rapidly spinning in the air, and when, due to the briskness of his movements, you cannot discern the precise outline of his body, which he is no longer able to control. He has surrendered it to dizzy spinning and found himself totally under the power of the laws of physics. His head is cocked to the side, and it makes swift circles in the air. His legs hang like whips and are so involved in vortical movement that there is no time for them to be pulled up; they smash against hard objects that come their way. Only his teeth and hands grip the rope as if in a cramp and hold it firmly. The twisting in one direction is followed, according to the law of inertia, by untwisting in the opposite direction. So it goes up to eight times, with the spinning so long and tiring that Joni gives in and falls down from the rope to the floor like a heavy sack.[7]

Hanging. Two of Joni's gymnastic tricks can be filed under the title "somersaults"; he attaches himself with his four little toes to a high wooden shelf and hangs head down, his hands also stretched down. Holding a few seconds in this position, he disengages himself and falls down on his hands, thus saving his body from being smashed.

For his next number, lightly holding to the same shelf with his toes, Joni rests his hands on a nearby pole, his head down. Suddenly, his toes unclench, and his body, rocking in the air, is ready to collapse flat on the floor, but his feet grab for anything steady or touch the floor before the hands let go of the shelf. The situation is saved; Joni is unscathed, but the ending is not always that happy.

There were occasions when, in such situations, Joni, falling flat on his back, smashed himself against the floor so hard, that the sound of his fall was heard in the next room through the wall. Nevertheless, as stated, no matter how badly Joni was hurt, he never cried, but only quieted down as if recovering. This mishap never lessened his desire for new hazardous tricks.

The chimpanzee's other play in motion involved jumping, somersaulting, and climbing.

Jumping. While with us, Joni voluntarily jumped down from heights no more than 1.5 to 2 meters. Even during his speedy descent from the ceiling of the cage in his room or from the house roof when he climbed down the poles to the ground, he jumped no earlier than when there was approximately 1.5 to 2 meters left. This is quite different compared to lower monkeys such as macacos and particularly to our monkey Desie, which lived with us before Joni; Desie could jump fearlessly from the ceiling, roof, or trees (i.e., from distances no less than 3 meters).

In the absence of other entertainment, intermittent climbing and jumping at the shelves, trapezes, and swings could fill the chimpanzee's entire day. Joni also loved to jump on one spot, especially on the springy surfaces of the sofas, which always prompted Joni to an uninhibited performance of his reckless tricks.

Joni manages to walk on all fours over the narrow (about 10 cm wide) plank at the back of the sofa, dragging a long piece of cloth in his foot. He comes up to the end of the plank and, not being able to turn around, plumps on the sofa, springs back up, starts jumping on his feet in the vertical position, or falls on all fours and jumps from feet to hands and back. Or, he somersaults over the sofa; stands on his head, whipping himself with a piece of cloth; or falls flat on his back and fidgets and spins, waving with his hands and feet. When allowed outside, Joni willingly climbs trees, fences, or roofs.

Climbing. In the country, climbing the barns, the terraces, or the house roof is one of Joni's favorite pursuits, and it engages him for hours [plate 15(1)]. While on the top of the house, Joni strolls over the roof slopes in all directions. He comes down the cornice, glances down, then comes up to the very ridge of the roof and walks along the top plank, back and forth, again and again. Joni enjoys these elevated walks so much that he refuses to come down despite our persistent and energetic calls. He pays no attention to various tempting objects shown to him from below.

Only such an extraordinary event as a staged attack on me can make Joni immediately and decisively abandon his high post; sympathizing with me, he burns with the desire to play the role of my defender and avenger. Then, he hastily, although carefully, descends the steep slopes of the roof, and walks along the fence. He avoids the tips of long nails on the fence (put there as a protection against the intrusion of children), climbs the gate poles, walks along the arch, rises to the roof, and comes down to the porch by the poles that hold the roof [plate 15(2)], runs down the steps and arrives at the scene.

In climbing as in running, the chimpanzee creates various difficulties for himself. He holds an oversize object (for instance, a big wooden ball) in his mouth, stretching his mouth almost to the point of bursting. Or, he turns a big iron key in his mouth or props his lips with a wooden stick.

While Joni is climbing, he often puts some object between his belly and his thigh or near his neck. He holds the object with his bent head and tries not to drop it on the floor during his aerial evolutions.

Joni willingly engages in somersaulting and climbing the trapezes, taking a piece of cloth in his foot or teeth. He dashes up and down with it, stretches the cloth over the trapeze's ropes. He gets enraged when it gets stuck and hampers his movements, and he applies considerable effort to get it out.

Sometimes, Joni covers his head with a rag, which he holds in his teeth so that it does not drop. He keeps climbing over the furniture, bumping into objects and hurting himself.

What is the purpose of all these complications, of these artificially created obstacles, and of the chimpanzee's ordeals associated with them? In these actions, as well as in the ones described elsewhere above, we see how the young animal fulfills the commandments of the great legislator—the old instinct which has brought the experience of the chimpanzee's ancestors from the depth of the centuries.

This instinct guards and nurtures the chimpanzee like a wise old loving grandfather would care for his grandson. It develops the infant and prepares him for future life playfully, gradually, invisibly, and without coercion.

Chimpanzee's Mental Activity

1. Tendency for Entertainment. The little chimpanzee is a very mobile creature not only physically, but also mentally. You can see clearly that his brain does not remain idle for a second. Joni finds something to entertain himself in even the most unfavorable and artificial conditions.

When he is being fed and his mouth is busy eating, his hands move incessantly. He will jerk surrounding objects with his fingers, fumble at my dress, pass his finger over my face, press my eyelids, or shuffle my hair [plate 41(2)] and stare at me as attentively as if seeing me for the first time. In addition to that, he gets distracted and turns around every second. While he is washed and combed, he gets into my pockets, knocks on the furniture, and tries to keep himself busy with something. Every tiny nail in the wall and every piece of paper on the floor are objects that rivet his attention for a long time and enticing play material.

As long as Joni is awake, he plays all the time until sleep knocks him down; then, he yawns widely and lies down, not having enough strength to remain standing on his feet. With his eyes half-closed, he still tries to entertain himself. He rolls from side to side, sticks his feet out into the air and looks attentively at them, or reaches the trapezes with his foot, rocking and pushing them until his eyes close entirely.

Even when Joni is exhausted with illness and hardly able to walk, he comes near my table and watches my every movement. When I put him to bed close to me, he does not take his eyes off me, and when he sees an object in my hands unfamiliar to him, he rises from his sheets, staggering from weakness, tries to climb on the table, and looks at this thing more closely.

A sick chimpanzee, unable to move independently, becomes especially aggravated with the necessity of staying in one place. Looking for diversity, he asks to be put on my lap, on the table, on the chair, or elsewhere all the time.

When there is nothing around worth his attention, Joni starts a thorough self-examination. He shuffles the hair on his body, arms, and legs, and bites his nails. He takes some tiny object, for example, a nail, in his mouth and moves it between his lips, trying not to drop it [plate 45(1)]. Or, he rotates his tongue or makes meaningless movements with his lips and jaws, opening and closing his mouth [figures 14(1)–14(3)].

Even after being blinded by the spotlight during shootings, Joni manages to find some entertainment for himself. With his eyes half-closed, he stretches pieces of string over his fingers, makes loops and tries to get through them, or tears pieces of cardboard, not remaining idle for a second. He is so active that those not privy to the whole situation would not notice that Joni does all these things literally without seeing anything.

2. Tendency for Changing Play.   Observing how Joni plays, if left to his own devices, we notice that his every play activity is very short-lived, and that the objects of his play and the ways of his self-entertainment change very rapidly. The chimpanzee's play clearly reflects his lively temperament, inconsistencies in his desires, unsteadiness of his mood, his impatience, his superficialness in using things, and, first, his careless curiosity.

Three examples of the chimpanzee's pursuits, described in minute detail and pertaining to a period of a few minutes each, solidly prove this idea. The chimpanzee was allowed out of the cage and put in a common room where he rarely has been before (no more than three times a week). First, he gets on a chair and looks at the street through the window; he sits down on the back of the chair and presses his face flat to the windowpane; then, he climbs onto the sill. He keeps looking through the window for a few seconds, knocks on the glass, and hoots every time he sees a passerby. There is an electric lamp on the sill, and as soon as Joni notices it, he immediately takes the cord, gnaws at it, and turns it in his hands, from time to time glancing through the window and knocking on the glass.

In a couple of minutes, Joni gets bored with the window. He moves to the middle of the room and jumps onto an armchair, takes a piece of paper from the armchair, and shifts to a nearby chair. The chair stands close to a dresser with several small objects, (bottles, porcelain knickknacks, etc.). Joni takes these objects and one by one pulls them into his mouth. It does not take too much time for him to do this, and he goes to another armchair near another window and looks through this window. From time to time, he leaves the window, rips the sparkling threads off the Christmas tree, and puts them between his lips. But,

Figure 14. The chimpanzee trying to find amusement and bored: (1) amusing himself by opening and shutting jaws; (2) and (3) amusing himself by rotation of tongue; (4) amusing himself by moving limbs; (5) the bored chimpanzee.

this becomes boring, too; he comes down under the tree, lies on the floor belly up, extends his hands and feet, and starts catching the fir cones hanging from the lower branches. He tears a cone off the branch, puts it in his mouth, and cracks it.

He leaves the tree, comes up to the window again, and stares through it, watching for about 5 minutes the men and horses moving on the street. Then, he comes back to the tree and occupies himself with shuffling its lower branches, taking some twigs in his mouth and gnawing at them. He leaves the tree again for the armchair. He stands on the arm of the armchair and looks through the window from a distance, extending his head toward passersby and following their movements.

At last, he lies down flat on the armchair and resumes his manipulations with the cones. He either tries to rip them off the tree, running the risk of tumbling it down on himself, or takes the cones in his mouth and persistently and force-fully gnaws at them. The entire observation period was no more than 5 minutes.

The second episode is in Joni's room. On the floor in the corner of his room, Joni finds a big aquarium without water, but filled with various small objects. He pulls the aquarium from the corner, drags it around the room, and then takes all the objects out on the floor. Suddenly, he gets into the aquarium and sits there for a few seconds; he gets out and knocks on the glass walls with his fist and drags the aquarium around the room again. Now, a small basin attracts Joni's attention; he puts it on the bottom of the aquarium. Then, he puts his hand in the aquarium, looks at it through the glass, and wants to touch it with his lips. He apparently is puzzled why he cannot do it because he repeats the same proce-dure a number of times. The aquarium is abandoned. Joni pulls a case with sawdust from the corner; he takes handfuls of sawdust and pours them all over the floor. Then, he stops doing this, too. He gets on the sink, takes a comb with which he usually is combed, stares at it at first, and then passes his fingers over the teeth of the comb. Now, Joni is intrigued by a new thing—a thermometer hanging on the wall. He takes and examines it, but soon puts it aside. He opens the lower door of the dresser and tries to get in. In a short while, he also abandons this.

He takes pliers and nails from the window sill and puts them on the floor, or he grabs a pillow and drags it over the floor. Then, he brushes off the sawdust that sticks to the pillow, puts nails on it, takes an iron hook, and strikes the nails with the hook.

Suddenly, it dawns on him to put pliers on his neck; grunting loudly, he tries to stretch the pliers by the handles beyond their possible limits. This also is abandoned. Joni grips the cage door, pulls his body up, then quickly comes down, takes a hammer, and hammers on the nails driven into the sawdust case.

Soon this also comes to an end, and Joni shifts to other things. He takes a piece of paper and an envelope from the table and scratches with his finger over the written letters. In another minute, he tries to pull a floor panel from under his cage. Failing to do this, he brings a case with nails and pours the contents on the floor. He takes the nails by handfuls and finds sunflower seeds among them; he picks the seeds out and cracks and eats them. This also becomes boring.

Joni takes some time to scatter the sawdust around, then quits. Suddenly, he finds a key, puts the key ring on his finger, takes the key ring off, and resumes strewing the nails. Suddenly, he starts thoroughly picking out some nails and fills his mouth with them. This also ends.

Joni comes up to the terrarium and takes off its lid; he then comes up quietly behind the stove and bites off a piece of lime from the wall. I turn around, and he takes off because he knows that picking the walls is forbidden. Now, he grabs a box, drags it around the room, or gets into it with his entire body. Or, he rolls it over the floor or, hitching it on a wooden bench, dashes around the cage leg with these things trailing noisily behind. He apparently enjoys it all very much, but gets bored with it anyway.

Joni takes the basin, turns it upside down, and drags it over the floor, bent down and with his arms resting on the basin's bottom. Then, he turns the basin back into its normal position, sits down inside it, and, pushing off against the floor, moves around the room as though sitting in a cart. This also loses its appeal. Joni takes a brush, passes it over the floor, pulls down a cover sheet from the bed, drags the sheet over the floor, and gets on the sink. He grabs a sponge, tosses it up, pushes it forward, drives it around the room, and chases it. The observation period for this episode was no more than 20 minutes.

In the third episode, Joni was playing in the cage. He pours sawdust around, throws it from one pile to another and covers a small piece of wood with it; suddenly, he stops doing it and runs to the swing, pulls himself up, hangs on his feet, and then falls to the floor. Quickly recovering from the fall, he hastily gathers strewn pieces of paper into a pile; pulling his cheeks inside, he tucks the paper under his body, sits on it, quiets downs somewhat, and sits rattling his lips.

In a moment, he already has changed his occupation. He takes the piece of wood again and begins fiercely gnawing at it, then he tries to pull out a panel from the cage floor. Then, he comes back to the swing and hangs head down again, trying to catch something on the floor with his hands. In a second, he throws himself on a high shelf and hangs on his feet, banging on the cage screen with his hands.

Now, Joni detaches himself, jumps down to the floor, pulls himself up to the shelf again, and then leaves it again. The next moment, he throws himself on the swing again, jumps down from it, and gets back up. He rocks the swing back and forth a couple of times, lies down on the swing, wallows with his belly up, and rocks slowly like in a hammock.

But this is also a short-lived; when I raise my eyes from the notebook, I can see his entire body pressed to the front screen of the cage. He freezes and stares through the window, but it lasts only a second, not more. He attaches himself to the swing again, and hanging head down, rocks in the air, opens his mouth wide, and tries to pick up a piece of wood from the floor. Then, he gets on the shelf again, pulls the swing over there, begins chewing the ropes, changes his mind, jumps onto the swing, and dashes up and down.

Now, he rushes back onto the shelf, lies down on it, and rolls from side to side. Suddenly, he jumps down from the shelf and peeks into cracks in the cage floor and tries to thrust his hand into them, but this also lasts no more than 2 or 3 seconds. Joni pulls himself up to the shelf and sways, hanging on his hands, then comes back to the swing and rocks, pushing off against the walls or trying to pick up a piece of paper from the floor with his feet to drag this paper during his aerial evolutions.

When he fails to do that, he comes down to the floor, throws himself on the paper, and rips it apart. Now, with the paper in his hands, he makes a somersault, or he crumples it with all four extremities, rolling himself into a ball.

As soon as I raise my eyes from the notebook, I see Joni sitting on the shelf. Then, he jumps down, grabs the piece of wood, and comes back up again. As if he has exhausted all possible ways of self-entertainment in the cage, Joni tries to involve me in play. He stares at me, bangs on the wall with his fist, and vigorously advances in my direction. The observation period was no more than 15 minutes.

Left to his own devices, Joni pays attention first to new objects, which he smells and examines thoroughly. Only then does he start to deal with the old ones; we are certain that nothing in the room will escape his eventual attention.

Joni sees a wicker basket for trash. He stands opposite this basket and makes a few ostentatious bows. Since the basket will not move, he pushes it with his hand and chases it while it is rolling.

He spends some time running after the basket, but soon he gets bored with it. He forcefully moves chairs and armchairs together and overturns them, but that also does not last long. He grabs shoes and, holding them by their shoelaces, rumbles them around the room, picking up other objects on the way. He gallops awkwardly from one corner of the room to another, gets stuck under the tables and chairs, and overcoming obstacles with difficulty, still moves forward. Sometimes, circling around the leg of a piece of furniture, he gets entangled with an object; he tries very energetically to break free to start new play.

In all chimpanzee's play patterns, we observe the same features mentioned above: the shifting of the chimpanzee's attention from one object to the next, short operation time with any single object, interruption of activities with an object, and reiteration of activities with that object after a break. The play process pertaining to a given period of time is not an integral process aimed at achieving a certain goal and subjugating all the component actions, but is a complex of disorderly, chaotic actions mosaically cemented together and involving all objects available in the room.

The choice of these objects is accidental and unmotivated insomuch that sometimes, when the chimpanzee aims at a particular play object but sees another one along the way, he immediately changes his mind and goes to the second object, forgetting about the first one. Even in anger, taking an object to threaten or throw at somebody, at the last moment Joni may become interested in the object and start playing with it.

Any new thing, after it already has attracted the chimpanzee's attention, soon becomes old for him and is substituted by another one and so on. In this infinite change of play objects, more than in all other forms of his behavior, the chimpanzee shows extreme inconsistency of desire.

What is the stimulus that urges and spurs his young mind toward this fleeting change of pursuits? First, it is an unappeasable curiosity that demands new materials for its satisfaction.

3. Curiosity (Tendency for New Stimuli).   Joni has an enhanced ability to imitate people's curiosity. If I stare at something, Joni immediately looks at the same spot. If I bend over some object, Joni immediately fixes his eyes on it.

Once, I took a simple little box with powder and, shaking it, started to examine the label. Joni bent over, rested on his hands, and fixed his eyes on the box, slightly extending his lips [plate 36(1)].

I began turning the box in my hands. Joni bent even lower and kept staring at the box, extending his tightly closed lips even further; his eyes were filled with curiosity and tense attention, and he followed my finger, which was tracing the label [plate 36(2)].

When I eventually tried to take the box away from Joni without opening it, he gripped my hand with his, grabbed the box with his other hand, and crying loudly, began to pull the box to himself [plate 36(4)]. When I tried to resist, pulling the box to myself, Joni bit my fingers and did not let go of my hand for some time [plate 36(3)]. Joni's curiosity toward the unknown contents of the box was so high, the desire to seize the box was so strong, that he resorted to such extreme means as biting.

Curious feeling, smelling, and examining of new objects bring Joni a lot of fun. When someone comes back from visiting other homes, he smells the person all over. New things brought from these visits are subjected first to smelling, then they go to his mouth. When we sort boxes or cases, Joni smells every extracted thing, every scrap of paper or cloth. Some odors seem especially intriguing to him; he repeatedly smells such things as raw meat, skeletons, his own dried excrement, live frogs, crawfish, and crunched small insects.

Not only new olfactory stimuli, but also auditory ones incite Joni's curiosity. On hearing an unfamiliar voice or unusual noise coming from the next room, Joni immediately stops whatever he has been doing, lies down on the floor, and peeks into the door crack. If he does not get access to the outside of his room and cannot satisfy his curiosity completely, he defiantly knocks on the door with his fist, demanding to open it. He shows the same curiosity toward new gustatory sensations.

One time, Joni received a box of colored marmalade. Tasting one candy, grunting from enjoyment, and licking his fingers, he nevertheless did not finish the candy, but gets another one, hardly takes a bite at it, and reaches for yet another one. In such a manner, he partakes of all 15 candies. Since not all the marmalades are absolutely equal by taste, Joni comes back to the abandoned first candies and eat them with the same grunting.

For Joni, there is nothing like getting a big bag with various edible products unknown to him. He takes time extracting these products (dried fruit, candies, cookies, bits of fruit and vegetables, nuts, sunflowers, etc.) from the bag and suddenly finds tasty surprises in every handful. Assuming possession of such a promising bag, Joni obediently enters his cage and without any resistance allows the door to be closed. He entertains himself for a long time with sorting, examining, and tasting the contents.

Joni is extremely intrigued by unexpected lighting effects. He looks at the lighting of a match with overpowering attention, as though being struck by the appearing light. He cannot see enough of a flame; he pitches in new dry matches, demanding the lighting be repeated after the flame has died.

You only have to give Joni a very simple, but new, thing; even a plain piece of paper, and he immediately gets invigorated and begins to play with it. He quickly satisfies his curiosity, though, and in a few minutes, the same piece of paper, if not torn to small pieces, lies totally neglected with other discarded materials. But, for a short time, this paper played its entertaining role. This passionate curiosity toward novelties, as any passion, is so overwhelming and insatiable in Joni that it can distract him from any other pursuit; overcome his tiredness, drowsiness, and stubbornness; urge him to strike bad deals; and make him fall for the bait of novelty, even at the expense of his freedom, which he values so much.

During the chimpanzee's most lively play on the trapezes, if he is shown a new thing from a distance, he stops playing at once and hurries, with his mouth wide open, to look at it. Sometimes, in the evening hours, you can see Joni yawning tiredly; he closes his eyes, half asleep, and is ready to drop and fall asleep any minute. But, if at the same moment you bring to him something unseen, he immediately is inspired, and the curious examination of this new object drives the sleep away completely.

It is often impossible to put Joni into the cage because of his resistance. But, as soon as you give him a new object, he voluntarily, submissively enters the cage and begins examining the object.

Sometimes, Joni sits on the roof of his cage, and no persuasion, entreaties, yells, orders, or threats will make him come down (because he knows from experience that such an action would be followed by putting him into the cage). But, as soon as someone pretends to have something in a clenched fist and looks attentively at that hand, Joni becomes very curious. He hastily comes down from the cage and tries to take a look at what is in the hand.

In spite of the fact that this bait is used primarily for the purpose of putting the chimpanzee back into the cage after he has satisfied his curiosity initially, Joni always falls for it, although the period of hesitation before making the decision to come down gets longer every time. But, curiosity eventually wins; Joni has never been able to resist the temptation to gratify his curiosity, or he has never rejected such an opportunity in view of some other benefits. By showing some new thing to the chimpanzee, you can frustrate the development of any affect, change the direction of any emotion, and set them on the track of curiosity. This curiosity develops under various circumstances and is especially strong when Joni is placed in new surroundings.

After Joni was brought to our home, while sitting on my lap, he looked around the room and fixed his eyes on some intriguing objects. He got used to the new place fast; now, he is unwilling to sit still and tries to break from me. He gets on an elevated place to have the whole room under his observation. Then, he comes down to the floor and, like a cat brought into a strange house, smells almost everything, touches some objects with his fingers, and takes objects in his hands and turns them over. He gets into every corner and looks under the tables, chairs, and sofas; in a very short time, he manages to examine, smell, and feel the entire room within the limits available to him.

Joni always tries to broaden the boundaries of his observations. The door leading to the adjacent rooms bothers him immensely; he wants to open it and pushes it, trying at least to look out for a second or to jump outside. When you take him from there by force, he bites and resists you.

Having been in the other rooms once, Joni concentrates all his efforts to visit them again. Since the door of the chimpanzee's room usually is closed, he sits by the threshold for a long time, looking for some delay, clumsiness, or confusion while the door is opening so he can sneak out. If Joni is not standing guard near the door when he hears some movement behind it, he swiftly runs up to the door and waits for the right moment to slide through the crack.

We described above the terrible scenes that occurred when we put the animal into the room and obstructed his access to the adjacent premises. We wrote about the various means and tricks to which Joni resorts to broaden the sphere of his reconnaissance.

The window in our living room that faces the street is an inexhaustible source of new curiosities, and Joni watches everything that is happening outside. As stated, sometimes he stands on his toes, turns his head in the direction of an interesting object, and follows it with his eyes from the moment of its appearance to the last moment before it disappears.

4. Self-Entertainment by Watching Moving Objects. Joni is especially intrigued by passing cabs and pedestrians carrying big packages; in response, he gives out his usual long sound "u-u-u," as in anxiety. Joni persistently and loudly knocks on the windowpane with his fist, trying to attract the attention of some passersby. Because the sidewalk is very close to the windows, the people turn around and smile. They notice the black face of a chimpanzee, to the great enjoyment of Joni, who presses his face to the glass so that his nose, flat as it is, becomes pale and flattens even more.

Running dogs and fighting boys rivet the chimpanzee's attention. His interest increases when the boys start playing with the dogs or attacking them. Joni gets worried, fluffs up, and grunts, participating emotionally in the event.

When we drive Joni in a cab along the streets of Moscow, he is so immersed in watching fleeting scenes around him that he seems paralyzed. He sits still on my knees and stares in front of him, not responding with a single movement when I call him by name.

He behaves similarly when being transported to the country. During the 6-hour ride, he intensely watches the changing landscape of forests, fields, and meadows and sits as if mesmerized. Only from time to time is this frozen state interrupted by his long and curious yell "u-u!" We hear this yell every time we meet another cab.

Sometimes, other cabs are right behind us; the horses almost touch our backs. We all turn around, and Joni recoils in awe at seeing the horse muzzle so close. He keeps glancing back fearfully and does not calm down until he sees a clear road behind him.

Riding in a closed automobile or railroad car, Joni looks through the window day and night, persistently rejects food, and stops looking only when it is too dark outside to see anything. This dominant watching is interrupted only by Joni's long and snappy grunting if he sees something especially interesting, for example, a human crowd during our railroad stops or roaming flocks of cattle. An express train thundering past makes Joni retreat from the window, but he soon is tempted by his curiosity to cling to the window again, in turn inciting curiosity and amazement in the passengers in the other train when they see such an unusual creature.

When allowed to the yard, Joni finds an inexhaustible store of new material to satisfy his curiosity. He peeks into the apartment windows that are close to the ground, runs over to a neighbor's porch, and examines every object there. He looks at the street, listens to the street noises, and climbs small trees. Coming down, he picks at the ground, plucks the grass, thoroughly examines and smells everything he can lay his hands on, and then smells his own fingers. In a swift and business like manner, Joni incessantly runs from one yard's corner to another, all the time finding new curious objects everywhere.

Put into a second floor room with a screened balcony, Joni spends hours there, hanging on the screen and greedily watching what is happening on the ground, as through unable to satisfy his insatiable curiosity. Every day, crowds of country boys and other curious people gather under this balcony to see a "clown" for free; Joni also enjoys seeing this joyous, lively, and noisy bunch. From time to time, he entertains the public by his sudden leaps over the screen; he shakes the screen and appears to be getting ready to jump outside. The crowd bustles, people dash from side to side, and there is squealing, crying, and laughter, which only enhances the pleasure of mutual watching.

This watching is very tempting for Joni. When closed in his room due to cold weather, Joni finds his way out to the balcony. He spryly unties the ropes holding the balcony door and goes out to accumulate new and changing impressions. When in the solitary confinement of his room or cage and without an influx of new things, Joni falls into a phlegmatic, lazy mood. He wallows on his sheets, belly up, yawning; he apathetically examines his hands and feet and completely ignores the abundance of his old, familiar toys.

5. Objects That Intrigue the Chimpanzee.   In a new room, after preliminary superficial reconnaissance and a brief examination of objects available for his observation, Joni directs his attention to opening hidden hollows. He peeks under the sofas, armchairs, or tables and opens window frames and takes various objects from there. He opens wardrobe or sideboard doors, looks at the contents, and touches every object; he pulls out drawers, baskets, cases, or boxes and starts taking things out of them.

Sitting in front of a case filled with scraps, Joni carefully takes them out of the case. He smells certain rags and puts them aside; he takes other rags in both hands, puts them around his neck, and makes a semblance of a tie (or a scarf), which he pulls over all the time from right to left and back. Suddenly, he becomes attracted by other rags. He takes the first ones off and puts the second ones on; these remain on his neck until he finds newer and more preferable rags.

6. Preferred Object Qualities.   I have noticed that, when he has to make a choice between a number of objects, it is the color that determines Joni's decision. Joni most often picks intensely colored red or yellow scraps. In other cases, he likes not so much bright, but glossy, silk or velvet cloths; he straightens them and smooths them with his hands. Sometimes, it is translucency and meshiness that intrigue his curiosity. He picks laces and puts his fingers into their holes; he finds veils or thin oilcloths, covers his eyes with them, and looks through them at the light.

Joni is drawn to the fire. He likes to watch candles or sticks burning. He gleefully turns on the electricity and stares at the lamp.

Often form and size attract his attention. He amasses especially long scraps of cloth, which he uses as either scarves or head covers. With these things on, he climbs on the furniture and runs around the room, holding the scrap with his teeth so that it does not slide down.

Sometimes, stimulated by his curiosity, Joni goes very far in his search. He opens stove doors and takes out coals or ashes; finding a samovar on the floor, he takes out all its contents of splinter and coals.

I have mentioned how willingly Joni examines, with a stick, even small cracks in the floor of his cage, driving out cockroaches from there. I am certain that there was no receptacle, no cavity, no hole in the room into which he did not take a peek.

Joni stares through the holes of the wicker chair seat. The opening depth, the transparent floor, seems to surprise him, and he gives out a long moaning sound. He tries to thrust his exploring finger into the holes, and when he fails because the fingers do not fit, he applies his efforts to tear the seat.

At another time, convex protruding, as opposed to concave, objects incite his unusual interest. Joni is especially intrigued by the protruding black keys of a piano; he touches them with his index finger, feels them, and tries to hit, or rather push, them down. When the sound emerges, he pushes the keys harder, up to his limits. As usually the case with him (as well as with people living through moments of extremely tense attention), his lower jaw hangs down, his mouth opens wide, and his look follows the movements of his exploring finger.

A similar pattern of behavior can be observed when Joni sees black stains clearly visible on white paper. Joni touches the stains with his finger, tries to pinch them (as though they were bas relief). When he fails to do that, he begins to scratch the paper with his nails, trying to detach the intriguing image. The less successful Joni is, the more persistent and ingenious he becomes.

Various books with illustrations also arouse Joni's curiosity. He loves to turn over the pages and look at the pictures; he treats pictures according to what they depict. Pictures of animals, particularly monkeys, are objects of special interest, especially if they contain contrasting shadows. Joni also tries to take the distinguishing objects off the pictures and becomes angry when he does not succeed. Photographs that depict animals with faces that have protruding eyes or bright specks of light in the eyes make the chimpanzee aggressive; he stares at them, fiercely scratches the faces, and bangs on them with his fists.

7. Chimpanzee's Reaction to a Mirror.   Seeing himself in the mirror for the first time, the chimpanzee opened his mouth in amazement and looked questioningly and curiously at the glass, as though asking silently but eloquently, "Whose is this face over there?" [plates 3(3), 39(1)].

When I pass a mirror in front of Joni's eyes, he does not move his body. He turns his head in tense curiosity, catching the image with his eyes; his mouth remains open all the time.

Soon Joni gets used to his image. He closes his mouth, tightens his lips and extends them forward, and keeps looking at the mirror [plate 47(2)]. But in a short while, this detached watching does not satisfy him anymore; he tentatively extends his hand toward his image, bumps into the glass, takes the mirror by the edges, pulls it closer, and, opens his mouth slightly, looking at the image intensely [plate 47(4)]. Seeing his image shifting, Joni tries to take the mirror from me, gnaws at its edges, and peeks behind its surface. If I do not let him touch the glass, Joni extends his hand behind it, trying to grab whatever it is there [plate 39(2)]. If I intentionally leave my hand behind the glass, he gets frightened at first and then presses my hand in his and tries to pull it to himself. If I resist, do not show him my hand, and hide it somewhere, Joni meets a new appearance of his image with stronger aggressiveness: Tossing his head back, he fiercely bangs on the mirror with his folded fingers, a wavy cramp rippling his lips and curving them near the canine teeth [plate 39(1)].

Then, Joni performs thousands of grimaces before the mirror, partly intimidating, partly curious and entertaining. Extending his tightly closed lips and humping his upper lip, he would give out a rattling sound, in the meantime not taking his eyes off the mirror [plate 39(4)]; or, opening his mouth, he would extend his tongue forward, writhe and rotate it in front of the lips, and shift it from side to side, from left to right, up and down. From time to time, his lips assume different configurations. He extends them forward in the form of a funnel (as in general excitability) [plate 47(6)], then he pulls them to the sides as if in a smile. He folds the upper lip in the shape of a promontory over the lower lip (as in intense attention), then he jerks, curves, or bares the upper lip above the canine teeth (as in anger).

We have mentioned the curiosity with which Joni looks at his image in the door and windowpanes and in other reflecting surfaces, for example, in the metal balls of a bed, in the black finished surface of a stapler, or in the shiny surface of a steel doorbell. It was reasonable to assume that stuffed animals and live and dead animals would arouse even stronger curiosity in Joni, and so it turned out indeed.

In the section dedicated to the emotion of fear, we discussed that, in spite of the fact that Joni was afraid of some objects (for example, a stuffed bear or masks of a human face and a wolf's head), Joni nevertheless made attempts to acquaint himself with them because his curiosity overcame his fear. Joni showed a similar intention to examine exciting objects when he saw a human skull, monkey skeletons, and dead birds or animals. Seeing these unusual objects, Joni initially used his exploring index finger, curiously touching the intriguing object; then, he smelled the finger, continuing to look at the object [plate 20(2)].

8. Chimpanzee's Reaction to Live Animals.   Live animals, with their ability to move (for example, guinea pigs), rivet his curious attention for a long time. In a dark terrarium, Joni immediately notices a slowly crawling caterpillar and watches its movement over grass blades for a long time.

A frog arouses even stronger interest in the chimpanzee. Joni bends over to the terrarium, raises and lowers his eyebrows, and presses his lips tightly, extending them forward and shaping the upper lip in a hump. The chimpanzee stares at the frog from different directions, following its slightest movements. Every time the frog freezes, Joni knocks on the glass as if urging the animal to move; when the frog becomes overly excited and begins to jump, Joni glances over the open top of the terrarium and tries to snap the creature with his fingers or hit it in some other way. A lizard that tries to find its way out of the terrarium and crawls over the screen stimulates long unabated curiosity in the chimpanzee [plate 48(5)].

Once, I showed Joni a live crawfish. Joni touched the paper in which the crawfish was wrapped, smelled the finger and then the paper itself, and stared at the crawfish. The crawfish started to move. Joni startled, curved his upper lip, and abruptly jerked the crawfish by the whiskers, then smelled his fingers again. Later, Joni became braver: He took the crawfish by the whiskers and began dragging it over the floor. The olfactory stimuli coming from the crawfish roused Joni's curiosity no less than the visual stimuli.

He touched the crawfish again, smelled his finger and then the spot where the crawfish had been, rubbed the wet stain, slapped the place where the crawfish had sat, and waved at the crawfish with his fist. When I put the crawfish into the chimpanzee's cage, he took it by the whiskers and threw it away from him and out of the cage onto the floor. The presence of such a creature too close to him irritated Joni.

All big cattle (such as cows, horses, sheep) incited Joni's unusual interest tinged with fear. I happened to see Joni many times hanging on the balcony screen and watching the flocks of cows and sheep as they passed to the pond.

Small domestic animals (such as pigs, cats, dogs, or hens) activate Joni's curiosity to such a degree that he cannot sit still; he runs after them to have more direct contact with them. When he is separated from them by a windowpane of the balcony screen, he attentively follows their movement.

People, especially newcomers, rouse Joni's curiosity more than his curiosity toward animals. When some new persons enter his room, Joni literally pierces them with his eyes. He has hardly looked at one of the newcomers, muttering his not too good sounding greeting (his worried hooting), when he looks at the door again, still hooting and extending his hand toward yet another new person. Opening his mouth, he now looks over the last person's head at the next person, and so on, until the door is closed.

9. Chimpanzee's Reaction to Newcomers.   Usually, Joni cannot stand passive watching of newcomers, but wants to get to know them better. If Joni's release from the cage is delayed while there are a number of new people in his room, he starts making an awful noise with the trapezes. He groans, whimpers, restlessly fidgets around the cage, and will not calm down until he is allowed out of the cage.

Released from the cage, Joni first aims his curiosity at new people. He runs over to them and tentatively touches their bodies with his index finger, thoroughly smelling his finger after that. He even touches some people with is tightly closed and extended lips, smelling the person at the same time.

Finishing this preliminary olfactory examination and making sure these new people are peaceful and harmless, the chimpanzee begins a more thorough study. He feels the faces; puts his fingers into the people's nostrils; shuffles their hair; slaps the bald heads; fumbles at their long beards; passes his hands over dresses and turns up the hems, lapels, and creases and inspects them; puts his hands in the pockets and empties them; looks curiously at the shoes; and touches spectacles, watches, and chains. Women's skirts that are long are especially intriguing for Joni; he grabs them by the edges, pulls the edges up, and peeks under the skirts. He touches the women's legs, to the utmost confusion of female visitors. The more a person resists such an examination, the stronger Joni's desire to continue this reconnaissance.

Joni usually prefers to play with strangers; he actually urges them to play with him. He lightly waves at them, slaps them on the back, pushes them, and looks happy when they respond to his calls. Joni gets used to the stranger very quickly and can play with the person for a long time, ignoring "his own" people.

One night, a carpenter with an assignment to make a partition entered Joni's room. Joni followed the man's every movement during his 3-hour stint. Joni clearly wanted to sleep; he yawned and had trouble keeping his eyes open, but nevertheless he attentively watched the newcomer until he eventually dropped and fell asleep.

Another time, a different stranger came to seal the window frames for the winter. This turned out to be big entertainment for Joni. The chimpanzee clung to the opened windows all the time and stole and tried to eat the putty. He stared at the new man, curiously and carefully tugged at the man's beard, smelled his face, touched him with his hands, looked at his high boots, and followed all his movements, from time to time interrupting his watch by teasing the man and urging him to play. Joni stood right next to the man, stomped his feet, knocked with his hand on the floor, charged and pushed the man vigorously, and ran to the opposite corner of the room to see what the response would be. If this maneuver also failed to involve the man in playful contact, Joni dashed around the room, overturning chairs. He fidgeted and tried to entertain himself independently.

We have to point out that Joni had an exceptional ability for observation and highly developed curiosity. He immediately noticed any new element and wanted to become acquainted with it more closely.

A young man (familiar to Joni) who had long hair previously, once came to us with his head clean-shaven. Joni noticed the change at once, came up to the man, and touched the man's shaved head, passing his hand in every direction.

Joni immediately notices my new dress, looks at it attentively, and smells it. He notices every little stain, every scratch on my face. He observes with curiosity a needle or a pin on the floor and every stain on the wallpaper. The senses of touch and smell play a more important role than vision in these examinations of new people and objects. For example, on entering a new room, Joni smells literally every object. As we have noted, Joni's behavior in this respect is similar to that of cats; once in a new house, they spend their first hours in incessant olfactory reconnaissance.

People and objects interesting to the chimpanzee are examined by him in a more tangible way—he feels them with his extended lips and touches them with his tongue. However, he is very careful with sharp things. He never cuts himself, although he deals with such things as nails, pieces of glass, pins, tacks, and thorns from flowers (roses) all the time.

Very often, the chimpanzee's lips play an auxiliary grasping role, completing movements that are especially subtle, such as when the chimpanzee wants to pick up a flat round piece of ceramic from the floor. Try as he might, he cannot take it with his fingers; he bends over to the floor and easily catches it with his lips; then, his hands take over.

When there is noting suitable for play at hand, Joni uses his lips, tongue, mouth, or hands for entertainment. Tightly closing his lips, he pulls his mouth first to one side, then to the other side [figure 14(4)]. He sticks out his tongue and passes it over his lips, slaps his lips with it, and extends it forward or pulls it back into the mouth[8] [figures 13(2), 13(3)].

Entertainment with Sounds

Sometimes, Joni opens his mouth as wide as he can, then quickly closes his jaws, and clangs more or less rapidly with his teeth. This sound amuses Joni, and he repeats it many times until his jaws get tired, until they are unable to open [figure 14(1)].

Often Joni entertains himself with sounds that he makes with his lips. He extends forward his tightly closed and bulged lips, slightly turning the lower lip inside out and humping the upper lip, [plate 39(4)]; he pushes air out of his mouth, giving a snappy cracking, sputtering sound. His lips become flat after the air comes out. Joni bulges the lips again and again, enjoying this "raspberry" for quite a while.

Another entertainment of the chimpanzee with his own sounds is no less comical. Once, Joni took his upper eyelid in his fingers and began pulling it away from the eye; a weak slapping sound could be heard. Joni immediately repeated this sound, pulling at the eyelashes of both eyes. Sometimes, he entertains himself by slapping and pinching the bare arms and legs of a human. He willingly claps his hands or rhythmically knocks on the walls or on the floor with his fist.

It happens very rarely that Joni exhausts all possible ways of entertainment. In this case, he extends his legs, takes his feet into his hands, tosses his head back, opens his mouth to its fullest, stares up at the ceiling, and keeps sitting in this silly and careless pose, swaying slowly back and forth [figure 14(5)].

I must say that usually the chimpanzee enjoys any sounds. He tries to reproduce them in various ways using any available object. Taking wooden balls, sticks, keys, metal plates, steel rods, and other metal objects, Joni strikes them against one another, rhythmically knocks with them on something hard, and produces a variety of sounds.

Taking a bunch of keys in his teeth by the ring, Joni shakes his head rapidly; the keys clang, and he likes it. He takes a belt with a metal buckle, a long stick, or a whip and waves these objects right and left on the walls and on the floor; the more noise he makes, the more he likes it. Often, he gets carried away, hitting himself on the head and on the trunk, but he does not stop, becoming more and more excited with every strike.

Rumbling with the trapezes and ramming them one against another and against the walls is one of Joni's favorite entertainments when he is alone in his room. You can hear such noise, clamor, and mayhem that you can think it comes from a dozen boisterous teenagers.

Opening and closing the cage door, the wardrobe door, or front door is Joni's common pastime. He cannot miss any open door without slamming it shut.

Joni uses practically everything to produce a sound. Taking a punctured rubber ball in his hands, Joni kneads it and listens to the screeching sound it makes. Putting a rubber ring on his face, he stretches it and lets it go; he hears the rattling sound and immediately uses this device for the purpose of producing sounds. Hitching the ring over his canine tooth, Joni stretches the other side with his toes; his fingers use the tightened rubber as a string. He keeps reproducing this sound with a smile until he tears the string with a sharp movement; then, he throws the ring aside as an unsuitable and uninteresting object.

He can be totally engrossed in rolling the cap of an inkpot over the table. He spins it like a top and holds it with his fingers by the handle. Spinning and stopping, the cap clangs, and this urges the chimpanzee to repeat his actions.

Experimenting Play   The "experimenting"[9] play represents a big part of his entertainment; it takes place when there are new, unknown, and unseen objects around. The varieties of this play are countless and versatile, as countless and versatile as the objects that surround the chimpanzee.

In this play, the infant chimpanzee familiarizes himself with the properties of the objects in his environment. He develops his five senses; he acquires the experience of dealing with things and the proper use of the things to satisfy his needs. His higher abilities, such as curiosity, attention, and patience, develop under these conditions in the process of experimenting.

Where does chimpanzee get the material for this play? An infant chimpanzee spends most of his life in drab conditions of confinement in a cage (with only a couple of benches, a wooden bed, sheets, and hanging trapezes); it is all new, unusual, and interesting when he is given freedom: There are water, soil, sand, rocks, grass. Various objects of everyday human practice are even more unusual to him, and he tries to explore them thoroughly, especially water, which due to its mobility, is an inexhaustible source of entertainment for the chimpanzee.

1. Play with Water.   You may assume that Joni suffers from unquenchable thirst because every time he sees water, first he wants to drink it. Even when he is being washed, he tries to lick my wet hand and to get one more drop, again resembling little children, who may drink even when they do not feel they really need it. If Joni does not drink, he rinses his mouth with the water. Then, he empties his mouth and splashes the water all over the floor.

If water is in a tall vessel (for instance, in a jug), Joni sticks his hand into the jug and tries to scoop the water; failing this, he takes out his wet hand and splashes the water around him. If water is in a wide vessel (for instance, in a bowl), Joni tries to catch the water with his hands and pour it from one hand into the other. He occupies himself for a long time with looking for various objects to put into the water and watching them immerse or float.[10]

Every time I washed Joni near the sink, Joni watched very carefully the spurt of water coming out of the faucet. Once, I left the water flowing and gave Joni the absolute freedom to do with it whatever he wanted. He immediately stood up, pulled his body closer to the water, put his hand under the spurt, and then pulled it over to his mouth; the water flowed off his palm, which was empty, by the time it was near his mouth. He sped up his movements, but in vain, he could not deliver water to his mouth.

Very attentively, Joni shifted to the other end of the sink and tried to catch the spurt from the other side. He pressed some water between his stretched fingers, as though gripping a hard object; water kept sliding away from him. He began running faster from one side to the next, coming yet closer to the spurt and bending his head as if trying to understand what was going on. Catching the spurt with his hands, he opened his mouth wide in time with his hands, put out his tongue, and turned the tongue in front of his lips and then pulled it back as if using it to help collect water.[11]

This procedure had taken 10 minutes of my observation time, but nothing helped. Finally, one time when Joni was very close to the spurt, he thought of catching the water with his mouth. Since then, every time he has wanted a drink, he comes to the sink, opens the faucet, and drinks, extending his lower lip in the form of a scoop.

In the country, Joni used to spend a lot of time near a creek that streamed from the hill. He tried to catch the stream with his hands; he dammed it with his fingers, splashed it, put his head and face under the freezing water, caught it with his lips, and got so soaked that we had to dry him for a while.

Often, if water is handy, Joni takes it in his mouth, empties his mouth, takes it again, splashes it on the cage floor, and smears it with his hand or with a rag.

2. Play with Loose Materials.   Soil, stones, or sand are Joni's favorite materials for entertainment. In the yard, Joni grubs about in the sand, finds stones, and takes them in his mouth or digs a hole in the ground with a nail.

When Joni fins a big stone, he scratches its surface with his nail, spits on the stone, smears his saliva over it, and then gnaws at it for some time. After he has taken the stone out of his mouth, he looks at it and finds a small hole in it. Bringing the hole to his mouth, he sucks up the contents of the hole, which consist of fine dust. Then, he uses the stone in his own, peculiar way. He puts the stone on the ground and lies down with his back on it; later, he puts it on his belly and breathes rapidly, as if trying to get from under the stone. The next moment, Joni takes the stone and is going to throw it at his image in the window pane.

Walking on the grass, Joni looks very attentively at every grass blade. He plucks one, ignores another, puts the third one in his mouth and discards it after tasting; he eats the fourth one and later keeps selecting this particular kind from among the population of many plants.

Sawdust, used to cover the metal floor panels in his cage before Joni got used to the material, serves for his experiments. Joni pours the sawdust from one hand to the other and bends his head low toward his hands. He looks at the sawdust, opens and closes his mouth, breathes rapidly, and from time to time quickly moves his lips; he tries not to spill the sawdust. Or, with his hands, he rakes the sawdust from remote corners of his cage into one pile, grunting all the time. He collects this pile on a piece of paper spread on the floor. Having gathered a small pile of sawdust, he pulls at the edge of the paper and drags the whole thing over the floor. The pile collapses; the sawdust spills on the floor. Joni gathers the scattered sawdust from the floor and puts it again on top of the pile. He carefully touches the top of the pile with his lips, picks up a small portion of the sawdust into his mouth, and chews at it for some time. Suddenly, he pulls the paper from under the pile, shakes the sawdust off the paper, puts the paper in a different spot, thoroughly wipes the sawdust off his hands, claps his hands to get rid of the rest of the sawdust, and then repeats the whole procedure from the beginning.

Often, having gathered a big amount of sawdust on the paper, Joni suddenly strews it around with his hand, throws it off by abruptly lifting the paper, or takes a handful of sawdust and tosses it up; at the same time, he bends his head down to prevent the sawdust from getting into his eyes.[12] Sometimes, Joni diligently rakes the pile apart, making smaller piles, which he places near the big one.

Getting into a big chest filled with sawdust, Joni sits for a few minutes all drowned in the sawdust, sifting through it and shuffling it.

3. Play with Transparent Objects.   The chimpanzee's reaction to transparent yellow medical oilcloth was very interesting. Joni smelled it, touched it with his lips, and looked through it; the transparency attracted the chimpanzee's attention. Joni liked to look through the oilcloth at the light; perhaps he liked the bright yellowness of the world around him. Later, he wanted to see everything colored in this unusual yellow and applied every effort to do that.

It was not an easy thing to do. Joni took one corner of the oilcloth in his teeth and slightly stretched his lips so that the oilcloth would not stick to his nose and would not hamper his breathing. He covered his face up to his forehead with the other end of the oilcloth and, tossing his head up to prevent the oilcloth from falling off, stared up at the screen of his cage and for some reason hit his forehead through the oilcloth. The oilcloth slid from the forehead and fell off. Then, still holding its one end in his teeth and holding the other end over his eyes, Joni dashed to the swing and kept looking through the yellow film at objects in his room. Holding the oilcloth near his eyes, Joni bumped into the walls in excitement, banged on them with his fist, spinned, jumped, leaped, and showed all the signs of a boisterous and joyous mood. During all of this, Joni made sure that the oilcloth would not fall off his eyes.

Later, the chimpanzee found other properties in oilcloth, which he also tried to use. He pressed the oilcloth through the holes in the cage screen; passed the oilcloth between his clenched teeth; laid it on a board and smoothed the wrinkled places; and peeled off, with his teeth, tiny scraps of the oilcloth and threw them away from him. Suddenly, lying on the floor with his belly up, Joni laid small pieces of oilcloth on his eyes and looked up through them, as if looking through glasses.

As soon as Joni stopped playing with a big piece of oilcloth, I wanted to take it from him, but he would not let me do it. Breathing hoarsely, Joni grabbed the oilcloth, bit me, and ran aside; he broke free when I caught him, hid the whole piece of oilcloth in his mouth, and fought over it as if it were a priceless treasure. When I gave Joni a piece of glossy but opaque paper on the next day, he tried a few times to look at the light through it as through the oilcloth.

I give Joni a long piece of rubber tube, and I am amazed how diverse his manipulations with this object are. Joni draws the tube to his mouth, smells it, thrusts his finger into the tube, stretches the tube while holding its end in his teeth, and sucks in the air from the tube; suddenly, he slings it over a hook attached to the ceiling, pulls its other end down, and tries to hand on it, but is unsuccessful.

Soon, Joni abandons this and thinks of something new: He presses an open end of the tube to his cheek, then intermittently bends and straightens the section of the tube that is in his hand. Then, he applies a new approach. He pushes the tube through holes in the ceiling part of the screen, and when the tube gets through the screen completely, he tries to reach it through the same holes. When he fails, he begins to push the screen itself; the tube jumps, and its end eventually gets into one of the holes and into the cage, at which moment Joni takes hold of it.

4. Play with Hard Objects.   The forms of examination and application of hard objects are entirely different. I gave Joni, for his full use, a big empty glass vial,[13] which earlier contained French turpentine. Joni hastily took it from my hands, ran into a remote corner of the cage, sat on the floor, and began to examine the vial, turning it in his hands. Then, he smelled the vial at its neck, put the neck into his mouth, and held its edges with his teeth; he pulled the vial by its bottom part, trying to tear off the neck. He did not succeed in this and soon left the vial alone. He took the vial out of his mouth again and began turning it in his hands; the vial opening attracted his attention.

He thrust his first finger into the hole, and when the finger barely made it, Joni raised his hand and stared at his vial-covered finger with keen interest. He repeated this procedure of putting in and taking out his finger. When suddenly he heard a slapping sound produced by the finger coming out of the vial, he became even more interested in the process. But this also did not last long.

Joni brought the vial close to his eyes: he noticed the transparency of the vial, and that gave rise to new manipulations. He turned the vial in front of his eyes, looked through it at surrounding objects, and then looked at the same object with the naked eye; he repeated this procedure with different objects and glanced at me from time to time. He tried to look into the vial through the bottom; he did this only for a very short time because the bottom was almost opaque.

Then, Joni looked into the vial through the opening. The next moment he put the vial on its bottom and poured sawdust into its opening, spilling more than filling the vial. Somehow, Joni delivered a small amount of sawdust into the vial. Then, he turned the vial over, pouring out the sawdust. Part of the sawdust struck to the inner surface of the walls and would not come out. Joni shook the vial with all his strength; the sawdust rumbled, and he listened to this sound, which obviously amused him. But, it also became boring.

Joni rolled the vial over the floor, stood it on its bottom, and sat on it and fell down. Then, he lay on the vial and rolled it on the floor with his back. When the vial happened to jump from under his back, he put it back again and continued rolling it.

His calm disposition turned into boisterous behavior: Joni's intention was to destroy the vial. He scratched the glass with his teeth and got mad at its defiance. He peeled off the label and tore it apart. He pressed the vial neck between his teeth and banged with it on the walls and on the floor. He dashed around the bed with the vial in his foot. He rocked on the swing, holding the vial in his teeth; he pressed it between his hip and stomach and walked in such a manner, trying not to drop it.

But, the inevitable end of all toys is destruction, and of course this happened to the vial: It fell and broke into pieces during one of Joni's most daring aerial evolutions. This finale was an absolute surprise for Joni. Filled with amazement, he looked at the fragments. When, fearing that he might cut himself, I began to collect the fragments, Joni grabbed one of them from my hands, put it into his mouth, and crushed it into tiny sharp pieces, which he shuffled in his mouth so rapidly that I feared they might pierce his lips and tongue like hundreds of small needles and turn his mouth into a gaping wound. But, as in the past, nothing bad happened as a result of this risky chewing, and the chimpanzee's mouth, to everybody's amazement, remained whole.

5. Play with Sharp, Piercing Objects.   Joni utilizes a variety of objects as toys: pieces of paper, sticks, straws, hay, nails, pins, vials, threads, and even hairs. Finding a piece of paper, Joni turns and rumples it with his hand and feet; then, he gnaws at it and tears it into tiny scraps.

Joni breaks sticks, straws, and grass blades with his fingers, makes small segments (about 3 cm) out of them, and props them between his lips [plate 45(3)]. He intensely tries not to drop a segment out of his mouth; he sits for a long time with his mouth open slightly, at the same time scratching himself in the groin, fidgeting, and jerking his trunk. His movements increase the chance of the stick sliding out of his mouth, which makes the chimpanzee more alert.

Joni also uses sharp vials, pins, and long nails as props. I was always surprised that there were no cases of Joni cutting or pricking himself; in playing with such objects, he was so careful and so perfect in the coordination of fine movements of his lips.

No less dangerous was Joni's play with nails. He put short (1–2 cm) nails into his mouth, chewed and shuffled them, pushed them from one cheek to the other, and smacked with his lips, as if sucking the nails. The first time he suddenly grabbed and put in his mouth a handful of these nails, I was appalled because I thought that he immediately would gag himself or scratch his mouth badly. I began begging him to give me back the nails. He would not do it for quite a while; he turned his head away from me and continued to chew. Only in response to my persistent entreaties did he dump from his mouth the whole bunch of nails, which were glued together by his saliva. I was quick to examine his mouth thoroughly and did not find a wound. Later, convinced that these nails were absolutely safe for Joni, I gave him full freedom in playing with them.

6. Play With Resilient, Thin, Long Objects.   Joni uses a long human hair to perform delicate play involving fine coordination of hand movements. Joni takes the ends of such hair between the first fingers of both hands, with his arms firmly resting on his knees and his mouth half-open. He touches the hair with his tongue and pulls the hair to the right with his right hand, while his left hand holds the hair in such a manner that it, sliding between the fingers and against the tongue, neither slacks nor tears, but remains taut all the time until the hair ends and falls from the left hand. Still holding the hair in his right hand, Joni brings it to his left hand, which takes the hair again, puts it into the left corner of his mouth and onto the tongue; then, he returns the right hand to the right, again using the fingers of his left hand as a mild vice to perform this resilient sliding [plate 49(2)].

Joni likes this process of tugging a hair and tickling his tongue tenderly, which brings him into the state of slumber. The chimpanzee closes his eyes and keeps pulling the hair, from time to time changing his hands, using either his left or his right hand to produce a moving force.

During my long observations of this procedure, I never saw Joni make a wrong movement and tear the hair. He often plays with a thread in a similar way, but in this case, he is less willing to play.

However, in the same type of play, Joni uses, very enthusiastically, sound-producing objects such as fine chains. Taking such a chain into his hands, Joni pulls it, as with the hair, between his teeth from right to left and back. The chain rumbles and clangs against the teeth, and Joni shows every sign of enjoyment.

Destructive Play   In observing the experimenting play of the chimpanzee from start to finish, we have found that they all end with actions that lead to the destruction of the object examined. Generally speaking, any gnawing, breaking, or tearing is itself a source of the chimpanzee's enjoyment. Of all things that he lays his hands on, nothing escapes his teeth; his destructive activity is the longest of his pursuits.

This inclination to destruction is sometimes so compelling, so overwhelming, that it can transform the chimpanzee into a completely deranged creature. Joni dashes around the room, takes everything that comes his way, and rips things off the walls; he tears, gnaws, breaks, and crushes them. It seems that he cannot stay indifferent when he sees a whole object and cannot help trying the strength of his teeth and hands on them; he will not stop until this object is ruined. He destroys not only the things that he is allowed to break, but also the things protected by my categorical restriction. Moreover, Joni attacks the forbidden things in especially fierce agitation; once he starts, he cannot be distracted from it by entreaties, threats, or force.

Left alone and sitting under the ceiling of his room, Joni bites at the protruding parts of plaster. He continuously rips off the wallpaper, gnaws at the lath that appears from under the plaster, and makes such an awful mess out of his room that when the owner sees it at the tenants' departure, he only gloomily shakes his head and vows to himself never to rent this house to anyone having such a rowdy child.

Sometimes, Joni tries to gnaw at the wooden frames of the glass partition in his pen to break them. Getting hold of the glass, Joni immediately breaks it with his fist; he looks at the fragments with unabated interest, starts to play with them, and carefully takes out the remnants of the glass from the frame and scatters them over the floor.

Joni uses the swing and trapezes first as additional material doomed for destruction. He tries to untie the ropes, throws the seats onto the ground, bites off all the ties. Only when he has failed to do that does he use them for the purpose for which they were designed. Joni is very enthusiastic about gymnastic exercises on the trapezes and rocking on the swing, but the passion for destruction prevails over all other pleasures. Eventually, you catch Joni assiduously untying the rings that hold the swing at the ceiling. He will not stop until his task is completed and the swing is on the floor. Without his favorite attractions, Joni obviously is bored; we rehang the trapezes. He resumes playing with them and then starts gnawing at the ropes again until the whole thing is brought to disrepair.

And so it was day in and day out. We labored every morning to restore the swing and trapezes to their places, and every evening, we found them on the floor, scraps of the tattered ropes hanging from the ceiling. The most ingenious hanging methods could not save the most solid ropes from destruction. All our tricks only lengthened the time the chimpanzee would have to spend on this work: The final result was the same. We sometimes found that the more unyielding to destruction the thing was, the more energy and persistence the chimpanzee added to this process. As always, the element of defiance in Joni's behavior urged him to continue.

Once, Joni found a way to bend the metal wire off the cage screen. As soon as he freed an end of the wire, he began unweaving the screen. He spent hours ripping the screen apart, making me do the useless work of patching the screen in the mornings when I knew that, by the evening, everything would be undone.

Unweaving and piercing baskets, breaking twigs, undoing loops and knots, and ripping off laces were Joni's favorite and constant entertainment. No soft toy in Joni's hands escapes destruction. Scraps and rags eventually are torn into tiny pieces; thick blankets appear in such a tattered and worn condition that you would think they have been under shell fire. The pillows are systematically rent; the feathers and down are raked out through the holes and scattered around the room.

I will never forget the scene that opened before my eyes when Joni was discharging the down from a pillow for the first time. It was dusk. I opened the door of Joni's small room and instinctively recoiled: Light white down was falling down and surging up all over the room like snow flakes on a windy winter day. Joni was sitting in the middle of the room, like a gnome from a fairy tale, his eyes staring into space: He was watching the movement around him. He was covered with down from top to toe; the down stuck to the top of his head and stood out like a white aura around his face. His white eyebrows and side-whiskers bulged in a funny way. Small flakes stuck on his eyelashes and in the corners of his eyes; he was blinking rapidly and was hardly able to see anything with his blocked wet eyes. Piles of down rested on his shoulders. There was a big ruined pillow, half empty, lying beside his feet, and new handfuls of feathers were taken out of it constantly and thrown up in the air. Like a swarm of white moths, they dashed up to the ceiling in the middle of the room and fell down on the floor slowly; they were met halfway by a fast upward swarm and were absorbed by it, then continued their up-and-down movement.

This was in the middle of the room, and the lightest bits of fluff were also hanging and dashing all over the place; they could not settle because of the constant airflow produced by Joni's movements. In a minute, I could not remain in the room any longer: The down had blinded my eyes and gotten into my nose; I wanted to sneeze. I saw Joni wiping off his face with the back of his hand from time to time, trying to brush off the most stubborn bits of fluff. It took a lot of time to bring the room and Joni himself back in order.

I was amazed when, entering his room in an hour, I found literally the same picture. The only difference was that now Joni could hardly be seen in the dark, and the white bits of fluff looked like dark flies. Afterward, no ban or severe corporal punishment concerning the discharge of down could stop the chimpanzee from enjoying this highly sought entertainment. Every time he got hold of a pillow, he immediately began rending it apart.

Joni showed similar behavior with small rags and scraps of paper. Not only did he throw pieces of paper up in the air, but also he tried to catch them. The very action of ripping paper apart, especially in the case of thick and rustling paper, gave Joni such pleasure that he went into his cage voluntarily and let us lock him up. Giving Joni paper for fun became a routine for me; every time I had to leave him alone for a long time, I gave him huge pieces of newspapers to brighten his loneliness, and I was sure that he would not be bored. The tiny paper scraps, in turn, were the material for various new applications and combinations of his play. Cartons or boxes given to Joni also are torn to pieces by him.

All new toys, after initial smelling, are tried to see if they can withstand destruction. For instance, metal, leather, rubber, or wooden balls first go into Joni's mouth; he gnaws at them, rubs them in his hands, presses them with his folded fingers, knocks with them on the table, and tries somehow to find a key to their destruction. When he is not capable of doing this, he throws these things up in the air, rolls them over the floor, and runs after them around the room. It may be assumed that round objects remind Joni of some fruit, such as oranges or apples, and he tries not so much to destroy as to taste them, but this hardly is plausible.

Other objects, that do not resemble any fruit are also subjected to the same examination. Once, I gave Joni a strange object, a metal puncher. Joni applied everything he could to destroy it. He examined the object from all sides, especially the holes. Holding the puncher with his hand and foot, the chimpanzee tried to bite it. He saw his reflection in its glossy surface, but was not distracted by it for long; he kept driving his teeth into the thing, bending his head over. Noticing his own face in it, Joni pulled his hand away from him and looks as if into a mirror; then, he drew the puncher back to his teeth again and bit it: The puncher was indestructible

Joni took it in his hands and tried to break it to no avail. He found the only yielding element, the spring, and pushed it with his hand. The spring yielded and then came back, but the object remained intact. Joni hitched the puncher over a metal hook and drew it to himself; he unhitched the puncher and knocked on the hook with it. With a sudden movement of his hand, Joni threw the puncher on the floor; it fell but did not break. Joni resumed gnawing at it and ripping it with his hands, but all his efforts were in vain.

He tried to hitch the puncher over the hook one more time by standing in a vertical position and reaching for the hook, but this attempt was faulty, and the thing fell off. Joni fiercely banged the puncher against the hook; the puncher was still whole. Then, Joni took the handle into his mouth, pressed it with his teeth, and with all his strength, pulled the puncher away from himself, holding it with his hands and feet by the base and scratching it with his teeth over the handle. But even such a heroic effort did not help him succeed in the destruction business. Only after long, vain destructive attempts did Joni change his strategy: He looked at his reflection in the surface and made challenging gestures toward his twin.

Once, Joni tried to disjoin the hammer from its handle: With apparent effort, he pulled the hammer off the handle. I brought the hammer back to its former status. Joni destroyed the bond again and offered me the parts as if asking for repairs. I restored the hammer again, and Joni took it apart again, demanding for me to put the parts together; he never wanted or never tried to reconstruct what he had broken.[14]

1. Play Based on Resistance (Chimpanzee's Will).   Generally, the willpower element is very strong in the chimpanzee's behavior. Sometimes, he does what he wants, bans or punishment notwithstanding; he acts like he is insane and unable to keep his actions in check, like a maniac submitting all his behavior to one obsession, the idee fixe.

Apart from the chimpanzee's passionate, persistent struggle for his property and his freedom, described above, his will strongly manifests itself in his struggle against his own fear, against self-imposed difficulties and hurdles (for example, in his mobile play), and against someone else's will. For example, Joni is banned from the destruction of his cage; he is yelled at, threatened, slapped on his hands, and driven away, but nevertheless, when you leave him, he immediately begins his destructive activity. And how joyful he is when he breaks from his cage without permission! He grunts, barks, runs over to every one of us, touches us with his hands or with his open mouth, and breathes rapidly; his whole body expresses enjoyment that is never as complete when he is *allowed* out of the cage.

A ban seems to urge Joni to counteract. It whets his persistent desire to do forbidden things, and that is what he does with extreme vigor. "Forbidden fruit is sweet" not only for the human, but also for the chimpanzee. For example, the chimpanzee broke a windowpane in the cage partition. He was punished immediately and relentlessly by whipping; as soon as we left, he finished the job by breaking the remaining two panes.

With the same persistence and in defiance of multiple bans and severe punishment (after the bitter experience of whipping), Joni peels off the wallpaper in his room, rips off the curtains, tears the pillows and throws the feathers around, and grabs and breaks the thermometer. Immediately after he has been forbidden to do so, Joni throws the clock off the wall, gets on the dinner table and runs over it, gets on the stove, plucks small objects from the desk, gets on the sideboard, and does many other forbidden things.

He is especially stubborn with respect to certain things and systematically shows utter disobedience in response to our bans. Apart from the destructive actions of gnawing and tearing, these stubborn activities include biting and grabbing things precious to me, climbing heights that are unreachable for me, running out of his room, picking the plaster, sticking his fingers into ink, and doing many other unacceptable and, therefore, forbidden things.

Joni uses every opportunity for trespassing. For example, during my short absence from the room, he immediately runs over to the folder with my notes, shuffles the papers, grabs a pencil and draws away in my notebooks. He opens the inkpot, puts his index finger into the ink, and then sucks the finger or smears the ink all over the paper. At other times, he gets onto the windowsill and pulls the insulation from the cracks; he reaches the abacas hanging on the wall, jerks its wires, and tries to throw it on the floor.

Once, when I was returning home after a 2-hour absence, I heard an unusual noise coming from Joni's room. Since I left Joni in the cage, I could not figure out what Joni could use to make such a knock.

I entered the room and saw the following. Joni had gotten out of his cage somehow; he sat in the middle of the room and, holding a jug by its handle, was banging on the floor with it.[15] The water that was spilled from the jug surrounded Joni on all sides. The sink faucet was open, and the water overflowed to the floor; Joni apparently tried to drink the water because his lips were wet and he was licking them.

The windowpane was adorned with milk. Joni put his fingers in the jug with milk. The soap was thrown from the sink on the floor; the comb that I usually use for combing Joni was taken out of the brush and also was thrown on the floor. The thermometer, usually on the wall, lay on the chest. One of my books was open and torn in several places; some pages were totally soaked in the spilled milk. The oven was open and ashes were partly raked out.

The cabinet with learning aids, forbidden for Joni to touch, was opened with a key left accidentally in the keyhole. A box with shapes was taken out, and the shapes were scattered over the floor. A second box with colored plates was also outside the cabinet, and all the plates were thrown around. Big white circles were separated from the group of colored plates; they were glued together by saliva and had apparently been in the chimpanzee's mouth.

A box with black wooden shapes was also taken out, but the shapes lay intact. From a box with gray (of different intensities) wooden blocks, one block was taken out and lay on the floor. Blocks of a neutral (gray) color probably did not interest Joni.

It must be mentioned that, as soon as Joni saw me, he dashed on top of the cage as if guilty. Then, at my first suggestion to him to come down, he hastily went into the cage and submissively let me lock him up (at some other time, this would have raised his wild protest), and he sat there quietly all the time that I spent bringing everything to order.

It would be a mistake to think that, in this case as in many similar ones, Joni did forbidden things because he did not know or remember that they were banned. Many facts prove that Joni is aware of his violation of the ban and apparently feels guilty for his disobedience.

For example, every time he runs out of his room without permission and is caught on the scene, he immediately runs back, goes into the cage and sits there not looking at me, with his head down and a timid, guilty look on his face; at other times, it is impossible to drive him into the cage even by force. Moreover, if Joni is punished by whipping, he accepts the punishment quite calmly as if he feels he deserves it. He quietly sits without trying to evade the strikes. In other cases (as in my experiments with Joni), even a mere waving gesture in response to his wrong choice of an object or a sharp disapproving tone on my part gives rise to his desperate cry.

Other instances prove the same: If I let Joni out of the cage and then try to put him back there in a while, this procedure would be accompanied with heavy outbursts of resistance on the part of the animal. If, however, Joni is at fault (he breaks a glass, grabs a forbidden thing, or overturns a bowl), or if he gets out from the balcony to the yard because he unlocked the doors, he would run back to the cage at my first request and would let me lock him up easily.

For large offenses, Joni accepts punishment quite stoically. I want to remind the reader of our punishment of him for biting children (he bit them after multiple verbal bans and after we used force to prevent him from doing this). As noted, after those bites Joni was severely beaten. He met every strike by curving the edge of his lip and shaking, but not in the least trying to take off and run away. (The beating took place on an open terrace in the middle of a big yard to which Joni could flee any moment and where, due to his fast running, he would be very hard to catch.) It is clear that Joni took the punishment as well deserved.

I have a reason to believe that Joni knows the strength of his bites. Getting carried away, Joni will bite one of us more strongly than the play would warrant. Joni immediately stops playing and looks the victim in the eye. If he sees a painful grimace, he gets worried, tries to locate the spot he bit, and turns the human's hand to look at the fingers, where the traces of his teeth can be seen. If he sees even slight damage, he quietly touches the spot with his fingers, fearfully and carefully touches it with his tongue, sucks at it (as he does with his own sore spots), and shows all the signs of concerned attention. If the victim utters a crying sound, Joni becomes visibly worried and gives out a long, low hooting sound.

As pointed out many times, if Joni burns with a desire to do something, threats, yelling, even corporal punishment cannot stop him because the excitement of his resistance makes him totally insensitive to the most severe strikes. Sometimes, the strike is so powerful that Joni fidgets and shrivels into a ball in expectation of a new strike, but he does not give in any way. In these cases, only mental fear (e.g., as a result of intimidating Joni by something scary) or shifting his attention to another object can make him give up his claims.

In some instances, a soliciting, soft voice is more effective than a forceful note. For instance, if Joni does not want to do something despite all the yelling and threats aimed at him, you ask him to do the same thing, but in a sweet voice; he immediately fulfills your request.

2. Whims.   In some cases, the will patterns in his behavior are such that they justly can be referred to as whims. For example, after Joni is given his usual evening meal, he stands at the doorstep of his cage and does not go inside, fearing that we may lock him up. He puts his fingers under the door so that we cannot close it, fiercely cries at our attempts to make him sit down, and resists us in any other way. The more insistent I am, the more violent is his stubbornness. Sud-

denly, something changes his mind, and although the circumstances have not changed a bit, he timidly leaves the doorstep and sits on the sheets in his corner. You would hardly believe that only a minute ago the very same act of entering the cage was accompanied by the chimpanzee's violent protesting gestures and body movements.

Similar phenomenon can be observed at my morning appearances. If Joni is closed in the cage while I enter the room, at my slightest intention to leave the room, he is anxious to get out of the cage and is on the verge of crying. After a few seconds of my inertia, my departure does not raise his protest again.

Chapter 5

Chimpanzee's Circumspect Behavior
(Deception and Slyness)

Joni often tries to do some forbidden things surreptitiously, inconspicuously. For example, I leave Joni in the cage with an open door, saying to him, "Sit here, you are not allowed to come out!" With me out of the room, Joni closes the door after me, runs outside the cage, and plays for some time, but when he hears my steps, he gets into the cage again, not realizing that the sound of a closing door and the puddle of urine in the corner will belie his disobedience unequivocally.

Joni is always forbidden to pick the plaster from the walls and eat it; after I have reminded him about it many times, I finally succeed: In my presence, Joni refrains from the destruction of walls, but as soon as I leave the room, he dashes to the ceiling, hastily bites off a piece of plaster, and tries to return to his place before I come back.

Moreover, he is in a haste to finish off the piece of plaster in his mouth to conceal all the traces. If he fails to do so and I enter the room unexpectedly, he intentionally keeps at a distance from me, sits with his back to me, does not respond to my insistent calls, and tries to win a few more seconds to eat the piece. He comes back to me no sooner than he swallows the last portion, totally unaware of the fact that his white nose and lips betray his secret exploits.

Sometimes, Joni takes entirely improper things in his mouth, for example, berries with seeds, small pieces of lime, tiny nails, small round plates, buttons, or vials. Fearing that he may gag, I ask him to give the thing back to me and to spit it out on my palm. If this thing does not have great gustatory or entertaining value for him, he dumps it from his mouth; if it does have a value for him one way or another, he is in a hurry to finish it off. He turns his head away from me, hastily trying to chew it up and to swallow if the thing is edible. If it is not and I keep bugging and begging him, he spits out only a part of it, hiding the rest under his tongue or behind his cheek; he does this so ingeniously that it hardly can be discovered.

Sometimes, I even open Joni's mouth and look inside, but I see no trace of the stolen objects. Furthermore, I explore with my finger all the secret spots in Joni's mouth and cannot uncover a secret cache of "illegal" things. I return to my daily routine, and in a while, I can see Joni stealthily chewing something that has apparently escaped my examination.

Often, Joni even resorts to deception. It usually happens like this: If Joni does not have anything in his mouth, and I keep holding out my hand and asking for more, he spits out some saliva on my hand, and I stop bugging him. Soon, he starts to abuse this trick, discharging saliva to get rid of me as soon as possible when he still has something in his mouth. When he succeeds, he surreptitiously extracts the hidden object from the depths of his mouth and resumes chewing it.

The chimpanzee's deceptive tactics represent the principle, "You cannot know what I cannot see." For example, when Joni wants to take a forbidden thing, he stares at me while his hands are stealing the desired object. Since he cannot see the object, he apparently thinks that others also do not notice his theft.

Sometimes, Joni uses much caution in performing an "illegal" action. For example, I do not allow Joni to take metal balls from my experimental set to play with them. Joni steals these balls, sits in a dark corner, and puts a pillow in front of him as a shield.

In another example, Joni finds a watchcase and gnaws at it. I say to him, "Stop gnawing!" Joni stops gnawing at the case, takes a piece of paper, and gnaws at it for a few seconds. Then, he leaves the paper alone and puts the case back in his mouth. I yell at him one more time, "Stop gnawing!" Joni takes the paper again and holds it between his teeth for some time; I do not prevent him from doing this. He looks at me, takes the case, covers it with the paper, and gnaws at it, now through the paper. I stop him again.

After some period of distraction, Joni begins gnawing at the case again. I slap him lightly on his fingers and say, "Don't do this!" Offended, he finds some fir cones and begins gnawing them. I allow him to do this, but this activity apparently does not satisfy him. Seeing the piece of paper previously abandoned, Joni picks it up and grabs the case; he turns both the case and the paper in his hands and takes both objects in his mouth. Looking at me, Joni lets go of the case, still holding the paper in his mouth. I go about other things, and when I suddenly turn my face to Joni, I see that the paper has shifted to his foot, while the case is again in his teeth. I yell at Joni; he pays no attention and keeps gnawing at the case. I slap him slightly; he runs away from me and hides under the table, holding the case in his teeth. Only when he sees that I keep chasing him does he leave the case; he grabs the paper and drags it in his foot or rumples it with his teeth.

Often, the chimpanzee shows heedful behavior, ingeniously conspiring against being put and closed in the cage. For example, if I cannot drive him into the cage, I use some trick. I put an especially tasty and intriguing (for Joni) object inside the cage, such as his favorite fruit (a pear) or a toy. Joni's cage has two doors. I put the pear near the left door, which is farther from me and is latched and padlocked; I sit near the nearest, right, door to have the opportunity to slam it shut after he has entered the cage.

Joni immediately becomes filled with the desire to grab the tempting object, but at the same time, he is afraid to be caught in a trap. Therefore, he applies various ruses. Sticking his body halfway into the door, Joni tries to reach the pear without entering the cage, but fails. Then, he sticks his body in farther, but fearing that the door might close, he holds the door by the knob or by the edge or puts his hand between the door and the door jamb.

But, since Joni also cannot reach the desired object from this point, he uses a new maneuver. He grabs, with his left hand, the belt hanging from the ceiling of the cage, sets his foot on the doorstep to prevent the door from closing, and tries to reach the fruit with his right hand. This does not help either. Then, he climbs on top of the cage, from where the distance to the pear is less. He shakes the screen and beats on the cage wall, but cannot make up his mind to come inside. Suddenly, he thinks of something else. He comes to the left door, the one closest to the bait, and spends a long time and much effort tinkering with the padlock. He opens the padlock, takes the padlock out of the hinge, and opens the door ajar. Now, both cage doors are open; Joni jumps into the cage through the left door, grabs the pear, and when I try to catch him, runs out the right door. He darts on top of the cage, and there he takes his time eating the fruit, enjoying the double pleasure of tasty food and freedom.

Sometimes, I do this differently. I close the right door from the outside and leave the left door open and sit somewhere near it. I place a much-craved object inside the cage near the right door in the hope that, entering the only open door and moving to remote corners of the cage, Joni will give me an opportunity to slam the left door quickly before he can escape. But, Joni does not fall for the trick; he opens the right door, that closest to the object, with the key. Looking at me with apprehension, he tries to grab the desired object; he moves into the cage a little, but is afraid to move too far from the door. As soon as I start moving, Joni recoils and retreats; when I sit still, he resumes his grabbing attempts, trying to come and go as fast as he can. We have described how carefully Joni chooses the right moment for fleeing the cage, how skillfully he uses this moment, how upset he is when he sees the impossibility of coming out (when the cage is padlocked), and how calmly he lets me go when the cage is closed only on a latch, which guarantees him easy and fast access to freedom.

Chapter 6

Using Tools

Circumspection in the chimpanzee's behavior manifests itself in a series of other actions, such as using all kinds of action aids. For example, Joni often tries to attack newcomers, but is apparently afraid of a direct encounter; rather, he threatens them from a distance using gestures or poses, or he raises a piece of cloth in the air and waves at the irritating object.

Sometimes, Joni uses a long stick for the same purpose. Also, Joni throws a stone at his reflection in the mirror.

Joni is more impudent and energetic during his attacks while he is protected. For this purpose, he puts a piece of cloth, paper, or his blanket on his head and then makes a full-swing charge at his "enemy."

In other instances, Joni uses tools as an auxiliary means of reaching faraway objects. For example, Joni sits on top of his cage and wants to reach an electric bulb that hangs from the ceiling, but the bulb is too far from him. He comes down, takes a long stick three times his size and brings it up on the cage with great difficulty; he also uses his head and shoulder to prevent it from falling off, and he finally pushes the bulb. At some other time, he uses a thin stick or straw to scare cockroaches from cracks in his cage or to get thick liquids from small vials.

Joni usually uses a metal nail for digging small pits in the ground. Joni applies a knife when he cannot bite the plaster off the walls. Below, we give many examples of his use of various tools as substitutes for pen and pencil.

Sometimes, Joni uses his own feet as tools. For example, the chimpanzee tries to break a stick, but his hands are not strong enough for the task. In this case, he places the stick on the sole of his foot and, grunting, breaks the stick in this way, reproducing precisely what a human would have done in this situation.[1]

When Joni wants to squash a bug but is reluctant to touch it with his hand, he covers the bug with a rag and smashes it through the rag. This circumspect behavior more correctly could be called "afterthoughtful" since the chimpanzee uses his previous experience in his subsequent behavior.

For example, I thrust the cone-shape lid of an inkpot through the cage screen. Joni takes the cone tip in his teeth and tries to pull the whole lid through, but the base part of the cone is too big and does not fit into the hole.

Tired of holding the lid with his teeth, Joni tries to pull his mouth away from the screen, but at the same time, he does not want to drop the lid. He sticks his finger into the nearest hole and from the outside helps his teeth to hold the lid. He also tries to pull the lid through the same hole with his other hand. One awkward movement and the lid falls on the floor outside the cage in front of a wide crack near the floor. Joni immediately takes advantage of the situation, quickly reaches the lid, and pulls it into the cage through the crack. Now, every time I put the tip of the lid through the screen, he invariably drops it and hastily picks it up. But before the lid accidentally fell, he could not think of such a short way of achieving his goal.

Chapter 7

Chimpanzee's Imitations

From the group of Joni's imitating actions definitely borrowed from man and associated with daily human hygiene, we can mention spitting, cleaning the nose, washing, and fanning. For example, after I had a sore throat, I rinsed my throat often, coughed, and spat in Joni's presence; he spit for days, shooting his saliva all over the room. It was just unbelievable how much saliva he could store. Joni's intention to imitate is so compelling that when, for example, he is given a bowl with food, he also spits on it, reproducing my spitting after rinsing my throat.

During the initial period of living in our house, Joni used to wipe the mucus coming out of his nose with the back of his hand; as is well known, such a method of wiping the nose is widespread among children of poorly educated people of all countries. I often observed it in lower apes living in captivity. But, I am inclined to consider this gesture of the chimpanzee not as imitation, but rather as a natural and genuine act because I could also observe it in children from highly educated families, for which there was no chance of imitating any adults.

Later, Joni acquired a new, obviously imitating, method of wiping his nose: using a small rag that he happened to pick up somewhere. In this case, he accurately imitated the using of a handkerchief by people around him, as well as my use of it when I was cleaning Joni's nose and other parts of his body. After smearing or spilling something on himself, Joni takes a rag or a scrap of paper and thoroughly wipes himself dry.

Joni easily learned how to use soap for washing. Fanning himself is another example of Joni's imitating actions. Joni would take a small feather and wave it like a fan. As a matter of fact, none of us ever used a fan in the house, but Joni could see one in action at his former owner's house; I consider this imitation and not something he did on his own initiative.

Joni undoubtedly imitates various manipulations associated with cleaning his room. Left to his own devices, Joni often takes a broom or brush and tries to sweep the floor, raking the trash in a pile. However, he does it so awkwardly and inefficiently, due to a lack of direction, that he spreads the trash over the floor rather than gathers it, and the floor is never clean as a result. Joni even moves the furniture, as it is done during a cleanup, although he often does not sweep the floor at the freed spot.

Sometimes, Joni takes a brush and tries to sweep the sawdust from the floor of the cage, but this action also can be characterized by his lack of accuracy, thoroughness, and effectiveness. Therefore, the results are questionable.

The same goal, cleaning of the floor of the cage, achieved more effectively when Joni rakes the sawdust he has just scattered around using only his hands. He usually closes his lips tightly and extends them forward somewhat, then meticulously gathers the sawdust, coordinating his movements in a more thorough and direct way. Sometimes, Joni opens his mouth and rotates his tongue in a funny way, shifting it from side to side according to the movements of his hands.

Seeing crumbs scattered over the table, Joni sweeps them with his hand, as I would do it. Once, Joni took a fine photographic brush and used it to sweep up buttons strewn over the table.

Very often and with good results, Joni also imitates the action of wiping. Every time Joni sees a puddle on the floor, he tries to wipe it. Joni urinates and defecates on the floor; in this case, he takes a rag or a piece of paper and cleans up the dirty spot, but as in most of his imitations, this work is also very sloppy, and he does not finish it. He evidently likes the process of wiping off his phlegm; he often takes water into his mouth, then sprays it on the floor and begins to wipe it off with a rag. Sometimes, he soaks a rag in water and sweeps the floor or the cage walls with it, imitating cleaning of the floor and walls, which he has seen before. When there is no water handy, Joni spits on the floor many times and then wipes his saliva off with a rag or paper.

Turning on an electric light can also be referred to as action of imitation. In this case, while flipping a switch, the chimpanzee intently watches the effect of his action, the appearance of light. This final result apparently urges him to complete the action.

For one more example of imitation, when I open the piano and hit the keys, Joni immediately does the same, but with a grain of fear and using one hand at first. Then, hearing the sound, he uses both hands and hits the keys with increasing force, a broad smile on his face.

The chimpanzee hangs the iron rings of the trapezes on a hook and takes them off, also imitating human actions. Joni is less successful at hammering nails. He often tries to hammer a nail into the floor panel of his cage, but it does not work for the following reasons: The chimpanzee does not exert enough force during the strike, he cannot hold the nail in a vertical position and the nail shifts around, or he cannot coordinate the movements of both hands and misses the nail. Therefore, even though he tried, Joni cannot hammer down a single nail. Nevertheless, his desire to hammer nails is so strong that he does not pass any protruding tip of nail without giving it a shot.

If there is no hammer nearby, Joni takes some heavy object and uses it as a hammer; if he cannot find even that, he strikes the nail with his fist, but this strike is very short and light because Joni apparently is afraid of hurting himself.

The act of pulling out nails protruding from some surface is an accurate and effective imitation. Joni takes a pair of tongs in both hands and squeezes it precisely as a human would. If there are no tongs around, Joni uses his teeth for pulling out nails that do not sit in the wood very firmly. Joni also tries to imitate our operations with the door hook and manages to open the door. Afterward, the chimpanzee often used this for self-release.

The more desirable for Joni are the final effect of his imitating actions, the more easily he learns and more precisely he reproduces these actions. I want to remind the reader that Joni learned how to open the outside door latch even from inside by sticking his index finger through the cage screen. Interestingly, Joni's initial imitating manipulations with the latch were entirely chaotic; the first opening of the door was quite unexpected for Joni, and he did not understand the meaning of this event. After unlocking the latch, Joni did not open the door, and when the door opened after an accidental push, he was visibly astonished and apparently saw no relation between the actions of unlocking and opening.

Only later, after multiple attempts to open the cage door, did Joni develop a more direct and accurate way of operating the latch, and he found a firmer relation between those two actions. Excluding some superfluous movements, he eventually found the shortest way of unlatching the door and gaining freedom.

When the chimpanzee is presented with the obstacle of the latch being tied to the hinge by a string and a belt, he also began imitating these tying procedures. More detailed observation revealed that there is direction in his actions toward opening the latch, and that there are also haphazard, inaccurate imitating actions with the string and belt, which reveal that initially he did not have the faintest idea how to get outside. Joni took the string into his fingers and made futile movements of putting its free end into the hinge from one side and then from the other. Soon, Joni changed the pattern: He put a concentrated effort and a lot of grunting in pulling the string toward himself and to the sides, sometimes tightening, sometimes loosening, the knot. He eventually loosened the knot by accidental jerking and often he even untied it, but after it is undone, he kept pulling the string near the hinge, unaware that his goal has been achieved.

Sometimes, sitting in the cage, Joni makes purely imitating movements not related to any goals. He takes the free ends of the string, puts them through the hinges, and awkwardly ties real knots, but he does it very rarely and without enthusiasm.

Joni, stimulated by his curiosity, spends a lot of time opening window latches and imitating our actions. He moves the latch with much effort until he hears the final click that signals the window is open. When I make an additional complication and lock the latch with a wooden stick and put a hammer, which I have just taken from Joni, into the cage to enhance Joni's motivation, he immediately removes the stick, unlocks the latch, quickly grabs the hammer, and rushes up to the ceiling of the cage.

I take the hammer from him, put it into the cage, and put an open padlock, (instead of a stick) into the latch hinge. Joni rushes to the lock with the same fervor, tries to open the latch, but fails. He takes a key, which hangs freely from the padlock, and makes some useless movements with it near the padlock. One of the movements accidentally pushes the padlock and it slips from the hinge. Joni immediately evaluates the situation, unlocks the latch, opens the door, dashes into the cage, and grabs the desired object.

Now, I lock the door in Joni's presence with a padlock that I close with the key in the keyhole. Initially, Joni does not pay any attention to the key; he turns the padlock to the right, then to the left; he pushes it up and down, but achieves nothing. He resumes turning the key in both directions and accidentally opens the padlock, but does not understand it has been opened and continues turning the key. After a series of useless movements, Joni performs some haphazard manipulations with the padlock; as a result, he accidentally pulls the padlock out, unlocks the latch, opens the door, and conquers the object.

In our fourth experiment with the closed padlock, Joni starts by jerking the padlock; later, he turns to the key and opens the padlock two times faster than in the previous instance.

Even in our fifth experiment, Joni's first movements are not with the key; he does not consider the key a crucial element in opening the door. He tries to rip off the rope attached to the key; having done this, he pulls at the doorknob or at the padlock and rocks the latter from side to side, again making haphazard movements that do not bring success.

Later, having accumulated experience, Joni starts the unlocking directly from operations with the key, skipping haphazard and useless movements with the padlock; but, this is true only for the experiments in which the key has been in the keyhole from the beginning.

The success and duration of this operation depend on whether Joni turns the key with his right or left hand. In the first case, he unlocks the door quickly; in the second case, he often fails. This happens because Joni turns the key clockwise with the key in his right hand and, therefore, opens the padlock. With the key in his left hand, Joni turns the key in the opposite direction, and the key becomes stuck. Only after a long time and fruitless attempts, Joni finally applies his right hand and opens the mechanism at once.

At first, when Joni hears the padlock clicking, he does not understand that the action has been completed, and he continues to tinker with the key until the shackle snaps open. Later, the clicking sound becomes associated so firmly with the opening that as soon as he hears that sound, he leaves the key alone, jerks the padlock, which speeds up snapping the shackle open, and takes the padlock out of the hinge.

Freeing the padlock from the hinge always happens in the same way, which Joni invented. He usually raises the dangling padlock and pushes the free end of the shackle with the index finger of his left hand through the hinge; at the same time, his right hand turns the padlock to the side, as if unscrewing it from the hinge. As soon as Joni takes the padlock out, the opening of the latch itself happens immediately, in less than a second.

Trying to make this operation shorter and more successful for Joni, I showed him (by placing the padlock and the key in his hand) that he has to take the padlock in his left hand and the key in his right hand. Joni learned this quickly and used the knowledge most of the time, shortening the time he needed to open the padlock by five times compared to his previous results (1 minute versus 5 minutes).

But, I went even further in making his task more complicated. I close the padlock, take the key out of the keyhole and leave it hanging on the padlock. Joni takes the key, but apparently does not know how to insert it with the groove up into the keyhole. After a few unsuccessful attempts, Joni leaves the key alone and tries to open the door in a different way, by pulling the padlock out of the hinge. He applies all his strength, but nothing helps, and he impatiently knocks on the cage wall with the padlock. Trying to help Joni, I give him the key in the proper position, with the groove up, and direct his fingers while he is inserting the key. After the key has been inserted, Joni turns it and opens the padlock.

At the second such attempt, Joni again starts from the key. For some reason, he puts it into his mouth, gropes for the keyhole with quick nervous movements, and scratches the padlock around the keyhole with the key.

Success depends on how Joni holds the key (i.e., whether the groove is up or down), and although I have shown Joni the right way of doing this, he often holds the key otherwise and, therefore, cannot put it in. Then, Joni turns to me and offers me the key as if asking for help.

If I have no intention of helping him and do not take the key, Joni resumes his own attempts, but because he has not paid attention to my instructions, he gropes at random again. If this does not work, Joni changes his hands when turning the key, takes the key into his mouth, or tries to put the key in while he is in a vertical position, which gives him more control over his movements. Assiduously groping for the keyhole, Joni rhythmically opens and closes his mouth, curving his lip on one side. The tenser the movements of his hands, the wider his mouth; the less successful his attempts to open the door, the more visible the angry wavy shaking of his lip. His facial expressions reflect all stages of the task; Joni reacts very emotionally to every stage of the process, especially to the final result of opening the padlock: A smile on his face shows that he is absolutely happy. Now, Joni is sure that the success is already his because pulling the padlock out of the hinge and opening the latch are not a big deal for him. In nervous haste, he pushes the shackle through the hinge, throws the latch open, opens the door, and enters the cage.

In time, Joni stops changing his hands while he opens the padlock. He omits putting the key into his mouth and covering it with his saliva, and he remains concentrated on the persistent movement of scratching around in search of the keyhole.[1]

Having learned how to open a padlock in his cage, Joni broadens the field of application: he tries to put every key that he happens to find into any hole, often not taking their respective sizes into consideration. Often, Joni takes a huge key and tries to open a small lock with it; or, taking the key of the small lock, he tries to put it into the tiny lock of a suitcase or into entirely different keyholes in dressers, wardrobes, and so on. He persistently tries to achieve his goal; in his long and futile attempts, he usually asks me to help by pulling at my dress or by touching my hands. If I do not pay any attention to him, he starts whimpering, imposes the key on me, and expressively looks at me and at the lock.

Speaking of the chimpanzee's other actions of the same nature, we have to mention taking apart a collapsible wooden egg that contains 10 smaller eggs. As soon as I showed Joni how to do it, he started to apply his hands and teeth in attempts to open the eggs, one after another, looking at every new egg with curiosity. The smaller the eggs, the more difficult it was to open them; but, Joni's teeth always came to the rescue. Although sometimes he damaged the egg surface, he successfully completed his task every time.

Joni's writing, or rather drawing, also pertains to this group of imitating actions. Joni can see that my pencil, my pen, and I are inseparable for days. He can watch the process when I take notes; as a result, he attempts to reproduce the same movements. While I am writing, Joni tries to snatch the pencil from my hand so he can also draw in the notebook. I resist and do not give it to him; he cries and keeps insisting.

When I give him the pencil, I also offer him a piece of paper on which to draw, but he wants to try his hand only where I have just written. He stubbornly pushes my hands off, gets angry, and cries.

Getting hold of the pencil, he apparently feels great joy: sometimes he pulls himself into a ball, bends close to the paper; resting on one hand and extending the other, he begins to "draw" assiduously [plate 50 (2)]. Sometimes, he draws by leaning back a little, bending his head, and holding the paper with the sole of his foot and with the folded fingers of his free hand.

Usually, Joni takes a pencil in his right hand, the way small children do, but in a short while, he changes hands every second, drawing thin lines here and there. Then, he gets carried away and draws longer and more visible lines, usually from the corner to the center of the sheet. He looks at the dark lines the pencil has left, stretches his lips in a smile, and begins drawing more vigorously and hastily, often moving his lips in time with his movements. He turns his tongue and opens his mouth wide as he does when he is engrossed in something; his lines are predominantly straight with short bends [figure 15 (1), original writing]. From time to time, Joni passes his hand over the paper and then resumes drawing with the same enthusiasm. He gets so carried away that, having filled a piece of paper, he draws away on anything: on the table, on the white walls of his cage, on the wallpaper.

If I also begin drawing on the same sheet and stand in his way, he immediately pushes my hand away, even slaps it slightly. If I persist, he may slap me on the forehead, push me off, or even bite my inquisitive hand.

It is hard to say which moment of his drawing he enjoyed the most, the movement itself or the rustling of the pencil, but it was clear that this action was so pleasant for him that I used it many times as an encouraging stimulus during my experiments. He often cried sincerely and bitterly when he was denied a pencil and a piece of paper. You had to see the comical scenes of our contention over a pencil to understand how the chimpanzee valued it and how desirable this

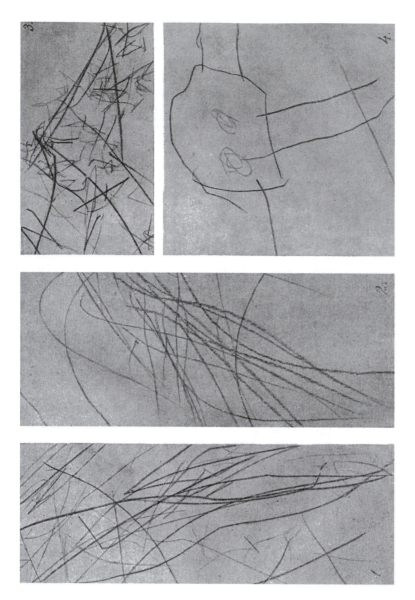

Figure 15. Samples of scribblings of chimpanzee and human child: (1) scribbling of Joni, first stage of drawing; (2) scribbling of Roody, first stage of drawing; (3) scribbling of Joni, second stage of drawing; (4) drawing "Mother" by Roody (3.0.8).

entertainment was for him. Almost every time I wrote in his presence, he tried to snatch the pencil from my hand and draw with it. If I did not give it to him, he got angry, attacked me violently, grabbed the pencil in his teeth, bit me, and pulled my hands away. In general, he behaved as he would if fighting for the most desirable things in the world. Finally, conquering the pencil, Joni runs away as far from me as he can, fearing that I might take this treasure back from him, and he starts drawing on everything that comes his way.

Sometimes, he draws on the hardwood floor, lying on his stomach; sometimes he "adorns" white wallpaper with his doodles. The greater the stretch of his drawing is, the more enthusiastically he runs his pencil back and forth.

Sometimes, I sit Joni down and try to move his hand with a pencil over paper, drawing simple figures (for example, crosses), but in this case, his enthusiasm for drawing vanishes at once. He sits like an enslaved student with a bored expression on his face.

He draws more willingly in a sitting position when he is given all the freedom he wants. He holds a pencil in his fist tightly [plate 50(3)], as preschool children always do. He does not take his eyes off the paper and draws away until the entire surface is filled with lines, and there is no more space on which to write.

Joni more willingly writes with ink, also imitating the action of dipping the pen into the inkpot, which he does with such strength that every time you clearly can hear the pen knocking on the bottom of the inkpot. Since Joni does not control the pressure while he writes, the ink spills from the pen all at once, the pen scratches over the paper, and he has to dip the pen again and again, spending more time on dipping than on writing and leaving more blots than lines. Sometimes, he dips the pen into ink for no apparent reason; he resembles a small child, particularly a boy I knew, who, having learned to print letters, felt that he needed to dip the pen into ink after every line he had drawn.

This kind of writing clearly does not satisfy Joni, and he resorts to a more effective way: He dips his finger in ink and smears the ink over the paper. If there is no ink around, Joni dips his finger into any liquid available: milk, jelly, water, or his urine. He smears it over the table. Often, he spits on the paper and spreads the saliva around.

In time, I noticed some evolution in his writing. He stopped changing hands as frequently and drew longer and more distinct lines without interruption. Sometimes he drew crosses, one after another [figure 15(3)].

Joni also uses long, thin sticks as writing tools, covering them first with his saliva. He scratches over the paper with a small nail or even with the nail of his index finger. Analysis of Joni's "drawings" shows that they can be filed under "first stage of complexity," which is not outside the limits of simple drawings by a human child.

As in other imitating actions of the chimpanzee, here transformation of actions and a change of tools have been also observed.

Chapter 8

Chimpanzee's Memory

Conditioned reflexes, habits. The strong and developed imitating ability of the chimpanzee attests to his good visual and motor memory, but he has auditory memory as well. For example, the sound of a door being shut and the subsequent silence in the apartment serve Joni as a signal that we have left and that for a long time nobody will come to see him. He constantly reacts to this sound by loud crying. But, if he hears a voice in the room next to his, this voice serves as a correcting signal for Joni. This means that not everything has been lost, somebody is still at home, and there is hope for communication with a human. Joni immediately stops crying.

After a period of silence, if there is the sound of a door being slammed and our loud voices can be heard, this is the signal of our homecoming. Joni hails us through the distance of three or four rooms with his loud, long hooting.

Thus, in the daily routine, Joni develops the following complex auditory reflex, which is tinged with the emotional overtones: slamming of the door—voices—Joni's excited and joyful hooting.

Surreptitiously observing Joni through the glass door of his room, you can see that, at the sound of the front door, he immediately abandons all his pursuits and freezes, listening for what would follow. At the next moment, depending on the additional auditory signal that clarifies the situation, he emotionally reacts to it (joyfully or sadly).

Joni developed a similar reflex with respect to the screeching sound of the inner door leading to the common room separating the chimpanzee's room from the ones in which we live. The screeching sound followed by our voices brings about the chimpanzee's joyful grunting; the screeching sound followed by silence makes the chimpanzee cry.

The chimpanzee's apprehensive reaction is usually caused by the electric ring and by the noise of the elevator. As soon as Joni hears these sounds, he becomes silent and attentively listens to find what will follow. If these sounds are followed by strange voices, Joni peeks under the door as if burning with desire to see the person who is coming. If he hears our voices, he gives out a joyful hooting sound ending with loud barking.

Joni also has developed a series of other auditory conditioned reflexes. The chime of the bell calling the construction workers for lunch prompts Joni to run over to the window and watch the workers coming down the scaffolding of the house across the street. If someone pronounces my name in my absence, Joni releases a gamut of hooting that ends with joyful barking. Joni can tell my steps easily from those of other people, and he gives out the same joyful sounds when he hears I am coming.

Once, I showed Joni a flying airplane; the noise of the propeller could be clearly heard. Joni tossed his head back and stared at the plane. On the same day, when he heard the noise of a propeller, he raised his head and his gaze followed the plane for some time.

There were multiple cases for which Joni developed a conditioned reflex in the sphere of visual perception. For example, if I enter Joni's room with books in my hands, the sign that I will stay with him for a long time, he greets me with happy hooting. After a while, if I begin to gather the books, this is the signal for him that I am leaving soon, and he begins whimpering and groaning. If Joni sees that I am entering the room without books, he becomes worried; he follows my every movement and does not grunt. As soon as I show him a book, which I have hidden, he immediately begins to hoot happily.

Similar grunting occurs when I move the table over, take a chair, or put an inkpot or notebook on the table. Every one of these actions unequivocally and definitely indicates to Joni that I am going to stay in his room for a long time.

If I come near the door on my way out of the chimpanzee's room, he starts to cry, but when he sees that I have left the inkpot on the table, he immediately calms down, knowing from experience that I will be back soon. When I leave in the evenings, I draw the curtain on Joni's cage (this is a usual preparation for the night); he cries. If I leave without drawing the curtain, he grunts happily because he is sure I will come back.

If I lock Joni in his cage before leaving, he fiercely cries because he knows from previous experience that he will not be able to get out. If I do not use the padlock and only latch the cage, he remains absolutely calm because he knows that he can open the latch without difficulty.

Every time our maid enters Joni's room with a big bowl, he gives out a happy grunting sound because he knows from experience that she will be washing the cage, and he will be given a long period of freedom.

The same joyful emotional reaction, accompanied by happy grunting sounds, follows bringing in of the firewood, which is also an unequivocal signal for Joni that he will be taken from his cage into other rooms while the oven is burning.[1]

Joni remembers how the vial with a "tasty" medicine looks after its first demonstration. He hoots and reaches for it when he sees from a distance that someone is bringing it. In contrast, he definitely rejects a vial with turpentine, remembering how it hurt him in the past.

The chimpanzee quickly and firmly develops the visual-tactile reflexes that involve pain. After burning himself with the candle flame, Joni does not try to touch the fire. Once, he burned himself when he started to bite a match that had just finished burning. Later, Joni carefully touched a match with his lips before putting it into his mouth.

During the initial period of Joni's life in our house, we had to use a whip to put him into the cage; later, he began to cry as soon as he was shown the whip, not so much fearing the punishment as being unwilling to enter the cage.

Joni's orientational, topographical memory is also developed highly. In a new, five-room apartment, he easily finds my room. Put into a new yard, at a distance of 120 meters, he easily finds the apartment from which he has just been taken and where I remain.

I have noticed the development of a time reflex in Joni. Only two times before Joni went to bed, I took him by the hand and invited him to go with me to the adjacent rooms. From that moment, when I entered his room in the evenings, he usually gave me his hand, asking me to take him for the tour. He cried if for some reason I could not fulfill his desire.

Here is one characteristic example of conditioned reflex he has developed. During photographic shootings on hot days, he quickly became overheated in the sun. Therefore, he began to groan, whimper, and cry, asking to be taken on my lap. If I did not take him, he became desperate. He covered his head with his hands, opened his mouth wide, shifted his hands from his head to his eyes [plates 30(1)–30(3)] and beat himself on the head.

I did not want to torture the chimpanzee and feared that he was indeed suffering from the overheat; I usually took him the moment he covered his head with his hands. He quickly utilized this maneuver. Even when he was in the shade, but anxious for me to take him, he opened his mouth, covered his head and his eyes with his hands, extended both hands to me, and remained in this pose, only uncovering his eyes from time to time. He looked at me imploringly, and if he did not see that I was responsive, he began screaming and crying and demanded more persistently that I take him.

As a result of developing these multiple and diverse conditioned reflexes, Joni developed an indirect conditional language for expression of his desires. For example, in wintertime, when he is taken from one room to another through a much colder corridor, we usually cover his head with something, fearing that he might catch cold. Every time when I am not very quick to take him out of his room, he grabs any piece of cloth, covers his head with it, and stretches his arms toward me, urging me to take him. If I am not going to bring him out and I take the rag off his head, he snatches the rag from me and stubbornly places it on his head again. He cries, gets angry with the rag, or tears it with his teeth. If I try to prevent the chimpanzee from doing this, he covers his entire body with the rag, reaches for me, grunts, hangs on my neck, and does not stop until I succumb and take him out.

The opposite reaction is observed when we are outside his room. Now, if I begin to cover him with a rag, he keeps pulling it off as persistently as he tried to put it on earlier. The only time he does not object to covering his head outside his room is when it is late and he is overpowered by sleep and ready to go back to his room. Another time when I am in Joni's room and take the same rag, he immediately stops all his activities, runs over to me, extends his hands toward me, and is convinced that he will be taken out soon.

Conditioned reflexes develop in Joni as fast as they fade away. After I treacherously used the trick of covering his head two or three times for putting him back in the cage instead of for taking him out of the room, he lost his trust in it, and the reflex was dulled. I showed him the rag from a distance, made inviting gestures, and pointed at the door—all in vain, the chimpanzee would not budge.

The following example is also characteristic of him. Joni usually enjoyed washing very much. I could make him come over to me from remote places very easily by pointing at the sink, but after I had used it a couple of times as bait for catching him, he no longer gave in to the temptation. The reflex arc (pointing at the sink as a cause of Joni's approach) was broken.

Thus, as a result of real tests, certain types of conditioned reflexes establish themselves firmly, while others fade away. For example, if nobody comes to his room for a long time or if somebody suddenly leaves his room, Joni becomes angry and makes noises with the trapezes, knocks with a wooden ball or with his fist on the wall, and fiercely throws various objects.

At first, this indescribable noise scared us, and we hurried into the room to see what was happening. That was exactly what Joni had wanted. When his meals came a little late (during the day or when he woke up), he knocked with any available objects on the walls and on the floor until the meals were delivered.

Such fast utilization of new experience was also observed with respect to intimidating stimuli. As mentioned, any new scary object (for example, a stuffed wolf's head, a picture of a chimpanzee, a mask, etc.) causes fear in Joni only at the first demonstration. After two or three contacts with these things, Joni understands that they are harmless, and he is no longer afraid of them. Similarly, my temporary inanimate substitute, the doll, lost its substitutional meaning as soon as Joni saw its pillowlike face.

Joni develops very firm conditioned reflexes as a result of systematic uniform actions. For instance, from the very beginning of our study, Joni has been forbidden to touch all the objects on the laboratory counter relating to the teaching process. Joni is so used to this prohibition that he considers this counter empty, although it is filled with the things that would be very tempting to play with.

These rooted habits allow us to see elements of deep conservatism in Joni's behavior. For instance, when Joni had a severe cold, he fell into the habit of wiping his nose with the back of his hand. Later, when the cold was already gone and there was no need to wipe the nose, he continued to use this gesture for a long time.

Usually, I hand Joni a mug with milk or allow him to drink by holding the mug in my hand and bringing the rim of the mug close to his mouth. If I put the mug on the table and give Joni the freedom to do what he wants, he does not take the mug in his hand, but drinks in a very primitive way. He stands on all fours, bends over to the mug, and drinks the milk from the surface if it is close to the rim. He has no idea what to do when the level becomes lower and therefore is inaccessible without tipping the mug.

After two or three training sessions, Joni starts to use a bed vessel. Once, when there was no bed vessel in Joni's room, he used a mug for urination. (During this training, as in establishing his other habits, gestures and body movements play a greater role than words.)

Also, Joni is very conservative with food. He usually is unwilling to taste something new and does it only after I insist. He drinks milk only if it is a certain temperature (29–30°C). If his milk is colder than that, he turns his head away after tasting the milk or takes some milk into his mouth and holds it there, apparently not willing to swallow until the milk reaches a suitable temperature.

Usually, after his morning meal, the chimpanzee is allowed to do some climbing over the cage. Therefore, when he is finishing his milk, he already has his leg raised, ready to get going. If you prevent him from doing this, he violently protests, gets angry, and cries until his trip around the cage is allowed.

Chapter 9

Conditional Language of
Gestures and Sounds

Communication with gestures and words. Gestures and images play an important guiding role in establishing complex visual-auditory conditioned reflexes. For example, I give Joni bits of a fresh cucumber. He eats only the core part of them and reaches for more. I point to what is left and say to him, "Eat this one!" He obediently eats the thrown-away part and later asks for more no sooner than he finishes what is in his possession.

When I look at the top of the cage, point in that direction, and say, "Get there!" Joni climbs up the cage. He sits down at the spot at which I pointed and said "Sit here." He takes the object at which I have pointed and said, "Give me this!"

Once, Joni was near the ceiling of the cage, and there was the mug from which he usually drank. I needed this mug and said, "Joni, give me the mug!" But, he did not understand what I wanted and only looked attentively at me and around him. Then, I took another mug in my hand; Joni understood immediately, took the first mug, and handed it to me.

In similar circumstances, he was at a loss when I said to him, "Give me this box!" But, when I showed him the top part of the box, he at once picked the box from a group of various things and threw it down to me.

Joni is quick to learn verbal commands accompanied by gestures; when he is ordered, "Sit!", "Lie down!" or "Make a tumble," he obeys. After 18–20 reiterations, Joni performs these actions in response to verbal commands alone (without the gestures).

"Understanding" words and phrases. Eventually, we have developed a conditional auditory language that helps us understand each other. For example,

I say to Joni, "Go to the cage!" He becomes sad, shakes his head, cries, extends his arms toward me in an imploring gesture, but nevertheless obediently goes to the cage.

I say to him, "Come over!" He takes off, rushes to me, and presses his body to mine. If I say, "Go away!", he groans but obeys the order.

I say, "Play with the ball!" He takes the ball in his hands or starts looking for it around the room if he cannot see it at once.

I say, "Pick it up!" He immediately picks the thing up and brings it to me.

I leave the room, and Joni cries. I utter the consoling words, "I'll be right back," and he grunts happily.

Joni does not want to come off the cage, but when I threaten him, "The wolf is coming!" and show him the stuffed wolf's head, he comes down at once.

Among other verbal commands that Joni obeys are the following:

"Climb on the cage!" He gets on the cage.

"I am catching you!" He runs away from me.

"Give me your hand!" He gives me his right hand.

"Play!" He gets on the trapezes.

"Back to your place!" He sits down on the chair near the counter.

"It's forbidden!" He stops doing whatever he has been doing.

"Leave this alone!" He stops touching something.

"Give me the ball!" He gives me the ball.

"A fly!" He turns around and looks for flies. If he spots one, he tries to catch it.

"Let's go!" or "Let's go for a walk!" He gives me his hand, inviting me to go with him to the adjacent rooms.

"Give me the rag!" He dashes to the corner and to his bed; he takes the rag and throws it on the floor in front of me.

"Give it to me!" He picks the rag up and hands it to me.

When I say, "Hot!" Joni becomes alert and uses extreme caution touching something.

Conditional gestures. Apart from the extremely eloquent language of instinctive sounds that accompany Joni's emotional experience, he has developed a series of gestures and body movements that cater to the expression of his desires.

When he wants to eat or drink, he starts sucking at my hands or my neck, or he puts my entire finger into his mouth. Considering that, at that age, under natural conditions Joni would still be sucking his mother's milk, we can understand his tendency to touch certain objects with his lips when he is hungry or thirsty. It is more difficult to decipher the touching with his lips and sucking of my finger in different circumstances, such as when I berate him for some mischief, or when he is especially happy and affectionate. Perhaps, here Joni is trying to enhance the scope of his requests, to arouse our empathy toward him in response to this affectionate touching.

It can be assumed that the initial instinctive sucking action, accompanied by the pleasant feeling of satisfaction from food, later evolved into the action associated with Joni's other joyful experience (for instance, the pleasure of seeing a familiar person or, generally, every time the animal is in a good mood). Later, Joni reproduces the same action under quite opposite circumstances. Seeing distinct signs of dissatisfaction expressed by scolding or spanking, Joni tries to arouse, as soon as possible, tender feelings in the angry person; his behavior runs ahead of his innermost desire, thereby speeding the arrival of the coveted event.

Other conditional gestures used by the chimpanzee extending one hand forward as a sign of request [plate 24(1)]; extending both hands as an expression of enhanced request or entreaty often accompanied by groaning and even a crying sound [plate 28(2)]; extending one hand and pointing at something, which means that Joni wants to have it as his own; shaking his head or turning his face away from something represents an action used to reject unwelcome food; curving his tightly closed lips to one side means a sign of dislike or aversion toward the food that we have imposed on him.

So, in the daily routine with Joni, a certain conditional language has been set up that enables us to understand each other. Joni has learned to obey my orders; however, he often completely disregards my requirements. This disregard occurs most often when Joni is in a state of anxiety or fear or in an extremely elated mood during play. At that time, he is deaf to my words or requests, even if I repeat them dozens of times.

Sometimes, Joni disobeys the orders when he is far out of reach for us, such as on the trapezes. Sitting there, he totally ignores our requests, calls, or threats and does what he pleases.

Chapter 10

Natural Sounds of the Chimpanzee

In summary of our observations regarding sounds emitted by chimpanzees, we have to emphasize that these sounds are additional attributes that predominantly, but not exclusively, accompany emotional states of the animal. Despite the certain versatility of these sounds, they are defined clearly by the form of their expression and by the certainty with which they can be ascribed to various types of emotions. They slightly remind us of unintelligible human sounds or sounds of familiar animals; they can be catalogued and described easily.

The least human is the sound emitted at the maximum of general excitability. This sound is represented by a long hooting that consists of alterations of two sounds repeated six times. The hooting has the tendency of heightening the voice after every two sounds and ends by the thrice-repeated sounds of the last third: "u-khu,uu-khu, uuu-khu, ua-khu, ua-khu." In some cases, this last representation is so sharp that it sounds like barking.

Translating this unusual auditory scale into the language of familiar animal voices, we have to say that, by their melodious arrangement, these sounds with their uniformly increasing intervals resemble the roar of a donkey; only with hooting substituting for the neighing timbre of the donkey's sound. In their final, more uniform stage, these sounds are very close to the barking of a dog.

This melodiously arranged hooting sound, emitted when the animal is most excited, ends with different types of barking that depend on the shades of the chimpanzee's anxiety. If this anxiety discharges into joy, the thrice-repeated barking is piercing, loud, high, and drawling (a variation of sound 1 to sound 2). Conversely, if the maximum anxiety is followed by an angry reaction, the thrice-repeated barking of the chimpanzee sounds like "u-au, u-au, u-au" (a variation of sound 1 to sound 3).

A moderate degree of anxiety that evolves into sadness is accompanied by a long single hooting, groaning sound "u-u" (sound 4). Anxiety with a shade of fear is combined with the appearance of short single hooting "u" (sound 5). Under fleeting angry anxiety, the chimpanzee emits either a sharp single hooting sound "ukh" (sound 6) or a hoarse grunting sound "khru" (sound 7), which is quite similar to the grunting of a pig.

Under longer anxiety, both sounds (6 and 7) are reproduced snappily and repeatedly as sound 6a "ukh-ukh" and sound 7a "khru-khru." Continuous, uniform grumbling grunting "khru-khru-khru-khru" (sound 8) often can be heard at moments of mental tranquility or when satisfaction follows anxiety. Anxiety accompanied by a threatening pose usually is combined with raising of a hand and emitting of a low, drawling-accentuated hooting sound "u-u" (sound 9, as in his attack against my offender).

In the state of worried surprise, Joni gives out a short, mooing sound, "m-m," directed inward (sound 10, reaction to meat), and a drawling "uuu" (sound 10a).

In reviewing the sounds associated with a sad mood, we should point out that these sounds (contrary to the sounds of anxiety) are very similar to those of a human and sometimes are indistinguishable from the sounds of a human child. These sounds vary considerably in form and intensity, depending on the degree of the animal's distress. For example, in the evenings, tired before his night's sleep and waiting to be put to bed, the chimpanzee gives out a weak squeaking sound.

During the initial stages of sadness, the chimpanzee extends his lips forward and emits a groaning intermittent sound "u," which could be compared with the groaning of a human, the whining of a child, or the whimpering of a dog (sound 11). An increase in sadness is accompanied by the increase in the duration and strength of groans and in the frequency of groaning (sound 12). At the next stage of increasing sadness, a yelling, drawling, rattling sound appears (sound 13).

At the subsequent stages of increasing sadness, we definitely can notice how a rattling groan evolves into an intermittent rattling scream (sound 14). At moments of an animal's despair, this evolves into volleys of a deafening scream. This scream is so loud that is exceeds the most intense crying of a child and can be heard coming from a closed room with wooden walls at a distance of 50 meters; or a shorter distance even through the thick walls of a brick house. When the chimpanzee is in the affective state of sadness, he repeatedly reproduces uniform volleys of this scream with such thundering strength that sometimes he gets carried away and loses his voice for a moment, as if choking; at the next moment, after a lull, he resumes the scream with the same strength.

Sad or unpleasant feelings of the chimpanzee are accompanied by diverse and extremely strong sounds; in contrast, pleasant and joyful emotions are almost silent. You rarely can observe a sad chimpanzee extending his lips in discontent without some groaning, but you often can see the silent smile of a joyous animal.

The chimpanzee's laughter, caused by light tickling or playing with him, is expressed by a wide open mouth and is accompanied by rapid and increased breathing (sound 16). I have noticed Joni's rapid breathing at moments of sudden joy. For example, if the animal has done something wrong and expects a punishment, but in fact, he is not punished, but is treated graciously, he breathes rapidly. In a second example, I came up to the door on my way out, and Joni was ready to start crying. Suddenly, I changed my mind and came back to the middle of the room; he breathed rapidly. For a third example, Joni was about to be placed into the cage; he extended his lips forward and was ready to cry. But, for some reason, the decision was changed, and Joni was brought to the next room—he breathed rapidly.

The same rapid breathing can be observed in the chimpanzee when he shows his affection toward certain people, when he presses his open mouth to their bodies, as if kissing them.[1] We constantly hear the sound of rapid breathing while the animal is sexually aroused, as when he presses his body to a ball and kneads it between his hips or when he tumbles over his head. In the first case, Joni breathes rapidly with his mouth tightly shut; in the second case (during tumbles), he smiles broadly, and his breathing is longer and deeper.

Also, Joni breathes rapidly in other cases; for example, when he attentively touches convex objects or when he examines himself, such as feeling a splinter or some sore spots on his legs or arms. In these two examples, the enhanced breathing apparently is associated with the chimpanzee's fear of breathing at the moment of tense attention during his touching (especially sore spots); at the next moment, Joni tries to compensate for this insufficient breathing by making it rapid, which in this case is not caused by any joyous experience on his part.

Pleasant sensations associated with tasty food give rise to three types of sounds: grunting, coughing, and barking. The chimpanzee's hollow grunting (sound 17) is absolutely indistinguishable from that of a child of the same age. During the chimpanzee's tasting of his favorite food, this grunting is almost continuous; sometimes, it is so loud that it resembles the sound of clearing the throat (sound 18). At the maximum gustatory pleasure, this clearing of his throat even evolves into loud barking (sound 19).

I heard a joyous grunting sound every time I came to his room in the morning and opened the dark curtains to let the sunlight in. Joyous grunting always accompanied when Joni took things out of boxes and there was something new in them.

The sounds associated with the chimpanzee's angry feelings were not formed clearly and often were represented by the transformed sounds of anxiety and joy or their combination. We could notice in our chimpanzee three specifically angry shades of these sounds: wheezing, barking, and shouting.

As stated, light tickling of Joni gives rise to his smile and rapid breathing, but if the tickling becomes more aggressive and strong, Joni bares his teeth, and his enhanced breathing becomes hoarse and can even reach total hoarseness (sound 20).

We can hear a similar hoarse sound when the chimpanzee's movements are limited, for instance, while he is being dressed, or during play when he becomes entangled or stuck and wants to regain his freedom. Joni's hoarse rapid breathing, which in our view is associated emotionally with the sense of discontent, occurs in the case of our contention with him over some object during play or during a spryness contest over it.

When it comes to a serious argument and the perpetrator also is a stranger Joni emits a short, snappy, twice-repeated, accentuated coughing, rather shouting, sound "a-a" (sound 21), followed by an abrupt offensive reaction and attempts to bite. I heard the same shouting many times when Joni was sick and was approached by someone other than myself; he growled and tried to push the person away from him.

Anger associated with fear, including elements of threat, often are accompanied by barking sounds. We already mentioned that low hollow barking follows thundering hooting under the angry excitability condition, but sometimes we hear only barking (sound 22). For example, when the chimpanzee throws smelly sheets away from him, he gives out snappy and angry barking. We heard the same barking when we showed Joni the picture of a chimpanzee and knocked on the back of the picture; Joni uttered piercing, ringing, or hollow barking, followed by long whining (sound 23) when he was shown the stuffed wolf's head.

We described in detail in chapter 7, Chimpanzee's Imitations, how quickly and enthusiastically Joni emulates a human when the human utters sounds borrowed from the chimpanzee, for example, the sounds of hooting, grunting, or shouting. We need to emphasize that, in joining the human, Joni increases and accentuates the intensity of his imitative sounds, as if trying to outvoice the person.

Joni imitates a dog's barking very well, but I have not noticed, in a 2½-year period, any attempts on his part to reproduce or imitate even a semblance of intelligible human sounds. The experimental data of Professor Yerkes in his book, *Chimpanzee Intelligence and Its Vocal Expressions,*[2] are in full accordance with our observations. This book indicates that, despite his 8-month attempts to involve the chimpanzee in reproduction of relatively simple syllables ("ba," "ko," "na") and despite the application of four different methods of imitative teaching, the experiments brought unequivocally negative results, although the chimpanzee (as determined by B. W. Learned, a musician, the coauthor of the book) was able to pronounce up to 300 musical phrases.

All the facts described show that the chimpanzee is an extremely noisy creature. Apart from the sounds that the chimpanzee reproduces with his vocal chords, he also can use additional sounds to indicate his emotional states. While in a good mood, the chimpanzee dashes around the room, rumbles, and knocks using every possible means. He slams the doors; he beats with his folded fingers or flat palm on the furniture, and the more resonant the knock is, the more time he spends producing it. he tap-dances. He throws different objects off the tables or hurls them across the room. He thunders with the trapezes, striking them against the walls with the utmost strength. He leaps from chair to chair and makes such an indescribable noise that you would think that a large crowd of boisterous youngsters were playing.

Joni produces a similar rumble when he is angry and excited. Assuming threatening or defying poses or standing on all fours, Joni jumps repeatedly from hands to feet, rattles with his lips, clangs with his jaws, clicks with his teeth, or rhythmically beats with his hand on solid objects.

The chimpanzee obviously likes to produce sounds. They amuse him and constitute an inseparable part of his affective states.

We also have to mention his snoring (during sleep); it is absolutely indistinguishable from that of a child. The guttural sound associated with deep yawning bears very close resemblance to the sound of human yawning.

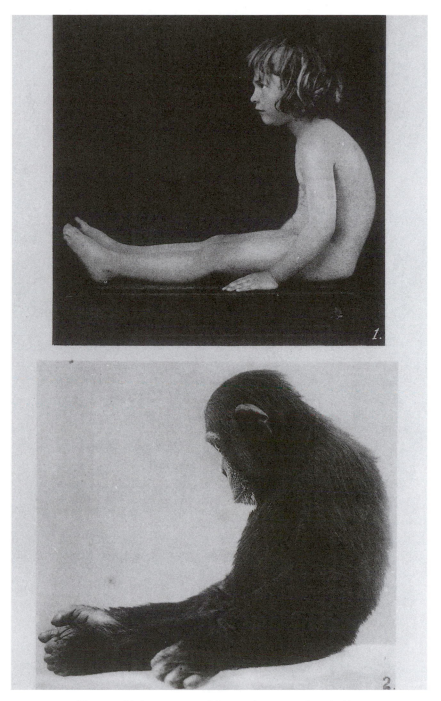

Plate 1. Sitting postures of human (at 4 years) and chim-panzee (at 4 years): (1) Roody sitting (profile); (2) Joni sitting (profile).

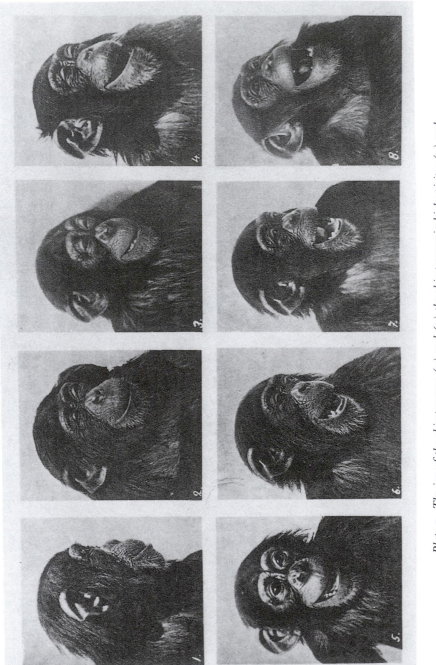

Plate 2. *The joy of the chimpanzee: (1) and (2) the chimpanzee in high spirits; (3) and (4) the narrow grin of the chimpanzee; (5) and (6) the broad grin of the chimpanzee; (7) the boisterous laugh of the chimpanzee; (8) the chimpanzee's laughter.*

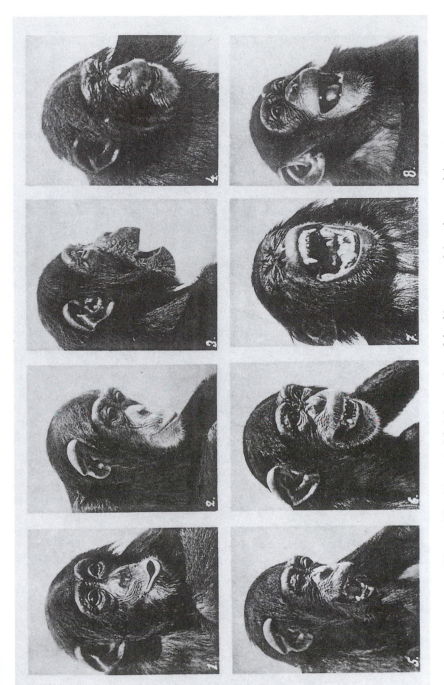

Plate 3. The eight typical facial expressions of the chimpanzee: (1) excitement; (2) attention; (3) astonishment (surprise); (4) disgust (repulsion); (5) anger; (6) fear (horror); (7) sadness (crying); (8) joy (laughter).

Plate 4. The chimpanzee's typical standing postures:
(1)–(4) typical standing posture, supported by arms.

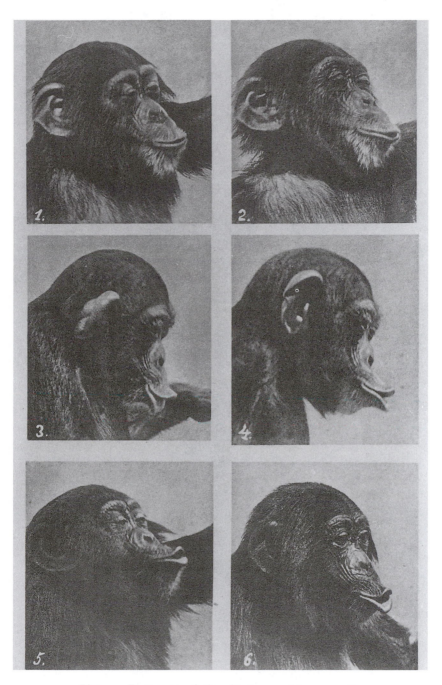

Plate 5. Excitement of the chimpanzee (six consecutive stages): (1) protrusion of lips; (2) humping of lips; (3) and (4) trumpeting of lips; (5) and (6) trumpetlike unfolding of lips.

Plate 6. Sitting postures of human and chimpanzee: (1) Roody (at 4 years) sitting (en face view); (2) Joni's (at 4 years) typical sitting posture (en face view).

Plate 7. The chimpanzee's unusual standing postures: (1)–(4) unusual way of standing, unsupported by arms.

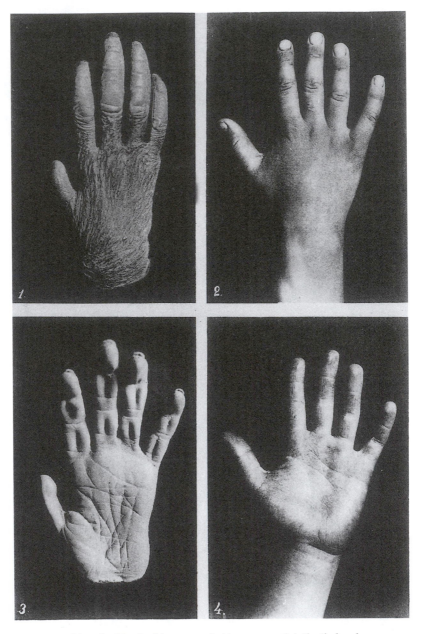

Plate 8. Hand of human and chimpanzee: (1) Joni's hand (seen from above, wax cast from dead animal); (2) Roody's (at 7) hand (seen from above); (3) Joni's hand (seen from below); (4) Roody's (at 7) hand (seen from below).

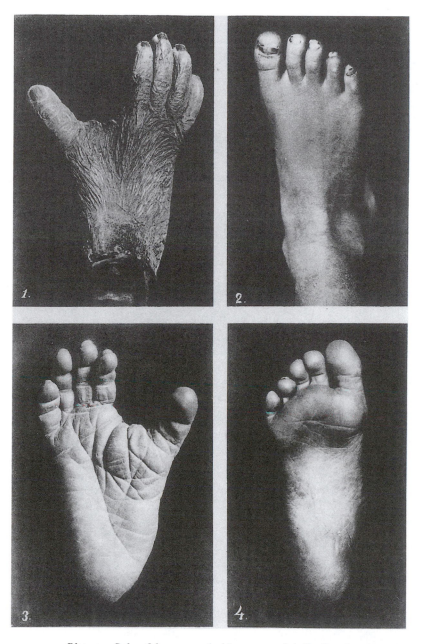

Plate 9. Sole of human and chimpanzee: (1) Joni's sole of foot (seen from above, wax cast from dead animal); (2) Roody's sole of foot (seen from above); (3) Joni's sole of foot (seen from below); (4) Roody's sole of foot (seen from below).

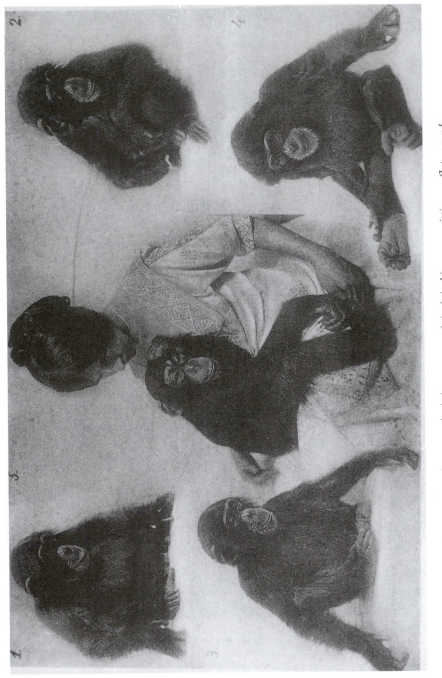

Plate 10. The chimpanzee's sitting postures: (1)–(4) chimpanzee sitting on flat ground; (5) chimpanzee sitting on knees.

Plate 11. Sitting postures of human and chimpanzee: (1) first attempts at sitting (Roody, 5 months); (2) Roody (3½ years) sitting on heels; (3) Joni squatting; (4) Roody (at 1 year 2 months) sitting on heels; (5) Joni sitting on bench; (6) Roody (at 3 years) sitting on bench.

Plate 12. The chimpanzee walking and running: (1) and (2) slow locomotion; (3) and (4) running (profile); (5) and (6) running (en face).

Plate 13. Assisted standing and walking of human versus chimpanzee: (1) Joni standing with support of human; (2) Roody (9 months 26 days) standing with support of human; (3) Joni (3½ years) led by hand; (4) Roody (at 3 years) led by hand.

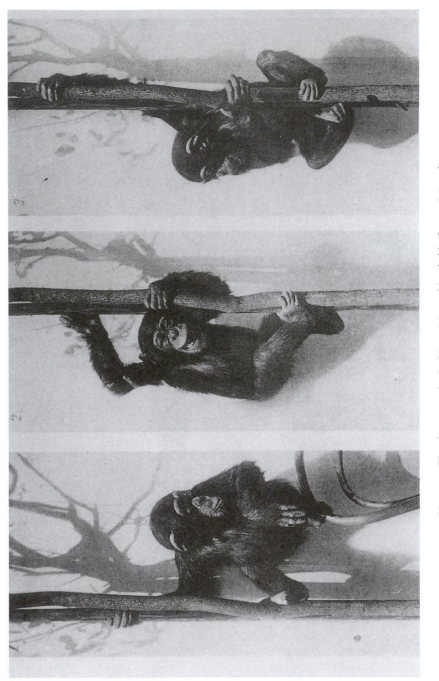

Plate 14. The chimpanzee's climbing: (1) catching hold of tree; (2) and (3) climbing up tree.

Plate 15. Negotiating altitudes, human versus chimpanzee: (1) Joni on roof of house; (2) Joni descending from roof of house; (3) Roody (4½ years) climbing up ladder; (4) Roody (5 years) on a trapeze; (5) Roody (2 years 4 months) climbing to the seat of a sleigh; (6) Roody (3½ years) climbing over chairs; (7) Roody (4½ years) ascending fence.

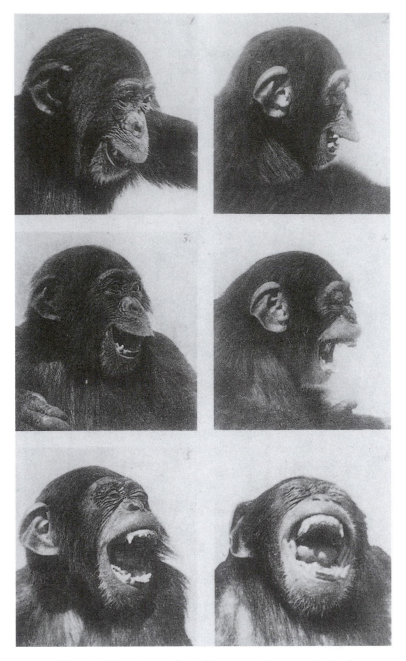

Plate 16. The sadness of the chimpanzee (six successive final expressions of the crying chimpanzee with varying expressivity): (1) and (2) initial stages of outcry; (3) and (4) roaring outcry; (5) and (6) the completely woebegone chimpanzee.

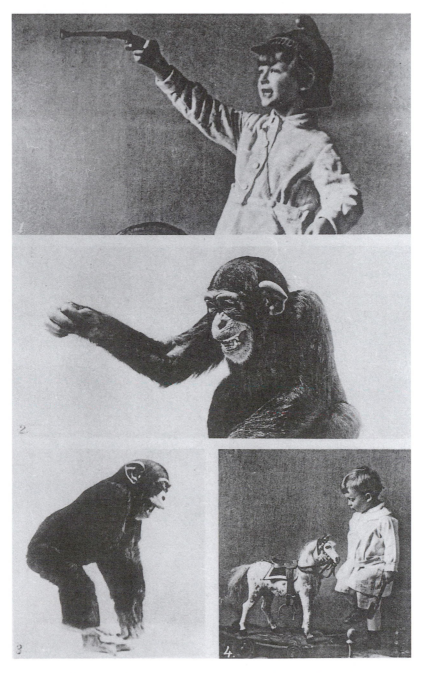

Plate 17. Expression of anger of ape and child: (1) Roody's (5 years) pseudoanger; (2) Joni's anger; (3) Joni's aggressive gesticulation (stamping foot); (4) Roody's (2 years 2 months) aggressive gesticulation (stamping foot).

Plate 18. *Aggressive reaction of the chimpanzee Joni to a stuffed chimpanzee: (1) standing erect; (2) banging with right foot and showing teeth; (3) throwing upper lip back— ready to attack.*

Plate 19. Combined or ambiguous facial expressions of the chimpanzee: (clockwise from top left) fear and anger; curious attention and fear; boisterous laughter; astonishment and strife; crying with anger; intense attention.

Plate 20. Expression of fear of the chimpanzee: (1)–(3) the chimpanzee frightened and excited (hair bristling) during a walk; (2) meeting the intimidating stimulus; (4) general excitation (slightly frightened).

Plate 21. Reaction of the chimpanzee to a dead hare: (1) first moment, frightened and squatting; (2) standing erect, waving clenched finger joints; (3) lifting up fist, pressed lips slightly curved sidewise; (4) threatening postures, ready to jump from feet to hands and vice versa; (5) waving right hand, shaking lower jaw; (6) attack on the offending stimulus.

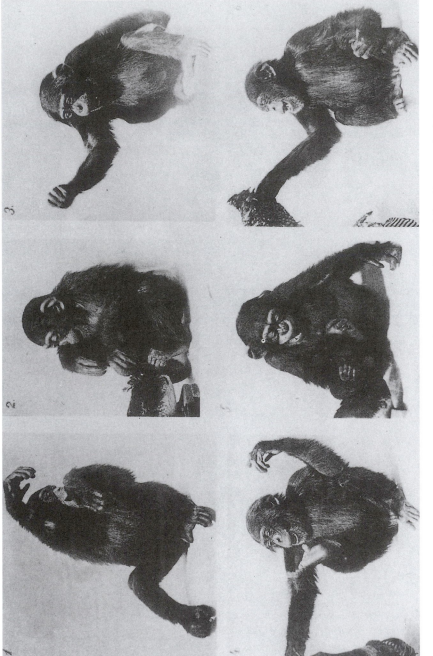

Plate 22. Postures of self-defense and attack (expression of anger of the chimpanzee): (1) and (2) protecting his face; (3) aggressive gestures with clenched fist; (4) self-defense with foot, attack with hand; (5) torturing his victim; (6) pinching his victim.

Plate 23. The six consecutive stages of general excitement of the chimpanzee: (1) slight bristling, straining of muscles, and protrusion of lips; (2) strong bristling, protrusion of lips, lifted hand; (3) partly rising to a vertical posture; (4) bending of body prior to making it erect; (5) making body erect and stretching out arms; (6) maximum stretching of body, lifted arms.

Plate 24. The chimpanzee specifically excited preliminary to the onset of various emotions: (1) excitement mingled with sorrow; (2) excitement mingled with anger; (3) excitement mingled with joy; (4) excitement mingled with fear.

Plate 25. Postures and gestures of the frolicsome, merry chimpanzee: (1) the chimpanzee in high spirits; (2) the chimpanzee is tickled; (3) The chimpanzee is tickled, and he laughs; (4) the chimpanzee in a boisterous and merry mood.

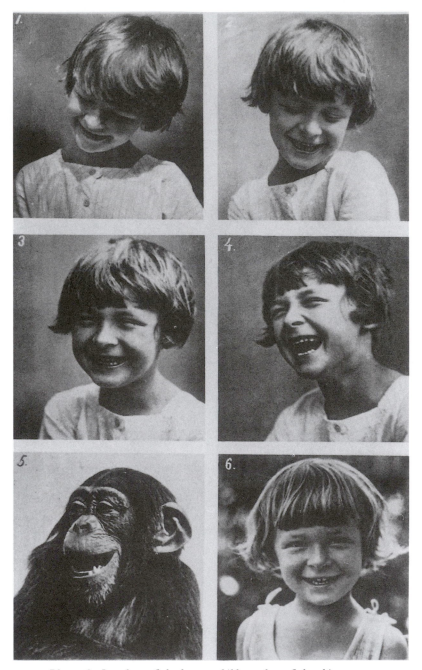

Plate 26. Laughter of the human child vs. that of the chimpanzee: (1) Roody's (6 years) narrow smile (second stage); (2) Roody's (6 years) broad smile (third stage); (3) Roody's broad smile (fourth stage); (4) Roody's laughter (fifth stage); (5) Joni's (4 years) broad smile; (6) Roody's (5 years) broad smile (third stage).

Plate 27. Sadness of the chimpanzee (six successive initial stages preparatory to crying): (1) the quiet chimpanzee; (2) the depressed chimpanzee; (3) and (4) the sad (groaning) chimpanzee; (5) and (6) the crying chimpanzee.

*Plate 28. Gesticulation and postures of the crying chimpan-
zee: (1) postures of the crying chimpanzee (overpowering
sorrow); (2) protrusion of arms; (3) hitting of hand; (4)
throwing up arms in despair; (5) raising arms.*

Plate 29. Gesticulation and postures of the crying chimpan-zee: (1) wringing arms; (2) and (3) raising arms.

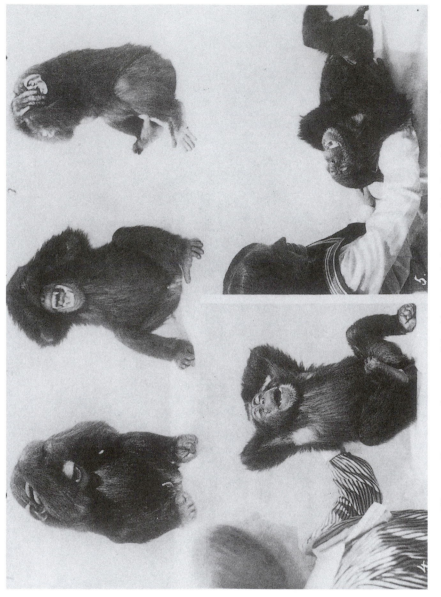

Plate 30. Postures and gestures of the chimpanzee in despair and in play: (1)–(3) crossing hands above head as sign of full despair; (4) and (5) tickling the chimpanzee.

Plate 31. Postures and gestures of the chimpanzee in different moods: (1) the chimpanzee in a good mood; (2) the chimpanzee in a bad mood (half sick); (3) the chimpanzee depressed; (4) the chimpanzee dejected; (5) the chimpanzee feeling despondent but excited; (6) the chimpanzee completely depressed.

Plate 32. Grooming of the chimpanzee: (clockwise from top left) cleaning feet; gnawing fingernails on hands; examination of feet; cleaning interstices of teeth; gnawing fingernails on feet; examination of lips.

Plate 33. Self-care of human child and ape: (1) Roody (5 years) removes splinter from finger; (2) Roody (2 years 4 months) examining his toes; (3) Joni examining the hair on his feet; (4) Joni examining the hair on his hand.

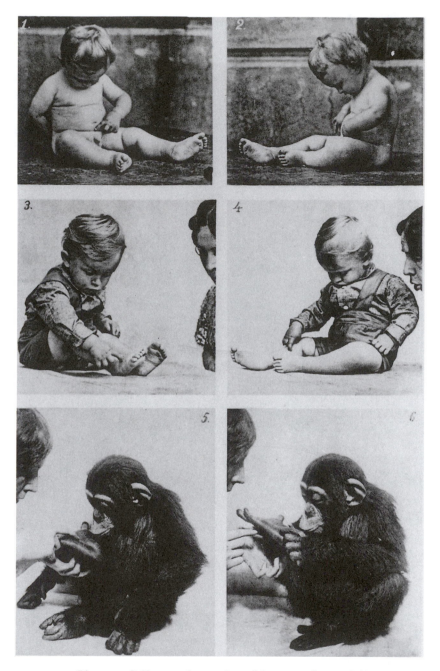

Plate 34. Self-centered attention of human and ape: (1) and (2) self-examination of Roody (1 year 2 months); (3) and (4) Roody (1 year 4 months) feeling and picking a pimple; (5) and (6) self-examination of Joni.

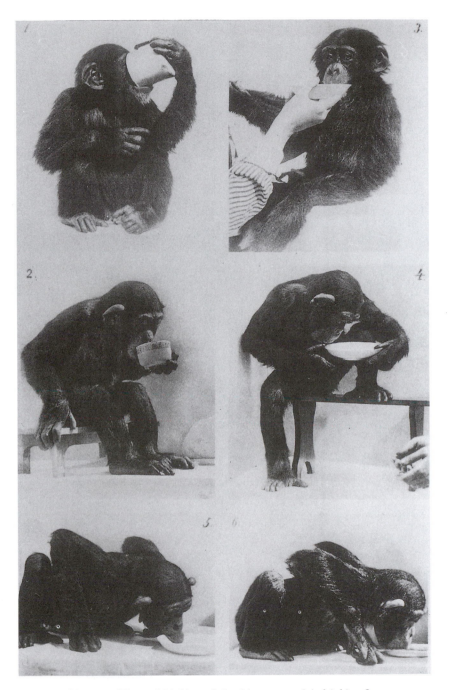

Plate 35. Ways of drinking of the chimpanzee: (1) drinking from cup (final stage); (2) drinking from cup (initial stage); (3) drinking from saucer with help of human; (4) drinking from saucer unassisted; (5) and (6) drinking from saucer "animal-like."

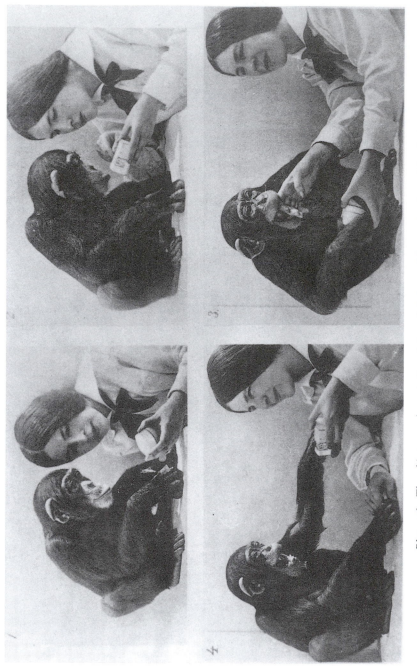

Plate 36. *The chimpanzee's curiosity: (1) and (2) careful examination of intriguing object; (3) resistance to returning taken object; (4) taking object away from observer.*

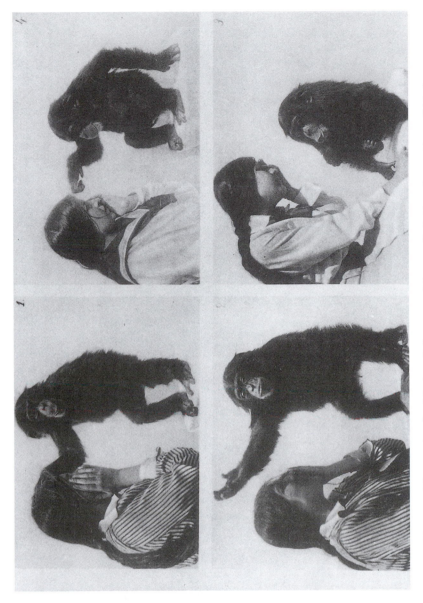

Plate 37. The emotional solidarity of the ape and human (imitating the emotions of his human friend): (1) expression of sympathy—putting hands on head of weeping human, whimpering; (2) and (3) general excitation on seeing a human cry; (4) invitation extended to weeping human to play.

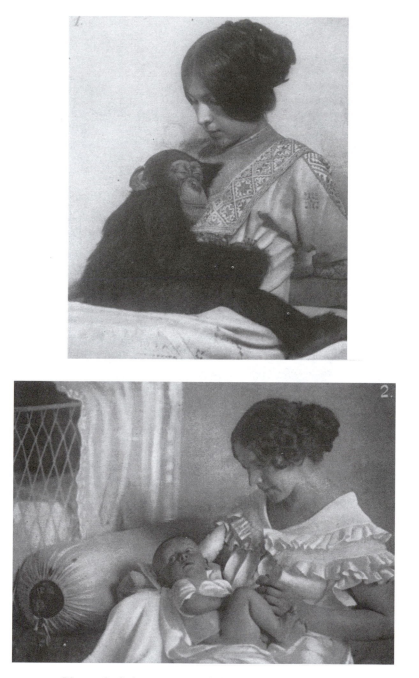

Plate 38. Lying postures of human and chimpanzee:
(1) Joni asleep; (2) Roody (3 months 21 days) lying down.

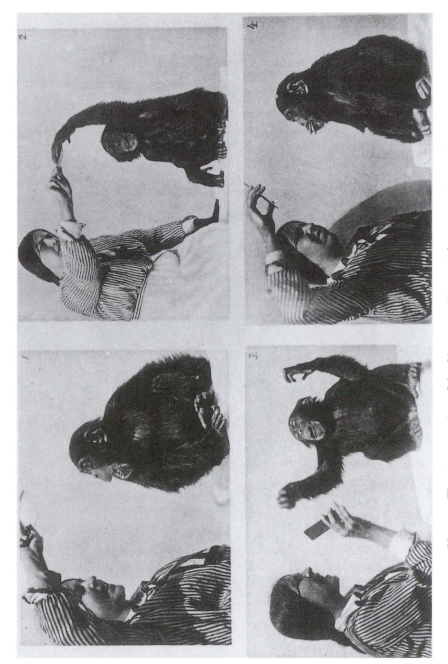

Plate 39. The mirror reactions of the chimpanzee: (1) astonishment at seeing own image; (2) seeking something beyond mirror; (3) threatening gesture directed against image, banging with fist; (4) grimaces in front of mirror (cracking of lips).

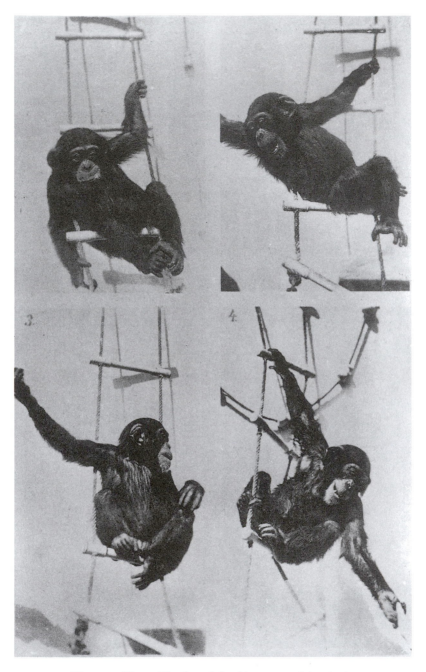

Plate 40. The mobile play of the chimpanzee: (1) swinging on a rope ladder; (2) catching objects with free hand in course of swinging; (3) sitting motionless on the rope ladder; (4) playful mood during swinging.

Behavior of Human Child

Comparative Psychological Study

Chapter 11

Human and Chimpanzee:
Comparison of Appearances

Part I was a long factual description of our study of the chimpanzee. His short life with its joys and sadness, fun and training sessions, interests and tendencies of the boisterous creature, flashed before the reader's eyes. To interpret all these facts and draw the broadest conclusions to enrich our knowledge and influence our comprehension of the world, it is advisable to conduct a rough morphological comparison of an infant chimpanzee and a human child of the same sex and age.

Let us start with the description of their appearances (figures 2A, 2B). Here they are, looking at me and at the reader. These two creatures are dear to me: I have been to them the closest person in the world: I have given them so much of my energy, feelings, and warm care.

Paradoxical as it might sound, I have to admit that, in my heart, both of them, Joni and Roody, take up an almost equal space. But, my exploring mind keeps pestering me to define the limits of their proximity in various forms of their appearance and behavior.

Face and Extremities Let us look closer at their faces (figures 2A, 2B).
in Stasis The chimpanzee's face is not the muzzle of a dog or of a baboon, and it is not even that of a lemur with its protruding jaws. It is a real face. A major part of this face has bare skin, as in a human child; it is slightly covered with silver velvety down. The eyes look sensible and are positioned as on a human face. In both infants, the eyelashes rim the eyelids. But, a couple of these features are basically everything they have in common; all their other features sharply diverge.

The face of the human child looks fresh, smooth, and plump at first sight. The chimpanzee's face, on the contrary, leaves the impression of being old, emaciated, and furrowed with wrinkles, as though not just 3 or 4 years are behind it, but a difficult life of 60 to 70 years [plates 6(1), 6(2)].

The more detailed differences can be found easily in the shape of various parts of the face, in the development of the hair, in the color of the eyes, in the creases of the skin, and in the pigmentation. A diagram that compares their main facial features reveals these differences.

While in a human child the forehead protrudes beyond eye level [plate 1(1)], the forehead line is straight, and there are no above-the-eye arcs. The chimpanzee's forehead is slanted back, the above-the-eye arcs strongly protrude, and the entire forehead is covered with short black hair, which is thinner in the central part of the forehead and longer at the sides and at the temples [plate 1(2)].

While in humans, the nose protrudes forward above the level of the forehead, the bridge of the nose is high, the tip of the nose protrudes most (in the sagital dissection of the head), the nostrils stand, and the nasal passages are aimed up and to the sides. The chimpanzee's nose is flat and hardly protrudes over the level of the forehead, and the bridge of the nose is low and impressed. The most protruding point (in the sagital dissection of the head) coincides with the middle of the upper lip. The nostrils lie flat, and the nasal passages are aimed inward and down [plate 2(1)].

The human child's cheeks are convex and smooth; the sides of the cheek are hairless. The chimpanzee's cheeks are concave and covered with thin dark hair; the sides of the cheek are rimmed with thick side-whiskers.

In the human, the forehead part, the nose part, and the jaw part of the face are equal. In the chimpanzee, the forehead is the smallest, and the jaw part is the biggest.

In the human, the upper lip is as long as the lower one; it is cut in the middle by a longitudinal, shallow, broad furrow that descends from the middle of the nose. Both lips are smooth; the edge of the lips has a red mucous turn. In the chimpanzee, the upper lip is much longer than the lower one and has longitudinal wrinkles; there is long white setaceous hair near the upper edge of the lip. The lips do not have the red mucous turn (only a slight pinkishness of the upper edge of the lips). The lower lip is covered with bumpy wrinkled skin that ends with a reddish thick edge.

In the human, the chin is outlined clearly; the mouth is rimmed with red lips, and the upper lip protrudes over the lower one. The length of the mouth equals the distance between the eyes or is slightly greater. In profile the upper lip extends forward beyond the lower one. In the chimpanzee, the chin does not protrude; it is bumpy and profusely covered with dense hair. The mouth is almost without lips and is almost twice as long as the distance between the eyes. In profile, the lower lip protrudes beyond the upper one.

Hair on the human head is long and highly developed. The eyebrows are in the form of almost solid stripes made of densely positioned, soft hair that clings to the skin. The whites of the eyes are light; the iris varies in color within the limits of a certain race. The hair on the chimpanzee's head is short; the hair is longer on the temples and on the lateral parts of the cheeks and resembles side-whiskers. The eyebrows are in the form of dark bristles that grow separately (at a distance of 1 cm) along the entire length of the above-the-eye arcs. The whites are dark; the iris is light brown.

On the calm face of a child, one can distinguish only three main furrows: two furrows that go from the nose wings to the corners of the mouth and one wide furrow that goes from the nasal partition to the middle of the upper lip. On the chimpanzee's calm face, one easily can see 90 furrows [figures 1(1), 1(2)].

When a human slightly raises the skin of the forehead, there are five thin parallel furrows that form in the upper part of the forehead, near the hair. In the chimpanzee, under similar circumstances, five thin parallel furrows form in the lower part of the forehead immediately above the arcs.

In contrast to the chimpanzee, human ears are absolutely hairless. The length of the ear by far exceeds its width. There is an earlobe (lobus auriculus), and there is a helix, which is uniformly thick along its entire length. The outer edge of the ear is not in the shape of an angle. The ears are semihard; they do not move involuntarily (for example, they do not flap when the human is running) or voluntarily [plates 51(1), 51(4), 52(5), 52(6)].

The chimpanzee's ears are covered partially with hair. The length of the ear is only slightly greater than its width. There is no earlobe. The helix is thinner in its middle area. The outer edge of the ear is broken in the shape of an angle. The ears are soft; they flap when the animal is running and move slightly when he is chewing or listening [plates 2(1), 2(4), 2(7), 2(8)].

The human leg is somewhat longer than his arm. The forearm is shorter than the shoulder. When the child is in a vertical position, the straight hanging hand reaches the middle of the hip [plate 53(1)]. The arm and leg are covered slightly with soft hair, which is more visible against light than under direct observation. The foot is longer than the hand. The hand is hairless; its length is twice that of its width. The palm length is larger by a quarter than its width.

The chimpanzee's arms are considerably longer than his legs [plate 39(2)]. When the animal stands, his hand reaches below his knees. The arm and the leg are entirely covered with black, dense, coarse hair. The hand is longer than the foot. The hand is longer than its width. The back of the hand is covered with hair to the level of the second phalanges, and hair covers to the main phalanges of the foot. The palm length is a third larger than its width.

In the human, the fingers are short and weak [plates 8(2), 8(4)]; the main phalanges are the thickest, with the first finger longer and thicker than the fifth one. The backs of the fingers have lightly colored down (in the middle of the first and the second phalanges). There are no callous bulges on the boundary of the main and middle phalanges. The fingernails are light.

Human legs are bare. The foot is smaller than that of the chimpanzee and also is bare; the big toe is parallel to the other four ones. The fingers are developed more than the toes [plates 9(2), 9(4)].

Chimpanzee fingers are long and strong (they are joined by a small membrane at the base of the fingers. The middle phalanges are the thickest; the first finger is thinner and shorter than the fifth one. The back of the main phalange of the hand is covered with hair. There are soft callous bulges between the middle and the main phalanges. The nails are black [plates 8(1), 8(3)].

The chimpanzee's leg is hairy. The foot is bigger than that of a human; it is hairy to the main phalanges [plates 9(1), 9(3)]. The chimpanzee's big finger and big toe are separate from the other fingers and toes. The toes are less developed than the fingers.

In the human, the hand lines [figure 3(2); plate 8(4)] are few and weak. Horizontal lines 1, 2, and 4 and vertical lines I, III, and IV are distinguished easily. The wrist lines (7 and 8) are more distinct than those of the chimpanzee. Vertical lines I and IV (going to the phalanges of the fingers) are less developed.

There are a number of very distinct lines on the chimpanzee's hand. Horizontal lines 1, 2, 3, 4, and 6 and vertical lines I, II, IV, and V are highly visible; the first horizontal and the first vertical lines are double. The wrist lines are less visible than those in the human. The vertical lines that go to the phalanges of the fingers are more visible than those in the human [figure 3(1); plate 8(3)]. Large individual variations in the lines of the hands of different species of chimpanzee have been observed [figures 4(1)–4(3), 4(5), 4(8)].

In the human child, the foot has an arch. The sole lines are less distinct and fewer than those on the palm. The second horizontal (at the base of the big toe) and the first vertical lines are totally absent; lines III and IV are weak and are visible only in the upper part of the sole [figure 3; plate 9(4)].

The chimpanzee's foot is not arched. The main sole lines are sharp, and there are more than in the human [see detailed information in the chapters that describe the chimpanzee and in plate 9(3)].

Static Poses

1. Sitting Poses. Comparing the positions of the body and extremities of the human child and the infant chimpanzee, we find sharply diverging features along with some features of unquestionable similarity. A normally developing child, like the chimpanzee, prefers running and walking to sitting still, and his sitting positions are no less diverse than those of the chimpanzee, but the typical chimpanzee's poses are totally uncharacteristic of a child.

First, the difference pertains to the position of the arms and legs. While in the infant chimpanzee the arms often rest on the ground when he sits [plates 6(2), 10(1), 10(3), 10(4)], in humans, the arms play a supporting role only at the initial stages of the child's development (from 5 to 9 months). We do not usually notice a snug position of the child's hands on his bent knees as is most common for the chimpanzee [plates 6(2), 10(2)].

The most typical pose of the chimpanzee [plates 6(2), 10(1), 10(3), 10(4)] is when he sits on flat ground and leans on his ischial bones with his legs bent at the knees and leaning directly on the feet, a pose quite uncommon for man. [Chimpanzees have callosity-like thickened skin on their "buttocks,"—ed.] In my long observations of sitting children, I have not noticed such leg position even once.

The legs of a child sitting on a flat surface may be completely straight, close to each other, with the rear parts of the thighs and shins resting on the ground so that the feet are directed forward and outward [plates 1(1), 54(1), 55(1), 56(1)]. Or, the legs are thrown to the sides, somewhat bent at the knees and spread out on the surface, touching the surface with the outer lateral parts of the thighs and shins and the soles directed inward to face each other [plates 49(1), 57(1)]. Or, one half-bent leg is placed on top of the other one, and the child sits with his legs crossed [plates 58(4), 59(4)], or the child places both bent legs close to each other and pulls them to the sides, aiming his shins and his feet in opposite directions [plate 6(1)].

Such tossing the leg up that I have described in Joni [figure 6(3)] I have never noticed in Roody because the mobility of Roody's legs in the hip joint is not as great as in his companion [plate 46(2)]. The most noticeable tossing the leg up in Roody is shown in plate 60(4).

On the other hand, the favorite pose of a child, the squatting position, is very well known [plate 11(4)]: The chimpanzee's body rests on the soles of his half-bent legs, his body balanced. He can stay in such a position for a long time, staring at something or engrossed in some activity. This pose can be seen here and there in the photos of my boy (at ages from 1 to 7 years) during his various activities, such as, when he was laying out or sorting things from boxes, when he was "working" with a hammer, playing with sand, and so on [plates 11(4), 57(4), 57(6), 61(4), 62(5), 63(5), 64(1), 65(1), 66(2), 67(3)].

Only in exceptionally rare cases (for example, when the chimpanzee was frightened) did I see him squatting, but this pose was very short-lived and not as steady as in the human child; it required additional leaning on his hands [plate 11(3)]. In the case of a child, the hands remain free and perform various moving and grasping actions.

These poses reveal the prevailing role of the legs in the support of a human body compared to that of his arms; in the case of the chimpanzee's sitting position, the arms retain an auxiliary supporting role. These examples clearly show that the child's legs are stronger and longer[1] than those of the chimpanzee; the child's bent legs are capable of balancing his whole body,[2] and they help him restore the vertical position without participation of his hands. When the chimpanzee stands from a similar position, he invariably uses his hands for support.

Due to the length of his legs, a human sitting in a squatting position does not reach the ground with his bottom; if he wanted to sit down, he would have to lean back somewhat. In contrast, the chimpanzee sitting in this position leans on the "buttocks," as though glued to the ground; his ability to stand is quickly encumbered.

Among various sitting poses of a child, one pose, which repeats itself over and over, arrests your attention: namely, when a child sits on his bent knees with leg directed backward [plates 61(3), 68(3)]. He either tucks one leg (usually the left one) under his body, ready to stand (as it happens during fast play) [see, for example, plates 56(6), 63(2)], or when he is involved more deeply in something (for example, when he examines a new toy), he may tuck his legs under his body completely, aiming his knees outward [plates 11(2), 69(5)]. I have even noticed that sometimes people carry this pose into adulthood, particularly those who used to write a lot in a sitting position and are tired of sitting but do not have time for a break. Of hundreds of photos that depict the chimpanzee in his natural sitting position, I could not find a single one showing him in such a typically human pose.

Comparing the sitting poses of the chimpanzee and human child under more artificial conditions (for example, on a human's lap or on a bench), one can find many similarities in both infants [plates 10(2), 11(5), 11(6)]. While Joni could sit on a bench as easily as an adult, Roody (when he was 18 months old) could not do it without falling on the floor; later (at 19 months), he used a more complicated way of sitting down. Coming up to the bench, he turned his back to it and moved backward, trying to touch the bench with the back of his legs. Then, he started to sit very carefully, touching the firm surface of the bench with his behind in an attempt to find a steady position. Only later did the child drop this preliminary search for a firm substrate during his attempts to sit on a bench.

When the infant child and chimpanzee sat on high places with their legs hanging freely, both of them showed the tendency for holding one foot with the other, apparently to prevent the legs from rocking [plates 53(3), 53(4)]. Sitting on a human's lap, both showed a tendency to press against the human, rest their heads snugly, and press their cheeks against the human body [plate 10(5)].

When both infants are tired, they use their arms as their head supports. But, while the chimpanzee rests his head on his folded fingers or even on the forearm [plate 7(4)], the human child[3] supports his head only with his fingers in the well-known and meaningful "thinker's pose." He rests the outspread first and second fingers on the temple and on the jaw and supports his cheek with his three other fingers.

2. Standing.   A comparison of the chimpanzee's standing poses with those of a human child reveals even less similarity. The typical standing pose of the chimpanzee leaning on his arms is quite uncharacteristic and even impossible for a child because of the shortness of the child's arms, if nothing else [plate 70(5)]. The auxiliary supporting role of the arms is observed, as is well known, only at the first standing attempts of a child (approximately at 6 months of age) [plates 13(2), 70(2)]. When a child is 9 months or 1 year old he can stand on flat ground quite steadily without the help of his arms [plate 70(4)].

A 1-year-old child yet positions his legs widely apart when he stands [plate 70(4)]; the same usually can be observed in the chimpanzee, particularly in his vertical position [plates 7(1), 13(1)]. But, even at this age, the human child rests firmly on the soles of his feet without the tendency to lean on the outer edge of the foot, as we can observe in the chimpanzee standing straight[4]; the trunk and the knees of a human child are absolutely straight, while the knees of a standing chimpanzee are half-bent. A 2-year-old child stands with his feet close to each other [plate 40(6)].

The straight-standing body position is a rule for children who are 1½ to 2 years old. The same position is an exception for a chimpanzee 3 to 4 years old. In the typical standing position of the chimpanzee, the axis of his body is inclined, while that of the human is vertical [compare plates 4(1), 4(2) and plate 71(5)].

Needless to say, the vertical pose of a 1-year-old child is more firm than that of a chimpanzee. The straight vertical pose is more advantageous biologically for human living conditions on the ground. It gives the human a wider range of visibility for subsequent movements and actions; it frees his arms for defense in case of danger or a threat from enemies or for creative work in a quiet situation during leisure.

3. Lying.   The poses of a freely sleeping child are similar to those of an infant chimpanzee [plate 51(5)]. Most often, Roody, like Joni, lies on his side, tucking one hand clenched in a fist under his cheek or under the pillow. During the night, their poses change; you can see Roody and Joni sleeping on their stomachs with their faces deep in the pillow, pressing their arms and legs to their bodies. Before going to sleep, the child (at the age of 5 months) rubs his eyes with his fists, rubs his face against the human who holds him, presses his face to the human's shoulder or to his elbow joint, turns away from the light, and sometimes puts his hands on his eyes and goes to sleep in this position. Roody, like Joni, tries to find a more comfortable position before going to sleep. He turns from side to side, turns his head, shifts up and down on the pillow, puts his hand under his head or takes the hand out, looks up, sees his mother, and smiles. Roody, even at the age of 2 months, goes to sleep holding the swaddling clothes; and when he sleeps on my lap (3 months old), he holds my jacket, his arm stretched vertically, reminding me of the sleeping Joni as he held the ropes.

For 10 to 15 minutes before falling asleep, the child yet tosses and turns; he raises his eyelids from time to time, opens his eyes and stares around him. Then, his eyes narrow somewhat, and one eye falls asleep, while the other eye remains half-open (in a 3-month-old child). The cramp rolls over his eyelids; his eyes close tightly, and he goes to sleep.

Roody (1½ to 3 years old) especially willingly sleeps holding my hand. If he wakes up and sees that I am not holding his hand, he asks, "Mother, give me your hand."

Like Joni, the child tosses and turns in the bed, throws his arms up [plate 51(1)], presses his hand to his neck [plates 51(3), 51(4)], and spreads his fingers widely; or, he crosses his arms on his chest (like the angels by Rafael), or he puts his hands under his head [plate 51(2)]. I have never seen Joni have this pose during his sleep. The sleep of a human child, like that of chimpanzee, is very sound. While in deep sleep, both infants allow me to turn them over or cover them with a blanket.

Poses in Motion   1. Walking. Comparison of walking, running, and climbing reveals even more distinct differences between a chimpanzee and a human of the same age. While the chimpanzee, as a rule, walks on the ground, leaning on his all four extremities and, as in his typical standing position, the axis of his body is inclined, the child's hands participate in his movements on the ground only when he is climbing or when he is 8 to 9 months old. As soon as the child starts walking in the vertical position, his support is only his soles [plates 72(3), 72(4)].

However, at the first walking attempts (at the age of 9 to 10 months), the child stands on his feet very unsteadily. He cannot yet straighten his back totally and holds his body at an angle; therefore, he may fall forward any second [plate 13(2)]. In making the first three or four steps, he instinctively grabs a person's hands or any object so that he does not fall, pushes a chair in front of him, or walks "along the wall" while his hands play a major auxiliary supporting role. But, when the child is 1 year old, he walks with his back straight, habitually extending his hands forward [plates 72(3), 72(4)].

At this age, Roody (13 months old) walks freely only after some preliminary training; to keep his balance, he often bends down and squats, but after a small pause, he straightens his body and continues walking. At this time, even such a small complication as a shallow and narrow rut[5] constitutes a heavy obstacle for him; not knowing how to step over it, he carefully steps into it (at the age of 15 months). After 2 months, the boy steps over the same rut very easily.

At the age of 1 to 1½ years, the human child is used to vertical walking to such an extent that he looks where to apply his freed hands. He willingly pulls or pushes various toy objects (wheels attached to a stick, horses, carts, or cars) [plates 73(1)–73(6)]. He even tries to carry light objects [plates 71(3)–73(6)].

According to Engels, freeing the hands from the necessity to support the body during standing and walking has been the main factor in transforming the animal-like human ancestor into man per se. When a 1-year-old child's hands are busy with something, he certainly walks less firmly; afraid to lose his balance, he often opens his mouth and extends his tongue forward [plate 71(3)]. But with each day and each step, his feet get used to the ground more and more, and a child 1½ to 2 years old, carrying a load commensurate with his strength, walks briskly, freely, and smiling joyfully, not suffering whatsoever from his hands being busy [plates 71(4)–71(6)]. The chimpanzee, even at the age of 3 or 4 years cannot walk verti-cally for a long time. He can hardly make three or four steps; to keep balanced, he has to use his arms [plates 7(1)–7(4)]. While he walks, his balance remains so unsteady that he cannot carry anything. As we have noted, leading the champan-zee by the hand does not make it unnecessary for him to lean on his free hand; for a human child, it is entirely superfluous [plates 13(3), 13(4)].

From here on I will refer to age with a shorthand indicating the number of years, months and days that have passed since birth, so that 1.2.20 means 1 year, 2 months, 20 days. As soon as the child masters the walking process, he demon-strates (at the age of 1½ to 2½ years) a spontaneous tendency to improve this ability. He (1.2.20) tries to walk in a more complicated way, which he has im-posed on himself. For example, Roody would try to mount a high place, or he would lay a board on the floor [plate 74(2)] and would attempt to walk on the board without falling off. He (2.4.27) would make, out of small planks, something like a narrow, shaky, and uneven bridge and step on it with one foot, keeping balanced with his raised hands; or he (1.7.21) would climb a high pile of sand, sinking in it on his way up. He would (1.9.14) complicate his walking with vari-ous tricks: he would go backward, walk only on his heels, toss his head up and walk without looking under his feet, intentionally curve his legs, walk on the sides of his soles, tap-dance, sway his arms, or rock from side to side (1.9.19). At the age of 2 years (1.11.9) Roody tries to imitate the marching step of a soldier; later (2.6.18), he does not get discouraged by obstacles he encounters on his way—he steps over ditches filled with water, goes around puddles and big stones (2.6.16). And, when he is about 3 years old and clad in a heavy winter coat, he can continue walking for more that 2 hours. He would climb small snow piles and sag in them, but keep going forward even more vigorously [plates 74(1), 74(3)]. It is as though the child instinctively tries to exercise and test his yet weak legs on various surfaces, applying his feet to different methods of movement.

The infant chimpanzee also tries to make his walking and running more complicated, taking different objects (rags, balls, or chains) in his hand or foot, going in narrow passages, and running with these objects around the legs of the tables and chairs. But, he continues his typical walking on all fours, never making an attempt to train himself to walk vertically.

The child also drags different objects behind him. Soon, he gets bored with rolling single objects, so he makes up long trains of rolling carts [plate 75(1)]. He builds so-called bridges and rolls the trains over them, although sometimes these operations usually end with a sudden fiasco [plates 75(2), 75(3)]. A 2-year-old child willingly drags heavy objects (for instance, a sledge) up onto a high snow hill [plate 76(5)].

There is one thing in which human child is undoubtedly inferior to the infant chimpanzee: running and climbing.

2. Running.   A 1½- to 2-year-old chimpanzee, using his four extremities and stepping on his soles and his semifolded fingers, ran so fast that a human adult was not able to catch him. He deftly climbed a fairly steep stairway and came down even more quickly; my boy, although he was very agile, liked to run and used to run down high hills in winter and summertime [plate 76(1)], but he was never able to develop the speed of the chimpanzee [plates 27(2), 72(6)].

3. Climbing.   Roody was worse that Joni in climbing; at the age of 1½ to 2 years, he could climb, only very slowly and awkwardly and also on all fours, on the stairs or on the seats of armchairs and wheelchairs. However, Roody used a peculiar way of climbing. He (1½ years old) put the bent knee of his left leg on a step and rested the sole of his semistraightened right leg on the floor. Then, he put the bent knee of his right leg on the same step, leaning now only on both shins. He straightened his right leg again and stood on its sole and put his left shin on the following step, and so on[6] [plates 77(1), 77(2), 78(1)].

Later, when my child was 2 years old, he learned how to climb the stairs in a straight vertical position. He ascended the stairs at that time only very apprehensively and hesitantly, leaning on his hand or vigorously balancing with his arms [plate 79(1)]. To speed up his climbing, he often tried to do it on all fours, but this time he leaned not on his knees, but on his outspread palms, transferring them from one step to the next [plate 77(2)], the way the chimpanzee had done it.

After a certain training period in climbing, the human child does not need balancing with his arms. A child 2½ years old (even in heavy winter clothes) briskly climbs the stairs [plate 79(2)]; only his outspread fingers belie a certain tension and lack of confidence. Having mastered the climbing, Roody (2.8.3) becomes brave and starts stamping his feet and waving his hands while he walks the stairs.

I never observed the chimpanzee climbing the stairs in such a way. Joni could walk up and down the stairs in the semivertical position on only three extremities when he was led by the hand and when he leaned, at the same time, on his other hand.

The human child is known to improve the ease and increase the speed with which he climbs the stairs; in a month or two, he is able to go up the stairs freely, without any help, including even treacherously slippery stairs covered with snow. If it were not for the unusually high lifting of his feet, nothing in him would have betrayed an insufficiently skilled walker [plate 76(3)].

Roody (1.4.9) used the same technique when getting onto a low armchair: he lifted one knee to the level of the seat, helping himself with his other leg [plate 78(1)]. Later, at the ages of 23 months and 2 years 4 months [plates 78(3), 78(5)], Roody already had mounted a relatively high place: he put one of his feet on it and then climbed with considerable effort, holding firmly to surrounding objects.

Coming down from elevated spots or stairs is more complicated for the child than for the chimpanzee. The human child can sometimes come up the stairs independently (1.10.19), but he needs help when he comes down.[7]

As stated, the chimpanzee comes down the stairs or inclined roofs the same way he goes up, that is, on all fours, face first; only when descending vertical poles does he come down backward [plate 15(2)].

My child at the age of 16 months to 2 years usually came down from an armchair [plate 78(2)] or from the sledge [plate 78(4)] face forward, holding firmly to something and carefully feeling with his feet for solid substrate. He finally came down only after he had reached the support.

One of my colleagues[8] told me that baby baboons, after climbing on the cage face forward, also come down face forward, while old baboons in the same circumstances descend backward. I myself observed the latter technique in 4- or 5-month-old baboons. It was fun to watch a little baboon coming face forward to the upper landing of the stairs and turning a few times to find a proper position for coming down, (i.e., back first).

I observed my Roody descending backward this way at the age of 2 years 4 months, when he once had to come down from a small toy car [plate 78(6)].

Obviously, in climbing the stairs on all fours or on feet with the help of hands, the face-forward pattern of descending is genetically earlier than the back-forward pattern. In the ontogenesis of the infants of lower and higher monkeys and the human child, the first pattern is earlier than the second. But, as we know, the human child later acquires a new pattern of descending the steps, unattainable for monkeys: descending in a vertical position.

However, at the age of 1½ to 2 years, in coming down the stairs, a child tries to hold the hand of an adult [plate 77(3)]. First, he puts one foot and then the other on each step. Later, when he comes down independently, he has to bend his body somewhat and balance with his arms, but when he is 2½ years and older, he can come down a few steps easily with his arms in any position [plates 79(3), 79(4)].

Although this vertical pattern of descending the stairs can be compared, by speed, with the pattern of a chimpanzee coming down the stairs and being led by man, by agility and by the height of climbing, the infant chimpanzee far exceeds the human child.

While the chimpanzee can climb easily up the screen of his cage and as easily up the smooth trunk of a tree [plate 80(3)], up a pole, and up a steep roof [plate 15(1)] and can freely walk along its ridge, my boy, even at the age of 4½ years, was able to climb a smooth tree only with significant effort (0.5 meter).

While the chimpanzee could hang firmly and confidently on such a tree [plate 80(3)], embracing it with his arms and legs and ready to play with his free hand, my boy obviously did not feel confident on a tree. He expected every minute to fall down. His mouth usually was warped with a grimace that reflected strong muscular tension; his lower jaw was visibly pulled down, and the lower lip was pulled away from the gums, baring his teeth. He was afraid to move lest he fall. Climbing a tree, he usually bit his lower lip and teeth with his upper lip and teeth, which also reflected great muscular tension.

Roody liked climbing: He had an opportunity to climb the stairs and trapezes in the garden; he willingly mounted low fences [plates 15(3), 15(4), 15(7)] and climbed the chairs [plates 15(5), 15(6)]. Nevertheless, climbing to considerable heights was next to impossible for him. Certainly, he was not able to climb on high roofs the way the infant chimpanzee did, easily and with perfection.

Chapter 12

Human Emotions and
Emotions of the Chimpanzee

Comparing the face of a chimpanzee and that of a human child in motion, we find in the child all eight main facial expressions of the chimpanzee: laughter, crying, fear, anger, surprise, attention, revulsion, and even the unusual expression of general excitability.

General Excitability   Once, when my boy was 11 months old, I brought a bowl of food to him but changed my mind at the last moment and took the bowl away; he immediately bulged his lips, apparently experiencing an unpleasant feeling.[1] Another time, Roody (at the age of 3 years 5 months) suddenly noticed something unusual on his hand (an ink stain). He extended his lips forward, making a short funnel at the end [plate 81(2)], and stared at this spot; the surface of this funnel was absolutely smooth and it did not show the creases and furrows characteristic of the chimpanzee in similar circumstances [plate 81(1)]. The child's hair did not rise as in the case of a worried chimpanzee.

The disgruntled mood of a child, as a reaction to a sudden upsetting circumstance that made him cry, usually manifests itself by extension of the closed bulging lips forward, lowering of the corners of the mouth, and a raising of the inner ends of the eyebrows. I could observe and photograph such an expression in my 2-week-old baby and later (4 years old) when unpleasantly surprised Roody was rejecting porridge without sugar (usually sugar was added to the porridge) [plate 82(1)]. This expression was strikingly different from the one that the boy had only 2 or 3 minutes earlier when he was eating a sweet porridge. In the first case, his eyes and the inner corners of his lips were down, the eyebrows were pushed together slightly and in a frown, and his lips were closed and bulging forward [plate 82(1)]. In the second case, his head, eyes, and lips were raised; his entire face wore a broad smile and expressed complete satisfaction [plate 82(2)].

Much later, I observed the same expression in Roody (7½ years old) when I was making an unwelcome suggestion to him to read an extra passage in a text-book. Sometimes, such expressions can be seen in adults. I saw an elderly energetic man who used to toss his head back and extend his lips, which were folded in the form of a tube with a short funnel at the end, every time he heard unpleasant news. These expressions were very reminiscent of those of an unpleasantly excited [plate 24(1)] or worried chimpanzee that was ready to cry [plates 27(4)–27(6)].

As noted, the emotion of general excitability in the chimpanzee is accompanied by a peculiar hooting or grunting sound (based on deep inhalation and exhalation), including the vowels "u," "a," and "e," and the consonant sound "kh." I did not see Roody giving any sound when, confronted with suddenly changing circumstances, he extended his lips forward. But, I can state with confidence that, when he was being pleasantly excited at the age of 5 months (for example, at my unexpected early arrival or when I showed him intriguing things, such as bright decorations), he made deep inhalations and exhalations that definitely included the vowel "a" and consonant sound "kh" (something like "kh-a-a-a"). Vatagin the artist told me that his 2-year-old daughter, looking at and touching various sculptures in his studio, inhaled and exhaled deeply all the time. Russian interjections used by adults, such as the ones expressing fear or sudden joy ("akh," "okh," "ukh") also include these vowels and consonants.

I want to stress that, like the hooting sounds of the chimpanzee, which are universal in meaning but similar in form, human common exclamations are also undifferentiated and short. The same interjection is used for expressing various emotions. Poets give us a number of eloquent examples:

> Ah!—she said—I did not recognize you,
> I thought you were someone else, at first.
> (Polonski, *Grasshopper-Musician*)

Here, the interjection "ah" expresses the emotion of unexpectedness.

In other cases, the interjection "ah" expresses pathos or the emotion of delight:

> Ah, my steppe, free steppe!
> How broadly are you spreading out!
> (Koltsov, *Poems*)

Sometimes, the interjection "ah" enhances the emotion of indignation:

> Ah, you are a glutton, you are a villain!
> (Krylov, *Cat and the Cook*)

The interjection "ah" is often used to express sorrow:

> Ah, how can I comprehend my fate?
> On those with souls you prey!
> Molchalins, that's who enjoy this world!
> (Griboyedov, *The Mischief of Being Clever*)

Thus, an adult in the state of anxiety or excitement lacks the distinct facial expressions typical of the chimpanzee or human child; the adult does not emit the expressive sounds of the chimpanzee, but we still can find the remnants of these sounds in the form of short exclamations in very emotional people.

These interjections ("akh," "okh," "ukh," "ekh") are like a mitigated echo of the sounds of anxiety and excitability of the chimpanzee. Apart from Russian, they are used in other languages, with the interjection "ah" being the most universal. With the increase of the cultural level and the person's well-roundedness, the intensity, frequency, and range of their usage decrease. It is well known that the high society standard in England and the United States awards the "title" of true gentleman to those able to withhold involuntary and sincere expressions of such emotions.

On the other hand, it is worthwhile to note that, among various sounds emitted by the chimpanzee, only the sound expressing extreme anxiety is designed melodically. I recall that the gibbon, with his peculiar singing, and other animals capable of emitting various long sounds reproduce them mainly in the state of anxiety. Many people start singing when they are nervous; I knew one such woman. "We save songs for a rainy day," says a famous Russian poem (Polonski, *The Road*). Indeed, an ancient Russian custom of "voicing" at weddings and funerals is well known, as are the Roman "women criers" usually invited to funerals.

Apparently, in all these cases, enhanced anxiety and excitement required the utterance of long crying sounds, which served to discharge this excitement and the emotions associated with it.

Sadness   The chimpanzee's and human's expression of crying are similar in general, but different in detail. Like Joni, my Roody, when he is crying, wrinkles the upper part of his face, lowers the skin of the middle part of the forehead and eyebrows, presses in the base of the soft nose, closes his eyes tightly, and opens his mouth wide, forming a cleavage (either crack shape, trapezium shape, or a wide oval), with the width dependent on the intensity of crying (plates 83–85). But, in humans, when the lips are pulled away from the gums as far as they can be, the gums are not bared, and there are no wrinkles that furrow the sides of the cheeks, as in similar expressions of the chimpanzee [plates 81(3), 81(4)].

On the face of a crying child, we can distinguish two arc-shaped large creases that go from the boundary between the bone and the cartilaginous parts of the nose and around the wings of the nose, and they pass at some distance from the corners of the mouth down to the chin. We can also single out a crease at the base of the nose that joins the tightly closed eyelids on both sides [plates 83(2), 83(6)].

When the child is crying, we can see tension of the muscles at the beginning of the eyebrows [plates 84(1), 84(2), 84(4)], a depression in the upper part of his face that starts from the corner of the eye to the lower eyelid and goes down to the sides of the cheeks; the second depression passes under the lower lip joining near the corners of the mouth with the large nose-and-lip creases [plates 83(2)–83(6)]. While in a crying chimpanzee we observe an area within the lip region that bulges forward and outward, in a human child, we see a large depression in the same region. While the face of a crying child usually reddens, sometimes very intensely, the chimpanzee's complexion only darkens a little.

As noted, the chimpanzee's crying is never followed by tears.[2] A newborn baby also cries without tears. In the case of Roody, from the first month (according to my observation of Roody), when Roody cried intensely, I noticed that the corners of his eyes became wet; at the beginning and in the middle of the second month, the tears flowed out of only one eye. Later, the eyes took turns issuing tears (one eye was wet, while the other was dry). Finally, at the end of the second month, the child began crying with real human tears, that is, the tears flowed from both eyes simultaneously [plate 81(4)].

Everyone knows that a child's tears can be so profuse that they drip from his cheeks in streaks or occur in frequent large drops. It is also known widely that these tears are fleeting: They vanish as easily as they appear. An intensely crying child suddenly may show a gleam of joy; his face lights up, and he smiles while tears still cover his face. These tears are very much like warm rain when the sun is shining! A cloud suddenly appears, the sky becomes dark, the rain starts, but the cloud is already waning, sunshine makes its way to earth, and although it is still raining, the rain is warmed by the light.

The tears and the sound of crying appear and disappear at the same time, and the intensity of crying corresponds to the amount of tears and to the width of the open mouth. But, as the child grows, we notice that he cries less and less, his mouth does not open as wide, and his tears are less profuse.

The external manifestation of crying is also interesting. Roody's opposition to my suggestion for him to read an extra paragraph, cited above, gave me an opportunity to watch his peculiar expressions. After he bulged the lips in discontent, he closed his mouth tightly, turned his lower lip inside out completely, and pressed his upper lip with the lower one.

It seems appropriate to discuss this peculiar shape of the lips in more detail. My 3-month-old baby, before starting to cry, often raised his upper lip and pulled the edge of his lower lip away from his gums [plate 85(1)].

Crying caused by pain also invariably began with turning the lower lip inside out [plate 83(3)]. The same reaction (turning the lower lip inside out and subsequent crying) was caused if I came to the door on my way out from Roody's room.

Darwin described and illustrated this expression in detail;[3] a very interesting book by Professor Dr. H. Krukenberg contains a description of remarkable expressions of depressive excitement of a mentally ill person.[4] Similar, but not so distinct, bulging of the lower lip was also visible in our chimpanzee when he was in a state of mental depression [plates 31(3)–31(6)]. I rarely observed this lower lip gesture in adults, but I vividly remember my friend in school (a girl 15 to 16 years old), who used to turn the lower lip inside out every time she was upset.

In the process of shaping the expressions of fear, the lower lip plays a more important role than the upper one (the initial stages of crying are characterized by jerking of the lower jaw, by uneven position of the lips, and by shaking of the lower lip, which is visible most clearly in babies. When a child is on the verge of crying but does not actually cry (either because the child is not upset enough or holds the tears back intentionally), the lower lip turns inside out and presses itself to the upper lip, which closes the mouth and prevents the lips from extending to the sides, as in crying.

In the ontogenesis of the human, there is a period when the person cries with tears without screaming or controlling the face or uttering any sounds. Finally, at the highest stages of a tragedy and when confronted with severe adversities, the person sometimes is unable to react by crying or with appropriate facial expressions. It is as if the person contains the sorrow within, suffering and crying invisibly with internal tears. Physicians know very well this dangerous form of suffering directed inward, and they usually try to ease it because their experience suggests that screaming and crying can discharge this powerful emotion. Experienced midwives advise women to scream louder during childbirth to alleviate their suffering.

If we analyze the expressive forms of sadness in the philogenetic aspect and supplement our observations with ontogenetic data, we obviously come to the conclusion that screaming is the primary manifestation of suffering because the chimpanzee, as well as a newborn baby, only screams. The secondary stage of the suffering manifestation is screaming with tears, with screaming dominant over tears (as in children between 4 months and their adolescent years). The third stage is tears and screaming, with tears dominant over screaming, the manifestation form typical for the adolescent period. The fourth stage is tears without screaming and is observed in adults until old age. The fifth stage is internal crying without tears.

Of the real human we expect logical completion of the highest forms of his reaction to tragedy—mental liberation from suffering; wise, philosophical, and stoical acceptance of suffering expressed in profound comprehension and in ideological and moral justification of the reasons for suffering. According to this internal purpose of a spiritually developed human, all external forms of manifestation of the suffering emotions, not only screaming and crying, but also invisible internal tears, are bound to vanish.

As stated in chapter 10 regarding sounds, the loudest scream of a crying child is by far less intense than that of a chimpanzee; also, the child's gesticulation that accompanies crying is not so significant. However, in the child, we find gestures not typical of the chimpanzee. For example, in connection with the irritating effect of tears and the child's desire to make his face dry again, we can see him pressing his hands to the face, rubbing it, and smearing the tears all over it [plate 83(4)].

I never observed crying children throw and wring their hands expressively the way the chimpanzee did a number of times (plates 28, 29). But, for example, the chimpanzee's habit in his fits of despair to place his hands on his head and to batter his own body with his fists [plates 30(1)–30(3)] is typical of extremely ill-behaved children. I even heard of a little girl who threw herself face down on the floor, crying, kicking, and shoving, when her desires were not satisfied imme-

diately. Even when a little baby cries violently, he may also make kicking movements with his legs, pressing them to and pushing them from his body. I also
observed my chimpanzee under similar circumstances throwing himself on the
floor.

M. A. Velichkovski, a very experienced observer working at the Moscow Zoo,
told me about a young female chimpanzee; the chimpanzee threw herself on the
floor when she was gravely upset and fiercely pounded around her with her hands
and feet. Once I saw a girl 4 to 5 years old beating herself on the head when she
was left with somebody else by her mother, who departed on a trolley. One
reliable witness related to me a tragic accident: A man run over by a streetcar
was fitfully clasping his head between his hands.

In this connection, I cannot help but mention that I have to witness invisibly,
silently, and on a daily basis the expressive manifestations of human tragedies.
The window of my study faces a morgue; countless times, I have to interrupt
my writing and look out the window when I hear the heart-breaking cries of
those escorting the coffins (especially with people killed accidentally or tragically).
I can see people mourning, most often women (mothers, sisters, wives); they
bend almost to the ground, under the wheels of the funeral carriage, and make
fitful gripping movements with their hands in the gestures of hopeless despair.

Humans, who used to consider the hands the best support in overcoming
various difficulties encountered in life, in moments of inconsolable grief often
resort to aimless, but expressive, hand gesticulation, as if discharging a grave
mental condition and alleviating suffering. The gesticulation of a desperate chimpanzee bears a striking similarity to that of a human, but our chimpanzee was
only an infant. In the child and chimpanzee, naive, unrestrained, ingenuous behavior free from conventionalities, the expressiveness of the bodily manifestations
in gestures and sounds corresponds with the strength of emotions. We can draw
the conclusion, based on the chimpanzee's contorted face, loudness of his sounds,
and expressiveness of his gestures, that the degree of his desperation has been
much stronger than that of the child.

The conditions for which the sadness emotions appear are generally the same
for the infant chimpanzee as for the human child. Any delayed or incomplete
satisfaction of their physiological and mental needs easily and quickly brings them
to sadness.

Only one characteristic difference must be underscored: While the human
child often cries from physical pain after falling down or hurting himself, I never
have observed the same in infant chimpanzee. Professor Yerkes, in his book *The
Great Apes*,[5] writes about an infant chimpanzee that did not cry after he had
fallen out of a tall tree and continued his usual activities in a few minutes.

On the other hand, expression of dissatisfaction toward Joni in the form of a
verbal reprimand or a waving aside gesture led inevitably and immediately to his
infinite grief, which was accompanied by crying. The same reprimand never is
perceived by a human child in such a touching and caring manner.

I observed many times that Roody (at the age of 8–9 months), when he inadvertently hurt himself or fell to the floor, did not start to cry right after the
accident, but after a few subsequent seconds of silence. There was obviously a

period of time needed for the pain to reach the consciousness; although this period between the perception of a shock and the reaction to it is very short, it definitely exists.

It is also meaningful that the child's crying (as well as the crying of the chimpanzee) can be stopped by immediate shifting of his attention to something else. As soon as you show even a baby 8 to 9 months old any new, intriguing object, he becomes silent instantly. This psychological peculiarity is used by babysitters to dissuade a child from crying (they often use so-called pain-killing rhymes, which sometimes may prove very beneficial). In the old times in Russian families, one such line was very popular. A nanny would say, blowing on a sore spot, "A little cat has his pain, a little mouse has his pain, a little dog has his pain, let little Petya's [the child's name] pain heal and be covered with fat." The verbal representation of animal images distracted the child from the pain, and he stopped crying. In our routine, bringing little Roody (at the age of 8 months to 1 year) to the adjacent premises of the Darwin Museum invariably stopped even very strong and inconsolable crying.

We have noted the anger and crying of the chimpanzee caused by his unsatisfied curiosity and by his unsuccessful imitation of a human activity (for example, when he was trying to open the padlock on the door to his cage). In the human child, we find different kinds of mental grief absolutely untypical of his shaggy counterpart. These are not only tears from physical pain, but also tears of compassion for people close to him, the tears of sympathy for "the powerless of this world," the tears of wounded pride and of frustrated ideas, all of which have nothing to do with the practical, material side of life.

When Roody was 9 months old, his grandmother was telling him a story about a little girl and began to imitate the thin little voice of the girl's crying. As a result, our baby was on the verge of crying.

When Roody was only 11 months old, I showed those around me a bulging bone at the base of my thumb. Somebody said, "This is a dislocation." They started pulling at my finger to readjust the bone. "Don't do that! I am afraid!" At that moment, the boy was going to cry; he frowned and began groaning, but as soon as I smiled, he calmed down.

Like Joni, in response to my pretended crying, Roody presses his body to me (1.7.8). When his grandmother complains about her pain, Roody comes up to her and embraces her knees (1.8.16).

But, the human child feels deeper sympathy than the chimpanzee does; I have never noticed that Joni would cry in response to my pretended crying, while Roody sometimes would cry bitterly out of compassion not only for me, but also for people who were less close. For instance, Roody cries (1.8.16 and 1.10.20) when he sees a bandage of the eye of his beloved uncle, when he hears people say that his uncle's eye is hurting or that he has abscesses on his fingers; Roody cries when he watches the nanny's grimace when she is taking a bitter medicine.

How touching and affectionate is the child's compassion for his favorite animals when they are in trouble (compassion not only for the real or toy animals, but for those described in books)! Roody feels sorry even for such creatures as insects like flies and bugs.

When Roody was only 2 years old, he heard a story about a big dog biting a small dog. Once, he saw on the street a big dog running up to a small one. His face became red, and he was about to cry, but I calmed him down by driving the big dog away. At another time (he was 3 years old), he saw a dog's leg cut off by a streetcar; he cried so bitterly that it took me a lot of time and effort to quiet him down. When someone read to him about the suffering of an animal, he always made a gloomy face. Once, he was shown a picture of wolves devouring a lamb, and he started to cry. Roody cried when he saw flies glued to the sticky surface of a fly trap and tried persistently to let them free. The feeling of compassion apparently was hard for him to experience; it brought up such unpleasant emotions that sometimes (in the cases when the suffering was not real) he instinctively shrank from them, as though sparing his young heart the pain of grief. For example, he often did not want to hear the bad end of a story. He either asked us to stop reading the story at a certain point, or chose that point in advance: "Mother, tell me about the little fly and the big fly, but don't tell me about the bumble-bee, she is bad, she bites."

Once, I told Roody a story about a little girl who played with matches when she was left alone in the house and got burnt in the fire; the next time, he asked me: "Mother, tell me about the girl, but not about the matches." When he listened to my story about the adventures of a baby wolf that got lost in the woods, in the most pathetic part of the story (the baby wolf became cold, hungry, and afraid without his mother, and he met a bear), Roody invariably whimpered. Before I started ready the story, he would ask me, "Mother, tell me about the little wolf, but don't tell me about that bear." "Why?" I would ask. "I feel sorry for the little wolf," Roody would say (3.0.4).

The child instinctively, by his pure, unblemished heart, feels that evil is unacceptable; he gets upset over any injustice and sympathizes with the offended, oppressed, and suffering creatures. Little Roody (1.8.23) heaved a sad sigh when he was upset, for instance, when he was unable to verbalize his desire or answer a question. Sometimes, when asked about the name of a certain object and not knowing what to say, he would give out a deep sigh in response. I have never noticed in Joni a sigh of sadness, but as its substitute in the case of a small worry, Joni would give a single sigh with the long groaning "u-u."

The child often cries over his failure to complete his own independently designed projects that have nothing to do with any practical goal. For example, at the age of 2 to 3 years, Roody cries bitterly at the sight of a collapsed toy house.

Once, I watched Roody (he was 3 to 4 years old) when he was hammering a nail, trying to attach a flag to a stick. He was unaware of the fact that the sharp tip of the nail had been broken off, and the nail flatly refused to go in despite the child's repeated efforts. With pent up tears in his voice, and continuing to hammer the nail, he kept lamenting, "What is going on? What is going on? It fell again! [about the nail] What is happening? [sounds of crying, a break in hammering] It cannot be so!" he sobbed, resuming the hammering, but with no success. He exclaimed then, with even more grief in his voice, "I cannot stand it! [crying] I simply cannot stand it!" [he sobbed and with new attempts of hammering].

At the age of 7 years, Roody tried to make up a so-called fossil man [plate 84(3)] out of irregular blocks he found. This project was hard to accomplish; it crumbled the moment the blocks were ready to bear any resemblance to man. The entire unsteady structure came apart at the slightest puff of the wind; the boy met this destruction with sorrowful grimaces and tears [plate 84(4)]. The child often cries because his pride has been hurt. Once, Roody (2.7.23) volunteered to recite a story, "Elephant Vambo," but after a few lines, he stumbled, forgot the words, and not knowing what to say next, began crying loudly.

This brief analysis shows that the crying of an infant chimpanzee apparently is associated with his unpleasant physiological and instinctual experiences, while the crying of a child is caused not only by physical reasons, but also by moral reasons.

Joy

Comparing the laughter expressions of both infants, we find hardly noticeable differences against a background of general similarities. During maximum laughter induced by tickling, the child and the chimpanzee open the mouth wide, [plate 81(6)], pulling the lips to the sides; at the same time, the child often bares the teeth of the lower and upper jaws, which we usually do not observe in the chimpanzee [plate 26(4)].

Our photographs [plate 81(5)] show that, in the chimpanzee, even during maximum laughter, only the upper edges of his canines or the upper edges of his lower teeth [plate 26(5)] are visible. Only if the chimpanzee in his vigorous mood has been filmed somehow from below [as in plates 25(3), 25(4)], can we see almost all of his teeth.

Another difference pertains to the eyes: Even a child's smile and, especially, his laughter invariably are accompanied by strong, uniform narrowing of his eyes [plates 26(6), 81(6)]. Sometimes, during maximum laughter, his eyelids close almost completely [plates 26(3), 26(4)]. In the case of a joyous chimpanzee, there are two patterns: He either opens his eyes wide, or slightly narrows the outer corners of the eyes [plates 2(4)–2(7), 81(5)].

In a smiling child, a furrow from the wings of the nose to the chin deepens significantly. The area from this furrow outward (to the cheeks) bulges distinctly; the area from his furrow inward (to the stretched lips) is depressed considerably and the chin protrudes sharply [plates 26(3), 26(4), 26(6), 81(6)].

In a laughing chimpanzee, these proportions are inverse. The lip area bulges significantly. The cheeks are depressed deeply and are covered intensely with round near-the-mouth wrinkles [plates 26(5), 81(5)].

In a laughing child, as well as in the chimpanzee, we often can notice sparkling eyes and two thin wrinkles, goose paws, in the corners of the mouth [plates 2(2), 2(6)].

In a smiling chimpanzee, the tongue usually is pulled inward, to the gullet. In the child, the tongue often is close to the teeth and even sticks out of the mouth [compare plates 2(5), 2(6), 81(5), 81(6) with plate 86(2)].

The chimpanzee's laughter is completely soundless; only in the case of tickling is it accompanied to rapid breathing. It is known widely that the human child of the same age laughs as loudly as an adult. This loud laughter appears in the child at the age of 4 months; before this age, we observe a series of stages that precede real laughter. Yet earlier, in the first weeks of life, there is an entirely soundless stage of laughter. In my notebooks, the slightest nuances of the development of Roody's laughing ability can be traced. I will cite only a few of them that coincide with the main stages of this development.

In a 2-week-old child, after meals, the lips narrow, and their corners are pulled up somewhat; no sound is uttered. In a child 2½ weeks old, in the same circumstances, we observe narrowing of the eyes, a broad smile, and the sound "akh," which is accompanied by closing of his eyes.

If in a good mood, a child at 3, 6, or 8 weeks utters increasingly diverse sounds; guttural and throaty sounds appear, and the sounds "a-a," "u," "gu," "agi," "gli," "reg," "gkh," "gl," "khm," and "gkha"[6] can be heard often. With these sounds, complemented with a broad or narrow smile, the child meets his first joys of life: his mother or objects associated with sucking, shiny or moving objects, a bath, or entertaining sounds.

I noticed in my 4-month-old boy the appearance of real laughter. Interestingly, Roody laughed in imitation of me. He reproduced snappy, coughing, aspirate sounds that ended with a melodious tone. Later (0.4.21), this sound was joined by a shrilling sound accompanied by jerking of the child's arms and legs.

Rapid breathing in Joni often was associated with his joyous mood (it invariably occurred when I was tickling him). I also observed it in my baby after meals when he (at the age of 3½ weeks) was wallowing on his back, waving his arms and legs.

I am not inclined to consider this sound as firmly associated with the emotion of joy, but rather with a tinge of anxiety because Roody (3 months old) sometimes showed the same rapid breathing before crying in response to sharp temperature changes (for example, when I rubbed his face with cool water) or, at an earlier age (1 month), when he tried to catch the feeding nipple with his lips or when he suddenly lost a pacifier.

Compared to Joni, Roody was more sensitive to tickling. Joni was ticklish only when I touched his armpits. Touching Roody's skin under the tip of his ear (0.4.24) or touching his chest or stomach (0.6.5) made him jerk his body and laugh. The sole of his foot and the lower part of his abdomen were especially sensitive; they could not be touched, even during bathing (1.4.27), because in this case the boy was ready to jump out of the bathtub. Both Roody and Joni tended to place their hands at the spot where they were tickled [plates 86(3), 86(4)].

In a 4-year-old child, as well as in the chimpanzee, it is easy to follow all stages of joy expression, from a narrow smile with an almost closed mouth [plate 26(1)], to a broad smile with a half-open mouth [plates 26(2), 26(3), 26(6)], to laughter with a fully open mouth [plates 26(4), 81(6)]. According to my photographs of Roody, the younger the baby, the less "smile-ish" his narrow smile;

while in a 9-month- and 1-year-old child, we observe only pulling of the lips to the sides and uniform shaping of the mouth [plates 85(2), 85(4), 85(6)]. When the child is 4, 5, and 6 years old, we see the corners of his mouth bending upward, which stipulates the semispheric form of his mouth [plates 26(1), 26(2)], as in the chimpanzee when he is in a good mood [plates 2(1)–2(5)].

Later, in mature adults, we sometimes again find this smile pressed to the corners of the mouth during held-back laughter[7] or in unemotional or reserved (as a result of their upbringing) people. But, this phenomenon is secondary, not to say artificial.

The joyous mood of the chimpanzee is accompanied by the expression of laughter; vivacious, irregular hand gestures [plate 86(5)]; foot stamping; ceaseless jumping; running from place to place; dashing around the room; and making noises by every means available (similar behavior is observed in the child).

A 2-week-old child, while in a good mood and lying on his or her back, usually wobbles the arms and legs, smiles broadly, shakes the head, utters unintelligible sounds, and jerks the shoulders. As the child grows, the diversity, strength, and boldness of these joyous movements enhance.

An infant chimpanzee, in a playful mood, makes noises by anything other than his voice, while a human child adds to those the sounds of his voice—shrilling, squealing, and whistling. To strengthen the sound effects, the child also adds the noise of various instruments (for example, drums, cymbals, whistles, trumpets, accordions, etc.). When Roody was 5 or 6 years old, he produced shrilling sounds of tremendous strength at moments of sudden joy.

An 8- or 10-month-old child, feeing happy, often waves his hands, slaps on different objects, or claps his hands in imitation of someone else's gesture [plate 85(4)]. The child willingly uses that gesture for expressing his joy; he claps his hands with his fingers outspread widely. I have recorded and photographed the moment when my 1-year-old baby clapped his hand after he had succeeded in standing up a toy deer; his face lit up with a smile, his eyes half-closed [plate 87(3)]. Haphazard, uncoordinated hand, foot, and body gestures observed in a playing child [plate 86(6)], remain until adolescence. The infant chimpanzee, in a frenzy of wild joy, also tends to slap on different objects with his flat palm; he beats even more persistently when the objects resonate in response. Professor Köhler found hand clapping in mature, joyous, dancing chimpanzees.

Growing children, at the age of 1 year and older, lose this childish manner of slapping various objects or clapping the hands. But, in adults, this gesture reappears as an atavistic habit, and it often dominates all other body movements related to joy.

Applause is the most widespread form of expressing the general approval or pleasant excitement of viewers at various performances. Also, sincere, emotional people often honor certain captivating moments of actors' performances with applause. I once witnessed, at one of the provincial theaters, not only a long (5-minute) ovation after the spectacular finale of a performance, but also rhythmic foot stepping.[8] Ancient oriental, as well as Russian, dances always contain rapid foot stepping and hand clapping.

Thus, human joy and that of a chimpanzee are always expressed in many sounds, but the human uses his voice more widely than the chimpanzee does.

A child's joy, as that of an infant chimpanzee, is brought about by physiological and emotional stimuli; the chimpanzee apparently lacks intellectual stimuli that induce laughter. Very early in human ontogenesis, we observe the appearance of the sense of humor in response to unexpected irrational stimuli, as opposed to habitual and common ones. When he went to bed, my Roody (4 months old) usually heard from me a certain babble that imitated the sounds of various animals; when, contrary to my usual repertoire, I inserted a "pif-puf" (gunfire), the boy started to laugh, although he was half asleep. Roody always laughed loudly when I suddenly appeared after hiding from him, when I opened and closed my eyes, or when I quickly moved my face before him. Once, I noticed that Roody (10 months old), lying on his back, was looking beyond the back of his bed; when he caught somebody within his field of vision, he laughed at the unusual, contorted sight of the familiar face. Another time (1.8.15), Roody started laughing when he saw my face pressed against glass; later (2.9.4), he laughed when he saw a skier fall suddenly.

It was unexpected for me that my 3-year-old child found the comicality in a circumstance of more refined nature. Once, when he was going to sleep and I was sitting next to him, he asked me, "Mother, what's a neighbor?" I said: "I am sitting next to you on the bed, I am your neighbor; your toy rabbit is next to you, he is your neighbor." At these words, the boy started laughing very loudly and then said, through laughter, "Oh, that's funny!" and laughed again. Apparently, he thought that attributing the serious name of "neighbor" to a small toy was funny; there was a conflict between the definition and the defined object.

Thus, a psychological comparison of the three major emotions of the human and of the chimpanzee—emotions of anxiety, joy, and sadness—indicates the following diversions. In the human child, the emotion of anxiety, by its strength, expressiveness, and frequency of manifestation, is much weaker than that of the chimpanzee; an analysis of sad emotions reveals in the human child the kinds of suffering that either are not present or are very weak in the chimpanzee. These are the suffering and crying from physical pain, crying in compassion or empathy, and often crying in response to the inability to complete a creative "project." Analysis of the child's joyous emotions shows that he has a sense of humor undetectable in the chimpanzee.

A sensitive and strong reaction of the human child to physical pain, which reflects his weaker physical endurance compared to that of the chimpanzee, indicates his worse biological fitness to an independent struggle for life.

This idea proves even more convincing in the process of comparative psychological analysis of a series of instinctive actions of the human child and infant chimpanzee, actions associated with the self-supporting instinct: eating, self-servicing, self-care, and self-curing.

Chapter 13

Comparison of Human Instincts with Instincts of Chimpanzee

Self-Supporting   1. Medical Treatment. In contrast to the chim-
Instinct   panzee, during an illness human children are very
suspicious and reluctant to any medical procedure and medicines; we all know
how much effort and patience it takes to care for a child and the tactics and
ingenuity you have to use to make even your own child available to an examination, to make him show his sore spot (a scratch or splinter) or to apply medicine
to it. The chimpanzee Joni "works" on a sore spot himself, picking at it until it
starts bleeding, pulling out splinters, and twitching from pain, he nevertheless
does not stop his manipulations. The human child is afraid even to touch the
spot because it may hurt. Also, he gets panic stricken when he sees blood. Once,
my boy (2.2.21) stained his finger with iodine; he cried with all his strength,
afraid to look at the finger and thinking that it was bleeding. "I am afraid!" he
cried (when he was sick with something far from serious) during his examination
by a physician. If the physician continued examining him, he screamed, "No, no,
no," rejecting this relatively harmless procedure. "I am afraid of powders," he
(2.4.16) used to say in response to the request to take medicine. On the other
hand, Joni took all medicine, including extremely smelly ones, very calmly.

Also, other manipulations with the child, such as putting Q-tips into his nose,
rinsing his eyes, or cleaning his nose or ears always made him scream. Only water
procedures such as washing his face or hands or bathing him in the bathtub
apparently were pleasant for him since he often smiled and never objected to
them; he seemed to enjoy the very sensation of warm water.

A 6-month-old child willingly allowed adults to do various manipulations with
him in the form of exercise. He would lie motionlessly for a long time when his
arms and legs were stretched to the sides or pressed back to his body, when he

was stroked on the back or on the stomach, or when he was rolled from side to side. He would not object to being pulled up by his hands into a sitting or standing position.

Chapter 12 noted how sensitively and compassionately the child reacts to pain experienced by people close to him. He willingly assumes the role of "doctor" treating a "patient"! He (1.0.6) welcomes an opportunity to treat the patients scratch with iodine, to put Q-tips in his nose, and to administer a spoon of medicine to him.

If the child's injuries are not too serious (for example, he breaks his fingernail or scratches his hand or foot), at the age from 1 to 3 years, he is inclined, like chimpanzee, to examine the sore spots with his fingers or toes, with his lips, or with the tip of his tongue [plates 33(1), 34(3), 34(4)]. In contrast to the chimpanzee, he stops these manipulations at the slightest pain; when older, he would rather conceal that he has a splinter than show it and allow us to remove it; certainly, he is not able, like the chimpanzee, to take the splinter out.

2. Self-Care.   The human child (1.11.30) very willingly examines the spaces between his toes [plate 60(2)] and cleans his nose (1.7.1); like Joni (despite persistent admonitions by adults), he puts in his mouth what he has extracted from his nose and is likely to eat it. Roody, like Joni, picks food residues out of his teeth and gnaws fingernails if he is not prevented from doing so.

Like the chimpanzee, a tiny man tries to keep himself clean; even at the age of 11 months, after having his fingers smudged with porridge, he tends to rub the porridge off. At the age of 2 to 3 years, Roody wipes his dirty fingers on his shirt, and wipes his lips with a napkin, bringing the napkin to his mouth after almost each spoonful of food. He also shakes off sand stuck to his clothes. At the age of 10 months, he already avoids stepping into wet spots and raises his legs when confronted with a puddle. A 9-month-old child rubs his fist against his nose, feeling that mucus is about to come out. Later, he (1.3.8), like Joni, wipes his nose with the back of his hand or applies a handkerchief; only much later (at the age of about 2½ years) does he start to blow his nose. It is funny to watch. Before bringing the handkerchief to his nose, Roody spreads it out and shakes it; he puts a lot of effort in blowing—his entire face wrinkles [plates 60(5), 60(6)].

Unfortunately, later, the child loses this habit of keeping clean. After 3 years, Roody neglects it totally: he smears his face and hands unconscionably during play and does not even try to restore his decent appearance. At any rate, he never paid as much attention to his looks as Joni did. Joni could not stand any mess under his nose or a dirty face, while you often had to remind the human child that he should blow his nose, wipe his lips, and so on.

The self-care habits appear in chimpanzee easily and quickly; in the human child, they develop after certain exercise. However, in the absence of a control on the part of adults, the child fails to put them in practice. This is especially visible when you watch his daily pattern of keeping clean.

A 2-year-old child knows relatively well how to clean his teeth with a brush; a 1½-year-old child knows how to comb his hair; and a 3-year-old child can wash his face, soap his hand [plates 60(1), 60(2)], and wipe himself with a towel. But, if you do not watch it done, the child can come back to you with his hands and face still wet. He often forgets to comb his hair or combs it sloppily, only to finish the boring business. Almost every day, the child has to be reminded that he needs to rinse his mouth after a meal or to clean his teeth before going to bed. Everybody who has watched a child during these procedures knows how hastily and ineffectively they are performed, especially by vivacious and mobile children. This cannot be said about the chimpanzee Joni; despite his extreme mobility, all his actions associated with his self-care are performed very seriously, slowly, and in a businesslike manner, although not as perfectly as by the human child.

The same hastiness also pertains to such actions of the human child as eating, drinking, dressing, and other vital procedures.

3. Feeding.   Similar to Joni, a small human child does not take his eyes off his mother when he eats; he follows her every movement and cries every time she leaves without feeding him. Like Joni, Roody (1.7.0) was very suspicious of new meals; he tasted solid food in tiny pieces and tried liquid food by immersing his finger in it and sucking the finger [plate 65(2)]. Certainly, the child never smelled the food before he tasted it, which was so typical of Joni.

Similar to Joni, my Roody ate better when eating was accompanied by entertainment. Even when he was 10 months old, I noticed that, sucking from the bottle, he would grope around, find something, and start shifting it from one hand to the other to enliven the eating process. Later (at the age of 1½ years), Roody's appetite was better if eating was supplemented with conversation, reading, when somebody else ate with him,[1] and when he ate near the window and could look outside. At the age of 2½ years, he demanded conversation or reading during a meal. Obviously, he was not as skillful as the chimpanzee in peeling and unseeding fruit. However, at the age of 2 years, Roody could peel a potato or tangerine, but he did it more slowly and awkwardly than the chimpanzee. We had to prepare a suitable food for him and watch so that he did not swallow the seeds. Nevertheless, we know that, despite multiple warnings and due to the negligence of those feeding them, children often swallow the seeds, and this can be fatal. Pediatricians are especially aware of the fact that a great number of children 2 to 3 years old accidentally swallow not only seeds, but also beads, thimbles, buttons, pieces of eggshell, and other small household objects that find their way into the children's hands.

My Roody was no exception in this respect. Once, at the age of 11 months, he lay on his back, playing with a small comb. Suddenly, I noticed that he was moving his mouth in a convulsive manner. Looking into it, I saw a tooth of the comb on his tongue. I was quick to pull this tooth out of his mouth, but it turned out that two teeth were missing from the comb (this second tooth was later discovered in his excrements). Another time, he managed to swallow unnoticed a glass vial, which fortunately did him no harm.

His little counterpart Joni was much more skillful in this respect. He could hold a large number of nails in his mouth or prop his lips with pins or vials, never swallowing them, hurting himself, or gagging.

Every human child, as well as the chimpanzee, has his favorite food and food he detests and rejects. Eating unpleasant (for instance sour) food, the child makes the "sour" grimace and wrinkles his face. Once, I gave Roody (0.6.22) a spoon of cranberry juice. He immediately closed his eyes, tensed his nose, pulled his lips to the sides, shook his head, and recoiled from the cup.

I observed an even stronger aversion reaction in Roody (7 years old) when he was given rhubarb powder [plate 89(4)]. He wrinkled the upper part of his face; as a result, the outline of his mouth became rectangular. His lips were tightened in the corners, and intense salivation occurred.

The boy had an entirely different expression when he enjoyed tasty, sweet food [plate 82(2)]. In such cases, he usually licked the spoon clean. And what a sight of pleasure his face was! His lips were closed, and they smiled; his eyes were half-closed. Untasty food caused eyebrows in a frown, bulging lips, dropped head, and cast-down eyes [plate 82(1)].

In response to a pleasant gustatory feeling, the infant chimpanzee utters a smacking/grunting sound. The same sound is typical of the human child under similar circumstances.[2] Adults of a sufficiently high cultural level try to eat noiselessly, but people with little education often champ, especially when they eat tasty food.

It is common knowledge that many children, after adults (or adults after children), mark the eating of delicacies and sweets with a smacking sound. It is extremely interesting that this smacking sound "ntsa" is used even by adults as a qualifier for nice things. I observed many times how people (especially from the southern and eastern parts of the country) visiting the "Beauty in Nature" section of our museum, looking at the exotic birds, utter this smacking sound as a sign of delight, their eyes shining with pleasure, their lips smiling, and their faces glowing with joy.

If some untasty food is in a child's mouth, he, like the chimpanzee, spits it out immediately. Once, I was giving Roody (at the age of 6 months) some minced carrot; he swallowed the first spoon at once, but suddenly his entire body started shaking, and his eyes narrowed. Hardly taking anything from the next spoon, he dumped the food on his lips. Before the third spoon was administered, he made the semblance of a vomiting motion. Rejecting unpleasant food, Roody (1.9.18) often spat it out.

The child usually swallows any new food with a grimace; he screams, jerks his face, and shakes his head. The following is a colorful description[3] of a 7-month-old baby after he has been given a piece of watermelon to lick.

> His face is contorted by a convulsion, his nostrils are dilated and whitened, his mouth is curved, his lips are tightened at the corners and droop, his eyebrows are pressed together and raised somewhat, he eyes keep opening and closing; the child shakes his head (from right to left), coughs, spits out saliva, cries and pushes the watermelon away from him.

An initial attempt to feed 9-month-old Roody mashed potatoes led to the following: After swallowing three or four spoonsful with a grimace, he did not want to open his mouth anymore; he turned his head away and threw his body back. Later, mashed potatoes became his favorite food.

The chimpanzee expresses his aversion to food less frequently and distinctly, if for no other reason because he smells it in advance and simply does not take unpleasant food (for example, containing butter) into his mouth; he turns his head away from it. In the case of the chimpanzee, aversion often is associated with the olfactory and tactile sensations. All of us know the disgust that children feel toward skin of boiled milk and the smell of castor oil and cod-liver oil, although aversion induced by the olfactory stimuli is weaker than that caused by gustatory stimuli.

There are foods that both infants preferred. The human child and the chimpanzee shared a passion for fruit, berries, sweets, and even for chalk and coal. My Roody (1.9.4) often pulled coals from the stove and chewed and swallowed them. Like Joni, Roody (1.8.15) craved salt, cranberry jelly and even lemon, although he often ate these things with grimaces and facial wrinkling and with his whole body shaking; nevertheless, he persistently solicited them from me. Perhaps Roody's craving for sweets was even stronger then the chimpanzee's. I often watched Roody take a candy and close his eyes in the anticipation of pleasure, he would grunt and smack his lips when eating the candy or utter the growling sound of a baby bear drinking milk. The human child, like the chimpanzee, cries bitterly when we refuse to give him the sweet things he desires.

For a child, as well as for the chimpanzee, milk is the main food. My Roody, like Joni, got used to warm milk and flatly rejected it when it was cold. Like Joni, Roody (at the age of approximately 1 year) drinks from a cup. At first (1.1.9), somebody else held the cup; in a year (2.1.12), he held the cup himself in both hands to drink [plates 90(1), 90(2)] the way Joni does [plates 90(3), 90(4)]. Later (2.3.21), he held the cup in one hand [plate 88(5)], which Joni is not capable of doing.[4] Like Joni, Roody (2.11.3), instead of taking the cup in his hands, often prefers to suck the milk in, bending over to the cup.

In the human child, as well as the chimpanzee, the habit of using a spoon develops only through long practice. Swallowing from a spoon occurs very early in a child's life (almost from the first months). However, my boy learned how to scoop with a spoon only when he was almost 1½ years old. The use of a spoon, complete with bringing it to the mouth, came only 5 months later (1.10.19), but even then, Roody, like Joni (wherever it was possible[5] and when nobody was watching), was more willing to use his hands instead.

Obviously, Roody at the age of 2 to 4 years [plates 88(1), 88(2)] used a spoon more skillfully then Joni, who often spilled the contents while scooping semiliquid food and bringing it to his mouth. Most often, Joni preferred to do without a spoon; he bent over and touched the flat bowl with his lips.

By the age of 3 years, the human child learns such things as how to drink from a saucer while holding it in both hands, how to pour liquid from a cup to a saucer, and how to use a fork[6] [plates 88(3), 88(4)], or a knife [plate 88(6)]. Very early in his life (2.5.0), Roody tended to reject any help in acquiring these skills. "Myself, myself," "Don't hold," or "I got it, Dad," he used to say when trying to do something on his own. In order not to use help, the child tries to eat in the primeval manner (with his hands) or to learn how to do it himself. In contrast, the chimpanzee always prefers the primeval manner of eating, does not want to learn otherwise, and does not object to human help.[7]

The egoistic feelings of Roody and Joni manifest themselves most vividly in eating. For example, when Roody (1.3.28) eats a cookie and someone asks him to share it, he turns his back to the person as though hiding the food from the person, goes somewhere else (1.4.9) to finish it, or hides the cookie between his knees (1.4.1). If the person is persistent, the child gives him such tiny pieces that they hardly can be picked up. Sometimes, the child collects crumbs and gives them (1.4.7) in response to a request. If the child has two pieces, he gives one of them more willingly; he would not give up his last piece for the world.

Once, Roody (1.4.0) gave a cookie to his friend, but in response to his second request, he took a stone and gave it to the boy, thereby confirming human egocentrism ("stone instead of bread"). If the child already has eaten enough cookies himself, he may share them with others. The chimpanzee, as we stated, is totally unwilling to share his food, particularly his favorite food.

The child (2.3.10) always leaves bigger pieces for himself, when he has pieces of different sizes, and gives somebody else the smaller ones; when the child takes something, he tries to take bigger chunks. Once, Roody (2.7.23) asked us for sugar. We told him, "Roody, ask your dad for a small piece of sugar." Roody said, "Daddy, give me a big piece of sugar." We repeated what we had told him before, thinking that he did not hear it right or did not remember, but he insisted on his version for the second and third time. Every time he was given things of different quality and size, he set aside the best and the biggest for himself and then was willing to share the rest with others.

The adage "Die Liebe geht durch den Magen" ["Love goes through the stomach."] is never as true as when it is applied to a child. Once, I heard from Roody (2.3.12) the following sincere words: "I love daddy, I love ham, daddy bought." Here is another example of his egoistic love: once Roody (2.9.28) said to me

"I love mother, I don't love nanny."

"Why do you love mother?" I asked.

"Because she gives me chocolate," he said.

"Do you love daddy too?"

"Yes, because he buys chocolate."

"And why do you love Gaga?"[8]

"Because she makes soup and potatoes."

"And why don't you love the nanny?"

"Because she doesn't give me anything," he said.

Like the chimpanzee, the human child enjoyed eating the food he had found himself (for example, berries); he liked to look for berries from the age of 15 months. But, in picking up berries, Roody did not show the same adaptiveness as was typical of Joni. For example, he was ready to swallow even poisonous berries and was not inclined to taste them at first and discard the inedible ones the way Joni did it. Roody (1.0.8) even put a small bug he caught into his mouth; Joni never did this.

When the human child does not want to eat, you only have to say, "I am giving your food to someone else," and he immediately starts eating. This is reminiscent of dogs, which often sit idly with a piece of food before them; they immediately devour the piece when they see another dog coming and there is danger of losing it. One credible observer told me the following story about a dog living in his yard. During a daily cleanup in the yard, the dog always finished off the food that was in her bowl, but until that moment remained untouched, because she knew from experience that the food would be gone after the cleanup. So she ate it, although it was obvious that she was not hungry.

Egocentrism, greed of the child, clearly is visible in his actions in acquiring food. Roody (1.4.9), for example, used to take more food than he could possibly eat. If he (1.4.7) saw a candy in someone else's hands, he was quick to finish off his own and get another piece.

The child sometimes does not want to eat, but is not willing to give up especially tasty pieces. Also, everybody knows how skimpy children are with respect to their belongings and toys and how reluctant they are to share them with their companions.

Ownership Instinct   Like Joni, my Roody showed his sense of ownership very early. This sense manifested itself in his vigorous guarding of his things, in collecting new things (acquisition of property), and in attempts to take things from others and make them his own.

Every time before Roody leaves kindergarten, he thoroughly collects his things, reminding me to take this or that toy. When he leaves his toys in our yard, he (3.1.5) strongly urges his grandmother, who remains in the yard, "Grandmother, guard my reins, my stick, and my knife, and my vial, guard my bowl and my board; and if somebody comes [to take them], spank him."

When the child (1.2.20–1.5.29) sees a toy he likes in somebody else's hands, he comes up to its owner and takes the toy. Later (2.1.23), he persistently asks for the toy he likes, although he already has the same one. He starts to collect the forbidden things surreptitiously and sometimes resorts to a lie (if the "legal" ways do not work). For example, he knows that I do not allow him to collect newspaper scraps near the garbage cans, but he does it secretly all the time anyway.

He tries to save his belongings (such as his cap or a light toy) from being blown away by the wind, and he cries out of fear of losing them. He (2.9.2) objects to taking his cups and bowls out of his room, reminding us, "It's mine." Once, when he saw me holding his father's hand, he asked me, "Why are you holding father's hand?" "So what?" I retorted. "You can't. It's my father," he said, expanding his sense of ownership over the people closest to him, whom he considered his "own" in the literal sense of word.

Roody (1.1.0), like Joni, collects everything in our yard he can lay his hands on: small stones, vials, chips of wood, matches, sticks, nails, and other waste. He (1.4.25) puts all his findings into his pocket and urges other people around him to do the same. He accurately remembers what he has collected. When he (1.2.8) comes home and dumps it all onto the floor, he gets very upset if something is missing. "Where is the stick? Where is the nail?" he insists and demands we start searching for the lost things. Later, this passion for collecting things grows, the number of collected things increases, the area of his search broadens, and the objects start to be differentiated by their quality.

My 4-year-old Roody, despite the abundance of toys we bought for him, could not pass a rusty, dirty piece of steel, nail, or fragment of a pipe without asking my permission to take it home with him. We had accumulated a huge pile of such waste, and Roody often did not use these things that much, but his passion for collecting things remained insatiable.

Once, we passed a pile of waste consisting of old utensils, rusty wire, nuts, and other metal parts. The boy said fervently, "I wish I could have it all." "You cannot, it has been collected for recycling," I said. He added sadly, "Lucky they are!" obviously envying the future owners of this junk. There are many examples of children from relatively well-to-do families who literally have inundated their rooms with collected junk in the form of stones, sticks, and the like.

When we went for a walk in the forest with 5-year-old Roody, his main desire was to collect. It did not matter to him what to collect and how. He was ready to collect virtually everything: mushrooms, pinecones, plants, berries, leaves, stones, twigs, and other innumerable objects of nature. All this was carried home, put in a corner, and often not touched again.

At a very early age, the child collects everything he comes across; later (after 1½ years), he prefers the things he is attracted to. We discuss the child's preferences in detail in the following chapters.

The child keeps all the collected objects in his room, in his coffer, or in his drawer. He carries most favorite, and portable, things in his pocket, holds them in his hands, and cries if he happens to lose them. Roody takes his favorite toy, a small rabbit, to bed with him and puts it under the pillow. In this respect, he resembles Joni, who also used to take his favorite toy (a small ball) to bed with him. Joni brings his "nest," color plates, and bright scraps of cloth he has stolen to his cage or to his room.

The human child wants to have his favorites with him all the time. Every time I am going to leave Roody (from 1.8.9 to 2.5.26), he cries. The child catches me, embraces my knees, pulls at my dress, tightly closes the door, tries to stop me, and like Joni, cries when I come near the door on my way out.

Roody's desire to have his favorite people around all the time can be illustrated properly by the following funny episode. Once, when Roody was 20 months old, he sat in the living room with four people he knew (including me) and one stranger. He came over to one of the "familiar" people, pulled at his dress, said "Come, come!" and pushed him toward the open door of his room. Then, he came back for the next "familiar" person and then the next. We obeyed, anxious to see what would come next. When the stranger was the only person left in the living room, Roody tightly closed the door of his room and, as if satisfied, started to play.

The more the child likes someone taking care of him, the stronger his desire to have this person around and the more he is afraid of losing that person. His words confirm this thought. Once, I showed Roody (2.8.6) my picture, hanging over his bed, and asked, "Who is this?" He said, "Mother," and added, "I want you here," pressing both hands to his chest, "I want to sleep with mother." Later (3.1.5), he said, "I'll never give you to anybody!"

Family Instinct   The child wants to go to sleep in his mother's presence. When I put Roody to bed, he often says, "I want to be closer to you." He takes my hand and holds it so that I do not go away. At the age of 3 years, he says, "Every day with my mother—at night, in the evening, and during the day." He even would like to sleep next to me; he gets upset when I do not go along with his request. Like Joni, Roody liked to fall asleep on my lap, cuddling up to me.

The child early (2.12.2) revealed the fear of losing his beloved person. "Mother, you will never die, will you?" he asks and, not waiting for my answer, "I don't want my mother to die!" Even at an older age (before 8 years), he is reluctant to let me go out in the evenings, not because he wants me to put him to bed, but because the presence of adults in the house at night gives the child a sense of peacefulness and well-being. He expresses this by the following announcement: "It's good when everybody's at home, when you put me to bed because I worry about you."

From his early age, the child feels his mother's care and protection against all the unpleasant, scary, and irritating things. It is quite natural that he, similar to the chimpanzee, wants to be under this protection when he is most helpless (i.e., during sleep). (Development of the family instinct is discussed below in more detail.)

The human child, like the infant chimpanzee, rarely is quiet during sleep. The baby usually moves his fingers, smiles, wrinkles his lips, or frowns with his eyebrows. I observed sobbing in a 6-month-old baby during his sleep; in a 9-month-old child during deep sleep, I observed moaning, whimpering, sobbing, and even babbling.

Roody (1.3.26) jerked during his sleep (Joni also does that often), cried (1.4.21), even uttered certain words (Daddy, Mother) (1.3.28). Later (1.6.30), I noticed that he made threatening gestures and said, "Give me, give me" (1.10.20). Once, he woke up from a scary dream and said, "The wolf woke me up; he ran over the bed" (2.9.3). Another time, Roody woke up crying, "The goat butted me with his horns"; the child could not quiet down for a long time. I never noticed that Joni emitted any sounds during his sleep.

A sleeping child, like the chimpanzee, often snores if his nasal passages are not in order.

Obviously, the human child protected by adults always has a bed ready for sleeping; like the chimpanzee Joni, he does not have to think about finding it, but his tendency to protect his sleep and to prepare his bed also pertains to his toys, which he animates. For instance, 3-year-old Roody, before going to bed, makes a place for his teddy bear and his doll to sleep; he carefully puts soft rags on the floor and covers his toys with blankets, revealing his budding family instinct.

I remember that my 3- or 4-year-old boy particularly enjoyed playing "nests." We made a semblance of a dungeon out of the blanket; we crawled in with our animal toys, and the boy assumed the role of the one who protected, defended, entertained, and supplied us with food. He did it with vivid enthusiasm, reproducing the function of a primeval man, who was the hunter and the defender of his home.

We already stated that the chimpanzee vehemently objects to being covered at night. In contrast, the child likes it when his mother covers and wraps him up with a blanket. "Oh, how good it is in bed," Roody would often say.

Such a difference in the behavior of both infants reflects the fact that the chimpanzee is afraid to sleep with his hands constrained because he wants to retain his strong ability to defend himself, while the human child ingeneously allows others to confine him because he apparently does not intend to defend himself on his own and relies in this respect on the people who surround him.

During one period in my boy's life (when he was about 5 years old), I noticed that, apart from his favorite toys, which he put under his pillow for the night, he began placing around him the spirited and warlike (in his view) animal toys (such as his teddy bear or cardboard soldiers on horses). He said, "They will guard me!" eloquently admitting his fear and lack of confidence in his ability to defend himself and his desire to pass the responsibility of defense along to the others.

Freedom Instinct (Love of Freedom)

The human child is as reluctant as the chimpanzee to put on his clothes. He is always uncooperative and unwilling to learn how to do this. Even a 6-month-old baby always cries when being dressed and undressed. Putting on even things as simple as a bib or an apron makes him whimper, scream, and cry; he stops crying right after you end the procedure. Putting a kerchief or a hat on the child or wrapping him up with a blanket are even more unpleasant for him. The child (7–8 months old) cries incessantly during these preparations for going outside. His face reddens; he cries to exhaustion, kicks, and tries to free his hidden hands (10 months old). A baby less than 1 year old even is cramped by such things as thin stockings or knitted hats; the child persistently tries to get rid of them.

Later (at the age of 1 to 1½ years), the child is unwilling to be dressed not so much because his movements are constrained, but because the whole procedure is boring; we can hear him crying, especially in wintertime when he has to wear plenty of clothes. For example, when we put gloves on Roody (1.10.6), he cries, "I am hurting!" and demands, "No gloves, no hat!" not because he really is hurting, but rather because he is less comfortable in clothes than without them. It is not surprising that, when this dressing procedure is accompanied by vivid conversations or casual entertainment (1.3.28) or his toys are involved in the process, the child calms down and submits his body for the manipulations needed.

Joni's tendency for self-dressing was developed weakly and manifested itself only in his throwing a rag on his head, neck, or back and covering himself with a blanket at night. Roody showed this tendency already at the age of 14 months. At that age, the child tried to put on his clothes himself and to help adults dress him; because of this, the procedure became less boring for him.

Later (at 5 or 6 years of age), he will learn these manipulations fairly well, begin to perceive the dressing process as a burden, and will haggle over what he is about to wear, trying to reduce the amount of clothes to a minimum. He will treat this as an indispensable, but unpleasant and boring duty.

In a baby 1 to 1½ years old, the attempts to self-dress are far from perfect and take much effort [plates 60(3), 60(4)]. For instance, when a child tries to put on a shoe, he touches it with his foot and stops in bewilderment, not knowing what to do next. He learns how to put on shoes properly only by the age of 3 years (my Roody was able to do that at 2.11.1) and only after long training. It is natural that he learns how to take off clothes more easily than how to put them on. For example, Roody (1.3.28) takes his hat off very easily, but putting it back on presents a big problem. As in any movement requiring coordination, Roody usually holds his lower lip with his tongue [plate 60(3)]. Even later (3.4.0), we can see how difficult it is for him to put on headgear [plate 44(4)].

The child likes being successful and gets upset over his failures. Roody shed a flood of tears before he learned how to take off his shirt (1.7.16), how to put on his stockings and pants (2.9.9), how to button and unbutton his coat or his shoes, or how to take off his galoshes. The reasons for this dissatisfaction are only too clear. The human child, no less than the chimpanzee, wants to have freedom for his activities and movements, and everything that obstructs this freedom irritates and upsets him.

The human child is overwhelmed by the joyous dynamic sense of freedom no less than the infant chimpanzee. Like the chimpanzee, the human child gravitates to unhampered physical and mental activity. Therefore, he persistently rips off not only his constraining clothes, but also fights the imprisonment in his room, his house, and his yard; he breaks the shackles of various bans imposed by adults and throws off the yoke of his responsibilities.

The child is freedom loving by nature. His proverbial disobedience, whims, and willfulness are virtually his robust reactions to the limitations of the freedom of his behavior. The initial methods of his upbringing are predominantly aimed at suppressing, limiting, and rendering a proper direction to the freedom-loving tendencies of the child. These tendencies reveal the true nature of a small "savage" who is violently protesting against acquiring the complicated, conditional, sometimes artificial, and sometimes blatantly unnatural habits needed to create a so-called civilized man.

Like the chimpanzee, the child looks for his own physical and mental space. He cannot yet talk or walk, but he already shows with his finger [plate 48(3)] that he wants to go outside his room (1.2.0). Once he has learned how to walk (1.4.0), he is anxious to leave the limits of his home, yard, or garden and is ready to walk tirelessly. The broader the arena of his movement, the more enthusiastic his run. Allowed to move freely outside (2.1.27), he enjoys uninhibited running and screams, "To run around, around!"

Those who, like me, have been to the steppe know that, coming to these open spaces where you can see only the horizon at an unlimited distance and there is only an unfathomable sky above you and a blustery wind near you, even a grown man is captured by the unrestrained desire to dash toward this boundlessness, not knowing where or why. For the same reasons, humans are attracted to mountain heights in the same way he is drawn to an abyss, powerfully, sometimes irresistibly. Climbers know the fascination associated with conquering mountain peaks, which can often end fatally. Everyone knows about the fear of heights and at the same time about the enticing power of an abyss.

A sensitive person, like a child, feels his association with Mother Nature. But, being detached from it and then suddenly confronted with its magnificence, he cannot resist its power and succumbs to it vigorously and passionately, perishing in its lap, thereby merging with nature.

In the chimpanzee Joni, his desire to come out of his cage, his room, his home, or the yard was very strong. It is interesting that, when set free, he liked most of all to go to high places, such as fences or roofs [plate 15(1)], rather than run far away, which illustrates his passion for climbing. The human child, left to his own devices, does not have this pronounced tendency for climbing; he satisfies himself at the ground level or climbs chairs, stairways, trapezes, or fences [plates 15(3)–15(7)]. Joni can sit or walk on the roofs alone for hours and come back only in response to our persistent calls. The human child at the age of 3 years is unwilling to part with adults and is not inclined to these lonely walks over the roofs.

| Self-Preservation Instinct | Let us conduct a psychological comparison of the emotions of fear in the infant chimpanzee and the human child. |

1. Fear.   At the initial stages of fear, which can be called *timidity*, the human child, as well as the infant chimpanzee, bends his head somewhat and looks up from under his forehead, his eyes open wide and fixed on the frightening object. He slightly raises his eyebrows, wrinkles the middle part of his forehead, and presses his lips tightly and extends them forward. A quite similar expression can be found in a timid chimpanzee [plate 19(1)], but in addition, his hair rises and his side-whiskers fluff [plates 19(2), 20(1), 20(3)].

As is well known, the rising of hair on the head of a human can be observed only in extreme circumstances: either in mental illness or in cases when fear assumes the form of an affect, when hair "stands on end." The initial stages of a child's fear are never accompanied by rising of the hair.

The human child expresses sudden fright by tossing his head back, by maximum widening of his eyes, by decisive tightening of his lips, and by pressing his fists to his chest. It appears as though he is ready to defend himself against any harmful effects [plate 89(2)].

As we have seen, strong fear, awe, in the chimpanzee is also expressed by maximum widening of his eyes and by the tense position of his lips, but while the child's mouth is closed tightly in such cases, the chimpanzee's mouth is open wide, his lips are pulled away from his gums, his teeth are bared, and his hair is pressed tightly to his body and does not bristle. The chimpanzee's body is bent; he rests on his hands and is ready to take off [plates 19(1), 89(1)].

Widening of the eyes is symptomatic. "Fear takes molehills for mountains," says an old proverb. Indeed, these widened eyes try to catch the moment of imminent danger, to preconceive the forms of possible defense. It is not by mere chance that, in the pictures of both infants, we see that their eyes are so dilated that they seem to be ready to pop out of their orbits, as in patients with thyroid disease [plates 89(1), 89(2)].

While a frightened chimpanzee, influenced by the self-preservation instinct, stays motionless [plate 19(1)] and is prepared to defend himself with his teeth, the human child clenches his fists and presses his hands to his chest. A photograph [plate 89(2)] was taken when my child (3 or 4 months old), sitting on my lap, was shown a big mirror reflecting bright light; he immediately assumed the pose described above.

At the age of 9½ months, my baby was extremely afraid of a piece of transparent dark brown curtain lace. Once, when we were outside, I showed him a piece of the lace, which was fluttering in gusts of wind. The baby made a wry face (raised the inner corners of his eyebrows and sharply dropped the corners of his mouth), threw his body back somewhat, and pressed his right hand to his chest in a gesture of defense [plates 54(1), 54(2)].

I observed the same defense gesture in Roody when he (at the age of 2½ years) was shown, for the first time, a live frog moving in water. He recoiled, crying, firmly held on to his clothes, and pressed his hands to his body. Later, at the age of 3 years and 3 months, Roody was shown a sparrow that had been shot, but was still moving. He seized my arm with his left hand, holding his right hand at some distance from his chest [plate 54(5)].

At the zoo, Roody (5 years old) once was led very close to an elephant. As soon as the elephant extended his trunk forward, the boy pressed his crossed arms to his chest, abruptly retreated toward the people he was with, and was ready to flee.

Every time Roody (at the age of 5½ years) fired his toy gun, he startled at the sound of the shot, but kept shooting. He held the gun in his right hand; his left hand was pressed to his body; his fingers were clenched in a fist; and his head was cocked to the side, as though in an attempt to defend himself against the dangerous consequences of the shot [plate 54(4)].

A frightened chimpanzee also has the tendency to make defensive gestures. More often, he covered his face with his hand [as in plate 54(3) and the photographs taken during the bird shooting], while the human child makes such gestures only in exceptionally rare cases [plates 22(1), 22(4)]. In contrast, a frightened chimpanzee reproduces this defensive gesture (pressing his hand to his chest) as an exception; it is typical of him that he presses his foot, instead of his hand, to his chest [plates 22(4), 54(3)]. And this diverging analogy is clear.

The infant chimpanzee, afraid of something, does not stay passive and idle. He is ready to stand his ground and does it despite his fear, defending first of all his most valuable body part—his head.

The human child, affected by fear, especially when he is among the people closest to him, seems to pass the task of his defense on to them. His fear represents a pure emotion, one of fear with no anger added to it. This fear apparently causes an unpleasant feeling in his heart, in his chest, and the child instinctively presses his hands to his chest, as though trying to get rid of this feeling.

We all know that sudden fear is felt in the chest as certain pressure and as a pulse that is higher than normal. In the case of frightening news, people faint, press their hands to the heart, and cross their arms on their chests as though trying to alleviate or hold back the heartbeat. In works of art depicting poses of repentance (the emotions always comprising certain elements of fear), crossed arms on the chest are frequent elements. Hence, the infant chimpanzee experiencing fear is more ready for self-defense than the human child.

The child's shyness, stemming from timidity, often is accompanied by looking downcast, turning his head away, or covering his face with his hands. I knew a 14-year-old country boy who, in response to a question about his age, hid his face in his bent arm and would not say anything for some time.

In the human child and in the chimpanzee, we observe such external symptoms associated with fear as shaking of the body, change in the complexion (the chimpanzee's face turns white, the human child's face[9] turns red), and the tendency to shrink, hide, run away, or give the self up to the protection of guardians. As a result of sudden fear, the chimpanzee often squats [plate 21(1)]. Generally, the conditions of fear that appear in the human child and infant chimpanzee are similar.

Here are some observations of Joni and Roody that invite analogy.

Frightening sounds. Abrupt loud noises scare the child as well as the infant chimpanzee. A newborn baby is not afraid of a knocking sound. My 2-week-old Roody did not react to the sound of a hammer while he was asleep. Even later (6 weeks old), he did not pay attention to abrupt sounds; once, he was in an arbor during a severe storm with earsplitting thunderbolts, and he did not even wake up.

Only later (3 months of age), I began to notice that sounds frightened him. Roody startled when a door suddenly slammed (0.3.0), when a heavy object fell on the floor (0.4.0), when I laughed during breast-feeding (0.4.0), when a motorcycle rattled (0.4.0), and when I suddenly screamed (0.6.15). The child started to cry when he heard strong laughter, noise, or loud talk (0.9.13). Once, I showed Roody a small rubber "devil." While the devil was silent, Roody was quiet, but as soon as he heard a piercing "udi-udi-i-i- i," he started to cry. But, in a couple of days, he no longer feared the devil; he grabbed it and pressed it to his body.

Roody (until 7 years old) always covered his ears every time he saw a train was coming because he was afraid of the locomotive whistle. He was also afraid when he heard the sound of a toy percussion gun, of a Christmas cracker, or of a bursting paper bag. Interestingly, he did not try to flee from the sound; he even asked people to reproduce it, but as if he was instinctively apprehensive of its extreme loudness (2.8.24) or of its harmful effect on his hearing and tried to mitigate it by covering his ears with his hands.

A child I knew (3 or 4 years old) was so afraid of the hooting of a steamboat that, every time he was brought to the boat, he cried loudly and for a long time.

Even a soft noise, when it comes suddenly and seemingly from nowhere, scares children. When it was dark and he heard the sound of a mouse or of something falling on the ground, my boy (3 or 4 years old) asked apprehensively, "What is it?"

Frightening luminous stimuli. The human child, like the infant chimpanzee, shields himself from strong light. As mentioned, my son was frightened by a mirror that was reflecting bright light. Much later (at the age of 3 years), he was afraid of lightning flashing in the dark, and we had to persuade him that it was not dangerous. (Joni also was afraid of lightning.)

Not only bright light, but also darkness and black objects scare children. Once, I bought Roody (2.1.6) two small metal cats, one dark blue and the other gold and glossy. He did not want to take the dark one in his hands and was visibly afraid of it, but he willingly played with the golden one. Another time, I noted that my boy (1.4.7) was afraid of a lady in a big black hat, of a man on the street wearing a black cloak (1.4.26); the child ran to me and clung to my legs. He was afraid of a black folder, blackboard, black coat, or black ball.

When Roody (1.8.0) walked with me up and down dark stairs, he often pressed his hands to his chest, saying, "I am afraid." Looking at dark illustrations in a book, he used to utter a sound of anxiety (1.6.4).

For what other reason than fear did the child (1.10.5) qualify dark, black animated or inanimated objects as "bad"?

My child (from 1.9.19 until 4 years of age) feared crawling into a dark corner, under the bed, or under the table. He cried every time he had to do that to retrieve a strayed toy; he always asked someone for help. We noted above his fear of a dark transparent piece of curtain lace (1.5.19). The child likes light and is afraid of darkness (1.6.27). Roody cries when some extra electric bulbs in the room are turned off, and he asks to turn them back on. When the electricity suddenly went off in the room, Roody started weeping (2.1.14). Once, when I went with Roody for a walk (2.7.11) and we crossed to the sunny side of the street, he said, "It's good here!"

As it mentioned in the first part of this book, Joni also disliked black objects and sometimes was afraid of them.

Frightening tactile stimuli. Sudden and especially painful touching frightens the human child as well as the infant chimpanzee. I noted in my diary instances of strong fright brought about in my 3-month-old baby by tactile and luminous stimuli. The boy was sleeping, and suddenly I abruptly jerked the rim of his diapers; he immediately woke up and opened his eyes wide. At this moment, he looked at the large illuminated surface of a white pillow. His eyes opened even wider and were fixed on one point; he started screaming, extended his hands to the sides, and clenched his fists. The child calmed down only when I began to breast-feed him.

Even in his sleep, a human baby is very sensitive to touch. He immediately startles and wakes up when a fly lands on his face or when you pull at his diapers.

Another time (0.4.4), he looked at a toy when his grandmother pulled at his gown. He sharply turned his head toward her and screamed loudly, but saw that it was his grandmother and calmed down immediately.

Frightening temperature stimuli. The child is afraid of temperature sensations. Even when the child is 2 years old and you start to wipe his face with a piece of cotton wool soaked in water at room temperature, he makes deep choking sighs similar to those made by a human dipping into cold water. He reproduces the same sighs under gusts of cold wind.

Frightening pain stimuli. We all know that children cry more from fear than from pain when they fall down. They are afraid of and try to avoid painful manipulations: filling tooth cavities, taking out splinters, or applying iodine to a scratch. Sometimes, they have not yet experienced pain during this particular procedure, but they anticipate that the procedure might be painful. Almost all children are afraid of doctors and their, often harmless, examinations.

My child cried when he heard somebody mention that the doctor would come. During the examination, he screamed "No, no, no!" and violently resisted listening and palpation by the doctor. This fear sometimes can be observed in children even at an older age.

Fear of new stimuli and new faces. An unfamiliar sensation sometimes frightens children no less than a sensation that is outright unpleasant, but familiar. The younger the child, the less experienced he is in communication with surrounding people. Therefore, he is most afraid of new faces, new circumstances, and new objects.

My boy (1.0.9) only after a lot of effort and 5 days of communication, finally made up his mind to sit on the lap of a new babysitter. Later (at the age of 3 or 4 years), he got used to a new nanny after a few weeks of familiarization. For a long time, he would not remain with the babysitter or nanny in the room when nobody else was around, and he did not want to go for a walk with them (for instance, to a nearby park). He screamed and cried when he was left with them and made up various excuses why someone from our household should remain with him.

When Roody was 9 months old, he tended to stay close to the people he knew when there were strangers around. He hid and turned his face away from them, and if they wanted to approach him (10 months old), he at once started to cry, held on to people familiar to him, did not want to play in the presence of the strangers, and calmed down only when the strangers disappeared from his sight. When Roody (1.7.11) encountered strangers in a public park, he would often press his hands to his chest and step aside, and his face would redden; he would get worried if they tried to talk to him (1.4.27), or he would hide behind his nanny's back. When strangers came up to him or sat down next to him on a bench, he (1.8.2) would not come off my knee, would not budge for about 15 minutes (1.6.8), and would not play (2.3.29) if they did not leave the scene. The child became shy when strangers wanted to talk to him (2.9.10); after they left, he unambiguously revealed the reasons for his behavior, "I'm afraid of them." When Roody is in a living room and hears the bell ringing, which means that some stranger is coming, he (1.11.10) hurries to his room.

As noted, the infant chimpanzee also treats strangers with apprehension and fear, but he gets used to them faster and is more relaxed than the human child. The human child, after repeated familiarization with new people, stops fearing them; he (2.2.10, 2.7.24) does not worry or blush when approached by them, speaks with them, and answers their questions. But, he behaves less arrogantly than Joni and does not invite others to play with him as aggressively.

While my Roody, (1.8.19) like Joni, is afraid of adults, he does not fear children. He comes up to them, calls on boys to join him (2.4.15), and plays with them (2.5.0). He remains shy to a certain degree for a long time (2.10.4), though. You could often observe that, during his first encounter with new people, he stands with his eyes cast down, his face blushing, and his tongue pressed to the inside of his cheek.

In the ontogenesis, the instinct of fear toward strangers varies, competing with the social instinct. We can notice that at first the child is afraid of strangers; then there is a moment (for Roody, this age was 2½ years) when this desire to communicate with people becomes so great that it overcomes his fear. The child sometimes gets so emboldened that he begins to approach people in public places, in streetcars, or on the street, pulling at their clothes or attracting their attention in some other way.

Getting used to the constant presence of familiar people, the child objects to being left alone not only in unfamiliar, but also in familiar, places. As mentioned, the child's intention to keep a familiar person near him when he goes to sleep is suggested by his fear of being alone in the darkness. For example, it took so much effort on my part to leave Roody in a small playground closed on all sides; he would run back home immediately. He did not want (until 5 years of age) at all to remain in the yard alone, especially when there were boys there unfamiliar to him. Later, having found out that the boys were harmless, he began playing with them, but avoided the new faces, and he ran away every time they happened to appear.

It would be fair to say that crying, which darkens the golden days of our childhood so often, is caused by a child's confrontations with stimuli that are frightening because the results are unknown to him.

Fear of new objects. The child is afraid of new things he encounters. My boy (1.8.2) did not want to put on new shoes and a new hat because he was afraid of them. He treated some of his new toys with apprehension. For instance, when I bought him (2.7.29) a big red balloon, he did not want to touch it at first. When he (2.1.6) received from me a toy horse with a bright saddle, he looked at it with suspicion, tried to avoid it, and admitted, "I'm afraid of the saddle." We have already mentioned the child's fear toward new food and toward new medicines: "I'm afraid of the powder."

In general, the child is afraid of any new experiences. When I was with him in a cab (1.4.1), he pressed his body to me. Another time, in similar circumstances, he screamed for several minutes until he got used to the cab.

New situations (for instance, swimming in a large pool) scare Roody as well as Joni. My boy started to experience this during his first attempts to swim. It was very difficult to persuade him to swim in a lake. Although he loved to sit by the water, make dams, launch toy boats, collect stones and shells, sit with his feet in the water, and splash his friends and be splashed in return, he could not make up his mind to go into the water. He began to enjoy swimming only after 18 swimming sessions; from that time, he urged me to go swimming with him and did not want to come out of the water.

The same gradual, sequential familiarization with a series of other stimuli that had frightened Roody at first later became routine and sometimes even attractive.

Fear of moving objects. Everything new, live, and moving, hiding unexplored possibilities, everything that becomes distinguishable against the background of a day-to-day routine attracts a child's attention and, first of all, frightens him. Once, when Roody was 4 months old, his grandmother brought cymbals close to him. The child screamed loudly, recoiled, and extended his hand forward as if defending himself against the toy.

Roody (at an age from 26 to 36 months) was extremely afraid of self-propelled toys, such as grasshoppers with shaking legs, cars, or mice running along ropes). Later, it was self-propelled toys that he especially enjoyed.

Fear of live animals. Roody was even more afraid of live animals with their abrupt, sudden movements.

Before the age of 1½ years, Roody did not show fear toward small insects or slow-moving crustaceans. When he saw a crawling ant, bee, wasp, bug, or crawfish, he examined them, touched them with his fingers, and put them into his mouth.

At the age of almost 26 months, the child is afraid of a tiny bug; he reddens when the bug crawls close to his bare leg. He squeals (2.1.26) when he sees an ant or a hairy caterpillar (2.3.22), a spider (2.5.3), or a fly (2.5.18) running toward him. He flinches from these creatures and is unwilling to take them in his hands. He is even more afraid (2.8.7) of such agile creatures as fish in the aquarium (he does not want "to touch" them even through the glass). But, at the same age, he touches a slow-moving crawfish, tries to tie and lead it on a rope, pours water on it, and obviously has fun.

Often, children, taking after adults, when they see a harmless animal, they are ready to touch it, but only if it not moving too abruptly. My boy is quick to take in his hand a uniformly fluttering fish, but adamantly refuses to touch a leaping frog.

Once, when Roody was 2 years and 6 months, he was shown a frog sitting in a bowl of water. He carefully came up to the bowl, tightly pressed his hands to his body, and with his lips closed, stared at the frog. As soon as the frog started to leap, his face immediately reddened, he retreated, jerked his hands back, and did not want to come any closer, screaming; "It's bad!"; he recoiled when the frog croaked. When he was urged to come closer, he made defensive gestures, tossed back and forth, cried and clung to the adults. When asked why he did not want to come closer to see the frog better, he replied; "Because it was leaping, and I was afraid."

The same fear Roody showed toward live mice, guinea pigs, and small live birds (bullfinches or martlets) [plate 54(5)], which were shown to him at the age of 3–4 years. It is natural that the bigger the animal is, the more frightening it is for the boy. Once, when I was walking with Roody (2½ years of age), we met a flock of geese. The boy was so scared that he did not want to come closer to them; he stood at some distance, cried, screamed, and resisted our efforts to make him approach the birds.

Roody (2½ years old) was very reluctant to touch a live cat [plate 64(3)]. Later, when he was shown tiny kittens (2 weeks old), he screamed, "I don't want, I don't want!" He did not want to look at them and clung to the adults. I want to stress that children need a certain mental maturity to experience fear. My 1-year-old son did not think twice about touching a cat he saw for the first time [plate 64(1)]. Later, the same cat brought about in him a delayed reaction, followed by the emotion of fear; then later (at the age of 5 years), after the child had gotten used to the scary object completely, he conquered the fear and willingly spent his time in the company of cats, inventing most diverse forms of communication with them [see chapter 14, Active Play (with Animals)].

When Roody visited the zoo for the first time (at the age of 5 years), he certainly was afraid not only of direct contact, but also of coming close to the animals, especially the bigger ones. We have mentioned that he was afraid of an elephant. When I offered Roody a choice to ride on a camel, a donkey, or a pony, he rejected the camel because of its size and preferred the donkey.

Not only live animals, but also their toy copies and sometimes their images, frighten the child. Fearing live mice (1.11.1), he is at the same time very apprehensive toward a toy velvet mouse. He does not want to take it in his hands despite my persistent persuasions. At the age of 3 years, seeing such a mouse, he keeps saying, "It's bad! It's bad!" He presses his hands to his stomach, screaming, "I'm afraid, afraid!"

When somebody mentions mice, the boy (1.11.9) blushes and worries. Roody hides his face, pressing it to my neck, when he sees the picture of a white mouse.

While in this case we may assume that the child does not know the difference between live and artificial mice, in the next example there is no doubt left that the child is afraid of the mouse image. Once, when he was 3 years old, he found a piece of carved wood (a furniture decoration). He pulled back from it, saying, "It's like a mouse." Despite my long attempts to reason with him, he did not want to touch the carving. Another time, when sorting out some junk, he found an oval piece of linoleum with a loose fiber coming out of it, and a gray whistle—things bearing a very fair resemblance to a mouse. The boy (3.0.12) immediately retreated, pulled his hand off, screamed, and stopped sorting the things out, and did not want to touch either object until I persuaded him to take a closer look at them.

Later, when Roody was 5 years old, the velvet mouse was one of his favorite toys, which he often kissed, put affectionately under his pillow every night, or sat down at the table with it beside him.

When Roody was 6 or 7 years old, even live mice were objects of his loving care. He often left scraps of food for them on the floor of his room and watched them coming and eating the food. With a vigorous passion and jubilation, he burned and destroyed mousetraps and protested and was genuinely upset when he heard people say that mice should be killed.

The child was afraid of small, but too active, animals. For example, Roody was very much intrigued by baby goats, but when they approached, he (2.2.17) stepped aside, blushed, and was reluctant to touch them. He was afraid of some stuffed animals and avoided them. Roody (1.5.9) was afraid of a stuffed turtle and wolf; he touched them with apprehension at their first encounter. He (1.1.28) was afraid of big stuffed birds, of fur on people's shoes (1.7.13), and of a stuffed white bear (2.5.9), saying, "I'm afraid of his whiskers." Roody recoiled from the dummy of Joni when is was shown to him for the first time ("I'm afraid of the monkey"); when I began to open the cabinet where the dummy was, Roody screamed, "Close it, close it quickly!" However, the boy got used to the dummy in a short while and started to examine it, touching its hands and feet and its eyes, ears, and nose. And when I took the dummy out of the cabinet, Roody even hugged it, saying, "To cuddle Joni." Roody (1.2.1) was afraid of big stuffed birds, but when he saw them for the second time 2 days later, his fear vanished completely.

Sometimes, Roody was even afraid of certain pictures. For example, when the boy was 3 years old, I bought him a book about a country boy who in the darkness took his own dog for a wolf. Having heard the story once, Roody flatly refused to read it again or look at the pictures. Most of all, he was afraid of a picture of the wolf with brightly burning green eyes. Roody (2.2.8) was afraid to look at the pictures of morels (mushrooms) with human faces, saying, "I'm afraid of morels." He (2.5.17) was afraid of the picture of a chimpanzee (and that was after he got used to the dummy of Joni); when I brought the picture close to him, he stepped aside and cried, "It's bad, it's bad!"

At the age of 6 years, Roody was afraid of the picture of a cyclops, a naked, single-eye giant sitting in a dark cave. This picture was apparently both intriguing and scary for him. He called me to look at the picture, but was not brave enough to look at it himself, and he covered it with his hand until I came; then, he hastily jerked his hand away, showed me the cyclops, at the same time closing his eyes or turning his head away. In answer to my question why this picture frightened him so much, he said, "A very big eye."[10]

Similar to Joni, the human child would try willingly to call forth a scary, but intriguing, stimulus again; it is this renewed familiarization that leads to the elimination of the fear emotion. In this case, we see again that the unusualness and magnitude of the stimulus are themselves the scary elements. The boy retains the same mental attitude until an older age.

The child's fear urges him to be inconspicuous. A 2½-year-old girl spoke only in a whisper the entire time she was in an unfamiliar house. The human child knows the unpleasantness of fear; nevertheless, like the chimpanzee, he is inclined to frighten others. Roody (2.7.20) enjoys frightening a sleeping cat; he suddenly assails the animal and laughs to see the cat wake up and dash away in fear. Roody frightens his doll with a roaring sound: He makes himself out to be a bear. Or, he (2.11.11) uses a toy dog to scare people, crying "Bow-wow!" He likes (after 3 years of age) to set an ambush in dark places and is glad when he succeeds in frightening somebody. He enthusiastically wears masks and various military paraphernalia to appear more scary for his friends, if not for adults. As mentioned, Joni often frightens people and small animals by covering his head with a rag or by suddenly charging at people from behind the furniture.

Comparing different stimuli that bring out fear in the human child and infant chimpanzee, we have to point out that the former entirely lacks fear stimulated by olfactory perceptions, as found in the chimpanzee. In the human child, the fear caused by luminous effects is not so distinct as in the chimpanzee. Taking photographic pictures of 2½-year-old Roody under the blinding light of flash-bulbs never caused him to panic; the chimpanzee literally fell off his chair at repeated attempts to use bright light. Also, Roody, in contrast to Joni, did not express uncontrolled fear toward the leopard and miniature snakes and turtles. On the other hand, Roody experienced more fear of heights. At 4 or 5 years of age, when crossing a bridge, he was afraid to look down, while Joni fearlessly traveled roofs as high as dozens of meters above the ground. Roody (at the age of 3 or 4 years), unlike Joni, was afraid of too active insects and small live animals. In contrast, Joni chased these same animals. Running free in the forest, Joni behaved more apprehensively than Roody; Roody was afraid to lose his escort and would not leave them even for a moment. Under the same circumstances, walking next to a human, Joni continuously turned around, assumed the vertical position, and examined the terrain every three or four steps. It was as if he did not trust the human experience completely and wanted to make sure that everything was safe before continuing on his way.

The following example is quite typical. The human child, fearing the picture of a cyclops, like Joni tends to reproduce his frightening experience. However, he reproduces it not for himself, but for other people. Afraid of looking at the picture, Roody insists that I look at it, blocking it from his vision. Thus, he tries to overcome his fear using outside help. The human child gets used to unfamiliar people less quickly, and he is not as relaxed with them as the chimpanzee.

Hence, this comparative psychological analysis of the self-preservation instinct and of the emotion of fear brings us to the conclusion of the stronger biological and mental adaptation of the infant chimpanzee relative to the human child. The self-preservation instinct of the chimpanzee significantly broadens the sphere and the range of frightening stimuli. It makes the chimpanzee afraid of more stimuli and express his fear more aggressively than is typical of the human child. However, the chimpanzee has a stronger tendency for overcoming fear by his familiarization and closer contact with a frightening object.

2. Anger. In the human child, as in the infant chimpanzee, the emotion of fear often brings forth anger, either in the form of protest against an unpleasant feeling or in the form of self-defense or revenge. A 3½-month-old child already

defends himself instinctively, closing his eyes every time he hears the sound of a rattle brought close to him. At the age of 4 months, the child tries to beat off the intruding hands in response to unpleasant contact or even light touching, such as the movement of a fly crawling on his face.

Once, my 2-year-old child was presented with a big toy horse covered with real animal hide. At first, the boy looked at it with apprehension, afraid to come closer, apparently thinking that this was a live horse. But, as soon as he realized it was not moving, he began to charge at it, stamping his foot [plate 17(4)] and yelling angrily.

The child's defensive gesture, his fist pressed to his chest, definitely can be considered a gesture that precedes his attack. Any angry facial expression of a human child is generally similar to that of a chimpanzee; both pull their lips to the sides, bare their tightly closed teeth and gums, wrinkle the upper part of their faces, and narrow the outer corners of their eyes.[11]

My boy, surrounded by constant care and attention, has never been driven to rage or even strong anger; but once (when he was 5 years old), I observed such an expression on his face under the following circumstances. Armed with a toy sabre and a handgun, he mounted a horse and aimed the weapon at an imaginary enemy, his lips pulled to the sides somewhat, his clenched teeth bared. But, since the entire offensive posture was make-believe, his eyes were smiling, his upper teeth were covered partially by his upper lip, and his face had an expression of facetious aggressiveness [plate 17(1)].

Comparing expressions of anger in the child with those of the chimpanzee, we easily can find that, in humans, the teeth are closed tightly, the corners of the mouth are pulled to the sides, and the gums are covered. In the chimpanzee, we see a gap between the teeth, the corners of the mouth are bent up, and the gums are uncovered [plate 17(2)].

In anger, the chimpanzee lowers the skin of the inner ends of his eyebrows, jerks up his nose,[12] and wrinkles the upper part of his face. We do not see this in the photographs (in this book) of aggressively disposed children, but I noticed this in Roody in response to a sudden unpleasantness (for instance, when I did not allow him to do something that he wanted very strongly, when he rejected food that was not tasty, or when he was affected by a sudden change in ambient temperature);[13] he invariably jerked the wings of his nose and wrinkled the skin of the bridge of the nose.[14] I often observed Roody frowning his eyebrows and speaking loudly, when he was angry and irritated.

Quite similar wrinkling of the bridge of the nose usually accompanied in Roody the sensation of gustatory aversion [plate 89(4)]. In this case, it was complemented by his expressive pressing of the corners of his mouth inward (shaping his open mouth in the form of a rectangle), by narrowing his eyes, and by profuse salivation.[15]

I have never noticed other attributes of anger observed in an aggressively excited chimpanzee, such as curving of his lips, baring of his canines, tossing up of his upper lip [plates 18(2), 18(3)], rattling of his teeth, shaking of his head, and drooping of his lower lip, or jumping from his feet onto his hands. However, I can assume that other observers of children and adults could draw similar parallels in aggressive movements and angry expressions of chimpanzees and humans.

We can find many similarities in the human child and the infant chimpanzee with respect to their aggressive gesticulation in response to frightening stimuli. The very intention to scare is no less typical of the human child than of the infant chimpanzee. All of us have seen, at one time or another, the intimidating gestures of children: waving or clapping of their hands or the use of rods, sticks, stones, toy guns, swords, or sabres to increase the effectiveness of a threat or attack. We all know how often children in a slight disagreement apply their fists and batter, pinch, scratch, and bite each other, then become serious and forget to stop. Most of the favorite children's games, struggles, fights, and wars invariably include, implicitly or explicitly, angry, distasteful, and sometimes ferocious feelings [plate 91].

Even my "soft-hearted" Roody (at the age of 5 years), after watching military exercises, was so inflamed with militant feelings that, for the next 2 years collected only military pictures, read books only about wars, enthusiastically played military games, asked for and made only military toys (sabres, swords, handguns, and endless types of rifles), adorned himself with military attributes (spurs, helmets, metal belts) not only at home, but also outside home. He surrendered himself to this passion with maniacal fanaticism until it was supplanted by another passion (collecting caricatures). I think that, in this case, the boy's aggressive and angry feelings, curbed by us, impeded the manifestation of the instinctive senses of strength, power, and superiority, which he had no opportunity to express, and it was only natural that such a spirited child as Roody discharged his feelings in various kinds of military exercises and games. Since their angry instincts are aroused when they do not yet know how to control them, children often harm each other severely, surrendering themselves fully to military games and getting carried away with them.

Similar to the chimpanzee, children and adults, in expressing uncontrollable anger, produce various knocking sounds, stamp their feet, make abrupt movements, throw and destroy things that happen to be nearby at the time, sometimes hurt themselves (bite their lips or beat their heads against the wall), fall on the floor, kill other people, and commit suicide. The sound "ah!" which usually accompanies the chimpanzee's angry excitement, can be observed as an exception, both in children and in adults. We all know that angry humans often speak loudly or scream.

Mothers know very well that children are prone to so-called angry crying, which includes certain trembling notes of irritation. This crying consists of uniform blaring volleys reproduced at equal time intervals with the same tone. I observed this crying in Roody when he (0.3.4.) was hungry; at that time, he made scratching movements with his fingers or later (0.5.25) waved his hands or jerked his legs.

When he was 7 months old, I often heard him give out a growling sound when, having brought his meal to his room, I lingered in giving it to him. He uttered the same growling sound when he was not allowed to touch certain objects. Roody (1.0.2) pushed aside his plate in an abrupt manner when we were persistent in feeding him something he did not like.

Stimuli that provoke anger. In general, stimuli that provoke anger are the same as in the chimpanzee: Everything that brings about fear, and consequently an unpleasant feeling as a response counterreaction, also causes anger. The child's angry crying often is associated with incomplete satisfaction of his physiological needs with respect to food, water, and sleep; his aggressive gestures and body movements are aimed predominantly at pushing aside the unpleasant or irritating object.

Our Roody (1.10.16), when we thwarted his desire to do something (for instance, prevented him from leaving his room), used to slap me or his nanny while we held him. Sometimes, under similar circumstances, he grabbed at our faces, waved his hands, and bit the sleeves of his shirt and tried to tear them with his teeth, becoming very serious when we tried to stop him (1.3.4).

Once, when I was feeding him (1.6.9) a hot cereal, I gave him, between the spoonfuls of cereal, a spoonful of soft-boiled egg, which he usually did not like. As soon as the boy took the spoon in his mouth and felt the unpleasant taste of an egg, he struck my on the face. When I pushed him away, he started to stamp his feet on the floor.

One time, Roody (1.4.27–1.8.12) invented and used biting and pinching when he was in an aggressive mood, in response to some of our restrictions concerning him, or out of revenge for his punishment. However, he bit and pinched not the person who had punished him, but another, uninvolved person who happened to be nearby. Once (at 1.5.21), he ran over to me and slapped me after his father had slapped him for some misdemeanor.

In general, I have to point out that my child's angry reaction was always based on the sense of revenge. For example, Roody (2.1.12) accidentally hurt himself against our stone porch; he started slapping the porch with his fist, saying, "I'll beat you up." Once, he hurt himself against a toy horse; he started crying and beating it with a stick. In response to my question, "Why are you beating the horse?" he (1.6.23) clapped himself on the temple, where the bruise was. Another time, Roody (2.2.27) painfully hurt himself with a stick; crying, he began beating a man who just happened to be there. Notably, Joni showed his anger not toward the man punishing him, but toward the whip, a means of punishment. Roody (2.3.25), started battering my after he had hurt himself against a bench.

Sometimes, a child, in response to being hurt, aims his unpleasant feelings at the surrounding adults, considering them guilty of his mishaps because he used to rely on their experience and their timely warning of the danger. Therefore, such a direction of his aggressive feeling is justified objectively. I once gave Roody (2.8.3) a glass roll and a metal ring on a string and showed him how to spin it. The boy, seeing the whole procedure, said with fear in his voice, "Mother, I'll hurt my forehead." I said, "Never mind, go ahead, play with it." He started to spin and, indeed, hurt his forehead; he immediately cried and struck me. He retained this aggressive slapping gesture for some time. For instance, when he was 2½ years old, he had to stay in his room with a boring young nanny who, unlike me, did not play with him at all. Every time he had a choice between me and the nanny, he expressively slapped her, pushed her away, and stretched his arms toward me.

Other manifestations of the child's anger in the form of stamping the feet and throwing things are no less known. Many times, my boy, already grown, was carving a piece of wood and not achieving the desired result; he abruptly and angrily threw his tool and his unfinished product aside. I also saw Roody whipping and yelling at a big toy horse because he was unable to move it. As noted, Roody at 2 years stamped his feet in reaction to the horse, of which he was earlier afraid. Later, I caught him stamping his feet and screaming at the nanny, who did not fulfill his request to the letter.

I am inclined to think that this foot stamping is not an imitative act, but rather an instinctive one because I observed it in tiny children who had not yet been able to see such stamping and imitate it. It is a well-known fact that a child 1½ to 2 years old who just started walking would angrily stamp his feet if not allowed to do something. The jerking of the feet of little babies when they suffer from colic is perhaps the stage preceding the stamping.

Once, I observed a 2-year-old boy stretching his hand to stroke a dog, but the dog kept running away from him. The child started to shake all over and to stamp his feet. I have mentioned the hot-tempered girl who used to throw herself on the floor and stamp her feet, cry, and scream if her desires were not fulfilled immediately. Here, the anger element certainly prevailed in her unpleasant feelings over the element of sorrow.

It is typical that the human child's anger almost always is justified and is not aimed at harmless innocent creatures, as in the case of Joni, although sometimes the child in his behavior targets the wrong objects. Another example is even more funny. Once, Roody (2.2.19) declared, "I'll beat up this book!" "Why?" I asked. "Because it's not good, I am not going to listen," he explained the reason of his angry attitude.

3. Compassion and Protection. I must mention that I have never observed Roody to enjoy torturing and beating small animals the way Joni did. However, we often tried to restrain his angry feelings even toward inanimate objects, even towards his toys, and we excluded stimuli that would be likely to cause such feelings. I tried my best to nurture in the child the sense of love and compassion toward objects he started to attack, particularly toward toy animals.

Our goal has always been that the boy would not offend a single animal, that he would not kill even insects. I have never seen him beating his companions, engaging in a serious fight, or even killing a fly. We have succeeded in that the boy, on his own initiative and guided by his kind heart, will try to help every time he sees an insect in trouble. As soon as he finds a bug or caterpillar in the house, he carefully carries it outside into the garden and puts it on the flowers. If he saw a butterfly beating its wings between the windowpanes, he begged us to catch it and set it free; flies were the objects of such a great care of his that there were even quarrels in our family over them. If Roody saw fly traps hanging from the kitchen ceiling, he persistently tried to free the flies glued to them; he protested, at first in a complaining voice and then categorically, against putting up the fly traps. He tried to destroy them and struggled with the maids when they did not go along with it. To attach value to his compassion, you should have seen the tenderness with which he saved flies that happen to get into the water. We mentioned that he cared about mice.

When he was 2 years old, he heard a story about big dogs biting a small dog he used to know. Soon after that conversation, when I was pushing the carriage with Roody along the sidewalk, he saw that wounded dog. While he was watching it, another, bigger dog ran over to it. My baby's face reddened, he eyes filled with tears, and he was ready to cry. I comforted him and drove the big dog away.

Certainly, this compassionate attitude of the human child toward animals is incomparable with that of the chimpanzee. There were many colorful examples of Joni's unjustified cruelty toward small and helpless creatures, despite my persistent struggle against these acts of torture (particularly toward children and live animals).

My mind, trying involuntarily to find mitigating circumstances for my chimpanzee, conjures up the thought that perhaps the infant chimpanzee was completely unaware of the fact that he was inflicting pain on the animals, the way many children are unaware. But, I immediately have to throw this thought away as unfounded because I recall that Joni was fully aware of the strength of his teeth and nails when he, in his playful attempts to bite and scratch me, constantly watched my eyes and stopped as soon as he saw a grimace of suffering on my face. But, despite the fact that the dogs desperately squealed under torture and the children who were bit screamed and ran away, he continued to chase and bite them. It was not always an element of play, but more often bore the features of a self-propelled cruelty.

I am convinced totally that by no means could I incite Joni's compassion toward practically harmless neutral creatures inferior to him. The most that could be expected of him, on the basis of my observations, was his compassion to his guardians, to the people who loved him; he sometimes even was ready to take revenge on those offending them. This was the apogee of his mental development; however, that was it and no more than that. On the other hand, my boy was ready to take revenge to defend an absolutely neutral inanimate object, for instance, a sculpture of a human. Roody (about 3 years old), on seeing in our museum the sculpture of a gorilla overpowering a human, said to me, "Spank him, he hurt the man!" He hit the gorilla sculpture a few times; then, coming back home, he bragged about it to everybody.

Later (at the age of 4 to 5 years), Roody energetically drove away, slapped, and chased the dogs that dared to attack our dogs; his aggressive feelings and actions could be justified by ethical purposes. These actions were in no way similar to those of the chimpanzee, who enjoyed torturing animals.

Again, we come to the generalization concerning the different directions in the behavior of the human child and infant chimpanzee and the degree of development of their social and moral feelings. Certainly, there are children whose cruelty can exceed by far the cruelty of apes, and there are apes who perhaps have not committed any cruelty toward creatures inferior to them. However, for me personally, the most crucial is the following inductive inference, which can be considered a postulate: The human child is able to raise himself to the stage of compassion toward his "smaller" brothers, and the infant chimpanzee is unable to do that. This postulate solves the principle question of ethical differences between humans and animals.

In summary, we must say that the infant chimpanzee is ethically at the stage of compassion toward his friends, while the human child rises to the stage of compassion not only toward his kin or people close to him, but also toward neutral creatures; this contains incipient elements of brotherly, social love.

Social Instinct   A certain similarity in development and manifestation of tender feelings of the human child and infant chimpanzee toward people can be found. Very early in the ontogenesis, the human child differentiates his relationships with people, favoring some of them and being neutral or sometimes disagreeable to the others.

1. Children's Communication (Expression of Tenderness and Attachment).   As for a chimpanzee, for a human child his mother is the creature closest to him; she feeds him and cares for him. A tiny 3-month-old child first recognizes his mother and welcomes her with a joyous smile. A 6-month-old child, encountering his mother after a long period of her absence, utters inwardly a sharp jubilant sound, smiles broadly, rapidly moves his legs, clings to her with his entire body, touches her face with his face, embraces her neck, and touches her neck with his open mouth, moving his lips and breathing rapidly (like Joni). It looks as though he wants to kiss his mother, but does not know how, although he has been kissed a countless number of times. Later, under the same circumstances, he presses his lips together; yet later (in Roody's case at 1.5.25), he produces a smacking sound, kissing for the first time [plate 52(6)].

Joni also knew how to kiss in a humanlike fashion, but such a way of expressing tenderness apparently was artificial for him [plate 52(5)]. I have never noticed that he used it on his own initiative or when he was under the influence of tender emotions. The chimpanzee used the body movements that were more natural for him when he showed his friendliness, sympathy, and tenderness toward the people he knew: He either carefully hugged his favorite person under the chin [plate 42(1)], or touched the person's face with his hands, clung to it, extended his lips toward it [plates 42(3), 52(3)], or touched it with the tip of his tongue and slightly licked it [plate 42(4)]. Sometimes, Joni pinched the skin of the person's face with his lips or carefully put the person's fingers into his mouth and slightly sucked them.

It would be appropriate to wonder whether this sucking can be considered an expression of tender feelings. But, my boy gave me, quite unexpectedly, the clarifying data. When he was 2 years old, I sometimes noticed that, when he was going to sleep, he would hold my hand in his, would kiss my hand or slightly touch it with his lips and tongue, or would lick[16] and suck it despite my restraining gestures. I think that this action reflects the infantile association of sweetly falling asleep with the nipple of his mother's breast or the pacifier in his mouth, the habit children sometimes retain until 5 or 6 years of age.

Kissing becomes for the human child (as early as 2 years of age) a natural and habitual way of expressing tender feelings [plate 52(6)]. Meeting his favorite people, the child smiles broadly, his face lights up, he runs to meet them, and he hugs and often kisses them. At the age of 2 years, he eloquently, unequivocally, and verbally expresses his love, "Dear mommy, I love you," Roody (2.1.17) says, hugging me after my 3-hour absence. After a longer period of absence, the child (2.6.9) meets me with the following words: "I miss my mother, I love my mother." As mentioned, the child likes to go to sleep when his mother is present; he presses his body to her, pulls her to himself, wants to be as close as possible to her, kisses her, and says (3.0.1) warmly and tenderly, "Mother, I want to be closer to you, I love mother," "Every day with my mother, every night, and evening and day!" Or, even more expressively, "I love my mother dearly, I love my mother every day" (3.1.3). With these ingenuous, simple words, the child expresses his deepest feeling, his clear thought, and reveals to us that he always, constantly loves his mother, that he wants to have her entirely for himself forever. This desire is expressed eloquently in the following powerful words that do not correspond with the weakness and helplessness of his age: "I won't give you to anybody!" (3.1.5), or later (2.11.2), "I don't want my mother to die!"

How strongly the child adheres to his bond with his mother, how afraid he is to lose it! I vividly remember how difficult it was for me to leave my child with his grandmother or his nanny when he was 1 to 3 years old. In such a case, I often had to resort (as with Joni) to persuasions, temptations, or some ingenious tricks. This attachment to mother sometimes turns into a passion. For example, Roody was always reluctant to let me leave home, saying, "It would be better if you stayed," "Don't go away," or "I don't want you to leave." I tried to sweeten him with tempting things, saying that I would buy him a chocolate or a book, but everything was in vain. He would say quietly, "Don't do that, please" or, with the sound of irritation and offense in his voice, "I don't need anything," unequivocally expressing his unwillingness to trade my presence for the world. When I was leaving, he affectionately said good-bye to me, waved his hand, followed me with his eyes, and said, "Mother, come back soon!"

In the section of chapter 12 devoted to the child's sad feelings, it was mentioned that he expressed warm compassion toward people he loved in reaction to their real or imaginary troubles; sometimes, he felt sorry for his mother that he was ready to sacrifice his strongest desires for her. When Roody was 3 years old, he liked very much running after his nanny to whip her with a rope. He applied so much force doing it that we had to restrain him, but he kept ignoring our words. Once, I offered myself as a replacement for the nanny. He ran after me, struck me with the rope a couple of times, and then said suddenly and sadly, "No, I can't. I feel sorry for mother." I tried to convince him that I was not hurt and asked him to continue the chase, but he would not. In other cases, he showed a touching compassion toward me. For instance, when I pretended to cry, like Joni he (1.4.25) tried to detach my hands from my face and to look into my eyes.

When I cut my finger or complained about some pain, he tenderly kissed the sore spot, remembered this for a long time, and asked me several times during the day, "How is your finger, mother?" When I got tired and he kept bothering me, I had only to say to him, "Roody, my head aches," and he immediately stopped his requests, calmed down, and became sad. I have mentioned that Roody felt sorry and expressed compassion toward other people and animals close to him, but his love toward his mother definitely stood out among his other sympathies.

A 3-year-old child does not want to make a single step without his mother. He runs to her when he feels a real or imaginary danger or when he is upset with something; he hides his face in her knees [plate 92(3)], leaning his head against her, and tugs at her dress. If a mother fulfills her great mission with respect to her child, the word *mother* is constantly on his lips and is used in all cases. This word starts his day, and he goes to sleep with it. No wonder that the greatest people of all times and nations have expressed their prose and poetry with beautiful, touching, and emotional words their feelings toward their mother. We can remember charming, moving lines from "Childhood and Youth" by Tolstoy, a touching poem "Mother" by Nadson, and Nekrasov's poems devoted to mothers. These fully reflect the unexplainable beauty, depth, and firmness of this bond. And we all know what makes this bond.

A mother gives her child so much: She gives him her body for development, her health for his birth, her youth and strength for his upbringing, her heart to his heart at the time of his distress, and her soul for his spiritual development. She gives all this so passionately, so unselfishly, so lovingly. It is as though she is deluged with a desire that may be expressed in the following words: "Give me the opportunity to serve you so that there will be joy in my soul."

The child, as we can see, pays with his love for her love. However, there is a common understanding that the child's love for his mother is rather selfish and not as strong as a mother's love to her child. My child in the dialogue above quite definitely showed the selfishness of his love. However, as we have just seen, there are moments when the child is ready to relinquish all his real possessions to preserve contact with his mother and is ready to sacrifice his personal desires and egoistic inclinations for her. "Love fosters love": On the basis of their initial mundane contact, thin but firm emotional threads are stretched between the mother and the child; these threads do not break when all material bonds between them have been destroyed.

Certainly, as long as a child is small, every member of the family is loved by him for some real deeds. The child often expresses his tenderness at the very moment he experiences joy. For example, Roody was anxious to go with me for a walk, and as soon as I told him, "Let's go for a walk," he (1.8.12) rushed to me, embraced my legs, or pressed his face to my dress. Every time I said to him that I would play with him, he (2.3.27) kissed my hand and caressed me. This was reminiscent of Joni, who under similar circumstances, touched my neck with his open mouth, breathing rapidly. In both cases, we can see the incipient manifestations of thankfulness as a natural response to joy given by a dear human being.

I could notice in Roody, as well as in Joni, elements of jealousy. I have mentioned that my child was jealous of his father and, another time, of a doll because of me. When I bought Roody a big doll (he was 1 year old) and played with it as though it were a child, he crawled over to the doll and struck it. Later in the evening, I put the doll, like a child, to bed, covered it with a blanket, and started to rock it; seeing this, the boy uttered pent-up crying sounds, and when I brought the doll closer to him, he belligerently slapped it.

Contrary to Joni, Roody granted his benevolence to many people around him. At the age of 11 months, he differentiated his sympathies toward members of our family, which could be seen easily in a simple experiment of leaving him at home with different people.

For example, my son for a certain period of time preferred me to all other people, and every time I had to leave him in the charge of somebody else, he cried; he always came back to me with joy. Later, during my illness, he spent a lot of time with his uncle, who took good care of him; as a result, in just 3 days my primary position was abolished temporarily, and my place became second.

During that period, his uncle's departures were accompanied by the child's loud crying; he would not trade his uncle for anybody else and sometimes cried inconsolably. We mentioned in another chapter how warmly he sympathized with his uncle when the uncle was ill.

But, if my boy happened to lose spiritual contact with his guardian, like chimpanzee, he tried to find a new one. Once, when Roody (1.0.1) was reprimanded by the same uncle for some misdemeanor; the child came over to his grandmother, gave her his hands, and imploringly said, "Grandma!" He surrendered himself to her protection, although before this incident, the grandmother occupied a secondary place in his heart.

Such shifting of the child's attachments also affected his going-to-bed preferences: He would rather remain with those of us who on that particular day had played with him longer. At the age of 2 years, Roody sincerely expressed his feelings about whom he loved and whom he hated. He began (2.3.25) to like some strangers coming to our house; he stopped fearing them and said, for instance, "I like the photographer" (the photographer's presence meant much fun to him). It is natural that a child's tender feelings very often focus on his friends, on little children, with whom he plays and about whom he cares.

2. Protective Communication.   We have noted that many children, like Roody, treat their smaller brothers with special compassion. It concerns not only those belonging to the genus *Homo sapiens*, but also animals and even inanimate objects, for instance, dummies or toys. The smaller, more helpless and oppressed the creature under their protection, the stronger the sympathy and compassion animal or object arouses in Roody. Multiple examples bear that out.

We have mentioned the compassion with which Roody treated suffering animals, including insects, although to experience the emotion of tenderness, the child needs a certain maturity that comes at different ages to different children. My Roody, for instance, longed for communication with live animals; however, he rudely grabbed a cat by the hair, which was obviously unpleasant to the cat.

But at the age of 2½ years he carefully stroked the cat; he also liked feeding a dog [plates 64(1)–64(3)].

Roody's plush teddy bear enjoyed his never-ending sympathy [plates 93(1)–93(6)]. He treated a porcelain elephant, plaster dogs and cats with touching tenderness. This sympathy was not limited to his new toys; it was extended toward old, half-ruined, frayed, and disfigured ones. "My beauty," Roody called the ugly head of a broken plaster dog, which confirmed many old proverbs proclaiming that love is blind. Roody did not want to throw away or give out a broken toy elephant and preferred rather to part with a new, shiny, but unwelcomed, animal, that an old, damaged one to which he was tied by the intimate bonds of his heart [plate 64(4)]. My observation notes (pertaining to the child from 1.7.38 to 3 years) contain numerous entries that prove this thought.

The child emotionally differentiates his toys from very early on; he (1.7.0) shows signs of affection toward some of them—kissing, hugging, stroking them; pressing them to his chest; and uttering grunting sounds. At that time, Roody's special sympathy goes to miniature rag or porcelain dolls and to filthy, half-broken, stained, shabby animals, the most frequent companions of his play. The boy always meets the reappearance of these toys after their temporary absence with joyful squealing; he is glad to welcome back his old companions from which he once was inseparable, and he covers them with kisses.

The child (1.10.10) expresses his sympathy even toward images of animals especially dear to him (for example, chickens, hares, squirrels). When he finds such drawings in a book, he brings his face closer to the page and kisses them, unable to restrain his affection for them. Roody's sympathy also extended to stuffed animals. While Joni was very aggressive toward a stuffed little chimpanzee [plates 18(2), 18(3)], and he attacked it, 2½-year-old Roody was afraid of it at first, but soon got used to it. He hugged it and pressed it to his body [plates 64(5), 64(6)].

Roody (2.11.7) showed great affection toward his numerous toy hares. When we went for a walk, he took his favorite hare with him, carrying it in his hand or in his pocket. He took it out of the pocket to show it interesting things that were happening around him. He did not part with the hare even during his most active play; he cried when he did not remember where he had put it and would not calm down until the hare was recovered. At dinner time, Roody sat the hare down on his knees and shared with him the tastiest bits of food. At night, he put the hare under the pillow or tenderly covered him with a blanket.[7]

The love of hares reached a point at which Roody did not want to eat a marmalade candy in a paper wrap bearing a hare's image; he asked not to touch the candy, saying: "I feel sorry for the hare." I could not persuade him to eat the candy by saying that we had no other candies and that the hare was not hurting. He was adamant in his decision, although it was a great sacrifice for him because he always had a sweet tooth and especially adored marmalade.

This last trait of the human child drew a distinct line between him and the chimpanzee, marking different directions in the development of their affectionate feelings. While affectionate feelings in the chimpanzee are associated exceptionally with those who take care of him, those who satisfy his egoistic needs (in essence, these feelings are egocentric and directed at powers that be; Joni behaves toward weaker creatures as an oppressor and despot). In the child, these feelings most often are aimed at the weak and helpless. Apart from these egoistic sympathies, the child also has altruistic tendencies. Sometimes, he seeks neither help nor protection on the part of inanimate objects he has granted his favors, for no other reason than he knows that these objects are not alive and cannot do anything for him. His love of the weak of this world is sometimes self-sustaining altruistic love, an antecedent of a higher form of brotherly love.

We are certain to point out that both the human child and infant chimpanzee possess a clearly expressed social instinct; their life ceases to be complete and turns out to be dull and sad as soon as they are left alone. If we look beyond the increasing elements of fear typical for every child left to his own devices and not altogether confident of his limited strength, we see that the feeling of loneliness deprives him of vibrancy and joyous emotions.

No wonder children brought up without the company of others become reclusive, somber, melancholic. They give the impression of little old men who lack the natural attributes of childhood that develop in a community; charming artlessness, briskness, and playfulness. We noted that Roody (2 years old), while on the street or in a streetcar, calls people, pulls at their dresses, and in any other way tries to communicate with them. He talks and plays with strangers, calls other children to come over and play with him, participates actively in common play and says (2.7.19) "It's boring to be alone" when none of his companions is nearby.

The human child, even to a greater extent than the infant chimpanzee differentiates his relationships with other people: He starts to discriminate between "his people" and "strangers" very early in his life; he willingly goes to the former and recoils from the latter. Also, his relationship with his people is individualized, not to say arranged emotionally in a hierarchy. At that time, it is based on the principle, "Wie du mir, so ich dir" ["I will treat you the way you treat me," or "Tit-for-tat"]. Roody (at the age of 3 or 4 years), in response to a question regarding which family members he loves more, speaks his heart with childish sincerity regardless of relative importance of certain people for him. After thorough examination, it turns out that he favors those who offered him more joy, affection, and love.

A 3-year-old boy, who was an object of Roody's (7 years old) affection, was asked whom he liked the most, his mother or father. He said suddenly, "I don't love mother, I don't love father, I love Roody."

3. Companionship.   The human child's reaction to strangers is less differentiated since it is more reserved. In this case, as in the chimpanzee, the age and the emotional factors determine the relationship to a greater extent than the gender factor; that is, children and adolescents and vibrant, active adults (men and women alike) attract him more and are his favorite company. Obviously, the child most wants to communicate with children of his age.

The human child is more reserved than the infant chimpanzee in communicating with adults. Not only Roody, but the most vigorous, not to say reckless, children were not so irresponsibly insolent as the chimpanzee. Certainly, the human child does not smell strangers, as the chimpanzee does, but limits himself only to visual examination. However, Roody (at the age of approximately 3 years) often started familiarizing with other children, not only by giving them the unceremonious once-over, but also by touching their faces, lips, and hair with his fingers.

The child tends to communicate to adults both his joyous and sad experiences. When he hurts himself, he not only hurries to inform those closest to him about it, expecting their help, but he also tells this to other people. Once, my Roody (2.8.24) pricked his leg on the horns of a toy giraffe and cried from the pain. To make the toy safe, we removed the horns. When Roody's father came, the boy immediately started telling him about the accident, accompanying the story with gestures (he made a quick pulling movements with his hand, showing how the horns were taken off). When I came, the child repeated the whole story to me.

Another time, coming home after a long walk, he (1.11.12) told everybody that his nanny had fallen on the snow-covered ground while fleeing from a car that was about to hit them. His story was very simple: "Car, nanny, bukh-tukh!"

Having succeeded in something, Roody wanted immediately to demonstrate this to adults. He was very glad when he gained their approval. "Here, I caught a fly, show to mother!" he said (2.0.27).

When he suddenly saw a tank passing by, he rushed into the house and persistently and passionately invited everybody to run outside with him and to look at the tank. If he valued something, he demanded the same love and respect toward it from others. This trait of his (3.0.16) can be illustrated by the following funny episode.

Roody liked very much to play with a rubber tube, which he called "the bowel." Once, he asked his father, "Daddy, do you like the bowel?" "No," his father answered. "I like it!" the boy continued; then, in a while, as though suffering from "ideological differences" with his father, he added very convincingly, "Daddy! Love the bowel!"

In experiencing some new sensation, the child invariably invites others to experience the same. We can often hear from our children, "Mother, smell it, try it, look at it." Once, Roody (2.10.15) put on new shoes; he smelled them and said, "Smells good," and insisted that others also smell the shoes.

The child is very responsive when adults try to share their interests with him. He emotionally empathizes with them, joyfully helps them, and is offended when they close themselves to him, concealing from him events of their life. For instance, my Roody (at the age of 1½ to 3 years old) was eager to fulfill our small requests, run errands, bring some things, or pass things to any member of the family. The child tends to imitate not only certain actions of adults, but also a series of actions, assuming the roles of different professions (newspaper reporter, sanitary worker, doctor, photographer, etc.).

Once, Roody (at the age of 14 months) performed a very difficult task. He stood a wooden lamb on its feet [plates 87(1), 87(2)]. He and all the people around him applauded him. When he did the same thing for the second time, the nanny did not participate in the applause. The boy looked at the crowd around him and abruptly said, "Nanny!" inviting her to join in common approval. The next time, to check whether our guess was correct, the grandmother did not applaud either. The child was very quick to shout, "Grandma!" The boy repeated this pattern with other family members.

As we have seen, Joni also expressed his intention to keep company with everybody, but it did not reach the proportions typical for the human child. The infant chimpanzee wants to be close to people mainly when he is afraid, angry, or upset, when he expects help from them. But, when he is in a joyous mood, he completely forgets about them and even prefers to communicate with strangers. The chimpanzee is infected by the emotions of people around him and imitates their actions; he shows elements of mutual service and defense, but only to a very limited extent. The human child (not only relatively soft-hearted Roody, but also other small children) never showed such hatred toward tiny (2- or 3-year-old) children that I observed in Joni.

Moreover, a grown-up child (Roody at the age of 4 to 7 years) assumes a patronizing position toward small children and willingly takes care of them the same way he has been taken care of by adults. I vividly remember that one time my Roody did not want to come back inside from the yard. When he was reprimanded for the disobedience, he said passionately, "But it was important! I was buttoning up Vova's pants; he doesn't know how to do it." And for a long time after that, Vova (a 4-year-old boy) was the object of Roody's very touching care. Roody gave him candies and toys, watched that he was not hit by a car, guided him through puddles, and in general helped him in every difficult situation of his life. Such a tender, loving relationship exists in many cases between a child and his younger brothers and sisters.

The bold and vigorous treatment of adolescents typical of Joni hardly was characteristic of the human child (until 5 years of age).

Hence, we must say that, in his communication with people, the human child is more reserved, timid, tender, and friendly than the chimpanzee, who is brave and sometimes arrogant with strangers.

The human child does not show this persistent tendency for retaining his leadership in active play with humans, as observed in the chimpanzee. It is difficult to say whether the chimpanzee does not trust people completely, instinctively prepares himself for the role of leader, or protests against any form of coercion (which is very typical of him). In any case, this tendency is so strong that Joni, unlike Roody, does not concede his leadership even to the people closest to him, does not compromise, and fiercely uses all his physical strength to come out of an inferior position.

My child (2.8.14) willingly communicated with a girl two times his age, obeying her initiative in arranging a play. He was so carried away with this relationship that, after she left, he used to say, "I want Nina!" or "I love Nina!" He hugged her tenderly every time they saw each other.

Joni, in contrast, at the age of 3 or 4 years, was eager to establish contact with people; he particularly liked active play with them.

We all know that a child, starting from a certain age, can initiate play with other children of the same age. For example, Roody (until he was 3 years old) limited himself to showing a familiar child his toys or examined the child's toys; he never tried to organize independently common play with his first friend, a girl of almost the same age. They played separately, although next to each other. They watered flowers or weighed something on a balance [plates 94(1)–94(3)]. Their cooperation required the complicity of adults. Later, this complicity was needed less and less. When Roody was 3 years old, he easily initiated contact with children of his age; he allowed them to ride in his toy automobile, his toy horse, or his bicycle [plates 94(5), 94(6), 95(5)] and showed them his other toys or stuffed animals [plate 94(4)].

Having experienced the joy of communication with his friends, boys of the same age, Roody could not sit still any more as soon as he saw them playing in the yard. It was impossible to take him away from playing with them since no other entertainment could compete with it in attracting his interest and attention.

I observed many times in playing children, particularly schoolchildren, the same challenging, teasing gestures that are also so typical of the chimpanzee; namely, pushing, beating with the fist, waving, hitching with the hands, pinching, and poking with the fingers—all body movements that invite a make-believe fight or chasing game.

The infant chimpanzee, as we have seen, also boldly challenges children. The chimpanzee's biting probably can be considered an awkward invitation to play. Like the chimpanzee, my Roody (2.8.7) ran from one person to another, hitching them, laughing, squealing, and screaming. He tried, by all means available, to involve them in common play or to urge them to chase him. However, he did not dare to do it when he was among strangers.

Chapter 14

Human Play and Play of the Chimpanzee

Active Play   1. Entertaining with Movements. The overwhelming pleasure in movement is as typical of the human child as of the infant chimpanzee.

Right after 9-month-old Roody awakened, he tried to stand up by holding the railing of his bed, to walk holding somebody's hand, to crawl, or to reach for the baby carriage. Even when he was in a carriage, he entertained himself with his own movements all the time.

A 10-month-old baby is a very mobile creature. In a state of excitement, Roody (at this age) did not stay in the same position even for a second. He stood up, fell on the floor, and tried to retain the vertical position by spreading his legs wide. As soon as he felt the firmness of his position, he started waving his hands vigorously as though trying to take off. He made a couple of steps forward, fell down, tried to stand up again by holding some objects, crawled here and there and up and down, fell again, and hurt himself, but he was not discouraged in his intention to continue moving.

An 11-month-old baby, who hardly has learned how to walk, entertains himself by running around the room for a long time; sometimes, he can run incessantly from one room to the next, from one side of the room to the next, and from one corner to the next.

A 1-year-old child runs after a rolling toy; he leans his body and his head forward and often falls on the floor. Sometimes, the child will throw things in front of himself and will chase them, moving in a meandering fashion.

Roody, even at the age of 3 years, liked to entertain himself by wallowing on the floor like a little baby bear, jerking his legs, as though remembering the babyish wallowing. "I want to tumble on the floor," he would say. We all observe, at one time or another, children dashing around the house, getting in everybody's way, hitching everything, and begging us to allow them "to run a little."

2. Running and Jumping.   A child can spend a long time simply running all over the place. "To run, to run! Around, around!" Roody shouted when he got into a spacious room. The child gets so carried away with running that he reaches a state of severe tiredness without even noticing it; he puts his hand on his chest and says, "It's hurting," but after a 2-minute rest, he resumes running. Roody (at the age of 1½ to 3 years) liked to spin to the point of dizziness, exclaiming joyfully after he came to a stop, "The chair is spinning, and the sofa is spinning, and bread, and milk, and grandma is spinning," apparently assuming that everything around him was involved in the whirlpool of rotation. Roody also loved to climb a high hill and run unrestrainedly down the hill, standing a chance to fall down and hurt himself.

Running accompanied by jumping especially inspires a child. My boy (1.6.14–2.7.6) took great pleasure in running from one room to the next, jumping on the sofas and bouncing up and down on them.

Later (at the age of 2 to 3 years), the child willingly jumped down the steps and leapt over a plush teddy bear, over the low barriers that he had made of wooden planks [plates 74(4), 74(5)], or over the rope. But, in all these exercises, the child obviously was less skillful than the chimpanzee. The chimpanzee that easily and fearfully jumped to the ground from gates or fences as high as two or three meters.

We all know that children like to play "squares" [hopscotch], the game that involves jumping on one foot or leaping over the squares drawn on the ground. Roody, at the age of only 4 to 5 years and with great difficulty, learned how to jump on one foot and only could do a short distance. Joni did not altogether show an ability to jump on one foot. But, both infants demonstrated an overwhelming passion for jumping in one spot; they were especially eager to do this on the springy surface of sofas or beds. I waged a long struggle with my child, who at one time became addicted to jumping on the furniture; I had to pull him from the sofas in fear that there would not be any whole springs left after his vigorous and systematic escapades.

3. Active Play (with Animals).   The fact that children vividly play with little puppies is common knowledge. It was a great pleasure for my Roody at 4 to 5 years old to "frolic" (his expression) with a little dog. He chased it, charged at it, pinched it, or wrestled with it as though it were a boy of his age. Similar to Joni and our little dog, both of which possessed an indomitable desire to chase any moving creature, little Roody (at 1.4.0 to 2 years) also liked to entertain himself by chasing sparrows, crows, stray cats, hens, dogs, or baby pigs.

Roody, like Joni, was willing to have live animals for company; he played with them even more ingenuously, recklessly, and vibrantly than with his friends of the same age. There was nothing in his behavior like the torturous treatment of animals that was so typical of Joni. While the chimpanzee, feeling physically superior to other animals, tried to exercise his power over them by chasing, pinching, and torturing them in any conceivable way, Roody (already at the age of 2½ years) tried to involve live creatures in his human activities. For example, when he was shown a live cat, he kept bringing to its nose a small toy cat, urging the former to kiss the latter. He (2.9.13) demonstrated a stuffed squirrel to the cat, gave it his rubber doll, put a pipe in its mouth, brought different toys to the cat, and explained how to play with them. Such explanations usually were quite imperfect. For example, when showing a pyramid ring to the cat, he would say, "This is for doing so," and he would use gestures to complement the explanation. After Roody heard no response from the cat, he answered for the cat to keep the "conversation" alive.

The child is especially willing to run with domestic animals; he arranges various active play, chases or catches animals, and flees from them or wrestles with them. As the child grows, the play with animals becomes more and more complex.

My 5-year-old Roody, having three kittens to share his company, could play with them for days, inventing various games. He would make a cage out of wooden planks and put the kittens in a so-called prison, he would take them for a ride in his toy automobile, or he would make them leap through a hoop, talking with them and for them all the time and making them participate in the mutual communication. I never observed Joni communicating with live animals in any form other than by chasing, cruelly pinching, or beating them.

When there was no live animal handy, a toy one could replace it as well. Of Roody's countless toy animals, I mention here only his favorite companion, a plush teddy bear—"half human, half animal"[1] according to Roody's definition.

There was hardly anything that Roody did not do with this bear. He (at the age of 2 or 2½ years) would put the bear into his toy car (or train) and drag the car [plate 75(1)]. Or, he (at the age of 3 or 3½ years) would start wrestling with the teddy bear and would push it against the floor as if defeating it; or, he would take the role of a doctor and put a rubber tube against the bear's chest and "listen" to its heart [plate 58(6), 65(6)]. Also, he would give it a "fish" caught in a "pond," actually a puddle. In summer, he would ride a bicycle or in a hand car with the teddy bear [plate 96(5)]. In winter, he would [plates 93(5), 93(6)] sledge with it. The boy would take the teddy bear with him when he would fly an airplane (built out of the sofa cushions) [plate 67(5)]. He would rock on the swing and climb the trapezes with the bear pressed to his body [plates 107(2)–107(4)].

At the age of 4½ years, Roody made up a complex game of catching the teddy bear; the game consisted of six stages [plate 97(1)–97(6)]:

1. The boy collects sweetbrier berries (food for the teddy bear).
2. He places a rope loop on the ground and puts the berries in it.
3. He raises the loop a little and, ready to tighten it, he asks somebody to throw the teddy bear into the loop.
4. After tightening the loop, he pulls the teddy bear to him.
5. He carries away the teddy bear by hanging it behind his back, like his spoils.
6. He ties the teddy bear to a tree.

Apparently, this play has a series of sequentially developing actions associated with the imaginary catching of the animal and represents a prototype of hunting play.

Later, at the age of 5 years, when Roody was fascinated with war games, he constantly made the teddy bear his companion-in-arms, supplying it with a gas mask and arming it with artillery shells, rifles, and cannons. Sometimes, he made the teddy bear fight on the enemy's side and, leaning it to the cannon, urged it to attack him [plate 91(5)]. Sometimes, he took the teddy bear to his camp and, embracing it as if it were his comrade, attacked his imaginary enemies.

Not only his favorite teddy bear, but also other toy animals (hares, cats, dogs, horses) and dolls were "animated" by him to such an extent that he tried to involve them in his entertainments. Since he liked to strum the toy piano (at an age up to 4 years), he made his favorite toys (the teddy bear, hare, and cat) beat the piano keys with their paws [plate 58(5)]. When playing with them, he constantly talked to them like to his friends.

The forms of play with animated toys become more complex as the child grows up. While at an earlier age his contact with an animated toy is limited to the representation of a single human action, at a later age (3 years), he performs a series of actions that illustrate a complex phenomenon and require the participation of a group of inanimate components. For example, Roody (at the age of 3 years) set out an entire chain of various toy carriages, placed dolls and animals in them, and "traveled" with them around the garden [plate 98(1)]. He put a number of animals on a truck and made me draw it; with a drum in his hands, he marched in front of the truck [plate 98(3)]. He made a so-called airplane, occupied the pilot seat, made me a passenger, and gave me almost all of his inanimate companions to keep me company [plate 98(2)].

Sometimes, the boy gathered all of his toy animals and dug them halfway into the sand or made a fence for the zoo [plates 99(2), 99(3)]; sometimes, he sat up his animals and dolls at the table, set out dishes, and poured tea into the cups; sometimes, he gathered a group of toy cats and showed them his favorite picture ("a cat running away from dogs through the window"), explaining to them in detail what the picture meant. He played, with aspiration and meticulousness, the leading role of a driver, a pilot, a courteous host, or a speaker.

You could often hear the child, leaving the yard, seriously admonish a toy cat remaining in the sledge with the rest of the toys: "Guard them! If somebody shows up, spank him and then with your teeth do this" (he brought his hand to his mouth as if he were going to bite it). Another time, he (2.11.1) said, "The chicken is crying, her tears are dropping," when he noticed that he had forgotten to sit the rubber chicken at the table where he was treating his toy companions with tea. Or, he (2.9.10) said, "The cat is crying," "The lamb is crying." At the age of 2½ years, Roody gave hay to a horse and stood, bewildered, unable to understand why the horse was not eating [plate 58(2)].

Roody (2.9.14) showed pictures in a book to his toy horse [plates 58(1), 58(3)] and tried to teach it to read. He said, "Well, read, little horse," pronouncing for her a series of incomprehensible sounds, apparently thinking that reading must be different for a horse. Another time, he recited rhymes to the horse and shared his candies with it.

Various dolls, animated by Roody no less (but, characteristically, not more, either) than his other toys, did some human things on the boy's orders: the 1½-year-old Roody tried to feed a doll [plate 58(4)]; he gave it a pipe and kissed the doll. At 2 years of age, Roody willingly drew a sledge or a cart with a doll in it [plate 100(1)]. The 4-year-old boy put a doll to bed, washed it, and put clothes on it. The child at 4 to 5 years tried to make clothes for a doll.

4. Active Play with People.   "Horses," the children's favorite game [plate 94(2)], reflects their tendency for unhampered movement. According to my observations, the child (younger than 3 years) usually prefers the horse part to the driver part; at a later age, it's vice versa.

When I started a chasing game with Roody and made him take the role of a catcher, he (at the age of 2½ years) did not know how to play that role; it did not make sense to him; therefore, he often stopped playing. Similar to Joni, the child preferred fleeing to chasing; he sometimes challenged you to catch him, ran away from you, laughed when he got caught, urged you to chase him again, squealed as soon as he saw you coming closer, dashed forward, and ran the risk of falling down; nevertheless, he continuing this engaging game.

5. Hide-and-Seek Game.   In the hide-and-seek game, Roody (like Joni) also preferred a more passive role (hiding) to a more active role (seeking); he sometimes did not know how to play the active role.

We observed more primitive hiding in the human child (from 1.5.22 to 3 years of age) than in Joni. The chimpanzee hides in places where he is not visible (behind curtains, in a dark remote corner of the cage, etc.). The child's hiding place is often ostensible; he would stand behind a wicker chair [plate 92(2)], where he is clearly visible (2 years 5 months), cover his face with his hand, hide his face between his mother's knees [plate 92(3)], or close his eyes. He also covers himself with a newspaper (1.7.10) or with his hat [plate 92(4)] or squats at the porch steps [plate 92(1)], covering his eyes with his hands (2.2.8) and turning his back to those seeking him; not seeing anybody, he thinks that he is invisible, when in fact he is in everybody's sight. Only after 3 years of age does the child start hiding more ingeniously under the bed, behind the wardrobe, in the arbor, or in other inconspicuous places. Sometimes, he is so good at hiding that the adults seek him for a long time and even begin worrying, while he sits pressed into some corner with a triumphant smile on his face, not showing any sign of his presence. The child also likes to bury things in the sand and then look for them until he finds them.

6. Games or Contests.   When active play includes the elements of a running competition, it is embraced by the child with special delight. I vividly remember how fervently my child played chasing games with me in our small garden, how

he strained himself to outrun me, how he exulted over coming to the finish first, how he could run an infinitely long time, to complete exhaustion.

The human child, like the infant chimpanzee, becomes carried away with games that include the elements of a contest in the agility of catching things. My Roody at 2 to 4 years of age loved to compete to see who was quicker to catch a ball thrown up in the air [plate 101(6)]. It was a great triumph for him to catch the ball in the air before someone else touched it. Since he did not always manage to do it, he was glad even when he succeeded in catching the ball on the ground. At the last moment, he developed such power in his attack that he spared neither himself nor his rival: He would fall down and hurt himself to be the first to touch the ball. This passion reminded me of Joni, although Roody was less arrogant and less vicious.

All kinds of play based on the speed of movement (running, catching live creatures or inanimate objects) in mobile games or on the speed of achieving the goal in immobile games, for which some objects are moved, excite the child's deepest emotions associated with his self-esteem and make his heart the arena of struggle between lively joy and cruel sadness. I remember how unwilling my Roody (about 3 years old) was to compete in running with his older friend because he feared he might lose. When finally he made up his mind and lost, he sincerely and bitterly cried. Joni never cried when he was defeated in a running contest, but became very angry.

I also recall with involuntary compassion how Roody, playing "Who is faster?" got an unlucky number and, according to the rules, had to start the game from the beginning. He sobbed so bitterly and was so anxious to catch up with his partner that he was a sad sight to watch in spite of the fact that the victory did not have anything in store for him except for the knowledge that he had finished first.

But how triumphant he was when he was the winner! His face lit up with joy, and his eyes shone vividly. He quickly placed his chip on the winning number, shouted loudly, and jumped up from his chair in joy.

As we can see, competition based on self-esteem and ambition makes an early and powerful appearance in the child's ontogenesis. This competition, more than anything else, develops the physical and mental strengths and abilities of the child.

Analyzing the mental purpose associated with competition, we must state that the chimpanzee also possesses more endurance than the human, who is usually much more upset than the chimpanzee in the case of failure. Roody (1.3.21) was extremely self-conceited. Every time he did not succeed in doing something, he started to cry. Sometimes, he calmed down very quickly or even laughed if any of us pretended that he could not do it either. He joyfully competed with me in the speed of amassing around him the greatest possible amount of toys (2.7.12) and was triumphant when he exceeded me in this.

The child would not miss any opportunity to demonstrate his (even imaginary) superiority. Usually, we could hear something like this (2.9.27): "Mother, I didn't cry when you left, but Andrey cries when his mother goes away" or

(2.10.12) "I cried a little and stopped, but Andrey keeps crying." Andrey was a boy of the same age.

Even in an earlier age (1.10.15), every time he was reluctant to eat, the moment I mentioned that some other boy ate better, he immediately would try to improve and would put more energy in eating. My 10-month-old Roody seemingly was offended when his favorite food was shown to him and then delayed for some time. When the food finally was delivered, Roody, instead of greedily grabbing it, writhed his body and wrung his arms, as if protesting the delay. Wasn't the child's vanity the reason for his crying (the child was less than 1 year old) in response to loud laughter of an adult? Later (at 1½ years of age), the same laughter bought up his aggressive feelings; the boy waved at the laughing person and struck him in the face; he probably suspected that this person made fun of him.

The child definitely suffers when he becomes aware of his "Minderwerthigkeit" [inferiority] in anything. As noted, Roody (2.7.23) wanted to retell for me a piece of complicated prose ("Elephant Vambo"). I expressed doubts that he really could do it. He started very briskly, but after a few lines, he faltered, could not remember the text, and cried, although I tried, by every means, to justify his forgetfulness, saying that it is a very difficult text even for me to remember. Later (2.8.24), the child, discouraged by this experience, still sometimes recited poems to the nanny, but in my presence, he either stopped short or recited without any intonation as if afraid to stop.

At 3 years of age, when he was among adults, he was confused not so much because off his timidity, but because of his self-consciousness, which was based on his self-conceitedness. In such circumstances, Roody usually stood motionlessly, pressing his tongue against the inside of his cheek and blushing when somebody tried to talk to him.

This higher mental vulnerability of the human child compared to the infant chimpanzee manifested itself in activities that were far from vital in importance to him.

Riding. Like the chimpanzee, the human child loves all ways of movement, but the fastest ones are the most preferable. For example, when Roody (3 years old) had a choice, he always insisted on going by bus as opposed to streetcar. In answer to my question of whether he would choose a bicycle or a motorcycle, he was always in favor of the latter. Since he had not had the experience of riding a motorcycle, he limited himself to staring at the motorcyclists passing by. He begged me many times to buy him a motorcycle and even decided to start saving money for this when he was told that I did not have enough money to buy it.

When Roody was 5 years old he immensely enjoyed riding in a pony-driven cart; he was ready to do it endlessly. Passive sitting in a baby carriage, sledge, or boat is not a desirable entertainment even for a child 2 to 3 years old.

Roody's urge to move is very strong. "In movement is my happiness, in movement," he seems to say; no punishment is so severe for him as limiting his movement. When Roody once was reprimanded for dangling his feet under the table, he said very eloquently, "Mother, I have to move something anyway!"

It is not accidental that the old school of child rearing (which in general brings out not very good memories) there always were various movement limitations as punishment for a too restless child, for instance, standing "like a pole" (in one place), standing in a corner (not only the ability to move, but also any mental activity associated with vision was impeded), standing "on your knees," and so on.

When a child is doomed to remain in bed due to some ailment (for instance, bone tuberculosis or a broken leg), it leaves a very sorrowful impression. In a very good book, *Solnechnaya*, Chukovsky masterfully describes how children, patients of a bone disease sanitorium, spend their time. Confined to bed, the children satisfy their strong desire to move by all means still available to them. They use threads with weights on the end or rings tied to a rubber rope to catch objects that are at some distance from them. Using these devices, the children play and fight with each other, compete in agility of snatching things, and bring everything to such unconstrained movement that from a distance you would think that the room is populated with boisterous kids endowed with all the natural gifts.

Sitting in a cart or sledge driven by toy or real horses, the child wants to be active by any means. An 11-month-old child driven in a baby carriage tries to stand on his knees, throws himself on the pillow, fidgets on his stomach, and throws things out of the carriage. The moving carriage stimulates his independent movements; he wants to hold the reins, although he does not know how to drive the horse. In a toy car, a child (4 years old) tries to turn the wheel, although he cannot yet do it properly [plate 94(5)]. A 2-year-old child sitting on a small chair pushes his feet against the floor, moving fitfully backward.

In a cab or streetcar, the child entertains himself by watching other people's movement. With his eyes and mouth open, he stares at the traffic and street crowds or watches balloons or wind wheels tied to children's sledges.

It was not interesting for Roody (3 years old) to sit on his toy horse motionlessly. Not knowing yet how to move it, he invented a new entertainment. After sitting on the horse for a while, he intentionally fell on the floor as though thrown off by the horse; in a few seconds, he stood up, mounted the horse, and fell again [plate 95(4)].[2] At the age of 4 years, the boy found a new way of movement on the same horse. He mounted the horse and holding it by the muzzle, pushed with his feet against the ground [plate 95(3)]. He also tried (age 4½ years) to tow his friend who was sitting on another horse behind him [plate 95(3)]. But a live horse, played by a man, was a special treat to my boy, as well as to the little chimpanzee. Getting on a man's back while he was standing on all fours, my Roody was ready to ride in such a way forever; he could bring the man to exhaustion and still urge him to move on.

And how ingenious he was in thinking up and building devices for movement (for more detail, see the section, "Constructive Play"). All ways of human transportation (riding a horse, a bicycle, a motorcycle, in a sledge, in a streetcar, on a bus, in a train, sailing in a boat, or flying in an airplane) are reproduced by the child with different degrees of perfection with a great variety of self-made devices that he uses alone and with his friends (plates 67, 98, 103, 104).

The more the child's initiative is involved in his movements, the more attractive they are for him. What a joy it is for him to slide down from snowy hills on a sledge, on a board, or simply on his behind. Roody (2.7.27) slid boldly down a large hill, persistently rejecting my attempts to hold him; he was even reckless, not holding the sledge. He was afraid; he even closed his eyes during the ride, but nevertheless, in answer to my question why he did not want to hold to the sledge handles, he said, "This is how it's done." He apparently thought that this reckless ride suited him more than a normal, passive, inert one [plates 76(2), 76(4)].

I have a vivid recollection of how joyfully my boy played the "railroad" game. I pulled the sledge, with him sitting on it, to various places in our yard, which I called stations. I remember that the most inspiring moment of the game was when the so-called train was departing, and Roody, the late passenger, had to jump on the running sledge. Roody was beside himself when he succeeded in thin. Like Joni, Roody was dissatisfied with the monotonous movement to the next station, so he invented the following trick. In the middle of the ride, saying that he had forgotten something, he suddenly stepped off the sledge and ran back to the starting point, and then he hurried back to catch up with the train. He acted in the same manner when he saw a big sledge passing by; he tried to jump on it. He reminded me of a kitten, which cannot sit still when it sees a rolling ball or any other moving thing. No wonder boys often hitch onto passing cabs or streetcars; doing that, they often disregard danger. I am convinced that no normal streetcar ride will bring a boy more pleasure than reckless rides on a streetcar bumper.

As though to meet the child's instinctive movement needs, such vehicles as a toy hand car have been designed that are put in motion by pushing and pulling the lever.

Roody at the age of 3 years and 3 months, already knew how to drive a hand car, but he only could move straight [plate 102(1)]. In half a year, he could turn the hand car with his feet; at the age of 4 years and 4 months, he learned how to drive a different hand car, one that required simultaneous pushing the driving lever with his hands and pressing the steering lever with his feet [plate 102(2)].

Needless to say, using such a vehicle as a children's bicycle is extremely exciting for a child. Getting on a bicycle for the first time, 4-year-old Roody rode somewhat tensely, awkwardly turning the handlebars and often falling off. In 2 or 3 months, he got used to it to such an extent that he whisked on the bicycle around the apartment; in addition, he put some handicaps along the way: thin sticks, planks, or pieces of steel [plate 102(3)]. He recklessly ran over them and became exhilarated when these "bumps" clanged, shook, thundered, and jumped to the sides. He did not get discouraged if he fell off the bicycle; he intentionally arranged the falls even under most benign circumstances [plate 102(6)].

In a year, his recklessness grew: There was nothing that he had not thought of in riding the bicycle. For example, he tried to ride it without holding the handlebars and leaving the control entirely to his feet [plate 102(4)]. Sometimes, he rode with a book in his hands, leafing through the book, looking at the pictures and not looking in front of him, skillfully turning the front wheel [plate 102(4)], and horrifying the onlookers, who expected a disaster to happen any moment.[3]

Like Joni, my 2-year-old Roody did not understand yet that not every rolling object was capable of moving independently. When, for instance, he received a small toy car as a gift for the first time, he sat at the wheel and waited for some time, apparently expecting that it would start moving; he was very disappointed when it did not [plate 96(1)] and lost interest in it almost completely.[4] He reacted similar to a horse-on-wheels, which could be moved, but could not move by itself.

8. Pulling Various Objects.   If the child cannot ride, he pulls objects behind him, and this satisfies his passion for movement. After Roody had hardly started walking (11 months of age), he willingly pulled his big carriage behind him, pushed his little carriage, and dragged a chair or a stool. Later (1 year and 4 months), he was delighted to push a small cart loaded with sand [plates 73(1)–73(3)]. In wintertime, he pulled a small sledge behind him [plate 93(5)]; this way of walking, in which he leaned on something, made his walking and running firmer and faster.

Later, he started dragging behind him everything that could be pulled or could roll: toy horses, toy cars, sledges, carts, rugs, and so on [plates 73(4), 73(6)]. As he grew, this ability expanded. At the age of 3 to 3½ years, he enjoyed moving carriages, which required significant muscular effort: a heavy sledge and carts loaded with snow in winter and in summer a heavy toy truck with a passenger (a little girl) [plate 94(5)].

Not only at the age of 2½ years, but also when Roody was 4 years old, he willingly entertained himself by pushing toy hares jumping in front in front of him [plate 73(5)], hares beating the drums, butterflies fluttering their wings, or windmills. Every time we went for a walk, the boy took a movable toy[5] with him to diversify his walk, which was already full of his movements.

Roody did not walk in a conventional way: Now he ran forward, now he returned; walking was just simply too slow for him. He used various tricks—he leaped and galloped. Sometimes, he ran like a madman, running the risk of colliding with passers-by. While Roody did not exceed Joni in the quickness of his movements, the diversity of his movements was at least as rich as those of the chimpanzee.

As in the case of the chimpanzee, the child likes to produce thundering noises while running. He enjoys dashing around the rooms dragging trains of toy carts, cars, or tractors behind him. The louder the sound, the more happy the child.

Like Joni, Roody loved to ride the doors, pushing against the floor. Because his hands were relatively weak, he could not hang for a long time like Joni did.

Similar to the chimpanzee, my boy would often set a polished board at an angle and roll down objects that could roll. Sometimes, he followed suit and tried to slide down in a sitting position. We all know how children love slopes made of wood or covered with ice [plates 76(4), 105(5)].

My 5-year-old Roody took enormous pleasure in sliding down a big and very steep sand hill; lying on the ground, he would roll down like a ball, from side to side and from back to stomach. Both human child and infant chimpanzee like to tumble over their head; the former does it more carefully than the latter. Sliding down the banisters is a favorite pastime of schoolchildren; they recklessly zoom past you in the stairways, neglectful of the danger.[6]

The child easily can be involved in walking and running of a more complicated nature. At the age of 3½ years, he already tried to walk on stilts using someone else's help. At the age of 3 to 4 years, he was very enthusiastic about skating and skiing [plates 106(3), 106(4)]. When he could not go outside, for whatever reason, he imitated skating and skiing at home: He would put big sandals on his little feet and, pushing against the floor with two sticks, would slide on the smooth stone floor as if skiing [plate 106(1)]; or, he would put wooden blocks under his feet and, pressing them to the floor with all his weight, would reproduce skating movements [plate 106(2)].

The child tries to increase his running speed, making his running more complex. The chimpanzee also complicates his running by holding objects in his feet or by making it less comfortable.

9. Rocking.   Any kind of rocking gives the child (at the age of 2½ to 3 years) immense pleasure. He sits on a rocking horse or rocking chair all the time [plates 95(1), 95(2)]. Sometimes, he would rather fall down from them and hurt himself than slow the tempo. The child also likes very much (3 years old) to sit on a human's feet while the human intermittently pushes them up, tossing the child up a little.

Obviously, the human child, as well as the chimpanzee, likes to rock in a swing. Since a swing was hung in our garden, it became my boy's favorite entertainment. Sitting on two swings next to each other, Roody and his friend would compete in their swinging amplitudes. It turned out later that they imitated airplanes and competed in the height of flying.

Very soon, simple swinging became boring for the child, and he, like Joni, diversified it with various tricks. He wanted (at the age of 3–4 years) to swing standing on the seat, he tried to hitch the adjacent swings and trapezes, or he started playing with a plush teddy bear, trying to catch it by his feet [plates 93(1), 93(2)].

The play with the teddy bear consisted of four stages:

1. Roody tied the teddy bear to a rope hanging next to him and sat in the swing.
2. Swinging, he tried to push the teddy bear at the same time so that it would swing alongside him [plate 93(1)].
3. While swinging, Roody tried to hitch the teddy bear by his foot and to pull it closer to him.
4. Roody hitched the teddy bear (after much effort) and put it on his knee [plate 93(2)].

10. Climbing the Trapezes.   For Roody, as well as for Joni, the trapezes offered inexhaustible opportunities for exercise and tricks. At the age of 4½ years, Roody willingly, but at the same time apprehensively, climbed a not very high wooden

ladder [plates 15(3), 93(4)]. He was delighted and loudly expressed it when he managed to reach the top of the ladder. At the age of 5 to 6 years, he already was dissatisfied with this achievement: He timidly climbed to the highest rods from which the swing hung and sat there tense and diffident [plate 15(4)].

A 5- or 6-year-old child can climb rope ladders fairly easily. Like Joni, Roody tried to place his feet on the rungs higher than his head [plate 80(2)]. But, despite his great effort, he did all these tricks much more timidly, tensely, and imperfectly than the chimpanzee. The weak hands of a human and lack of tenacious support on the part of his feet did not provide him with such a confident, easy, brave, and unrestrained ability to move, which were so typical of the chimpanzee.[7]

Roody at the age of 6 to 7 years learned (without our help) how to twist and spin on the trapezes and could engage in this for hours. Certainly, Roody could not hang only by his feet with his head down; Joni could do this, although very briefly. Roody was also considerably worse than the chimpanzee in other exercises involving exclusive use of his hands, such as hanging on wooden beams or ropes or in the agility of climbing the trapezes.

A desire to crawl through something is very strong both in humans and in the chimpanzee. Roody (at the age of 4 years) noticed a hole in the rope net on the side of his bed. From that moment, he got into his bed not by stepping over the side rail as he used to do, but through this hole. He got stuck in it, breathed heavily, and squeezed his body through the hole; nevertheless, he preferred this complicated way to any other. I recall that, in my childhood, I tried to stick my head between the vertical rods of the bed's back, although I could not always free myself from this self-imposed imprisonment.

Comparing the play of the human child and the infant chimpanzee in motion, we can draw the following conclusion. Although in both infants a strong tendency for active and fast movement is observed, the infant chimpanzee is better in terms of energy, speed, and diversity of all kinds of movements. Both infants diversify their movements, making the movement more complicated by erecting obstacles along the way. In the case of the human child, these complications are associated less with physical suffering than for the chimpanzee; they are directed at spiritual development rather than physical endurance and nurture braveness, courage, and agility, but the ape learns to suffer through pain stoically. The human child to a larger extent trains his ability to walk, and the chimpanzee trains it to climb, hand, and crawl through.

Like Joni, Roody often complicated his gymnastic tricks by taking with him one of his favorite toys (most often, his fairly heavy plush teddy bear). But Roody, unlike Joni, never took anything in his mouth for this purpose.

11. Entertainment with Easily Movable Objects.   Like the chimpanzee, the human child, when prevented from moving, tries to put in motion everything he can lay his hands on. Already at the age of 7–9 months, the child tried to push a clock pendulum or pictures hanging on the walls, flip a notebook cover or cupboard door open and closed, turn faucets, jerk ropes, and throw objects to watch their flight.

For my Roody, throwing objects in front of him, scattering his toys around the room, or waving with objects in his hands were the most common entertainment. He could not sit still and watch objects on the table; he threw them off the table by an abrupt movement of his hand and watched them fall on the floor.

Roody (1.5.17) entertained himself by spinning a ring or a thimble on his finger. At the age of 2 years, he used to rock a small wooden ape or other moving toys. Of all the toys that I bought Roody when he was 6½ months old (a collapsible pyramid; three colored balls; a rubber rattle; a wooden owl with moving wings, ears, and nose), he especially liked the owl because of its moving elements.

Obviously, the child's favorite toys are the ones that are easily moved and capable of rolling. We can understand why a ball (particularly a rubber ball) is a human's toy for his entire life, from the moment a 4- to 5-month-old child, who cannot yet sit up or stand, reaches for a small ball hanging above his crib and through his adulthood, when he plays volleyball, tennis, basketball, and even airball, which he can set in motion only with partners. It seems that no toy can compete with a ball in the diversity of its use. The child cannot yet walk, but he wants to move. He is given a ball; the ball falls out of his hands and rolls on the floor. The child reaches out for it and eventually catches it; he throws the ball in front of him so he can run after it and catch it again. The child's potential activity has found an application.

A ball was Roody's first desired toy: The child willingly rolled it over the floor, threw it, and caught it. He carried the ball with him, although it took him much effort to do so (he opened his mouth and stuck his tongue out in a gesture of diligence [plates 71(3), 101(1)] because he could not yet coordinate the movements of his feet easily when his hands were busy doing something.

Roody (2 years old) already tried to imitate playing football by pushing the ball with his foot [plate 101(2)]. Later (at the age of 2–4 years), he invented numerous plays with the ball: He would roll the ball over the floor, aiming it at his companion, and the next moment meeting the ball coming back. Or, he (at the age of 1.3.8 to 2 years) would put up bowling pins and knock them over with the ball [plate 102(5)]. Also he would fling the ball against a wall and try to catch it when the ball bounced back, although he was not good at it at the time; or, he would throw the ball for others to catch [plates 102(4), 102(6)]. He (2.3.7) also would lay the ball on the ground and drive it with a stick. Until he was 3 years old, my child could not catch a ball in the air [plate 102(3)] and was limited to catching it on the ground; nevertheless, even such play brought him much joy. Joni also was not good at catching a ball in the air.

At the age of 4 to 5 years, the boy already liked to play basketball and volleyball; he rather would have played real soccer if he had not been forbidden to do it. Watching a soccer game was one of his favorite entertainments, particularly when he was allowed to bring the errant ball back to the players.

At the age of 3 years, Roody liked very much to roll small balls or toy cars down an inclined plane [plate 105(5)]. He could spend hours watching objects sliding downhill and coasting over the floor.

Four-year-old Roody was already able to play cricket and was fairly good at it. Rolling wooden wheels or hoops [plates 105(3), 105(4)] over the ground with a wire guide was so captivating that Roody simply could not go out of the house without something to roll in front of him.

Like a puppy or a kitten, which are driven by any moving object, the human child cannot resist the powerful instinct that urges him to run and to follow everything that flees from him; the child can think of many entertaining ways of moving the objects. We all have observed children trying to catch a spinning top [plate 105(6)] or how children in wintertime kick pieces of ice in front of them to make a long trip less boring.

A flying balloon [plates 59(1)–59(3)] or a soap bubble enthrall the child, intrigue his mind, and make him happy. How glad is a 3-year-old child to receive a colored balloon [plate 59(3)]? He joyfully runs around the room, holding the balloon by the thread, letting it go, and catching it again. An ephemeral soap bubble captivates him even more. The child inflates the soap bubble through a straw with great difficulty [plates 59(1)]. He is proud of this achievement; he passionately wants to share his joy with adults, but he is afraid to take the straw out of his mouth. He says hastily, while his teeth still hold the straw, "Look at this bubble, how big it is!" The bubble suddenly bursts, blown by a gust of air. The child wants to make an even bigger bubble and resumes his attempts. He watches a new bubble form; it displays iridescent flashers, hangs in the air, and descends gradually. He attentively watches its flight [plate 59(2)], enjoying this view, but the bubble bursts again. This seems to tease the child and challenges him to make a bigger bubble in order to savor the sight of its flight for a longer time, to see it rise higher in the air, to run after it around the room, and finally to catch it.

Inflatable rubber "devils" [plate 107(2)] peep loudly when deflated and combine all the best features of a toy: They move (inflate and deflate), they make sound, and they have a spherelike shape.

Other types of toys that move are an inexhaustible source of entertainment, including those that jump (grasshoppers, frogs), crawl (self-propelled mice, bugs, butterflies, ladybirds), rock (monkeys, parrots), and jerk (dogs that toss their legs up, puppets, shaking monkeys, balloons that dance in the air, birds with fluttering wings). All rotatable objects strongly attract the child, and he immediately tries to set them in motion.

A 2- or 3-year-old child persistently seeks permission to turn the wheel of a sewing machine [plate 105(2)] and is upset when he is not allowed to do it for an indefinitely long period of time. He is very happy when he gets a toy sewing machine or a windmill in his unlimited possession [plate 69(3)].

After 2-year-old Roody saw a spinning stool, he could spin it back and forth for hours [plate 105(1)]. In the beginning, he was so engrossed in this that his mouth was open constantly. Sometimes, he put one of his toys on the seat; the toys shook and fell off the stool at a certain speed, and that was the part he enjoyed the most.

12. Watching Objects That Move.   When the child is unable, for whatever reason, to participate actively in movement, he entertains himself by watching it. A 4-month-old baby already stares at people when they comb their hair, put their

clothes on, or make other movements, and vice versa: When people around him are not moving, he gets bored and starts whimpering.

My 5-year-old Roody, like Joni, was willing to watch street traffic through the window for a long time (people walking, horses and cars moving), emitting an unclear mooing sound (similar to the sound Joni uttered in similar circumstances). Sometimes, he turned his head while watching an especially interesting object.

The 8- or 10-month-old child intensely watches a spinning top and tries to grab it. Smiling and joyfully shouting, he watches scraps of paper carried by the wind, toys floating on water, and dogs running around the yard.

Moving objects (a clock's pendulum, leaves fluttering in the wind) attract the child's first active attention (in Roody's case, it was at the age of 0.2.18). Later (at the age of 1.5.0 to 3 years), the child never misses the opportunity to watch closely children who are playing or marching soldiers for a long period of time. A suddenly appearing group of skaters or skiers on the street is a special treat for the child: He becomes so captivated with such a sight that he seems ready to "swallow" them with his eyes. At this age (1.8.0), every time Roody was on the street with me, he was so eager to see a car that when there was no car around for a while, he began whimpering and demanded, "Du-du, du-du." I observed him many times (for instance, at 2.10.0) staring at slowly moving tanks. If the tanks happened to stop, he would exclaim, "I wonder what they are going to do?" After the tanks left, the child would say sadly, "I am bored without them!"

The child (1.4.0), hearing the sound of a propeller, greedily watches a soaring airplane. Later, every appearance of an airplane or a balloon is a joyful event for him.

Watching animals (another kind of moving object) is very interesting for the human child. Everybody who has observed and empathized with children's delight when they are watching live animals at the zoo or in animal stores will not need an additional proof of that.

In the case of more accessible animals, Roody (like Joni) did not limit himself to watching and entered into more active communication (as noted in the section on active play).

Child's Mental Activity   We all understand very well why children engage in active play so passionately. The child's rapid physical and mental growth stimulates his unceasing activity. Everyone who undertakes the joyous effort of observing a 3- to 5-year-old child will immediately find how creative the child is in inventing a great variety of entertainment even within the constraints of his room. The human child, to an even greater extent than the chimpanzee, demands to be entertained endlessly.

At the age of 4 months, Roody already was bored if there were no changes in the immediate environment; I could observe him shifting his eyes from one object to the next and starting to whimper when he saw nothing new. He immediately calmed down when he was taken into another room and was engaged in play.

If a child lies in a deep carriage from where he can see only sky, he soon gets bored with it and shows visible signs of uneasiness and sorrow. He cries and whimpers. In contrast, when he is carried in a person's arms and can see a changing picture, he feels satisfied and therefore can stay calm for a long time. When the child is awake, he needs to change his activities and objects of entertainment all the time. A 10-minute record of my 11-month-old baby during play clearly shows how his attention shifts from one object to another.

March 19, 1926

Roody (0.11.15) touches and turns over a small wooden figure of man, then a scoop; takes a toy chimpanzee, walks with the toy in hand to his uncle, goes around him, goes to a chair at the other side of the room, drags the chair, comes up to me, jerks at my dress, whimpers, looks at me, comes up to the chair again; suddenly, sticking out his tongue and stretching his arms to the sides, runs to a chest of drawers at the other side of the room, cries; stares at me, giggles, runs over to the carriage, takes a teaspoon, bangs on the table with it, takes a napkin, drops it, takes the cover from a can. Suddenly he runs, grabs a small chair, drags it, turns it skillfully when it gets stuck by the bed, walks to the corner, takes a wooden wheel, runs with it to the chair, grabs the chair again; with the wheel in his hands, comes up to the door, comes back to the chair, takes it; holding the chair and the wheel in his hands, walks to the wall, goes away from the wall; now he comes up to the carriage, takes it by the wheel, knocks with a big wooden wheel on the carriage, drops the small wheel, grabs the chair by its back, drags it for a second, suddenly picks up some toy from the floor, drags the chair again; with the toy in his hand goes to a dresser, comes up to me, looks at my shoes, goes away from me, takes the carriage by its wheels again; bends to me, looks at my shoes; bends lower, looks under the bed, reaches for a bowl under the bed; takes the chair again, knocks it over, drags it over the floor by its legs, drops it again, comes up to the bed, looks under the bed at the bowl; goes to another corner, takes the carriage, drags it back and forth, approaching it from one side and then from the other; leaves the carriage, takes the chair for a second, comes up to the bed, looks under the bed again, touches my shoes, sticks his finger into one of the shoes; bends over, looks under the bed again, suddenly comes up to the carriage, takes it by the wheels, comes over to the wardrobe, comes back to the bed, stares at my shoes again, touches them, hitches to the bed screen, looks at me, comes up to the dresser, takes a spoon, a can cover, drops them, runs into the dresser, cries, tucks his head in my lap, comes up to the wardrobe, touches a toy.

During these 10 minutes, the child changes his movements over 80 times and touches more than 20 objects; as can be seen in the record, not only his actions repeat, but so do the objects that attract his attention.

The same uneasiness showed later in his life. While a 10-month-old baby only tries to reach various objects, a 2- or 3-year-old child left to his own devices runs, grabs, examines, and touches everything he sees (ground, stones, plants, roots, leaves), turning from one to another.

Similar attention shifting pertains to the child's food. Once, I gave Roody a box of candies. He started tasting them all one after another (just like Joni) and came back to the first candy at the end.

However, 3-year-old Roody displayed more attention concentration than Joni, for example, during creative activities, imitations, or when somebody read a story to him.

The child entertains himself even when he is sick.[8] He tries to enliven with play all mundane and boring procedures associated with going to sleep, eating, getting dressed, and even fulfilling his common physiological needs. My 3-year-old child, in bed with influenza and with a temperature of 38°C, persistently demanded that somebody read to him or bring pictures and toys for him to look at and for play. Entertainment often makes the child forget about pain. Roody at the age of 9 months severely hurt his head against the metal back of his bed; he cried so hard that it seemed nothing could calm him, but as soon I brought him close to our clock and began knocking on it, he stopped crying.

Everyone knows how reluctant a child can be to go to bed at night. The child will make up excuses to delay going to bed even for a couple of minutes, saying something like, "I have not played enough." When the child is already in bed and is about to go to sleep, he nevertheless tries to talk. When we interrupt him and urge him to sleep, he asks, with is eyes closed; "Well, what else?" as though hoping that there still might be a chance for additional entertainment.

Enormous effort usually is spent on teaching a preschool child to sit still during meals or when he is reading or writing. For example, while at the dinner table, Roody would knock with his feet on the table legs, knock with a knife or fork on the plate, or invent other indescribable entertainment.

It is well known that when school-age children sit at the table without the control of adults entertain themselves by kneading bread crumbs and throwing them at each other. Sometimes, there is a real "bread battle" at the table. It has been mentioned that Joni also ate more willingly when eating was accompanied by additional entertainment.

I previously have described at length[9] how difficult it was for me to teach Joni to sit still and how hard this immobility was for him. My lessons with the little chimpanzee were especially successful when they assumed the form of play, that is, after every correct answer, I engaged him in active play as a reward.

Children are always reluctant to put on or take off their clothes. Roody (before 1½ years of age) often cried during these procedures; later (1½ to 3 years of age), he rattled or played on his lips with is finger to amuse himself a little. He did the same thing when he sat on a chamber pot. Sometimes, it seems that there is no regular activity a child will not try to enliven by his play. For instance, when Roody was washing himself (at the age of 3 years), he deliberately splashed water at people around him or at the walls; he hung the towel on the hook and took it off. He played with a comb while he was combing his hair (for instance, rattling the comb's teeth, etc.).

The human child, like the infant chimpanzee, follows his instincts for mental growth and development, greedily grabbing new experiences from the environment. This tendency for the new and entertaining is as typical of him as his desire to move and his interest in everything that moves. His own words prove this idea better than any commentary. At the age of 4 years, the boy said to me during our long walk on the streets, "I could walk and walk like this if there were new things to see all the time." Later, at the age of 7 years, he made a similar remark when he was looking at pictures in a book.

At the age of 2 to 5 years, this tendency was expressed in his constant craving for new toys; every time someone came carrying shopping bags, the boy would say, "Did you bring a new toy?" although he already had hundreds of toys. When we lived in our summer house in the country, Roody did not have a flow of new things for him to examine, and he bothered me for days: "Show me something new."

If there are no new impressions for the child to experience, you can often hear from him, "Mother, I am bored." The child burns with curiosity to an even greater extent than the chimpanzee; very early in his life, he shows enormous mental initiative and searches for new objects that he can watch and examine.

As early as at the age of 2½ months, when a child is brought to a different room he looks around very attentively, shifting his eyes from one object to the next. Sometimes my 3- or 4-month-old boy would cry, and we had no idea how to calm him. As soon as we brought him to a different room, he would stop crying immediately, look around the room, and open his eyes wide. And as soon as we brought him back to his room, he would resume crying. At the age of 8 months, Roody screamed joyfully and jerked his legs energetically when he was brought to a different room.

The older the child, the stronger his desire to expand the limits of his curious observation. At 1½ years old, Roody passionately wanted to go outside the limits of his apartment, house, garden, yard, and street, which proved that Schiller was right when he wrote, "A baby is comfortable in his crib; he grows up and the whole world becomes not large enough for him."

All new situations that broaden the human child's (or infant chimpanzee's) observation sphere are desirable: looking out the window, visiting unfamiliar houses and streets, moving from one apartment to another, watching street scenes, and at a later age, watching theater performances and movies.

Those who, like me, observed children watching a puppet theater would never forget the enraptured and shining eyes of the children, their open mouths, and their tense poses, which revealed their entire fascination with this sight.

It is well known that as long as there is a demonstration, a funeral train, marching soldiers or children, a train of tanks, or travelling actors with street organs and trained animals on the street, every window, gate, sidewalk, or square will be filled with curious children drawn to the sight as flies are to honey. Roody (at the age of 4 to 5 years) was fascinated with military attributes: every time he saw a tank or an armored vehicle on the street, he ran into the house crying widely, "Tanks, tanks are coming!" inviting us to participate in the watching.

We all know how delighted children are to move from one home to another. My 5-year-old Roody, during our move to our summer house, for hours would look out the train window and ask everybody questions about what he saw at the moment. The earliest child's recollections (as noted in autobiographies of many famous people[10]) are associated with a change in the environment such as moving to another city, village, house, or the like.

At 1½ years old, Roody was anxious to get into the adjacent rooms occupied by our museum, where he could find an infinite number of new objects. He (1.8.20) tried by all means (like Joni) during our evening walks to look into the windows of other people's apartments. When we rode a streetcar he (1.10.0) intently looked out the window.

The human child is as interested in meeting new people as the infant chimpanzee. Every time you visit a family with children for the first time, you immediately become an attraction. Bolder children will examine you from head to toe, get on your lap, and like Joni, touch all intriguing objects (watches, brooches, beads, rings, buttons). More timid ones will limit themselves to peeking through a door crack at the newcomer. My boy as early as at the age of 2 months smiled when he saw a new face. Later, at the age of 3 years, he expressed little interest in regular visitors, but was anxious to see foreigners (those who were German, British, Chinese, black).

A 1-year-old human child (like the chimpanzee) expresses interest in any new object coming his way, grabs it with a smile on his face, and often puts it in his mouth. I often showed Roody (younger than 1½ years old) something new to stop his crying. Showing new objects to the human child (as well as to the chimpanzee) distracts him from anything he is doing at the moment.

Curiosity and desire to see new things and experiment with them were my boy's very conspicuous features. Roody at 11 months of age could entertain himself with any trinket for a long time if only it was new to him. Once, I gave him a tiny striped piece of cloth; He took it in one hand and then the other, stared at it for along time, picked it up when it fell on the floor, and refused to give it back. When a child of this age sees something new, he persistently reaches for this object [plate 48(1)], utters a long sound "a," points his finger at it, and will not calm down until he gets it.

At home, the child would willingly sort out drawers filled with junk that was new to him [plate 62(5)].

Unlike Joni, whose sense of smell was as significant for him as his vision, Roody's vision played a predominant role. The human child smells only things with a very strong odor (for instance, perfumes, vials with medicine, fragrant flowers, etc.).

While for Joni, at a later age, smelling and touching retain their importance along with vision, in Roody the sense of smell gradually retreats to the background; touching with the hands and lips plays a more significant role compared to vision only in his early years (before 1½ years of age) [plates 108(1), 108(2), 108(4), 109(3)–109(5)]. Later, it is only of secondary importance.

I noticed in my boy almost until he was 7 years old an atavistic habit of touching various things (even things that were picked up from the ground), his fastidiousness notwithstanding. This tendency for touching things sometimes remains strong even in adults. Those working in museums know it very well. In a museum,[11] there are signs "Do not touch, please" everywhere; visitors are usually very eager to get permission to touch.[12] This tendency is especially intense in children deprived by nature (as in deaf-mute children). Once, I gave a group of such preschool children the opportunity to touch some of our stuffed animals and birds; they were so fascinated with this that there was a real danger that they would take these objects apart. Obviously, the tendency for touching is even more pronounced in blind children because touching compensates here for the absence of vision; this fact is well known, but it stands somewhat outside our topic.

Entertaining with Sounds   The child is intrigued with and amused by new sounds. I have the following facts registered. At 3 months, Roody was crying loudly and long; everybody thought he was ill. Suddenly, he heard a whistle coming from the street; he stopped crying as though he wanted to listen to the whistle for a second and did not resume crying at all. Another time, when 4-month-old Roody cried very loudly and nothing seemed to calm him down, I uttered an unusual hissing sound; the child quieted down immediately and soon went to sleep. When a child hears smacking or rattling lip sounds for the first time, he smiles; he visibly likes to hear these sounds again.

The child likes spinning objects and objects that make sounds immensely. The child at 1½ to 4 years old enthusiastically spins wooden or metal humming and buzzing tops that he can set into motion on his own (3 years of age) [plate 105(6)]. With the same pleasure, he spins homemade humming toys, metal rings that slide up and down and back and forth along a wire. Sounds by themselves amuse the child no less than the chimpanzee; both infants sometimes reproduce them in a similar way and under similar circumstances.

Early in ontogenesis, the child enjoys sounds and tries to reproduce them. A child from 7 to 12 months plays joyfully, takes a toy in each hand, and knocks one against the other. The child often beats with his palms on smooth table surfaces, knocks with metal spoons on solid objects, throws ringing tinware on the floor, and bangs with hammers and toys on solid surfaces. The child is especially good at clanging metal objects on the metal rods of a bed. With a smile, he rattles with a box of matches or a bunch of keys. Incidentally, it was a bunch of keys that was the first toy that made sounds that attracted Roody's attention at the age of 4 months and entertained him later (until 1.7.0).

The child (1 to 3 years old) tries to make sounds when he walks. He drags thundering carriages or scraping shovels, he intentionally shuffles his feet, or he passes a stick over the poles of a fence, producing rattling sounds.

I recall how glad 2-year-old Roody was when he found out that his new shoes screeched.[13] He enjoyed the screech when he was walking or running; he was genuinely upset when we wanted to eliminate that sound. Another time, I watched how Roody (1.4.0) was delighted to run over rattling autumn leaves.

All kinds of toys that make sound are used by a child with great enthusiasm. Roody at 2½ years old loved to fire his toy gun, passionately burst paper bags (1.7.0), and later fire his percussion gun. When 5 years old, Roody used to throw a big chunk of lead against the wall, relishing the ringing and deafening sound.

We constantly can observe countless ways that children of the ages of 1 to 7 years use sound-imitating and sound-reproducing play. To produce a sound, the child uses whistles, metal tubes, wooden or clay pipes [plate 107(1)], keys with holes in them, combs, rattles, handbells; if he has no such things in his possession, he whistles and rattles with his lips.

Like Joni, my Roody at 1 to 2 years of age often entertained himself by repeatedly knocking on solid surfaces. Roody (1.6.0 to 3 years of age) enjoyed toys that made sound: cymbals for which sound was produced by special hammers [plate 107(4)], a toy piano [plate 107(6)] and, first and foremost, drums [plate 107(3)], which were his favorite even during his school years. The more creativity that is involved in the process, the more he likes doing it. For example, the child likes accordions [plate 107(5)] and guitars better than mechanical musical toys, which have a monotonous procedure of extracting sound. The object that makes sound and that can also roll is really something else. My boy (3 years of age) was absolutely fascinated by a small musical wheel; he begged us to buy it for him, and we finally did. He could run around our garden for hours, pushing this wheel in front of him, enjoying the way it moved and sounded.

The child at the age of 1½ years already is willing to listen to a piano or wind instruments; when he sees a brass band marching along the street, he (2.2.11) usually smiles and asks for permission to follow the band. When he heard a band playing at some distance, he (2.1.25) said, "Good music." Once I took Roody (1.9.20) to a skating rink. It was illuminated brightly, and music was playing. The next time when we went there, there was only illumination, but no music. The child was so upset with it that he started to cry.

The child does not miss any opportunity to utter a sound. Roody at the age of 3½ years liked very much to push the buttons on a typewriter, press the handle of a puncher, and beat the keys of a piano keyboard. At the age of 1½ years he enjoyed the sound of the piano keys as much as he did the sound of the piano lid opening and closing.

When Roody heard the clicking of a photocamera, he begged the photographer to allow him to play with the shutter. He kept playing with it for a very long time.

All children love to scream and to make various sounds with their tongues, lips, hands, and feet (more details are in the section, "Imitation of Sound"). My boy, for example, before the age of 6 years, liked to shriek. He could not do it during the day because of the classes in the next room. He asked my permission to cry every morning before 10 o'clock (the time the classes began).

I often observed that, when Roody (about 10 months old) was bored, he started immediately to rattle his lips and blow saliva bubbles like Joni did when he was bored. Later, when Roody was 2 to 5 years of age, he rattled his lips every time he put on or took off his clothes or during other boring procedures or long waits. Children, when they are bored with long journeys in streetcars or buses, also rattle their lips. This habit sometimes is rooted so deeply that is carried over into adulthood: I observed a 17-year-old girl who rattled her lips while doing some daily chores.

The human child, like the chimpanzee, likes to clap his hands not only to express his joy (as it was noted in the section on Joy), but also for entertainment [plates 85(4), 87(3)]. After Roody learned how to clack with his tongue, he kept doing it dozens of times, demonstrating this ability to everybody around him.

A 10-month-old child is eager to listen to rhymes recited to him; he likes drawling declamation and new unusual sounds and words. Once, when Roody was going to sleep and I was imitating domestic animals, I asked him, "How does a gun fire?" and I answered myself, "Bang, bang!" The boy started to laugh. When Roody was 10 months old and I read him a rhyme about the purring cat, he invariably laughed at the sound of the final line. Roody at 1½ years old listened intently to long fairy tales; he liked them very much, demanded that we read to him, and cried when we did not do it (1.9.0).

When the child (at the age of 4 months to 1 year) is in a good mood (after meals or after sleep), he utters ringing, shrieking sounds that visibly amuse him. Later, he can entertain himself for hours with babbling that consists of diverse, single, unintelligible or intelligible sounds that constitute the basis of his future language. Awakening in the morning, 9-month-old Roody babbled to himself for 10–12 minutes, endlessly pronouncing separate syllables, for example, "a-poo-ba-ba-pa-ka-bo-bo-kkh-ta-da-ta-dia-ta-to-pa-pa." Sometimes, Roody (3.3.22) uttered a series of unintelligible words rhythmically connected with each other, for example,

kamenone-lamenole   iltenoen-emenone
kamentaen-lamentor   kelbororen-minemore-kamentor

Comparing Roody's entertainment with sounds with that of Joni, we may conclude that both infants use various ways of sound reproduction, but in the case of the chimpanzee, I have never noticed the tendency to entertain himself with the sounds of his own voice as it was the case with the human child.

Experimentation
Play

1. Play with Solid Objects. For experimentation play, the more plastic and flexible the objects are, the more attractive they are for the human child. The more dynamics are in the toy, the more the child is anxious to use it. In contrast, an object that offers no alterations, after a short examination is ignored totally as soon as it ceases to be new.

Play with solid objects, such as stones, pieces of paper, vials, nails, and buttons occupy an important place in the child's life. Like Joni, Roody found materials for experimentation all the time; everything Roody happened to find, he utilized in his play. His mental activity was expressed in his play with such materials as chips of wood, matches, rubber and wooden sticks, water, fire, sand, small stones, and more.

This play was very important for the child's development because he familiarized himself with three natural elements: earth, water, and fire. Everyone knows how babies like water and how they like to be bathed.

2. Play with Water.   While in water, 3-month-old Roody lies with a cheerful expression on his face; he smiles, extends his hands forward, and bends his back. At 7 months, Roody allows us to wash his face; he opens his mouth and catches every drop of water dripping from his face, rinses his hands in the water, brings his hands to his mouth, and sucks his fingers. While in the bathtub, he usually beats at the water with his hands, smiles, closes his eyes, splashes himself with water, and chokes, swallowing water. Roody at 9 months watches closely as the bath is prepared; he waves his hands, reaches for the water, and screams, anxious to get in the tub. Taking Roody out of the bathtub invariably is accompanied by awful crying, kicking, and other protesting body movements.

At the age of 1 year, the child already is drawn so strongly to water that it is difficult to drag him away from it. As soon as the child sees a wet spot, like Joni, he tries to spread it with his hand or with a rag the same way the infant chimpanzee does [plate 110(2)]. If Roody (1.6.0) notices a puddle in the yard, he invariably steps in it, whips it with a stick, throws a stone at it [plate 55(4)], or sometimes simply dips his foot into it, splashing himself and everybody around him. When he (2.1.25) is taken away from the puddle, he gets irritated and beats at his babysitter; his face reddens, and his eyes fill with tears. The child is delighted, after a rain, to watch cars and buses splash water from the street, showering the pedestrians. If you give a big bowl full with water to a child [plate 110(1)], you will see how joyfully he splashes in the water until there is no more water in the bowl.[15] Later, the child (1.6.6) takes water in his hands, floats toys on the surface of the water, catches the toys when they sink, sinks the toys himself, and scoops water with cups and pours it, raising his hand with the cup as high as he can [plate 110(3)]. He collects water in rubber toys and squeezes it out.

Leave a 1½- or 2-year-old child in a bathtub, and he will know how to have fun! The child (1.5.0) opens the faucets and catches the pouring water in his hands.

If you let a child swim in a pool or pond, after he gets used to the new circumstances, you will have difficulties taking him out of the water because it offers him so many different forms of entertainment. Roody (at the age of 7–8 years) often played in huge puddles in the yard, floating wooden boards and making some kind of a raft on which he tried to stand.

Children at 3 to 4 years old, smudged with clay or covered with sand, like to run into water and wash themselves. They build dams, float boats [plates 63(1), 63(2)], and splash each other and all those around them. "Water is alive, it runs," and this mobility provokes the child's quick mind and senses and stimulates his initiative. One time, Roody (1.2.17) did not want to stay in one part of our garden and kept asking me to move to another. Answering my question of why he wanted to go there, he said, "I can see water from there," meaning a big rain puddle that was rippling in the wind. At home, all activities associated with water are especially long. The child, with visible pleasure, waters flowers [plates 94(3), 111(4)]; the 1½- to 3-year-old child throws various objects (stones, wooden chips) into water and watches them float or sink. Once, I caught Roody (2.10.9) really experimenting: While putting objects on the surface of the water, he was murmuring to himself, "The lid is floating" (wooden lid). "Will this one float?" he asked himself, putting another lid (metal) on the water. "Why doesn't metal float?" he asked his father, seeing that the metal lid sank. "Stone doesn't float, but teddy bear does."

Roody liked to pour water from a teapot into toy cups, wash a doll's clothes in a bowl (2.6.0), and put toys into water and take them out, enjoying their shine and was reluctant to wipe them dry [plate 110(5)]. He filled toy tanks with water, dragged them around the room, or poured water from one vessel to the next (1.9.26).

But, no toy associated with water was so desirable for my 3-year-old child as a common rubber syringe. This object became an inexhaustible source of entertainment. The child filled it with water; dripped water from the syringe on his hand [plate 110(6)]; aimed the spurt at the flowers, at the wall, at the people; or, immersing the syringe into water, he listened to the bubbling sound. Playing fireman was the ultimate fun because it involved water that needed to be taken out of a barrel, poured into a watering pot, and then applied to the so-called house on fire [plate 65(2)].

At 7 years, Roody simply could not just wash himself at the sink. He blocked water from the faucet with his finger, breaking down the spurt, or like Joni, he caught water in his hands and splashed it around. Even at the age of 7 years, he enjoyed filling his mouth with water and blowing it around. We all know that, after massive rains, puddles swarm with children, who are using every opportunity to have fun with water.

Roody, like Joni, was afraid of larger reservoirs and was extremely careful while in them.

3. Play with Loose Materials.   Such play materials as soil and sand are the second most popular ones after water. Like the chimpanzee, a 2-year-old human child likes to sort sand and look for small stones [plate 61(4)], to pour sand from one hand to the other, to rake it into piles, to sift it through a sieve, to throw it around, to fill forms with it [plates 49(4)–49(6)], and to put it on carts and drag the carts. Roody (2.3.11), like all children, used to hide his toys in the sand and find them (sometimes, he could do it up to 50 times in a row). He also liked to wallow in the sand, dig himself into it, or slide down a sand hill.

At home, 1- to 2-year-old Roody (like Joni) got into stoves and raked ashes or cinders out of them. He broke the cinders with his hands, bit them, tasted them with his tongue, and rubbed them with his fingers (1.7.3).

For the child 2 to 4 years old, wet sand was the most convenient material for making various shapes. However, a 2-year-old child more easily destroys what he has just created because he cannot yet use the forms, but soon he learns how to do it [plates 110(4)–110(6)]. Roody at the age of 3 years could make, out of sand, a semblance of a house with a chimney. Like Joni, Roody played with sawdust scattering it over the garden paths.

4. Play with Fire, Metal, Transparent, and Shiny Objects.   We noted Joni's strong tendency for acquiring and examining shiny and luminescent objects. A

similar tendency is observed in the human child; the bright and moving light coming from fire is the most attractive for him.

In my 4-year-old baby, I observed his attention concentrated on an object and on reaching it. This was when he was intrigued for the first time by shiny rattles, white flowers in a vase, and a shiny clock pendulum.

Fire fascinated 8-month-old Roody. Once, he was crying very loudly; as soon as we brought him close to the fire, he calmed down. At 9 months, Roody stared for a long time at the oven fire or at the flame of a kerosene heater.

A child 1½ to 2 years old tries to look closer at the fire, catch the flame in his hand, blow the fluttering candlelight [plates 59(5), 59(6)], or strike a match [plates 59(4)]. Roody (2.5.27), after he had blown out the candle, wondered where the flame had gone and started looking for it under the table and chairs.

The child is overjoyed to receive a flashlight. I remember how my 2½-year-old Roody enjoyed his "hand fireworks," giving a host of sparks as a result rubbing small flints. But Bengal lights [blue lights or flares for signaling] were his favorite (2.8.24); he burned an unlimited amount of those.

The effects of all kinds of light fascinate the child completely; a reflection of a sunbeam on the wall, a searchlight in the dark evening sky, a holiday illumination stimulate his intense positive emotions.

We noted how delighted Roody was to use his percussion gun because of its reverberating sound and bright light effects. Play involving real flame (for example, building a fire) are as attractive for a child as play with water. Many fires, indeed, are caused by improper use of flammable materials by playing children.

Roody (1.9.26) liked to watch a colorfully illuminated skating rink so much that taking him home was very difficult. Like Joni, Roody was intrigued by transparent objects: oilcloth, colored glass, combs, eyeglasses, magnifying glass, and so on.

Pressing yellow oilcloth or a broad rim of transparent red comb to his eyes, Roody (at the age of 2 years and 3 months) liked to stare at the ground, at the flowers, or at the sky [plates 112(1)]. The surrounding world, appearing in a new light, was especially intriguing for him. Once, looking through such a comb at the sky, he exclaimed, "Oh, the sky is so red, so hot!"

When 4-year-old Roody was given a magnifying glass, he would not stop looking at various objects through it [plate 112(3)]. A toy stereoscope with its infinitely changing colorful pattern was one of the child's favorite toys [plate 112(2)].

5. Play with Elastic Objects.   Such things as a rubber tube for the child, as well as for Joni, are most coveted objects and exceed any other toy in attractiveness. Roody was very creative at finding new tricks that we could do with this tube. He (2 to 3 years old) would beat at the floor with it, watching its waving movement, he would fill his mouth with water and subsequently fill the tube from his mouth, or he would use it as a "telephone receiver," putting one end of the tube to his mouth and the other end to his ear [plates 113(4), 113(5)].

Sometimes, the boy used the tube for listening to his teddy bear, imitating a physical examination [plates 58(6), 65(6)]. Sometimes, he (2.3.11) put the tube on the spout of a watering can and watered flowers through it. He also wound it in a ball, tossed it up, and watched it unwind and fall down; or, he girdled himself with it in imitation of a fireman. With another boy, he blew water from one end of the tube to the other.

At that time, the rubber tube became an inseparable part of his activities; he went out on the street with the tube in his hand. When we tried to dissuade him of that because other children made fun of him over this tube, he did not understand the reason for our objection and repeated stubbornly, "I like the bowel" (he called this tube the bowel).

This universal use of a long elastic rubber determined its attraction for the boy. Professor W. Kohler noted that such a universal object for his chimpanzee was a stick. For my boy (and for Joni), a stick did not enjoy such exceptional success as the rubber tube; although, among solid objects, a stick could compete in the diversity of its application only with a ball. Other flexible objects (belts, ropes, threads) never carried much weight with Roody. He used them considerably less than the rubber tube.

However, thin long bands also amused 1-year-old Roody. He (2.2.0) would hold the band by its ends, pulling it from side to side behind his neck (Joni also often did that); he would make a loop out of it and try to put his head through the loop; or he (1.4.27) would drag the band crying, "Hey, hey!" The baby could also play with soft long scraps of cloth, spreading them out (1.6.2), hiding them in the sofa, pulling them back, and crying impatiently when it was difficult for him to pull out the whole piece quickly.

The child (1.1.7) shuffled his diapers in his hands, put them from one place to the next, pulled them by their ends, folded and unfolded them, raised them, put them down, or waved them.

Like Joni, 9-month-old Roody liked to take a cord or a rag by one end and wave it. He did it so vigorously that he often fell on the floor, unable to keep his balance when his movements were too forceful.

At 10 months, Roody could examine, shuffle, or pull a thread between his fingers and from one hand to the other for a long time. Even common human hair attracted the little child's vivid interest. At the age of 11 months, he ripped a hair from my head and started looking at it very closely. At the age of 1 year and 4 months, Roody, like Joni, could pull a hair and very attentively watching it slide between his fingers [plates 49(1), 49(2)].

6. Play with a Stick.   A long, solid object of universal application, such as a stick, was used by the child almost as often as a ball. As soon as the child (1.1.7) learned how to run, he took a stick or a twig in his hand and, waving it in the air, ran more energetically. Later (at the age of 2 years and 2 months), he drew on the ground with a stick [plate 100(2)]; he was glad to leave a visible trace there. The child used a stick to dig a hole in the ground, to poke everywhere [plate 100(4)], or to knock on solid objects with it. A stick was used when the child (1.2.22) did not want to touch something with his hand: a kitten, a big doll [plate 100(1)]. With a stick, the boy threw aside things that he was reluctant to touch because they aroused disgust or fear for for some other reason. At the age

of 2 to 3 years, Roody used to throw sticks over high barriers or over fences. He destroyed things with a stick [plate 55(3)], scared off the dogs [plate 100(3)], and shooed pigeons. A simple stick substituted for various kinds of weapons (for example, a gun, a saber, or a sword). A stick was his companion on forest trips; he beat at tree trunks with it, or knocked off inedible mushrooms, leaves, or tree branches.

This addiction to sticks was so overwhelming in my boy that he could not pass any stick on the ground without taking it. "This is a good stick," he used to say. He always found an excuse why he needed a particular stick. Although this need did not always materialize, and he had scores of such sticks already, his desire to pick up sticks did not diminish.

A slightly bent stick was used to make a bow; long and thin sticks served as arrows; a forked twig was a catapult (such twigs were highly valued). A stick was also used in various games for which the accuracy of throwing it or beating with it determined the score [plates 55(6), 114(1)–114(3)]. At the age of 3 years, the boy invented a special game with a stick. Holding a smaller stick in his left hand, he beat it up into the air with a larger stick he held in his right hand [plate 55(5)]. We mentioned the enthusiasm with which children play various ball games [plates 101(1)–101(6)].

The human child sometimes arranged play, without any adult participation, similar to that of Joni. He (1.6.10) propped matches across his palms (Joni put them between his upper and lower teeth). Like Joni, he pricked his hand lightly with a broken vial and pulled the hand back when it hurt; or he put his head through a hole in the bed screen and pulled it back.

Experimenting play reveals how intense a child's mental activity is, how anxious he is to become acquainted with the environment, how he enriches himself with new impressions. The world around the child is completely new to him; therefore, he spends practically all the time at the beginning of his life curiously watching it.

Play for Familiarization with New Objects   1. Surpise. I noticed that the child reacted to a new or sudden event by sighing deeply. Like Joni [plates 115(1), 115(2)], he opened his mouth wide, apparently experiencing surprise. When Roody (1.1.6) was given a watch, he pressed it to his ear, his mouth half open [plate 113(1)]. But then he pressed the watch to his eye and to other parts of his head, as if checking whether he could hear the ticking sound with something other than his ear.

Once, I gave 10-month-old Roody a thermometer he was reaching for; he immediately took a deep breath. In approximately 2 weeks, I heard a similar sigh when the doors of our wardrobe were opened suddenly and he saw the big dark space. Earlier, when Roody was 5 months old, I noticed that he sighed deeply when he was shown new things or when I suddenly appeared before him after along absence. The expression of surprise in the form of the open mouth was recorded many times. When the boy was 6 months old, for the first time he saw his grandmother wearing glasses with a light-color shiny rim. The child opened his mouth, stared at his grandmother, and at the first opportunity grabbed the glasses. For other examples, I showed 1-year-old Roody a shiny vase; he opened his mouth and stared at the vase [plate 115(1)]. Roody, like Joni, showed a similar reaction when he saw a blue medallion on my neck [plate 108(1)].

I also observed Roody opening his mouth when he (1.2.0) was looking at a human face [plates 41(3), 41(6)]. Yet later, at the age of 2½ years, he did the same when he was looking at peonies in full blossom, and at the age of 7 years, when he was shown a huge toy bow.

It is interesting that a human retains this habit of opening his mouth even in his adult years. During my work as a museum tour guide, I often observed people (especially emotional ones) open their mouths when they were shown unusual objects.

A 1-year-old child's typical gesture of surprise is spreading his arms with his palms up. My boy reproduced this gesture when he (at the age of 1 year and 1 month) saw the bright spout of a metal watering can [plate 108(3)]. A 1½-year-old child often utters an "a! ba!" sound when he is surprised; Roody uttered this sound when he looked out the window and saw snow for the first time.

From an early age, the child displays a strong tendency for examining, smelling, and careful touching of new things with his lips and tongue and for greedily accumulating new impressions. A baby, as early as at the age of 4 months, starts looking at his hands and at his outspread fingers.

My 5-month-old baby kept putting his fists into his mouth [plate 109(2)] or sucking his index finger; sometimes, we had to take his fists out of his mouth up to 15 times in a row. The corners of his clothes, his diapers, or everything he could lay his hands on also went into his mouth. When Roody was between 2 and 2½ years of age, he would suck his index finger while listening to a story [plate 109(6)].

Roody, at the age or 4 to 5 months, tried to put his toes into his mouth and suck them while lying on his back. He put pieces of paper, cotton wool, toys, and metal objects in his mouth and sucked them, salivating profusely. He especially liked metal [plates 109(3)–190(5)], glass, or wooden objects and showed a grimace of discontent toward rubber and things covered with oil paint.

The 6-month-old child embraced the shiny metal ball on the back of a bed and tried to touch it with his open lips and tongue. At 7 months, Roody wanted to touch and stroke my velvet dress, a marble table, and stove tiles; he passed his hand over soft fur, pinched it, and pulled it to his mouth. The 8-month-old child gladly put his fingers in somebody's hair, shuffling it. But, he was unwilling to touch anything that bristled, for example, a brush, because he was afraid of pricking himself [plate 49(3)]; he always took a brush by the smooth handle.

The child puts any new toy in his mouth and slightly scratches it with his teeth. At 7 months, Roody looked attentively at his fingers and toes. He noticed such tiny objects as a relief mark on his pillow case and made scratching movements over its surface with one hand, then with the other; he touched the mark with his mouth, exactly like Joni while he was examining the embroidered anchor on a sailor's shirt that he wore. Roody at 9 months carefully touched my bare foot, slapping, stroking, pinching, and scratching it. When 10 months old, Roody reached persistently for various objects, such as buckets, shoes, carriages, and toys, he uttered a log "e-e-e" sound. Every drop of water attracted his attention. He bent toward the intriguing thing and did not stop trying to reach it until he finally was allowed to touch it with his hand. He touched heating pipes, tried to turn valves, or touched the valves with his lips and tongue. He took a metal pot, examined it, scratched its bottom, slapped it, turned it from side to side, knocked on it with a wooden toy, or pressed the pot to the wall.

A baby at the age of 9 to 11 months is interested in unaccessible spaces; he tries to look behind the back of the bed or pulls out a drawer and looks inside. He sticks his finger into the hole in a pyramid ring many times [plate 62(1)]. If he finds a hole in cloth, he rips the cloth apart [plate 62(3)].

When Roody (before he was 1 year old) walked around the room, he usually had a toy in his hands somewhere near his mouth. He bent every second to pick up pieces of paper, tiny specks of dust, fragments of toys from the floor, and despite constant admonitions on our part, he pulled them to his mouth. There seemed to be no object he would not want to feel with his lips and tongue. When Roody was sitting at the piano, he at first pushed the keys a couple of times, and then he started scratching the piano edge with his teeth.

Sometimes, he made a grimace while tasting an object with his tongue or trying to scratch even hard things with his teeth. At 1 year, Roody touched new leather shoes, then put the shoe's nose into his mouth; he pinched the pom-pom and pulled the laces. My boy retained this tendency to touch objects with his lips almost to the age of 8 years.

At the age of 9 months to 1½ years, he put the most inappropriate things in his mouth: rattles, metal caps, utensils, and the like [plates 109(3)–109(5)]. Once, Roody (13 months old) saw a watering can glistening in the sun. He stared at it, spreading his arms; the next moment, he grabbed the spout and began touching it with his lips [plates 108(3), 108(4)]. Later, I literally became tired of preventing the boy from stuffing his mouth with metal objects, sticks, wooden toys, and more. Too fastidious to use his spoon for the second course if it had residue of the first one on it, he would forget about this completely, pick up something from the floor and put it into his mouth, or he would lick his dirty finger, which he had just used for digging a hole in the ground [plate 109(6)]. Even at the age of 8 years, he used to come back from the yard with his chin and lips smeared with soot, which indicated that he had picked up things from the ground and had touched them with his mouth.

In the forest, in the garden, or at the riverbank, the child finds an inexhaustible storage of things to satisfy his curiosity. We mentioned (in the section on the ownership instinct) how greedily the child collects, examines, and acquires everything we sees around him: unusual tree branches, mushrooms, moss, berries, flowers, stones lying in the road, insects, and countless other objects. Even walking in a tiny garden, the child comes up to every flower, touches it, and asks about its name. During closer examination, the child usually closes his lips tightly and extends them forward [plates 115(5), 115(6)], exactly like Joni did under similar circumstances [plate 115(4)]. Sometimes, Roody felt the flowers; sometimes, he plucked them or broke the stems, looking very attentively at the "fruits of his labor" [plate 115(5)].

Thus, the child familiarizes himself, by way of experiment, with pleasant and unpleasant, harmful and useful properties of objects. He touches a very hot oven, jerks his fingers off, and then touches it again, but this time more carefully (1.7.17). He touches a bristling coat, throws it aside, scratches his hand, and wipes it off on his apron as if trying to eliminate the pricking sensation (1.10.23). As noted, Roody was more willing to touch soft objects. We noted that, to become acquainted with an object, he used his vision and sense of touch first. He looked at and examined the object with his hands, lips, and tongue. He applied olfaction, unlike Joni, only to objects with a strong odor.

To familiarize himself with an object, Joni started from smelling it, which revealed the importance and fineness of his olfaction. However, it can be stated that the human child (at the age of 1½ to 3 years) can differentiate olfactory perceptions, discern rather weak odors, and qualify them by the degree of their pleasantness. Once, Roody (2.6.26) asked me, "Mother, what did you bite?" "What do you think?" I asked him. "Chocolate," he answered, smelling the odor of chocolate candies. Another time, he felt I had eaten sausage. Roody (2.9.5) said to his father, who had just returned from feeding the guinea pigs hay, "I smell hay." Smelling his uncle, who had been in a smoky area for a while, he (2.7.8) said, "I smell smoke, put the smoke away to the wardrobe!"

I observed Roody at 1 to 2 years of age extend forward his tightly closed lips [plates 34(3), 34(4)] when he was looking attentively at something (for example, at his hands or feet). Similar expressions were observed in Joni [plates 34(5), 34(6)]; he was so engrossed in this (1.2.21) that he did not respond when we called him by name. When I observed Roody (at the same age) listening to something, I noticed that he tilted his head in response to faraway sounds and opened his mouth in response to nearer sounds (for example to the ticking of our clock).

2. Keenness of Observation.   The child, even before he is 1 year old, notices new objects immediately. At 10 months, Roody noticed that I wore black shoes instead of the usual white shoes; the moment I entered the room, he stared at my feet and then bent and touched the shoes. He immediately noticed (11 months old) my new white apron with a dark rim, rumpled it, and touched its rim, sleeves, and pockets with his mouth.

Roody (1.9.0) noticed new suits or shoes we put on him. In this case, he was more willing to surrender himself to the dressing procedure; he touched and stroked the new clothing, uttering a smacking sound.

The child showed very keen vision. Here are some examples. Roody (at the age of 9 and 11 months) picked up hardly noticeable tiny crumbs from the floor [plates 61(1), 61(2)]. Later (1.4.27), he saw an airplane flying so high that no noise was heard; he stared at the airplane and watched it until it disappeared. Roody (2.1.24), examining other people's hands and his own legs, found and touched tiny pimples (plates 34(3), 34(4)] and dark dots made by a pen (1.10.23). Roody (2.1.24) noticed a very thin hair, which had stuck to a piece of bread, and took it off before he ate the bread.

When we moved pictures from one wall to another in the boy's absence, he (2.6.11) immediately noticed the change and asked, "Why this picture? It wasn't there," and he showed where these pictures used to hang. At the same age, the boy noticed a slight dent in a metal teapot and said, "Bent." Once, Roody (2.8.9) saw a dark horse with dark blinders and asked, "What is it on his eyes?"

3. Sensory Illusions: Vision and Hearing.   I noticed that Roody, in his indomitable desire to examine everything at close range, tried to grab things at some distance from the place they really were. The hand he brought to his mouth turned out to be empty; only at that moment he realized his mistake, and he immediately reached for the same thing again. Roody at 11 months uttered a long "e-e-e-e" sound and sometimes reached for a chandelier, a street light, or a white sheet that was hanging on a third-floor balcony of the house across the street, obviously unaware of the inaccessibility of these objects. When 1 year old, Roody extended his hands toward the moon; seeing people on the roof of a house, he (2.2.13) extended his hands toward them and said, "To reach the men." It is also obvious that when the child (at the age of 2.2.13) looks at a lamp through a transparent comb, he sees the lamp closer than it really is and tries to grab it.

In 1-year-old Roody, like in Joni, I observed many times his illusions associated with his perception of stereometric images. For example, when he (1.4.26) saw a good picture of a carrot, he tried to take it; when he did not succeed in this, he turned to me with a begging sound "e-e," inviting me to help him. Later (1.9.22), he tried to grab the image of a bagel or of a doll, saying, "Give, give." Once, Roody (2 years old) tried to dip a big chair in a small bowl, obviously not understanding the futility of his attempts [plate 110(4)].

Roody (2.0.23) finally understood that grabbing things from drawings was impossible and expressed it verbally. He came up to a picture of a man, pretended to be taking the man off the picture, came back to me as if to lay the man on my lap, repeated the entire sequence of these movements, and said with disappointment, "No way." One time, after he (1.8.4) had walked sideways in a narrow aisle of the museum, he continued walking in the same manner in a wide aisle, as though avoiding the shelves, completely ignoring the fact that this time there was ample space around him.

At the age of 1½ years, Roody tried to catch the shadow of his hands. The child (1.10.7) mixed up transparent and glossy objects; for example, he took a glossy postcard and tried to look through it as though it were a transparent film or a comb through which he used to look at the fire or at the sky.

I also observed auditory illusions in Roody. For example, he misperceived the volume of some auditory stimuli. Seeing his father standing outside behind a double-pane window, he tried to speak with him in his regular voice, although he could not be heard even if he cried with all his lungs. At the age of 3 years, he tried to speak with us from the street, also not understanding that his words could not be heard through the glass of the window.

4. Self-Attention, Self-Perception.   The child, possessed by a desire to examine everything, sometimes directs his attention at himself and people around him.

Roody (1 year old) was examining his naked body [plate 34(2)]: he saw the navel, touched it, and stared at the area around it. He passed his index finger along a deep above-the-forehead crease of his skin [plate 34(1)]. He examined his arms and legs; suddenly, he noticed a pimple [plate 34(3)], extended his lips forward (exactly like Joni used to do under similar circumstances) and began picking at this spot with his finger [plate 34(4)]. He touched his nails with his fingers and then with his lips. At the age of 10 months, Roody put his index finger into his ear and mouth. At 8 or 10 months, lying on my lap, Roody noticed my ear was not covered, as usual, with my hair; he laughed and put his finger into the ear. Another time (at the age of 14 to 16 months), Roody started to examine my face as if seeing it for the first time; he touched my nose, chin, and ears [plate 41(3)], hair [like Joni; plate 41(2)], or my teeth. Sometimes, the child passed his hand over a strange face as if familiarizing himself more closely with it [plate 41(6)]; he put his index finger into the nostrils, ears, and corners of the eye and under the lips. If you opened your mouth, he looked inside and laughed when you moved your tongue, or closed and opened your jaws. The child immediately noticed that his grandmother was wearing a new glistening brooch and touched it with his fingers. He (2.1.24) looked at kittens with great interest and concentration and touched their eyes, ears, and noses, naming these body parts. He apparently thought that the noses were sore because they were pink, and he said that he was sorry for the kittens. Now (at the age of 2 years), he tried to name things when he examined them.

At the age of 1½ years, the child liked very much to look at his image in the mirror.

Reaction to a mirror. In different periods of the child's life, his reaction to a mirror is different. Looking at the mirror at 4 months of age, Roody smiled at his image. At 5 months, Roody predominantly looked at his own image in the mirror, even when I was next to him. When he was brought close to a mirror, he stared at it, squinted his eyes, touched it with his index finger,[16] smiled, and uttered certain sounds. At 6 months, Roody, looking at his and my image in the mirror, extended his arms toward the mirror, passed his fingers over it, turned his head to me, looked at the mirror again, and so on. Sometimes, he stamped his feet and smiled while looking at the mirror. I observed that Roody (at the same age) looked at his image in the glossy lid of a pot, bringing his face closer to the lid. Roody at 7 months struck his image in the mirror, as Joni did after he got used to the mirror. An 11-month-old Roody laughed when he was brought close to the mirror and struck his image with his palm. At 1 year old, Roody touched the mirror with his lips, then looked at himself and again at his image as if looking for someone. Roody (1.3.24) beat his image abruptly with his fist; the child (1.4.27) reached behind the mirror as if looking for someone, like Joni did [plate 39(2)]. Roody (1.5.22), like Joni, spat at his image in the mirror, made grimaces [plates 47(5), 47(6)], and rattled his lips. When asked who he saw in the mirror, he answered, "Mother and a man" (he called himself "man"); he often touched the mirror with his lips.

At 1½ years, Roody made various grimaces[17] and gestures in front of the mirror. Seeing his grandmother in the mirror, he laughed, turned to the grandmother and then to her image, saying, "Grandma." He giggled when he saw me in the mirror.

One time, I found Roody (1.8.17) looking at the glossy ball at the back of his bed; he was raising his hands and calling himself "man." Roody (1.9.2), like Joni, made grimaces [plates 47(5), 47(6)] while looking at the mirror. Later (1.9.1), I saw him kissing his image in the mirror, looking behind the mirror, and then looking again in the mirror. Sometimes, he brought his face closer to the mirror and kissed[18] his image. When we asked him who was in the mirror, he answered, "Apa" (his pronunciation of his own name) and "ntsa" (good).

At the age of 2 years, Roody made threatening gestures with a rag or stick toward his image. Seeing himself naked in the mirror (2.2.9), he started doing physical exercises.

Roody (2.2.22) looked at the glossy lid, calling himself by name; he (2.3.21) made exaggerated chewing movements in front of the mirror. He (2.9.26) smeared his face on purpose and made grimaces in front of the mirror; he extended his lips forward and opened his mouth wide.

As we can see, the child's reactions to the mirror are fairly close to those of Joni under similar circumstances. Chronologically and by their exterior manifestations, these reactions unfold in the same sequence as in Joni:

Stage 1:   looking at the mirror

Stage 2:   smiling

Stage 3:   touching with a finger

Stage 4:   beating with hands, reaching behind the mirror

Stage 5:   touching with the lips, spitting, lip rattling

Stage 6:   grimaces, gesticulation

Stage 7:   threatening gestures

Joni's behavior lacks the following three elements: self-recognition, comparison of real objects with their reflections, and affectionate treatment of his image. Apparently, Joni does not comprehend fully the relationship between the thing and its image; he does not understand that the chimpanzee in the mirror is his reflection.

In the curious attention of the child, we easily can find his ability for recognition, generalization, finding similarities, curiosity, preferences, and imagination.

5. Recognition.   The child as young as at 2 months of age can recognize things and differentiate familiar from unfamiliar ones. Roody invariably smiled at me when he saw me after a long absence; when somebody else took him, he immediately started crying. At 3 months, Roody smiled at all the people in the house when they were within the limits of his vision; he did not smile at strangers. The 4-month-old child smiled in recognition of the things associated with his feeding. He hailed these objects with a smile, like good friends. At the same age, the child recognizes his mother's voice and distinguishes it from other voices. Hearing his

mother's voice, but not seeing her, he keeps crying until she appears and takes him in her arms. The child does not want to be taken in other people's arms (0.3.13) and bursts into tears.

Not only did 10-month-old Roody recognize some objects, but also he developed some sort of generalized perception. For example, when asked, "Where is the button?" he pointed to various buttons that differed by color, shape, material, and size, and buttons placed against different backgrounds, on different clothes, and on different people (red buttons against bright background, large black shiny buttons, little black shiny buttons against black background; striped, orange-gray, green cloth buttons against green background; pearl white buttons against white background; white linen buttons blending with the background cloth) [figures 16(1)–16(9)]. Roody at 11 months pointed at eyes and a mouth on a rubber toy with an imprinted picture of the human face. The 1-year-old boy answered correctly the question [plate 41(4)], "Where are the doll's eyes, hands and nose?" At that time, not only did the child recognize people around him, but also he called everybody by name (Father, Mother, Nanny, Uncle, and so on). Roody (1.6.15) recognized caps on various wooden toys that portrayed people without anybody previously showing them to him.

Roody (as early as at the age 1.6.15) recognized animal toys and knew their names: a lion, tiger, brown bear, and giraffe. He differentiated certain domestic animals and also knew their names. He recognized me and his father on pictures from 16 years ago. But, he did not recognize his father at the age of 12–14 years; he called him "Uncle" (the same as he called strange men). Roody (at the age of 1.10.20) recognized himself, me, his father, nanny, doll, and toy horse in a family photo and called everybody in the picture by name. At the age of approximately 3 years, the child, looking at a family picture (painted in oil by a not-so-good painter) said, "These are Mother, Father, and Apa," and went on to say, "Father is not like my father, mother is like my mother, Apa is not like Apa." He added, "Mother's face is not good, Father's face is not good, Apa's hair is not good,[19] Apa himself is good." Roody (2.6.7) looked at pictures in a book and picked up a similarity to people he knew; he noticed various facial expressions and asked about one man's face, "Why does the uncle laugh?" Roody (2.10.9) discerned a slight dissimilarity in the expressions on a two-face toy, saying "This toy is laughing, and this is not."

The child (1.8.21) easily recognized and sorted various rings and pyramids, calling some of them old (the faded ones) and some of them new (the brighter ones). Roody (1.8.1) recognized a cat from a long distance (about 50 feet from his window); another time, he (1.4.29) noticed a crow sitting on the roof of a three-story building.

The child (1.9.18) recognized the whole toy by a part. When I took apart some toys (a goose or a bell) and showed him their parts, he recognized correctly which was which. Roody (2.4.18) identified images so well that he could play lotto [plate 116(5)]. He (2.9.16) recognized in the picture and could name 85 objects on the lotto cards. Roody (1.7.17) recognized an airplane in a picture on a matchbox.

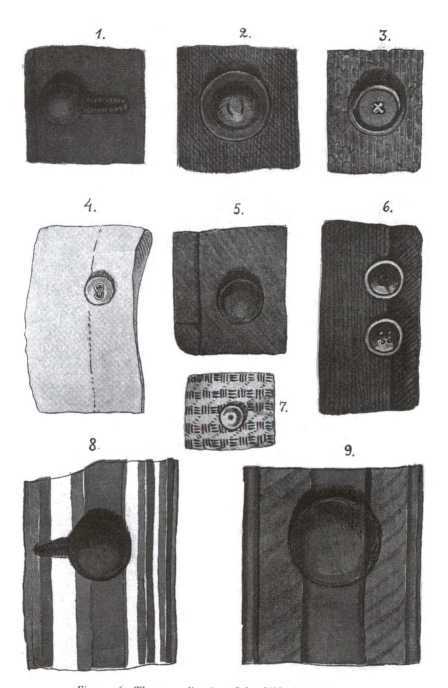

Figure 16. The generalization of the child: (1)–(9) recognition of nine buttons widely differing with respect to color, size, and material. The child (10 months) has mastered the generalized notion of "button."

6. Generalization.   The child (1 to 1½ years old) recognized objects by generalization. For example, he called any strange man "Uncle," any strange woman "Aunt" or Gaga (our maid's name), any old woman "Grandma," any child "Katya" (the name of a girl he knew), all black birds "Karr," and all cats "kkh." We already noted that Roody used the generalized image of a button, pointing at various buttons that differed by size, shape, and color.

The child easily found similarities in many objects. Answering the questions, "Where is the mouth (ear, hands, eye, head, or hair)?" he (1.2.3) pointed not only at his body part, but also at mine. He pointed at a horse's legs in a picture, then touched his own legs. He (2.1.15) found eyebrows on a toy and then pointed at his own eyebrows and at the eyebrows of all people around him, adding suddenly, "It has eyebrows," and pointed at the black rim of a white porcelain cup.

Observing a child (at the age of 1.4.25 to 3 years), we see that his ability to find similarities broadens. He finds similarities in individual characteristics such as color, shape, and size and trains his imagination, fantasy, and abilities to recognize and abstract, identifying with enhanced accuracy many images and generating concrete concepts.

7. Finding Similarities.   Of numberless examples of Roody's (at the age of 1½ to 3 years) verbal expressions and pictures registered in my notebook, I cite only several well-illustrated instances [figures 16(1)–17(6)].

Similarity by color. Roody (2.11.16) calls a piece of yellow medical oilcloth "oil." He (1.11.4) calls a piece of meat-red rubber "ham."

Similarity by color and shape. He (1.11.14) calls an orange bowling pin "carrot." He (2.1.0) calls a yellow corncob "weight."

Similarity by shape. He (at the age of 2 to 3 years) calls shapes according to what they depict: "car," "boat," "Aunt," "Uncle" [figures 18(1)–18(18)].

These terms include "moon," a piece of marmalade bent in the shape of a horseshoe (1.10.13); "axe," a jagged piece of cheese (2.7.10); "house," a cookie (1.11.24); "tears," stains on the wallpaper (3.6.0).

Similarity by transparency. He (2.6.0) calls a glass funnel, glass vial, or pipette "ice." A transparent stone in a ring is a "bubble" (2.6.0); matte buttons that resemble medicine are "tablets" (2.6.0).

Similarity by elasticity. He calls a long piece of apple peel "rubber" (1.10.27) and uses the same term for a measuring tape made of oilcloth (1.11.14).

Similarity by size. Roody calls big screws in his bed "Mother," and small screws "children." Roody did not always find similarities in terms of main characteristics of an object, but sometimes he could discern significant qualities correctly among a great number of them and make a genuine abstraction. I followed most thoroughly his abstraction of an airplane, which was imprinted vividly on his mind. Even before he started to speak, Roody called an airplane "gkh," imitating the propeller sound. One time, he (1.6.27) found a chip with a shape that apparently reminded him of an airplane; he lifted that chip and said "gkh" [figure 19(1)]. Later, if he saw objects with two perpendicular lines, they always reminded him of an airplane [figures 19(1)–19(7), 20(1)–20(7)].

Figure 17. Analogization by color: (1) a bottle with cod liver oil; (2) a piece of oilcloth—"fat" as defined by Roody at 2.11.16; (3) a carrot; (4) an orange ninepin, defined as "carrot" by Roody (1.11.14); (5) a weight from a clock; (6) a corncob, defined as "weight" by Roody (2.1.0).

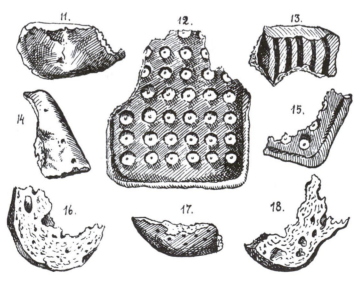

Figure 18. Analogization by form using bitten pieces of cheese and biscuits identified with different objects by Roody at various ages: (1) axe (2.7.0); (2) motorcar (3.2.0); (3) woodpecker (2.8.0); (4) motor car (2.8.0); (5) cat in car (4.0.18); (6) uncle in motor car (3.10.6); (7) ship (2.8.0); (8) carriage (2.9.0); (9) rooster (2.7.0); (10) crow (2.7.0); (11) motor car (2.6.0); (12) house (1.11.24); (13) motor car (3.0.0); (14) aunt (3.8.0); (15) shoe (1.11.21); (16) moon (1.10.13); (17) boat (3.0.0); (18) uncle in sledge (3.8.0).

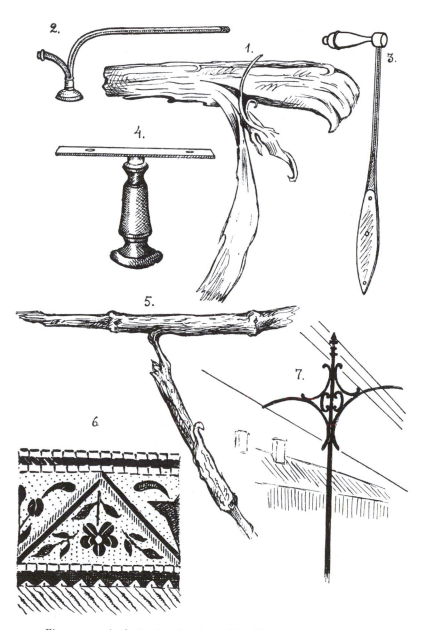

*Figure 19. Analogization by form—identification with "airplane":
(1) a splint, "gh" (Roody's reaction meaning airplane, age 1.6.27);
(2) a pulverizer (Roody's reaction at 1.8.16); (3) a doctor's hammer
for auscultation (Roody's reaction at 1.10.15); (4) a stamp handle
(1.11.24); (5) V-shaped twig (1.11.24); (6) pattern on tablecloth
(2.2.1); (7) tramway wires crossing (2.2.1).*

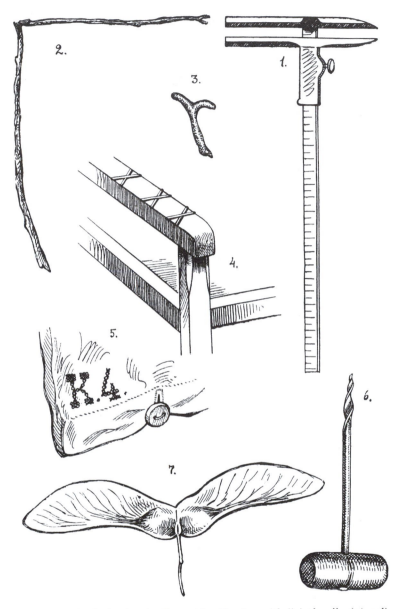

Figure 20. Analogization by form—identification with "airplane": (1) calipers, "gh" (Roody's reaction meaning airplane, age 1.7.3); (2) a bent twig, "gh" (Roody's reaction meaning airplane, age 1.8.14); (3) a piece of coral, "gh" (Roody's reaction meaning airplane, age 1.11.24); (4) charred decorative pattern, "gh" (Roody's reaction meaning airplane, age 1.8.27); (5) figure 4 sewn on a pillowcase "gh" (Roody's reaction meaning airplane, age 1.8.0); (6) a hauser, "gh" (Roody's reaction meaning airplane, age 1.2.2); (7) flying seed from maple tree, "gh" (Roody's reaction meaning airplane, age 2.2.4).

These are some "airplane similarities" that Roody indicated by the sound "gkh" or later (starting from item 31) by the word "plane."

1.	A chip (the first object that reminded him of an airplane)	(1.6.27)
2.	A pattern on cloth	(1.7.3)
3.	Calipers	(1.7.25)
4.	Number "4" on a pillowcase	(1.8.0)
5.	Number "7" on the bottom of a plate	(1.8.4)
6.	A piece of paper with a slit in it	(1.8.10)
7.	A bent piece of wood	(1.8.10)
8.	A metal rod	(1.8.10)
9.	A bent twig	(1.8.14)
10.	A sprayer	(1.8.16)
11.	A small hammer	(1.8.16)
12.	An airplane model	(1.8.20)
13.	A cardboard plate	(1.8.24)
14.	A pattern in the form of a cross on a chair	(1.8.27)
15.	A gnarled stick	(1.8.27)
16.	A doctor's hammer	(1.10.15)
17.	A piece of coral	(1.11.24)
18.	Pictures of airplanes on matchboxes	(1.11.24)
19.	A handle of a signet	(1.11.24)
20.	A forked twig	(2.2.1)
21.	A flat chip	(2.2.1)
22.	An ink stain on cloth	(2.2.1)
23.	A pattern on a tablecloth	(2.2.1)
24.	A half-split piece of wood	(2.2.1)
25.	A picture of an airplane in a book	(2.2.1)
26.	A piece of plywood	(2.2.1)
27.	A bore	(2.2.1)
28.	A piece of a toy	(2.2.1)
29.	A crust of black bread	(2.2.1)
30.	A branch of a tree	(2.2.1)
31.	A streetcar bow	(2.2.1)
32.	A flying seed of a maple tree	(2.2.4)
33.	A piece of a broken toy	(2.2.9)
34.	A piece of chocolate	(2.2.9)
35.	Letter "T" in a word	(2.7.12)
36.	A bowling pin	(2.7.12)
37.	Patterns on a cup	(2.8.10)
38.	A jagged piece of cheese	(2.9.0)

Thus, the child found similarities in terms of one characteristic of 38 objects that differed by material, shape, color, or other qualities.

Sometimes, we observe that one object (for instance, a jagged piece of cheese) evokes many comparisons in the child's mind. After Roody broke off small pieces of cheese, he described them differently each time according to his immediate associations, which demonstrated his ability of elementary abstraction. At the age of 2 to 3 years, Roody defined pieces of cheese as follows [figures 20(1)–20(10)]: baby carriage, axe (2.7.0); automobile, woodpecker (2.8.0); fish (2.11.24); crow

(2.7.0); airplane (2.9.0); seal (1.11.24); chicken (2.7.6); boat, man in a car (3 years 8 to 10 months). The child can be very thorough in indicating, for instance, where the "chicken's head" or "wing" is. In this propensity to find similarities, the child's imagination and fantasy clearly are visible.

To me it was most evident during my analysis of Roody's definitions of plants. Leaves, flowers, fruit, and gnarls, which he examined, not only were real objects for him, but also were a rich source of similarities. For example, Roody named the following objects [figures 21(1)–21(6), 22(1)–22(6)]:

"Gkh" (airplane) (at the age of 2 years):   A dry leaf

"Plane" (2.2.4):   A flying seed of a maple tree

"Caterpillar" (3.1.0):   Gnarls of pine and linden

"Seal" (3.2.0):   A birch twig

"Crab" or "porcupine" (3.0.9):   A part of a fir branch

"Crawfish" (3.0.28):   A twig

"Deer" (3.1.0):   A twig

Within the same time frame (1½ to 4 years), the child liked to look at the pictures in a book and traced the most vivid images with his finger.

At the age of 2 to 3 years, the child was very interested in sorting various small things he found in a chest of drawers [plate 62(5)]. He did not limit himself at this point to brief touching and looking at them; he examined them very thoroughly and asked, "What is it?" [plate 116(4)].

New toys also underwent his thorough visual and tactile examination.

8. Curiosity.   The older the child, the deeper he tries to comprehend things. He is not satisfied with just looking at things; he intends to find what is hidden in them. Curiosity has appeared.

Now, the child destroys or breaks a thing to understand it better. This propensity for destruction is overwhelming and, unfortunately, is not always interpreted correctly by adults.

Depth intrigues a child very early. Roody, at the age of 8 months, saw a hole in the seat of a chair, stared at it, and put his fingers into it. At 9 months, Roody stuck his fingers into a bottle or burned-out cavities in a table. He touched relief patterns on cloth with his fingers (like Joni) and then with his lips and tongue. Seeing holes in linen or pillowcases, he put his fingers in them and tried to rip the cloth apart [plate 62(3)].

The child willingly picks holes in soft furniture with his index finger, boring them to the very bottom. At 11 months, Roody used to take a ring of a collapsible pyramid and put a finger in its hole; then, he put the same finger in his mouth and tried to turn the ring [plate 62(2)]. He also used to put his entire hand in a pot (or pitcher) and feel it from inside. Sometimes, he tried to stick his finger into a tiny hole (for instance, in a teapot lid); he was not easily discouraged when the finger did not fit into the hole.

Roody's interest in examining hidden spaces, similar to Joni's, was expressed very clearly. He (0.9.9) tried to open the door of a bookcase with his finger or by

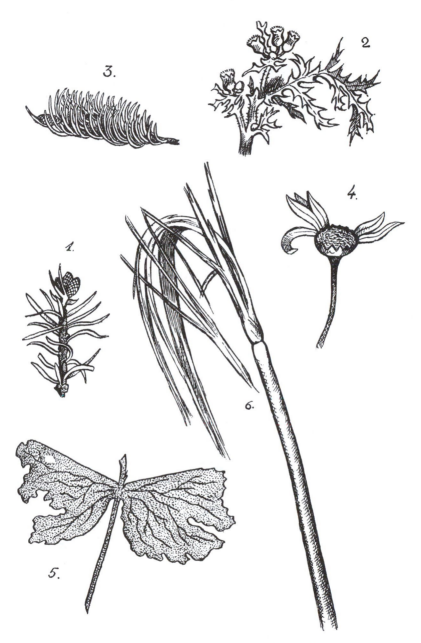

Figure 21. Identification of plants with different objects: (1) a crab—a spruce bough with cones (age 3.0.9); (2) mortar—flowers (age 3.0.9); (3) "hedgehog—flowers (age 3.0.9); (4) "dragonfly"—a daisy with most petals torn off (age 3.0.9); (5) airplane ("gh")—a faded leaf (age 2.0.0); (6) "snail"—a half-broken stem (age 3.0.0).

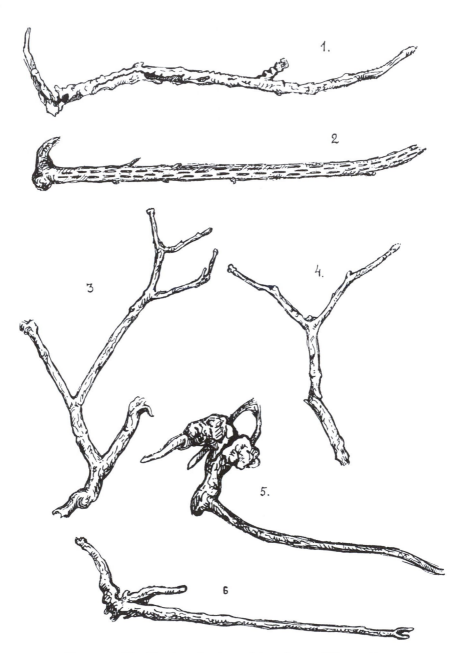

Figure 22. Identification of twigs and branches as different objects:
(1) caterpillar—a birch tree branch (age 3.1.0); (2) caterpillar—
a pine tree branch (age 3.1.0); (3) Crabfish—a ramified branch (age
3.0.28); (4) stag—a dichotomic branch (age 3.1.0); (5) goose—a dry
twig (age 3.2.0); (6) seal—a dry twig (age 3.2.0).

pulling at the key [plate 62(4)] to look inside [plate 62(6)] and browse in every drawer until everything had been unearthed. Another time, the boy (0.11.11) pulled a picture from the wall by looking behind it; he looked under a bed and into a bucket. Everywhere his examining hand went, his index finger preceded his eyes.[20] The child at the age of 1 year tried (like Joni) to look under skirts or other clothes; he stuck his fingers into pockets and searched them. No wonder that such toys as bright and colorful eggs or "matrioshkas," which intrigue children with their hidden contents, are their favorite.

The child usually is very persistent in his attempts to open the lid of a pocket watch or take a watch apart down to the last screw if he is allowed to do so. He likes very much to unwrap packages; he is drawn to stoves, trash bins, dumps, or yards, where he can find unexpected things. He examines all these, rejecting some and appropriating others.

Like Joni, my Roody (2.3.0) liked to look at the world around him through a yellow transparent oilcloth, a comb, or pieces of colored glass [plate 112(1)]. At the age of 4 years, he like to look through binoculars or a magnifying glass [plates 112(3), 112(4)]; he would go around the place for hours and shift his glance from one object to the next, enjoying objects enlarged or diminished. He loved to look into his toy stereoscope, which gave a new intricate colorful mosaic pattern at every turn [plate 112(2)].

9. Aesthetic Tendencies of the Child—Preferred Qualities of Objects.   The preferred qualities of objects can be determined by finding which things most strongly attract his attention and interest, which ones he chooses for play, and how he characterizes objects verbally.

We have stated that everything new attracts the child's attention and enriches his developing mind. As early as 1 year, he immediately notices his new suit or cap, touches them with his index finger, and reaches for the cap to try to pull it off. He notices and tries to rip off new buttons or brooches. Roody (1.5.0) called his new boots, shoes, or shirts "ntsa" (good), reserving the word "byaka" (bad) for the old ones. At the age of 2 years, he did not want to put on old aprons and reached for new ones, "Old—no!" He (2.8.28) enjoyed putting on a new suit and rejected the old one. "This is a good dress," he said (2.3.10), seeing his new apron.

I also noticed another tendency of his: He liked new things for himself, but did not always like when he saw a changed image of somebody he loved. One day, I changed my hair style, combing my hair away from the forehead. Roody (2.1.17) said "not good"; when I returned to the way my hair looked before, he said "Good."

He showed a similar reaction when he (2.7.8) saw me in an unusual long dress (vogue of the 1920s), which I put on for a special occasion. He said, "This dress is bad. Take it off, mother!" Or, he insisted that I take off a kerchief from my head, saying, "Take it off, mother. This is bad." One time, he (1.10.10) tried for some reason to pull a striped bright gown off me, calling it "bad."

Sorting out things and finding new ones, the child (1.6.11) always tried to take them into his possession. Roody (1.9.29) did not want to listen when his old books were read to him and demanded new ones. One day, I gave him a pile of 20 books from which to choose. He chose 9 books; those were the ones that had been read least to him or newer by the date of purchase or by the familiarity of their contents to him.

The child's qualitative discrimination ability, which is the embryo of his aesthetic feeling, tends in his visual perception toward everything bright and shiny that attracts his attention. His choice is usually an intensely colored object.

Gloss attracts the child's eye very early. The 2½-month-old baby smiled and uttered joyful sounds when he saw shiny metal things (balls in the back of his bed, a clock pendulum). The 3-month-old child stared at a bright chandelier and golden rattle or watched a moving light (0.4.0). The 6-month-old child predominantly reached for shiny objects, such as silver spoons, keys, utensil lids, metal clamps, faucets, candlesticks, and the like.

Roody (1.7.19) uses the term "ntsa" (good) for a glossy wrap of chocolate; he also calls it "beauty" (2.6.1). "Ah, how pretty," he (2.8.16) says as he admires a golden fish made of cardboard. "How pretty are these stars," he (2.9.22) says, looking at snowflakes. "Oh. Good, shiny," he (2.11.7) says about a hummingbird.

We noted above that the child preferred to play with shiny and luminous objects. He could not remain indifferent when he saw metal objects in the trash and resisted when I tried to persuade him to throw them away. "It is bright," he said, explaining his desire to have this thing.

Like Joni, my Roody clearly preferred the colors of the first half of the spectrum,[21] particularly showing his fascination with red. It is registered in my notebook and experimentally proven that, confronted with the choice of objects of different colors, Roody (1.3.9 to 3.0.0) invariably preferred red objects.

The child marked the following objects with the word "ntsa" (good): bright yellow sunflower (1.3.9), red blanket (1.4.27), red pants and shirt (1.4.22), bright gown (with blue and red stripes) (1.4.24), red flag (1.4.27), bright blue apron (1.5.0), colorful carpet in which the color red prevailed (1.6.20), yellow-red top and red polished brush (1.8.5), red pencil. He yelled gleefully when he drew lines with a red pencil; at the same time, he called a black pencil "bya" (bad) and did not want to draw with it (1.9.9). Roody (1.10.4) called red buttons "ntsa" and blue-green buttons "bya" [figures 23(5), 23(6)]; red and pink pieces of paper were "ntsa", purple or green pieces were "bya" (1.11.2). A red blouse and orange or pink flower were "ntsa"; a green or blue flower was "bya" [figures 23(1)– 23(4)]. Moving his dresses from place to place, he only called red dresses "good"; the others did not receive his (2.1.17) comment.

When turning over the pages of a calendar, he said "good" when he saw red numbers and "bad" when he (2.11.25) saw black ones. Pointing at a green basket he said assertively, "Blue, I don't like, afraid." "This I like," he (2.2.0) said, pointing at a red basket (2.2.0). Seeing black horses and white horses, he (2.2.28) called the black one "bad" and the white one "good."

When he played with colored beads, he (2.3.22) asked, "Where are the red beads? I don't like the black ones." He was reluctant to pop an orange paper bag (2.3.22), but he popped gray bags without a second thought. Once, I said to

Figure 23. The child's color preferences, preferred (1, 3, 5, 7) and refuted (2, 4, 6, 8): (1) an orange flower made of velvet; (2) a green flower made of velvet; (3) a pink flower made of paper; (4) a pale blue flower made of paper; (5) a red button; (6) a greenish-blue button; (7) a yellow hat; (8) a green hat.

Roody, "Look, a cloud is flowing." He looked at the cloud and said, "Biaka" (bad). I asked, "Why?" "Because it is black," he answered.

Roody (2.4.9) picked only red plates from a pile of colored plates. Seeing a strange black cat, he said, "The cat is bad, black, I don't like black, I like white, red." He said, "Good dress," when he (2.5.21) saw my dark red velvet dress. He called a green hat "bya," a yellow hat "ntsa" [plates 35(7), 35(8)]. He (1.11.18) called a red wooden ball "ntsa," a grey rubber ball "byaka."

I observed only once that my son (1.11.6) chose seven blue plates from 64 plates of different colors and started playing with them (as Joni did), but he did not call them "ntsa." In a few days, he also at first chose 12 blue plates and then some orange and red ones, which he called "ntsa." A conclusion can be drawn that, although the human child sometimes (like the chimpanzee Joni) chooses different objects (for example, blue objects) for play, this does not mean that he likes this color better; it is possible that blue objects attract him by contrasting with his favorite color, with which he is bored at the moment. Things that the child likes get his qualitative rating, and the more definite this is, the older the child; he wants to keep the things he likes (examples below clearly demonstrate the child's desire to have a red airplane, adorn himself with a red tie, or put on red clothes). These are rudiments of certain aesthetic tendencies, which can be characteristic even of a 3-year-old child. My Roody was very confident when he said, pointing at a nicely made paper flower, "What a beauty!"

When, for instance, I gave Roody a number of cloth scraps, he (1.6.26) called bright red satin "ntsa," separated these scraps from the pile, spread them out, looked at them intently, and put them on his head [Joni usually put them on his neck; plate 44(3)], as though adorning himself with them. Having 24 sticks of eight colors in his possession,[22] Roody (1.11.6) picked for himself only especially bright sticks (from the first half of the spectrum: red, orange, and pink). Another time, he (3.0.7) picked a large number of white sticks and said, "I collect them, I want to have a lot of them."

Having the choice of two pens, green and red, Roody (1.10.16) took the red pen and called it "ntsa"; he gave the green one back to me and called it "bya."

Roody (2.0.16) did not want to put on a gray shirt, but reached for a red one. Another time, he (1.10.9) refused to put on white pants and yelled "bya," but willingly put on striped pants, which apparently seemed more attractive to him.

Once, Roody (2.6.0) saw a red tie and demanded it, using expressive gestures, saying, "Put on, mother!" In a toy store, he (2.3.7) saw airplanes of different colors and said, "Red plane buy, green plane not buy."

The tendency for self-embellishment is expressed in the child stronger than in the chimpanzee. I observed the following instances associated with his aesthetic preferences:

Somewhere, Roody (1.5.0) found a collar of bright yellow feathers; he immediately put it on his neck, smiling happily.

He (1.7.12) found a string of blue beads and put it on his neck.

He (1.9.5) put pieces of rope and tape and ran around the room.

He (1.9.22) put colored beads on his neck.

He (1.10.20) put a bright piece of cloth on his neck, attached a piece of bright velvet to his shirt, and joyfully ran around the room.

He (2.9.2) put on a black glistening chain and paced around the room.

He (2.7.2) asked to put a light-green silk ribbon around his neck; later, he asked to pin a coral brooch to his shirt.

Speaking to the sizes of objects as criteria for the child's choice, we can clearly state that he, like the chimpanzee, preferred small objects to big ones. The human child, like the chimpanzee, likes miniature things.

The child's attention is attracted to tiny objects of the animate and inanimate world. The 8- or 9-year-old child already showed interest in tiny drops of water, black printed letters on paper, or threads in cloth and touched them with his finger. He diligently collected tiny crumbs or small pieces of paper from the floor [plates 61(1), 61(2)]. Just starting to walk, he looked for small stones or vials in the yard, bent down or squatted to pick them up, and gave them to adults [plate 61(4)].

Once, we drew two birds of different sizes for Roody (1.11.8); he called the bigger one "byaka" and the smaller one "ntsa" and kissed the drawing. Later (2.1.6), when he was shown two pictures of woodpeckers, he called the bigger one "bad" and the smaller one "good." Also, he (2.1.6) played with small bowling pins more willingly than with bigger ones. We mentioned how the child likes tiny live and toy animals and small balls (compared to large ones); he calls them tender names and sometimes even kisses them.

This certainly does not contradict the fact that, when choosing, for instance, a candy or a weapon (stick), the child prefers large objects because he looks at these things from the standpoint of their utilization; therefore he values their greater usefulness. However, if the object serves solely for entertainment purposes, his choice is directly opposite.

I remember that, at the age of 2 years, Roody could not take his eyes off tiny crawling snails [plate 61(5)]. Everything tiny evoked his tender sympathy. If you gave a set of wooden eggs to the child, you might be confident that he would choose the smallest one and look at it with touching affection in his eyes [plate 61(3)].

Roody, like Joni, of different geometric shapes, preferred spheres, especially small balls. He (2.3.13) said, "I love tiny."

Answering my question about which ball he liked better, Roody (2.10.6) said, in a tender singing voice, "Sma-a-a-al." He asked, "Why is mother's watch small and father's watch big?" When I asked him which watch is better, he said softly, "I love tiny, tiny crumb," and kissed my watch.

Before 5 years of age, Roody liked, among his many toy rabbits, the smallest one (length 2 cm); this rabbit was always with him. It was this rabbit that he armed with a cannon during his "military" games, as if he was aware of the rabbit's total helplessness against enemies [plate 61(6)]. We all know the sympathy children feel toward animal babies; children prefer playing with them to playing with adult animals. Young animals in the zoo are usually fun to watch not only for children, but also for adults.

We noted how joyfully children ride on small animals (donkeys, ponies, or goats). It appears that the process of riding does not change whether a small or big animal is harnessed to pull the carriage; nevertheless, a small animal apparently brings additional joy to a child.

We sometimes can see (I observed this in my child, too) the tenderness with which a child 5 to 7 years old treats younger children; we can see that he is more relaxed playing with them than playing with older children. This is entirely opposite to the tyrannical treatment of "the small creatures of this world" by the chimpanzee; this is the rudiment of the child's future humane feelings.

Also, among his aesthetic tendencies, we have to mention the human child's tendency for symmetrical arrangements of objects. Roody (2.7.7) arranged bowling pins as the radii of a circle; originally, he put the last pin across one of the radial pins, but later he corrected himself and also arranged that pin radially. Another time, he (3.0.18) corrected a flap on my slippers that had turned inside. Once, he (3.0.9) noticed that one of the horns of a wooden deer was broken; he immediately insisted that we also should break the second horn. I never noticed that Joni showed a sense of symmetry when he constructed something, but when he was dismantling symmetrical structures, he tried to keep them symmetrical until the end.

We stated above that the child preferred spherical objects to any others.

In regard to the olfactory perceptions of the child, we have to emphasize that he had some preferences. Roody (1.7.9–1.9.0) willingly smelled a handkerchief sprayed with perfume; smelling the odor, he inhaled with a sound and smiled broadly. Sometimes, he clearly relished the aroma of soap; he picked at it with his finger and said "ntsa." Roody liked to smell boot cream (2.3.23) and cheese (1.7.23); in the latter case, he grunted as he always did when he ate tasty food. He greedily smelled tangerine (1.11.2), orange, kerosene, and mint toothpaste. He definitely liked the odor of some flowers, such as sweet peas or phlox, and the odor of ether. He (2.5.27) analogized some odors unknown to him with something familiar; for instance, smelling anise-ammonia drops he said, "Smells like apple"; smelling some ointment, he said, "Smells good like gas" (lighting gas). Roody clearly disliked some odors. Smelling formalin, he asked, "Mother, what do you smell?" I said, "Do you like it or not?" He (2.7.8) said, "This is bad, it stinks." We mentioned that the child did not like the odor of smoke; he said, "I smell smoke, put it away to the wardrobe."

I never noticed that the human child feared or utterly disliked any odors,[23] which was so characteristic of the chimpanzee. On the other hand, I did not find in Joni any special attraction to repeated smelling of any inedible things with an odor. However, he used to smell a tasty fruit after every bite; Roody never did that.

Obviously, the human child and the chimpanzee did not qualify different gustatory sensations equally. As noted, both infants showed great liking for sweet products, such as sugar or candies (Roody more than Joni), and for fruit and berries. It must be emphasized that the human child did not have such a dislike for butter and boiled meat as Joni.

I am not inclined to go into the details of the preferred gustatory sensations of the human child and the infant chimpanzee since I noticed that the sphere of these sensations is the least steady. Not only is this sphere extremely individualized according to the infants' different eating habits (and, consequently, is determined fully by their living conditions), but also it varies in different periods of their life. For example, my Roody in the early years of his life (1 to 1½ years) liked lemon; later he (2.9.29) summed up his preferences as follows: "Oh, I like tangerine!" "I like orange a little, but I don't need lemon." The child often becomes bored even with tasty food; what he liked yesterday he may reject tomorrow.

As to the infants' tactile preferences, we can state that both of them liked smooth, soft, velvety things; they qualified coarse, prickly, and rough things negatively.

The child (1.9.26) said "ntsa" while the stroking soft velvet cloth, even though it was black. He (1.11.9) called a soft down jacket "ntsa." He (2.2.1) stroked a soft satin blouse and called it "good"; although it was dark green. He (2.2.1) called a coarse jacket "byaka" although it was red. "It's prickly, bad," Roody (2.5.9) said about a hard straw hat; "this is good," he said about a soft felt hat. "I like these pants" (of cotton, soft), and "I don't like those" (of broadcloth, coarse) (2.8.25). "I wore these pants, but I don'r like them all the same," "Oh, this is prickly, that's the thing," he (2.8.25) said about a coarse winter hat. "This is good," he (3.0.0) said, stroking my bare leg, and he asked, "May I touch your other leg?"

Joni never demonstrated an ability to differentiate finely between various tactile perceptions, nor did he prefer some of them over others, probably due to the coarseness of the skin of his hands.

Very early, the child qualifies things in terms of their aesthetic, ethical, or idealistic nature. He (1.9.9) subjectively characterized even people of our household, calling some of them "ntsa" and others "bya." Roody (1.10.3) differentiated pictures of animals, calling some of them "ntsa" (dogs, donkeys, horses, cats, cows, pigs, geese, and chickens) and others "byaka" (goats, hens, and ducks).[24]

We mentioned that Roody (2.4.30) treated different toys differently. Roody was also selective when he chose the books we would be reading to him: He (1.9.16) chose some books and persistently rejected others. Moreover, he wanted to listen to particular pages in a certain book, calling them "ntsa" (1.9.28), and he flatly refused to listen to others, calling them "byaka." For instance, he did not want to listen to the first four pages of general descriptive nature in the book *Four Colors*, calling them "byaka"; as soon as I come to the fifth page, which described girls going for a walk (pages that were more dynamic), he said "ntsa" and later made me read these pages three or four times in a row. Sometimes, he did not want to listen to the last (sad) page of the book.

Although, I found the chimpanzee had differentiated perceptions in terms of visual, olfactory, tactile, and gustatory characteristics and partially of auditory characteristics, as well as emotional preferences and rejection of some concrete objects or images, I obviously could not find in him even traces of differentiation in terms of idealistic characteristics, which was expressed so clearly in the 2-year-old human child.

10. Imagination.   A primitive aesthetic sense often is determined by the unusualness of the stimulus. My 7-year-old boy was very interested only in things, books, and drawings that were uncommon. He begged me to read him only about things that "cannot be found in the world." He wanted to see pictures in a book, but he always asked, "Is there anything in this book (i.e., are there any drawings) that can be found in the world?" If he received an affirmative answer, he immediately became uninterested and did not want to take the book. He looked at the images in a book on Greek mythology with great enthusiasm; he listened with unabated interest to fairy tales and legends filled with fictional elements.

I assume that the fanatical and protracted interest of my child from 6 to 8 years old in collecting caricatures was based predominantly on his passion toward everything unusual. I think that when Roody's first curiosity toward real objects that surrounded him was satisfied, not knowing yet how to penetrate the depth of things and lacking new experiences, he broadened the limits of his perceptions by shifting them into the region of fantasy, thereby developing his imagination.

We all know that the more unbridled the child's ability to substitute imaginary things for real ones, the more joyous the child. I vividly remember how my baby was delighted with a "doll' I made swiftly for him out of a handkerchief, tying knots at the corners; he preferred this home-made improvised toy to a more "real," humanlike doll.

I brought Roody even more joy when I included in our play a simple stick with a girl's face drawn on it. He was also delighted when, for lack of any other toys, I used my five fingers animated as different people speaking and acting.

The child at the age of 4 years is no less enthusiastic to ride a stick that substitutes for horse than the "real" toy horse [plate 100(5)].

If you monitor the day of a child 1½ to 3 years old, you will find that he spends 50% of his time in the "unreal" world with imaginary things and creatures, for which he speaks and acts.

Roody liked to communicate with live animals and animate toy animals or images of objects, even parts of his own body. I happened to hear Roody (2.4.19–2.8.30) treating his leg with a cookie, saying to it, "Take it, eat." When he would hurt his finger, he would come up to me and say quite seriously, "Mother, my little finger is crying." I observed many times Roody (about 2 years of age) trying to feed wooden geese; he bent their heads to the water and said, "Drink, drink, eat cabbage." He sometimes fed a wooden horse, giving it grass and expecting that it would start to eat [plate 58(2)]. He fed children in a picture, saving real bread for them. When he had no such things as bread and water handy, he always could imagine them; his play did not become less interesting and absorbing because he did so.

Roody (2.9.30) became so carried away with his imagination that he was able to describe the details of a doll's behavior: "Rebecca has choked, I should give her sugar." He (2.5.11) accidentally stepped on a doll and said seriously, "Excuse me," as he would say to a person. However, the boy considered the doll somewhat unequal to him by its mental abilities; therefore, he tried to guide it, ascribing "incorrect" actions to the doll. Once, I found Roody (2.11.1) talking to a doll: "I want to show something to doll Muni," "Muni has not seen Apa sawing," (speaking of himself, Apa). He started sawing correctly and said, "Muni says that Apa is sawing wrong"; then, he resumed sawing by holding the saw flat. I said, "What did you say, Muni? This is not right." Roody laughed, corrected himself, made some other faulty remark on the part of the doll, and laughed again. The sawing finished, he said to the doll, "Now you know how Apa works, don't you?"

We mentioned that the child entrusted his toy animals with guarding him during sleep, and that he introduced toy animals (a teddy bear or rabbit) into his world of entertainment. Roody brought a book to the eyes of a toy horse and said, "Read, little horse, what is written," and he (2.9.16) read on the part of the horse, turning the pages (he used some clearly distinguishable words, but not words of a human language).

Roody (2.8.30) enthusiastically played with imaginary water. He declared with great confidence, pointing at the floor, "Here is water," and added, "I want to jump into water." He fell on the floor and started making swimming movements. The child's imagination added some properties lacking in inanimate objects. "Let it be an apple," Roody said, taking a wooden ball; then, he gave the ball to a toy monkey, saying, "The monkey will eat."

I cut out a cardboard butterfly. Roody (2.6.13) asked, "Why doesn't it fly?" I said, "You have to attach a thread to it." He said (enthusiastically), "To attach a thread to it, it will fly around all the rooms and around Apa's room; it'll spread its wings" (he spread his arms). "It is spreading the wings," he cried, running around the room with the butterfly dragged on the thread. But, Roody was not confused, "I'll take it by the end of the thread, and it will fly." "How will it fly in the sky?" he asked, seeing that the butterfly, indeed, did not rise into the air.

The child's bright imagination is determined by the liveliness and brightness of his perceptions. "Oh, oh," Roody jerked his hand off the picture of knives, as though he had been pricked or cut with them. This feeling was imaginary.[25] I said, "Touch them, they won't cut you." He touched the picture, jerked his hand off, and said, "Oh, it pricks."

Sometimes, the child was fully aware that the action was not real, but this did not discourage him. He often added, "Make believe," but this make believe did not darken or diminish the joy of his experience; it was as if it did not reach his consciousness so strong was his imagination and the colorfulness and liveliness of these images.

This very imagination is the source of creative fantasy of a human; it sometimes embellishes the person's routine, day-to-day, narrow, lonely life, usually limited to his room and his family.

Here, we can see that the human child mentally converts the real world to a brighter, more colorful and sonorous fictitious world; we have not found its elements in the chimpanzee, if we do not take into account Joni's imitated fight or assault and his intentional clashes with handicaps he himself had put up. We do not have sufficient data to make a judgment about whether this was a substitution of an imaginary object for the real one or not. Judging by the vigor with which Joni engaged in such play, we can assume that at the beginning of the play, Joni perceived these objects as real, but then he became carried away and lost the sense of reality, taking inanimate objects for animate, imaginary for real.

Destructive Play   Perhaps none of the groups of play makes the human child and infant chimpanzee so akin as destructive play.[26] Even a 5-month-old baby during breast-feeding pulls at his mother's clothes, making gripping and ripping movements.

An 8-month old baby joyously engages in tearing and scattering papers and knocking over standing objects. When somebody stands a toy in front of him, he uses his hand or something in his hand to overturn the toy. Sometimes, the child uses his teeth to destroy or scratch things, for example, to rip buttons from clothes; he does it so vigorously that you would think that he may break a tooth. Breaking sticks, destroying toys, or smashing sand structures presents (at the age of 1½ to 3 years) unlimited pleasure for a child.

At the age of 1½ years, the child cannot yet build structures from wooden blocks, but with one swift movement of his hand, he most energetically destroys wooden or cardboard structures made for him by adults [plates 55(1), 55(2)].

My 3-year-old son often asked me, "Mother, give me something to tear"; when I gave him a sheet of paper, he would tear it to small pieces, scatter them on the floor, and throw them up in the air. Later, I saw my 5-year-old Roody in the company of other children breaking glass bottles and vials or smashing pieces of glass on the ground with stones. Those who have seen toys used by children know that, in most cases, these toys are a collection of junk mangled by a child's hands. The child's hands always move; they always are driven by the passion to destroy.

We mentioned that the child cannot remain indifferent when he sees a hole; he always must make the hole bigger. Roody was no exception in this respect; as soon as he saw a hole (in a linen sheet, pillowcase, or stocking), with sadistic pleasure, he tried to enlarge it to the limits [plate 62(3)].

Everyone who has watched children playing outside knows how vigorously they pluck grass and tear off leaves and branches without any concern for how they are going to use these objects; they enjoy the process of destruction per se. It is also known how much effort it takes to save fragile plants in public places from destruction; children, and even adults, out of sheer mischief break metal rails guarding the trees. Psychologically, such destruction is based on a tendency to unrestricted (in this case, wanton) behavior.

The first clear examples of the child's actions are associated with his pulling and pushing things, with the latter action usually more vigorous. At 5 to 9 months, Roody tried to throw things off the table, including the tablecloth; he intently watched these objects as they fell. He (0.8.23) tried to reach a thing he had just thrown aside and cried when he could not do it; he smiled when he managed to reach it and then threw it again on the floor. At 9 months, Roody used to push things down an inclined plane and throw them away from himself. A tiny baby who can hardly hold an object throws it away from himself, reaches for it, and throws it again endlessly, clearly enjoying his first self-expression in action.

The child's will initially is expressed in such throwing movements and destructive play. A 1-year-old baby throws things and makes loud sounds with special joy. Roody (at 1.2.20) could aim a ball at members of our family when asked, "Throw the ball at Father, Mother, Nanny, Grandmother." When I asked him to throw the ball at Apa (himself), he pressed the ball to his chest. At the age of 2 to 4, Roody assumed various poses before throwing the ball: He tossed his head back and bent his body back so far that he seemed to be close to falling [plates 101(4), 105(4)]. Roody (at the age of 1½ to 3 years) often tried to throw objects (stick, stones) over high fences. He usually could not do it in one shot; therefore, he repeated this action many times. In my notebooks, there are indications that Roody (at the age of 1½ years) showed an unbridled tendency for throwing things. In the yard, he threw stones [plate 55(4)], sand, or soil into the puddles; in his room, he scattered his toys around. Getting into his bed at night, he threw out diapers, pillows, and blankets. Sitting in the bathtub, he threw handfuls of water at me or at the nanny, laughing and undiscouraged by the fact that he was losing water between his fingers.

Once, I gave Roody (1.9.18) a chest of drawers filled with things just for him. The baby began scattering the things around the room with great enthusiasm. When the entire chest was emptied, I said, "Now, put everything together." But the child put only two or three things back into the chest and expressively said, "Mother!" thus requesting my cooperation. When I refused to cooperate, he said "Grandma!" When she also refused to join him, he started to appeal to his toys one at a time, pleading for their help. When I said, "Apa!" for the second time insisting that he do it himself, he again offered me everybody except himself.

The unhampered tendency of the child at 2 to 3 years for throwing and destroying things is expressed in knocking down bowling pins [plate 101(5)] or skittles [plates 114(1)–114(3)], for which aiming and throwing objects with destructive results are the main elements of play. Further development of this tendency, although in different forms, can be observed in a child 4 to 5 years old who, armed with a catapult, breaks windowpanes, shoots arrows with a bow, fires a gun [plate 91(2)], doing all this more willingly the more destructive the end result.

It is only natural that all kinds of military games (for example, tin soldiers) or battle games attract the child so strongly.

Modern children easily pick up the newest things in military weaponry: They use cannons and machine guns, arming their real and toy companions with them [plate 91(5)]. For example, Roody heard about military exercises and gas attacks and tried to demonstrate a gas attack in his own way. He invented a peculiar method for this purpose. He filled a metal jug with sand and threw the jug with all his strength in the direction of a "hostile aircraft" [plate 91(3)]. As a result of the jug crashing on the ground, the sand hurled up and to the sides. This gave the impression of a smoke and gas wave.

Play Based on Counteraction   1. Child's Will. The child shows great willpower in his movements of gripping and throwing. If I did not allow Roody (10 months of age) to take the things he was reaching for, he stared at me and uttered a shrilling yell, repeating it many times until this thing was given to him. If I resisted longer, the child started to cry. Roody, like Joni, showed a strong tendency for counteraction; refusal seemed to enhance his desire to get what he wanted. Someone else's action often generated his counteraction. For example, if you spread a piece of cloth on the seat of his chair, he immediately started throwing it off; he (0.10.15) used all his energy, cried, and despite all the difficulties, continued this action until it was complete. As soon as one of us covered his or her head with a kerchief, hat, or cap, the child tried to pull it off; if you put a toy in front of him, he threw it, or if you stood it up, he knocked it down. Once, I put a stick under a tangled electric cord; 11-month-old Roody managed to pull the stick out with great difficulty; he did it many times, crying when he failed to do it quickly.

As soon as an adult did something, the child immediately tried to do an opposite action. I sat a rubber toy on a stick repeatedly; every time, Roody threw it down. I spun a top; Roody caught and stopped it. I put a toy before him; Roody threw it to the side. We put stockings on him; he pulled them off. We put a kerchief on his head; he yanked it off. We put a piece of oilcloth on the chair; he peeled it off. If he was not allowed to touch an object that he was reaching for, he persistently tried to touch it every time he passed it; if he was initially allowed to touch this thing, he would not touch it any more.

When I tightly fitted a pyramid ring on a stick, 1-year-old Roody took every effort to pull the ring off [plate 56(1)], his face reddened, his teeth were clenched, and his head shook from the tension. As soon as he succeeded in taking the ring off, he put it back on the stick. The ban on an action seems to excite him (like Joni) to perform it. For example, Roody (1.3.28) was forbidden to pluck the corollas of flowers; nevertheless, he did this. He (1.6.16) was not allowed to take such things as glasses and scissors; nevertheless, he tried to grab these objects. If you said to him at the table that he was not allowed to touch any food on the plate, he (1.11.12) immediately touched it. If I asked him to give me his right hand, he (1.9.23) stubbornly offered me his left hand.

2. Stubbornness.   Sometimes, the child did not want to give his hand to the people who asked for it, yet he willingly gave it to other people who were vague about this intention. Sometimes, the child (1.11.13) was selective in fulfilling requests, for example, to bring cookies to certain people. In this case, however, this selective attitude was determined by his feelings toward these people; at any rate, his own free will also was a factor here. Once, Roody (1.9.26) clearly revealed his defiance under the following circumstances. He came back home from

the street, took some snow from his boots, and put it in his mouth. We told him, "You cannot do that." He immediately reached for his galoshes and tried to lick them. We also stopped him this time; he came up to a table and pressed the edge of the table between his lips. In response to our negative reaction to this, he lay on the floor and started licking the carpet.

Another time, Roody (2.3.8) put his index finger in his mouth and started sucking it; when we objected, he did not take the finger out of the mouth, but put it even deeper. When we insisted he stop doing this, he almost put his entire hand in his mouth until he started to choke.

Such disobedience of his was indicated in my notebooks even later (3.0.18). The child was forbidden to run over flower beds in the garden, but he did this on purpose. I remembered a remark made by Huxley about his little grandson, Julian: "The boy must have been about 10 years old. I told him not to go on the wet grass again. He just looked up boldly, straight at me as if to say, "What do you mean by ordering me about?" and deliberately walked on to the grass."[27]

Like Joni, Roody sometimes showed his discontent when his demands were not satisfied immediately. For example, he (1.4.22) asked for something; if we gave it to him after a pause that lasted a second or two, he would take it and throw it on the floor.

When the child's (1.7.16) wishes were not met, like Joni, he turned his back to the offender and did not respond to our calls. Once, Roody (2.1.15) accidentally hit his father on the head with a ball. His father reprimanded him very quietly, but the child's face blushed nevertheless, and he was on the verge of crying. Another time, he was throwing sand, aiming at the nanny, but he missed; he turned his back to her and stood motionlessly and silently for a long time.

The child at this age is already self-centered. Once, I gave the boy a pear; when he asked for more, I cut another pear in two unequal parts, gave him the smaller piece, and said, "I am giving you a small piece." The boy was offended and whimpered, "You did not say big!" He started crying and refused to eat the fruit.

Another time, Roody (3.1.7) disobeyed his uncle, and his uncle did not help the boy dress. Later, when the uncle appeared, the child avoided him and did not look at him. When I asked Roody, "Why are you avoiding your uncle?" Roody said, "He offended me." "How did he offend you?" I asked. "He didn't want to help me put my clothes on," he answered. In the evening, when we all sat at the table and drank tea, the uncle put a piece of raisin bread on his plate; Roody liked this bread and ate it every time his uncle gave it to him. This time, Roody said, "I don't want the raisin bread, it's bad, my bread is good," although his was a dry piece of common white roll. It was absolutely clear that the child's hurt feelings included even his favorite bread, revealing his predisposed attitude.

3. Slyness and Deception.   Sometimes, the human child, like the chimpanzee, uses slyness and deception to circumvent bans on some of his activities. Roody (1.9.8) was forbidden to gnaw a wooden block; he went behind the back of a high armchair and gnawed the block there. He was not allowed to take a match in his mouth; he obeyed, but as soon as adults left the room, he started biting a match.

I did not permit Roody (2.0.7) to peel paint from toys with his teeth; he either hid behind my back or behind the furniture; there, he did what he wished. His grandmother did not allow him (2.3.13) to pick berries in the garden; he walked behind her back (which is unusual for him) and picked the berries. Sometimes, this deception was naive, to say the least. For instance, Roody (2.2.6) did not want to drink milk. I said very severely, "Drink!" The child put his hands behind his back and said, "No hands," or, "Afraid of milk." I insisted and asked him directly whether he wanted to drink or not. He did not say anything. I said, "Why don't you answer me?" "I am not talking," he said. Apparently, he did not want to drink, but did not want to say yes and was afraid to say no. When I tried to squeeze an answer out of him, he totally disarmed me by saying, "I have no tongue."

Another time, Roody's uncle scolded him for disobedience and said, "Tell me you will do what I say." Roody (2.3.9) answered, "I don't have a tongue." When Roody sometimes did not want to do what he was asked, he used these deceptive pretexts. We said to him (2.2.23), "Give me your hand." "I can't hear you," he answered. We said, "Water the flowers, please." "I can't see them," he said, although we pointed at the place he was asked to water. Roody (2.2.20) does not want to eat white bread in a sandwich, only ham. We asked him, "Why aren't you eating bread?" He said, "To choke," "To cough," or, "Bread's sour." When Roody (2.5.6) was urged to eat soup, he put his hands behind his back and said, "No hands."

Earlier in his life (1.6.20), I observed that the child used deception to put off the moment of going to bed; he either wanted to go to the bathroom and spent a lot of time there doing nothing, or he (1.7.11) ate his evening meal slowly, so that you were about to lose patience. In all these examples, we unmistakably can notice the child's planned behavior, in the choice of a detour to complete some desired action or avoid something undesirable.

The human child, like the chimpanzee, knows the consequences of his actions and designs his behavior accordingly. Like Joni, confronted with failure, he does not know yet how to conceal his real goals and clearly exposes the illusiveness of his lie.

When Roody, like Joni, hid his hands or feet, since he did not see them, apparently he thought that others also did not see them. He showed a similar pattern of behavior (at the same age) in the hide-and-seek game.

As noted, Roody (1.5.22) hid by covering himself with a kerchief or cap [plate 92(4)], covering his face with his palm, or simply by closing his eyes. Later (at the age of 2 years), the child hid behind a transparent wicker armchair [plate 92(2)], standing at the same spot every time, or (2.2.8) squatting near the porch steps [plate 92(1)], or pressing his body to a wall, covering his eyes with his hands, turning his back to those looking for him [plate 92(3)], shielding himself from the outside world, and probably assuming that he also was hidden from our sight.

But, while destructive tendencies strongly develop in a child's life (at the age of 1½ to 4 years), we must state with great satisfaction that constructive actions form along with them, initially expressed in sheer imitation and later even in creative activity.

<table><tr><td>Imitative
Entertainment
of the Child</td><td>The human child, having broader access to the material world than that possible for the chimpanzee, has a more developed and sophisticated imitative</td></tr></table>

ability. Like Joni, Roody easily became infused with the mood of the people around him. Once, his nanny told the story about how she was scared by a drunk man; the child immediately became silent and serious. "What is it?" I asked. He (2.6.14) said, "I am afraid." He (1.2.20) heard a baby crying on the street; his face became sad, and he was ready to cry. Even more easily, the child became infected with such active emotions as joy or anger.

Roody often reproduced the gestures and expressions of people around him. For example, feeding Roody (1.8.24), I took a spoonful of milk from a cup, lifted the spoon, and poured milk back into the cup to cool it down; the baby also repeatedly lifted his hand and put it back on the table. I blew at a saucer filled with hot tea; he immediately also started blowing. Roody (1.5.21) saw in a museum the dummies of a wolf and of a monkey with an open mouth or (1.8.0) the picture of a girl with an open mouth; he immediately opened his mouth, too. In a picture, he saw a dog with its tongue out; he (2.0.9) opened his mouth and put his tongue out. Roody (1.6.26) saw a picture of a Boy Scout with his hand raised; he imitated this gesture. Roody (1.11.30) saw students exercise in the yard; he tried to reproduce these movements in his room. He (1.6.22) saw a man limping on the street, and he began limping. He (3.0.16) saw a man walking with crutches; he took two long sticks, walked limping, and said, "My legs ache." After watching marching soldiers, the child (1.4.2) walked by raising his feet high and stamping the ground. Suddenly, we noticed that he spat incessantly, saying, "I will be spitting." "Why?" we asked him. "Because nanny spits." "Smoke, smoke," he said, he (1.10.7) asked for a match, put it in his mouth, and pretended to be smoking [plates 117(3)].

Roody (1.8.10) saw boys in the yard throwing snowballs; he started doing the same. After he watched people skating and skiing, he (2 to 4 years of age) put on big shoes, ran around the room, and occasionally slid on the floor, shouting, "Skates!" [plate 106(1)]. Another time, he put two sticks on the floor, stepped on them, took two other sticks in his hands (2.8.4), tried to move, and became very disappointed, saying, "Why isn't it moving?" Later, he (2.9.20) put two wooden planks on the floor, stepped on them, and moved, pushing with two sticks against the floor. "I am skiing," he said [plate 106(2)]. Roody (2.11.19) saw the nanny put a new apron on a little girl; he said, "I want a new apron too," although he wore the same kind of apron, but an old one.

It can be stated with a high degree of confidence that there is no such activity, profession, or role that child would not want to imitate.

His growing soul, unsteady in its present condition, not rooted in its brief past, craves greedily for acquiring new experiences; it cannot resist the power of imitative attractions. In the process of imitation, it exercises, learns, develops, and perfects itself. The flexible soul of a child, like dynamic plaster, possesses the propensity for solidification, as if waiting to be shaped. Like liquid plaster, which quickly settles and assumes a new form, the child's soul is shaped by a surrounding medium, by patterns of day-to-day life, and especially by the psychological attitude of those in contact with him.

Apart from a series of actions associated with the child's instinctive imitations, satisfaction of his physiological needs and his learning of day-to-day procedures (we mentioned some of those in the previous chapters), there are no vitally important imitative actions associated with hand movements.

The child tries to develop his motor skills very early in his life, usually rejecting adults' help. Once, 10-month-old Roody tried to reach for a piece of wood, but in vain; his father saw these efforts and gave this piece of wood to Roody. He took it with a disconcerted expression on his face, threw it away from him, and tried to reach it again. Sometimes, the child did not want such help altogether and did not take the object given to him by adults.

It was observed that Roody predominantly used his left hand, and only after a number of corrections did he (2.2.1) start using his right hand most frequently.

Roody (1.6.11), like Joni, used to imitate actions associated with cleaning the room and preparations for his bath. He put firewood into the stove (1.6.11), arranged chairs next to one another, swept the floor with a broom (1.10.11), wiped wet spots on the floor with a rag (2.6.0), and cleaned the furniture with a brush (1.9.28). In the yard, he tried to sweep the ground with a broom (1.4.4). In winter, he took a huge shovel (1.9.3) and scraped the sidewalk with it [plate 111(2)]; or he took a smaller shovel [plate 111(5)], picked up snow and threw it aside (1.10.0).

The child started using a watering can very early in his life (1.4.9) [plate 111(1)]. He tried to take a very large can, which he could not lift; therefore, his attempt at imitation was useless. Later (2 years 1 month of age), the boy watered flowers with a small can [plate 111(4)]. He (3.0.18) watered flowers through a rubber hose, which he attached to the spout of the watering can.

It seems that there is no action the child would not want to imitate. "To turn like these men," Roody (2.2.22) said, rotating potter's wheel. "I am painting the house," the boy (2.3.22) said; he took a brush, dipped it in water, and made movements that imitated those of the workers painting a house across the street [plate 117(1)]. "To dig like these men," he (2.5.10) took a shovel and imitated shoveling movements; at the age of about 4 years, he took a big scythe [plate 111(3)] and tried to cut grass with it. Roody (2.11.4) took a long nail and scraped along a crack in the floor, saying, "I am cleaning the tracks." "Who does this?" we asked. "A streetcar," he said. Roody (2.8.3) stuck a twig into a crack of a wall, saying, "To take mud out of the tracks that streetcars can go."

Imitating the adults, Roody (2.1.25) looked at the clock on the wall or at his father's wristwatch (3.4.0) and said [plate 113(2)], "Ten minutes to one," or "Ten minutes to six," repeating something he had heard because he did not know how to tell time. It appears sometimes that the child cannot resist his desire to reproduce adults' actions. For example, he (2.10.12) saw through a window a horse falling on the street; he immediately knocked down his toy horse, saying, "The horse fell." Roody (2.4.2) put a long file (which reminded him of a thermometer) against the wall and asked, "How many degrees?" He (1.7.5) watched older children eat the seeds directly from a sunflower; he brought his hand to his mouth (an imaginary sunflower) and then spat the husk out.

Obviously, his attention was riveted to telephone conversations; because he was allowed to talk on the telephone, Roody either put to his ear the round lid of a shoe polish box (1.7.9) and imitated a telephone conversation or took a rubber tube [plates 113(4), 113(5)] and, along with his companion (2½ years of age), led long and lively conversations, sometimes using (3.0.18) expressions from a poem "Telephone") by Marshak.

The work of his parents also captured Roody's imagination, and he tried to reproduce its individual elements. Once, Roody saw a group of people had come and said (2.9.3), imitating his father, "I am going to lead a tour." Another time, he (2.9.27) said to me, "Mother, let's build a zoo. You will be the tour." Later, his (3.0.5) desire to follow in his father's footsteps was expressed in the following way: "When I grow up to the ceiling, I'll put on a white laboratory coat, I'll lead the tours and tell stories to naughty boys." At 4 years old, Roody, in a business-like manner, demonstrated his toy museum to a friend, a girl of the same age [plate 94(4)].

Social events leave strong impressions in the child's mind. Once, Roody (3.0.29) put a number of his animal toys into a toy car, saying, "They are going to take part in a demonstration." Walking on the streets during a demonstration, he wanted to carry a small red flag and a red badge similar to those carried by adults. The child (3.0.6) tried to put into action something he had read in a book; for example, he took pieces of paper, soaked them in water, and stuck them on the door, calling them "announcements." After he read a book, *Lost*[28] [plate 69(2)], he (3.3.0) put up announcements all over the doors of our flat, trying to reach as high as possible. When I asked him what his announcement said, he repeated a sentence from this book, "Residents are not allowed to let their cats out to the hallway, or the cats will be put to sleep."

Sometimes, the child (2.5.0) tried to imitate an action he had seen in some pictures. For example, he took a broom and swept a path in the garden, remembering the picture of a boy sweeping the ground.

Analysis of a child's imitative actions shows their diversity, effectiveness, and accuracy compared with those of the chimpanzee. There was no activity that the child would not want to repeat after adults.

When his much older friend (F. E.) scattered sheaves of straw for drying, 2-year-old Roody immediately positioned himself next to him, also took a sheaf of straw with great difficulty, and tried to scatter it [plate 66(1)]. In the wintertime, when F. E. dragged big baskets filled with snow from one corner of the garden to another, 2½-year-old Roody repeated the same operations on a somewhat smaller scale [plate 66(5)].

When F. E. swept snow from the sidewalk, Roody (2½ years old) did exactly the same. If F. E. carried bundles of hay on his back, Roody (3 years old) demanded that a bundle of hay also be hung on his back. I began to rake leaves in our garden; the child also began raking them [plates 66(3), 66(4), 66(6)]. The boy did not give up even if the task was too difficult for his limited physical abilities. When F. E. put up the dummy of a gigantic African elephant and kept coming down the ladder to pick up various things he needed for the work, the boy stayed near the man and wanted by all means to participate in everything, saying, "I want to help you." He took big chunks of clay and, straining himself, tried to reach the man; he followed F. E. everywhere, trying to be helpful. If F. E. took a big chisel to make holes in a metal sheet (the elephant's ear), the boy (2.5.2) took a big nail and a hammer, sat at the other side of the sheet, and started hammering the nail [plate 66(2)]. If F. E. took a long metal rod and got ready to make holes in it, the boy (2.5.2) took a similar rod and followed him. If F. E. placed the rod on a stone, put the chisel to it, and began hammering it, the child reproduced all these movements. No wonder that, when someone later asked in the boy's presence who made the elephant, the child said, quite seriously, "This man and I did."

At the age of 3 years, Roody watched with delight his father stamping papers and typing on a typewriter. The boy tried to imitate the entire sequence of actions: pushing the keys, sliding the carriage, shifting the lever. This "typing" was so ravishing that you hardly could pull him off the "typewriter" [plates 69(5), 69(6)].

The child often watched the photographic process and imitated it with great enthusiasm; as he grew up, the imitation became more and more complicated. Initially, Roody (2.2.25) "photographed" by holding two long sticks (crossed and then separated) at the level of his eyes [plates 118(1), 118(2)]. Later, (3–4 years of age), the boy took a real photographic tripod, placed it on the floor, adjusted the screws in its legs, and then "took a picture" [plate 118(6)].

In both instances, the child was impressed by a tripod more than by anything else associated with the photographic procedure. At the age of 3 years 6 months, the child paid more attention to the camera; he took a round can, held it in front of his chest with its opening directed forward, asked someone to hold an object to be photographed (for instance, the dummy of a squirrel) for him, and clicked the imaginary shutter [plate 118(3)].

At the age of 4 years, Roody's attention was directed not only at the camera in general, but also specifically at the lens, for he now used an electric flashlight, which he held in front of his eyes [plate 118(4)]. This time, he looked above the flashlight at the object (a stick) he had positioned for the picture; he seemed to bring the image into focus and pushed the flashlight button, which gave a clicking sound resembling that of a shutter [plate 118(5)].

Roody wanted to imitate various professions. He often saw a newspaper salesperson running and shouting on the street (Roody at the age of 4 years was asked to buy papers), and he wanted to sell papers. He gathered old papers and posters, turned them into a roll, and offered them to us, demanding that we take them [plate 65(3)]. Roody (3.10.0) saw a junkman in the yard; the boy asked for a sack, put various things in it, and went around the room, shouting, "Shurum-burum" [plate 65(4)]. Suddenly (at the age of 2–3 years), he wanted to become a driver; he (3.11.1) took large gauntlets, put them on [plate 96(2)], and said, "I am a driver." He started dabbling around the wheels of a carriage (3 years 10 months) or toy car (2 years 3 months), arming himself with some tools and pretending to fix the car [plates 96(5), 96(6)]. Soon after visiting a physician (at the age of 3 years), he imitated that profession. He listened, with a rubber tube in his hand, to the grandmother and his teddy bear [plates 58(6), 65(6)], putting the free end of the tube now to the head, now to the chest. Roody (1.0.10) wanted to put cotton wads into someone else's nose after he underwent the same procedure himself. The roles played by the child voluntarily and enthusiastically were numberless.

After Roody had seen anglers on the banks of Moscow River, the next day he chose a bench in our garden near a puddle. He threw a stick into the puddle, placed next to him a bucket for so-called fish, put up a toy rabbit to guard the fish, and sat motionlessly for hours, like a real angler, once in a while raising the fishing rod and taking the imaginary fish off [plate 63(4)]. After Roody was read a story and shown the pictures about a shepherd pasturing cattle, he (4 years old) placed his animal toys all over the grass, took a whip and a horn, and imitated a shepherd [plate 65(5)].

Roody's thrill when he saw firemen passing by equaled the delight with which he imitated extinguishing a fire. The more accurate he was in reproducing adults' action, the more satisfaction he got out of play. Roody did not limit himself to just pouring water on the fire, he put on a helmet and arranged a so-called fire truck, in which he put all his dolls. He fastened a rubber hose on the spout of a watering can, took a toy ladder, and with all this completed, he rushed to "the scene," a house he built especially for this purpose, and started extinguishing the fire with great enthusiasm.

After visiting a store or a pharmacy, Roody reproduced buying and selling activities. Having seen a road being paved, he (4 years 7 months) tried to pave a path in the garden, driving stones into the ground with a hammer [plate 65(1)].

We mentioned how Roody was captivated with the army after he saw military exercises.

Sometimes, Roody tried to be like people around him even by his clothes. For example, he (3 years 3 months) saw his friend (an adult) attach his medals to his coat; the boy pinned some old insignia to his shirt and, so decorated, paraded for hours. Suddenly, he decided he wanted to look like a military man; he came out on the street wearing a metal belt (a piece of a thin copper grate) and walked without so much as feeling uneasy (and perhaps with an element of secret pride) because he attracted attention. Enraptured by the sight of people marching in gas masks, he begged us to allow him to go outside in a real gas mask; when we did not give him permission, he found goggles and wore them for days, refusing to take them off even during meals.

The child's imitative tendencies were sometimes of an adverse nature. For example, Roody saw that all children in the yard wore caps; he was unwilling to put on a hat and begged me to give him a cap, too. He vehemently resisted my attempts to make him wear a tie because other children in the yard did not wear ties. He protested if I attempted to tie his red tie in the form of a bow; he wanted it tied in the "young pioneer" fashion. My son also showed the same tendencies much later.

Fascinated by cartoons (about 6 years of age), the boy made caricature painter Efimov his idol; Roody drew in his notebook in the same pen as in Efimov's book. He was beyond himself when, at the age of 7, he got a pen and pencil of the same type of those of Efimov.

It is commonly known that teenagers become so carried away by imitation of the heroes of fictional journeys as to attempt these most risky trips themselves. Also, it is very well known that children are infected easily with bad habits, particularly smoking. My Roody (17 months) would find a stick, put it in his mouth, and hold it like a cigarette; he (1.10.17) also said, "Smoke, smoke" [plate 117(3)]. At the age of 2 years 4 months, he imitated smoking with greater perfection: He held one stick between his teeth with a smile on his face, while he brought the second stick to the first stick as if lighting it with a match [plate 117(4)].

Also, other less accessible and more serious pursuits, such as reading papers [plate 69(4)], writing [plate 50(4)], drawing [plate 69(1)], telephone conversation [plate 113(3)], typing on a typewriter [plate 69(6)], playing the piano [plate 107(6)], watering flowers, spreading paint [plates 117(1), 117(2)], digging or drilling [plate 119(6)], sawing [plates 119(1), 119(3)], chopping or raking [plate 119(5)], cutting [plate 119(2)], and so on, which were part of day-to-day human activities, were picked up and reproduced by the child with various degrees of effectiveness and accuracy.

Drawing. The child (1 year of age) takes a pencil and tries to imitate adults' actions [plate 50(1)]. But, even brief observation of his actions shows that the child reproduces only the appearance of this process; in his attempts to draw, he raises the paper from the table and sometimes does not even look at what is produced. At this time, the child cannot learn the most common and proper way a pencil needs to be held. He grips the upper part of the pencil in his entire hand, the way Joni did [plates 50(2), 50(3)], which also does not help make a detailed picture. Roody (1.2.21) sometimes took a pencil upside down, not leaving anything on the paper; while he "drew," he often looked around and not at the paper (1.3.3).

Roody, like Joni (1.7.1), demanded an ink pen and tried to draw on paper, on a book, on the table, or on his hand. His drawings at that time resembled those of Joni [figures 15(1), 15(2)]. But, the human child at the age of 1 year 9 months employed a slightly different way of holding a pencil: Although he held it by the upper end, he entrusted the leading role to his index finger and positioned it along the pencil axis, somewhat forward with respect to the other fingers, and pressed from above [plate 50(4)].

At this time, while drawing, the child usually concentrated on this process, his head bent, and his lips closed tightly. His drawings of that period undoubtedly revealed greater complexity than those of the chimpanzee. Roody's (1.8.14) drawing entered a new stage. Instead of haphazardly positioned, predominantly horizontal lines, we observed crossed lines; the child paid special attention to the crossed lines, calling them "gkh" ("airplane" in his terminology, although these hardly bore any resemblance to an airplane). In a month, he (1.9.1) drew, perhaps accidentally, two parallel lines and called them "wo" (they probably reminded him of reins). The child's imagination helped him with definitions. Once, he (1.10.23) covered an entire sheet of paper with scrawls; answering my question about what he had drawn, he said, "Smoke."

We observed in Roody, at the age of 2 years, the predominance of round and circuituous drawing, while the chimpanzee at the age of 3–4 years, even after extensive exercises in drawing, did not go beyond drawing straight, sometimes crossing, lines haphazardly scattered on paper, which were so characteristic of the first two stages of the child's drawing. In the 2- to 3-year-old child, this tendency for drawing round lines was extremely compelling. Whether he drew with a pencil on paper, with a piece of chalk on the blackboard, or with a stick on the ground, everywhere we found scrawls, sometimes in the form of irregular ovals [plate 69(1)]. Joni, despite multiple exercises with the pencil during his almost 3 years in our custody, did not reach such a stage in drawing.[29]

More accurate analysis of Roody's and Joni's drawings reveals that for a very long period of time, Joni's single tendency was reproducing perpendicular lines; we can find hosts of tiny crosses on his sheets of paper [figure 15(3)]. Joni's drawings generally are monotonous, while Roody's reflect rapid progress and diversity. Sometimes, Roody (1.11.1) took a pencil in each hand and drew with both hands simultaneously; his right hand drew more lines than his left hand.

Roody (2.1.11) was never at a loss to define what he was drawing. Making irregular strokes, he called them, for example, "airplane" or "pike," although it was sometimes very hard to find even partial resemblance to these objects. Roody (2.1.28) to a greater extent succeeded in reproducing an airplane, this success being rather involuntary.

Roody (2.1.28) took a stick and drew separate short lines with it on the sand and then said suddenly, "I'll make an airplane"; he drew some additional lines, and a primitive airplane emerged. The child (2.3.14 and 2.7.11), now at the stage of drawing ovals, then made more accurate definitions: circle, snake, ring, moon, and wheel. He (2.7.25) called a group of circles "big metal wheel." He called another circle, "This is sun"; here, we could find partial similarity between the drawing and the object.

He (2.7.5 and 2.3.14) found outlines of a snake and scythe in some long lines. In some crossing lines drawn on the sand, he saw a cross or an airplane (2.2.16).

Roody (2.4.28) started drawing groups of images, three on each sheet of paper, giving each picture a name. As pictures were drawn, they and their definitions often changed drastically. The child (2.5.3) drew two perpendicular lines and said, "Airplane"; in a minute, still drawing, he added, "Some bird"; and yet later, he made a more accurate statement, "Flamingo."

Both drawings and verbal definitions were in abundance. He called his not very strongly differentiated scrawls little wheel, snake, hook, pin, worm, frog, teddy bear, horse, doll, "This is a long doll," scythe, and seal. Sometimes, the similarity was slightly visible.

Now, the child tried to qualify his drawings. Looking at his two consecutive drawings of airplanes, he defined the first one by saying, "This airplane is not like an airplane," and the second one by saying, "This is an airplane," "Good airplane" (2.7.29).

He drew three long lines, called one of them "worm," and called the two others "also worms." Suddenly, he regarded them as having a family relationship, calling them "Mother-worm," "Father-worm," and "Little boy-worm." Here, all three lines were of equal size, but soon Roody (2.8.11) singled out, in his drawings and verbally, objects of different sizes. He said, "This is a big ostrich, and that is a little ostrich."

Sometimes, the child (2.8.20) did not know what he was drawing, he would say, "I don't know." Suddenly, he would stop for a moment and give some definition, more or less suitable.

Roody (3.0.8) drew his first picture of a man in the horizontal position. He drew an oval (head) and something within the oval, then two long lines, which he crossed with a line at the other end. "What did you draw?" I asked. He said, "Uncle." Then, he looked at the picture and said, "Where are his arms?" and added two lines at the sides. I said, "Draw your mother." He made a second picture, this time a little more thoroughly, but also in the horizontal position.

The stages of his drawing are as follows:

1. Roody draws an oval.
2. Roody draws something within the oval and calls it eyes.
3. Roody draws two long lines.
4. From these lines, Roody draws two short bent lines, saying, "Legs."
5. For arms, two additional lines are drawn at the sides [figure 15(4)].

Roody (2.8.11) imitated artists who used to work in our house; he spread paint with a brush, but he did not name the pictures. Apparently, he was captivated with the colorfulness and brightness of the pictures, and this did not require additional interpretation.

The drawings of the child at 3 to 4 years old are undoubtedly more refined in their design and quicker in the speed with which they develop. Analysis of these in essence naive and primitive drawings of a 3-year-old child reveals the following intellectual processes:

1. Differentiation of drawings by the child.
2. Tendency for reproducing real objects.
3. Identification or analogization of reality with the pictures (by the picture itself and by additional verbal explanations).
4. Qualification of the drawings based on the degree of proximity of the drawing to reality.
5. Fantasy—a colorful verbal addition to make the drawing more expressive.
6. Discerning characteristic features of the object.
7. Child's perception of real objects, which turn into images in the child's mind.
8. Comparison of the images with the real objects, complementing and perfecting the image to attain better similarity.

I did not observe in the chimpanzee all these mental attributes of the imitative activity of the human child, and they probably did not exist. This is true in spite of the fact that Joni had been no less enthusiastic in drawing than Roody. Joni often cried for a pencil; it could be taken from him only by force. He drew with keen and lively interest and looked at the objects with great attention. However, he did not show visible improvement in the drawing technique [plates 50(2), 50(3)].

Like Joni, Roody used substitutes for the pencil as a drawing tool: a stick, his own finger, a nail, dry macaroni. The more intense the traces he left on paper, the more engulfed he was in the drawing process. While Joni usually drew with his finger, which he had previously dipped into ink, Roody drew with red chalk or painted with a brush dipped into dense black paint.

Using Tools   As mentioned, the child at the age of 1 year and older imitates a series of actions using various tools: a hammer (1 year 1 month and later) [figures 4(1), 4(3)–4(6); plate 65(1)]; pencil (1 year 1 month) [plates 50(1), 50(4), 50(5)]; brush (3 years 10 months) [plates 117(1), 117(2)]; broom (2 to 2½ years); shovel (1 year 10 months) [plates 111(2), 111(5)]; watering can (1 year 5 months; 2 years 1 month) [plates 94(3), 111(1), 111(4)]; pliers (2 years 5 months; 4 years 4 months); brace (4 years 5 months) [plate 119(6)]; saw (2 years 1 month; 2 years 4 months; 3 years; 4 years) [plates 119(1), 119(3)]; pitchfork (4 years 1 month) [plate 119(4)]; rake (3 years 1 month) [plates 66(6), 119(5)]; scythe (4 years) [plate 111(3)]; broom (2 years 6 months) [plate 66(3)]; scissors (4 years) [plate 119(2)]; knife (2 years 11 months) [plate 88(6)]; fork (2 years 7 months) [plates 88(3), 88(4)]; cup (2 years 5 months; 3 years 3 months) [plates 88(3), 90(1), 90(2)]; spoon (2 years 1 month; 4 years 1 month) [plates 82(8), 88(1), 88(2)]; soap (3 years 2 months) [plate 60(2)]; towel (2 years to 3 years); comb (2 years to 3 years); handkerchief (2 years 6 months) [plates 60(5), 60(6)]; toothbrush, bed vessel, key (2 years 6 months); and awl (2 years 5 months).

In getting used to operating a tool, the child strives for independence. For instance, Roody (2.2.8) took a watering can, wet the flowers, and repeated, "Apa himself, Apa himself," rejecting any help. He acted in the same way when using spoons and cups, "Don't hold, don't hold," "Apa himself," he shouted (2.5.18).

The child often reproduced only the outward form of the action. I was cutting bread with a knife, and Roody (1.8.25) tried to cut bread with the handle edge of a spoon, obviously not succeeding.

The child started very early in his life to use auxiliary objects for reaching something. For example, Roody (1.0.14) reached for an egg on the table, but it was too far for him; then, he pulled the tablecloth toward him and took the egg. After failing to reach an object from the upper shelf, he brought a stool and stood on it.

At 2 to 2½ years old, Roody, like Joni, liked to imitate sweeping the floor or sidewalk, but like Joni, he did not do it very effectively. Often, the end result was not perfect, but he was apparently more interested in the process than in the results.

In the case of Joni, there were relatively rare occasions when he used an auxiliary tool (spoon, stick, nail, knife, or key); given the choice, the chimpanzee preferred to do without them (for instance, when he was offered to eat jelly from a cup with or without a spoon, he always chose the later option). The human child, like the chimpanzee, in all instances when he can do without a tool prefers not to use it to make his task easier. This is visible most clearly with respect to his use of a spoon or fork. Roody had learned perfectly (by the age of 2½ years) how to use a spoon and fork, but each time, he preferred to take food with his hand when he did not feel he was watched by adults. To even a greater extent, he was inclined to simplify his actions with a comb or handkerchief. Like Joni, he always favored wiping his nose with the back of his hand than using a handkerchief. He tended to ignore altogether such boring procedures as combing his hair, brushing his teeth, and washing himself with soap because he did not care about the end result.

Speaking about the effectiveness of Roody's imitative actions, we must state that, according to my observations, except for sawing with a hand saw, which he tried too early (2 years 1 month) and which was totally beyond his capabilities, he performed all other actions quite accurately and brought them to completion. He most enthusiastically made holes with a brace [plate 119(6)] and peeled the bark from birch logs with tongs; he fairly easily rotated the brace, hitched, and pulled (the bark), achieving visible results.

He put great effort in learning how to use scissors. Initially (before 4 years of age), Roody held scissors with both hands, which is the common way of using large garden scissors. Only after long experiment did he learn to hold scissors with three fingers of his right hand [plate 119(2)].

Consecutive mastering by the human child of actions associated with tools, if put in chronological order, reveals gradual development of his skills.

Roody (at the age of 1 year 1 month) took a hammer in his hand, but frequently missed the nail [plate 57(1)]. Later (1 year 6 months; 2 years 1 month) [plate 57(3)], we caught Roody hammering small nails into a bench or into the wooden wheels of his toy freight car; although the direction of the hammer was correct, he could not hammer the nail all the way. The result was determined by the resistance of the material the nail was entering; in that case, the material was too hard [plate 57(4)].

If the material were soft enough, the success was easy; the child (2.4.2) entertained himself by hammering dozens of nails into a wooden plank. Roody (2.11.24) said, "I'll be working," and pounded with a hammer, chipping away ice from a garden walk. In a little while, he sat down, saying, "Oh, I've worked, and I am tired," and then he resumed hammering nails.

At the age of 4 years 7 months, Roody coordinated very easily the movement of both hands; he held the object (being hammered) with his left hand [plate 57(6)] and the hammer with his right hand when, for instance, driving stones into the ground [plate 65(1)].

At 5 years, Roody used a hammer very skillfully. When he could not drive the nail in directly, he took an awl, made a small indentation in the wood, and beat on the awl with the hammer, easing the future hammering of the nail [plate 24(5)].

The child at the age of 1 year 10 months tried to use a small shovel (for example, for making a snow pile). Sometimes, however, his strength did not measure up to the task. He took a huge shovel, but was unable to lift it and dragged it helplessly over the ground.

I observed his inability to evaluate his own strength adequately when he was using a watering can. At 2½ years, Roody could use a small watering can, but he often tried, in vain, to lift a big one (the size of a bucket) [plates 111(1), 111(4)].

Sometimes, the child (at the age of 1 to 2½ years) did not know how to use a tool altogether; nevertheless, he imitated operations with it. I observed this in my son when he tried to use a pencil (1 year 1 month), hammer (1 year 1 month; 2 years 1 month), tongs and awl (2.5.5), saw (2.4.1), and key (1.7.10). Sometimes, the child only pretended to be writing, hammering, opening, and the like. For example, Roody (2 years 5 months) pulled a toy automobile with a little girl in it [plate 94(5)]; from time to time, he stopped, bent over the wheels, and pretended to be fixing them [plate 96(6)]. He really enjoyed playing with his toy railroad car, which had a space where he could put his tools. Often, while riding on this car, he would stop, take out some of his tools, and pretend to do something with the wheels [plate 96(5)]. Although the ride was real and the work was imaginary and unneeded, nevertheless, the work part was no less entertaining and desirable for him than the ride.

"To saw a bench," Roody (2.4.1) said and, taking a hand saw, made sawing movements at the edge of the bench [plate 119(1)]. Roody (2.5.5) saw an adult fixing a toy train; he took tongs and poked with them, quite senselessly, at the train here and there. Another time, he (2.1.26) pretended to be opening a door, although he did not know how to insert the key and how to turn it. The child was not satisfied with his ineffective imitations; he was on the verge of tears when he could not saw for real, and he became extremely excited when finally he succeeded in sawing off a piece of wood. When Roody (2.1.26) managed to saw off a piece of wood on his own, he ran over to me, showed the piece, and exclaimed, "Mother, I sawed this off!"

The child uses various objects as a means of attack or defense very early in his life. It is natural that the child greatly surpasses the chimpanzee in terms of the diversity of these objects and their forms of use, which obviously influences the effectiveness of the child's actions.

As stated, Joni, afraid of a certain object, usually took a rag or anything else that happened to be nearby and made threatening gestures or threw these things at the object.[30] Roody did the same thing in similar situations as early as at the age of 11 months, but later, he enhanced these weapons, emulating the adults.

Roody (1.5.18), while running, liked to make a swishing sound with a stick; he would whip "his horses" and shout, "Go, forward." The child (1.10.4) used a stick to speed up a spinning top.

Once, he (2.4.3) even threw a stick in the air, which was swarming with mosquitoes, saying, "I'm shooting the mosquitoes"; armed with a stick, stone, and ball, the child (2.4.23) chased stray cats out of our garden. At the age of 2 years, Roody took a long stick and beat the objects around him with it, as if training himself how to use it (I wrote about this in greater detail in describing the child's military games).

Like Joni, Roody used tools when he did not want to be in contact with an object due to his fear or fastidiousness or when the object was not within his reach. For example, Roody (0.9.2) took a spoon and used it to knock down standing toys. The child (2.1.11), using a stick, extracted things from under the furniture. He (2.4.2) took a twig and touched a caterpillar with it because he was afraid to touch the caterpillar with his finger.

Constructive Play   The human child from his very early days without any doubt excels over the chimpanzee in constructive activity, heralding his future vocation as a world reformer. His creative tendencies developed with progressively increasing speed.

While a 1-year-old child, only with a great difficulty, can change the position of a thing such as putting up a toy animal from the lying position [plates 87(1)–87(3)], after a year, the child already can assemble the rings of a wooden pyramid [plates 56(2), 56(3)]. After another year, he puts these rings on one another without an axis, making a fairly tall tower [plates 56(4), 56(5)]. The child easily puts together and takes apart collapsible bowling pins (1 year 11 months) [plate 56(6)]. But, he does not limit himself to elementary design; very early in his life, his activities become truly creative.

Roody's first creative project was building an airplane [figures 24(1)–24(9)].

1. Roody (1.7.10) takes a pyramid ring, inserts a pencil into its hole [figure 24(7)], and says "gkh," which means airplane.
2. He puts two matches perpendicular to one another and calls them "gkh" (1.7.13).
3. He puts a feather on a stick and calls it "gkh" (1.8.6).
4. He bends a twig at an angle (1.8.10).
5. He puts a brick on another brick (1.8.14).
6. He puts two pieces of dried bread on one another (1.9.5).
7. He puts a cookie on a piece of dried bread (1.9.9).
8. He sticks a wooden chicken on a pencil (1.10.22).
9. He finds a twig on the ground, bends it at a right angle, raises his hand, and says, "gkh" [plates 67(1), 67(2)].
10. He takes a small pillow, raises his hands, and says, "gkh" (2.0.27).
11. He finds a piece of wire, bends it, and says, "Airplane" (2.2.9).
12. He puts two long sticks perpendicular to one another on the ground and calls this airplane [plate 67(5)].
13. He arranges a pattern of blocks and calls this airplane [plate 68(3)] (2½ to 3½ years of age).

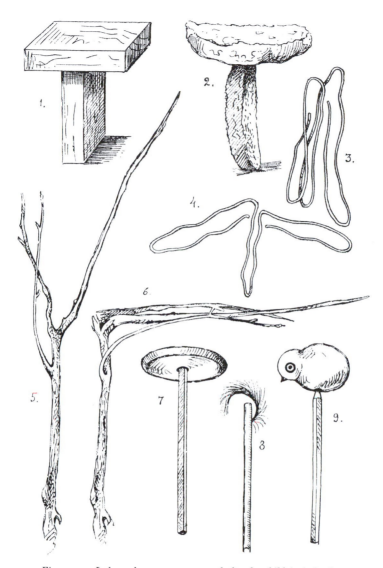

Figure 24. Independent structures made by the child in imitation of an airplane: (1) Roody (1.8.14) having put one brick on another, said "gkh" (airplane); (2) Roody (1.9.5), having put one biscuit on top of another, said "gkh" (airplane); (3) a bent wire found by the boy; (4) the same wire bent by the boy and termed "airplane" (2.2.9); (5) twig found by the boy (1.11.0); (6) the same twig bent by the boy, termed "gkh" (airplane); (7) wheel pierced by pencil, called "gkh" (airplane) (1.7.10); (8) feather placed on top of stick, called "gkh" (airplane) (1.8.6); (9) wooden chick perched on pencil, called "gkh" (airplane) (1.10.11).

14. He puts two sofa bolsters perpendicular to each other on the floor and calls this "gkh" [plates 103(1)–103(3)] (3.6.0).

In the case of the last airplane, Roody sat on it himself and invited me and a teddy bear to join him [plate 67(5)]; in a few minutes, he intentionally fell on the floor, imitating a plane crash [plate 103(3)].

At the age of 3 years, Roody made a "balloon"; he put a toy soldier or sat in its basket, saying, "In the balloon" [plate 67(4)]. At 2 to 4½ years, he became so engulfed in his creative activities that there were practically no materials he did not use, and there were no devices he did not try to reproduce.

As noted, Roody at the age of 2 years willingly played with sand, making various figures. He scooped sand, kneaded it in the forms, and put it on top of the figures [plates 49(4)–49(6)].

At the age of 2 years 1 month, Roody arranged wooden blocks in a long twisting line, calling it "worm" [plates 120(1)–120(3). At the same age, he created other edifices: a bridge from piled-up toys (2 years 3 months); a well out of sticks (2 years 11 months); a boat (after he was boating on a lake). As the photo shows, he made this boat differently each time: He sat in the boat (2 years 3 months) and even used "oars," or he (2.5.11) sat on a bench, put his feet in a case, took two sticks, and began to rock as through rowing.

Roody constructed a great variety of boats. At the age of 2 years, Roody made up a chain of small objects [plate 104(5)], such as a toy truck, small chair, small cart, watering can, etc.; he sat in the middle of the chain, took sticks (oars) and the boat was ready. At the age of 3 years 11 months he arranged half-torn bags in a row, calling this "boat" [plate 63(5)].

At that time, he tried to build a house, Zoo [plates 99(2)–99(4)] and telephone [plate 113(3)]; a certain progress in this (obviously without outside help) could be observed during a period of several months.

Before 2½ years of age, the child called "house" a series of standing or lying blocks next to each other [plates 68(1), 68(2)], at the age of 3 years he put out blocks perpendicular to each other [plate 68(3)] around a square. Later (3 years 4 months), he concentrated blocks near a certain point [plate 68(4)], or piled them on one another [plate 68(5)], or built a semblance of a tower or gate. At 3 years of age and later, he created multistory tower-type structures [plate 68(6)].

Comparing the drawings [plates 68(1)–68(6)] of these buildings, we can see the development of the child's creativity and how superior he was in this respect to a chimpanzee of the same age.

Imitation of Sounds   There is no imitative activity other than imitation of sounds in which the chimpanzee would be most inferior to the human child. We have stated that not only is the chimpanzee reluctant to reproduce articulate human speech, but also he is unable to do it even after long, intensive training sessions. In contrast, the child's proclivity to imitation of sounds is indomitable, which clearly distinguishes him from his chimpanzee counterpart.

Language, words, and speech first and foremost make a human "human." Even brief superficial observation of the child's language development reveals the point at which the bridge between the infant chimpanzee and the human child cracks and how early this proclivity to sound imitation starts to develop. I observed that my 3-month-old baby uttered pleasant melodic sound in response to my soft singing. At 4 months, Roody emitted something resembling a laugh in response to my laugh. At this age, the child can imitate adults' "lip rattling." A 10-month-old child often gives a superficial cough after he has heard the real cough of adults.

As soon as I uttered "ah-ah" in Roody's presence, he repeated this sound after me several times. My 10-month-old Roody repeated right after me the words "yes," "uncle," and "aunt."

I have the following entries in my notebook depicting Roody's (1.5.0 to 3 years of age) more or less accurate sound imitations. Roody imitated snoring (1.5.27), nose blowing (1.8.1), sneezing (1.0.10), coughing (1.9.5), hissing and sniffling (1.9.15), loud laughing (3.0.17), the inarticulate sound of cabdrivers (2.0.7), adult conversation (imitation by loud yelling) (1.7.0), crying baby (2.2.2), soldiers' singing (by yelling) (1.5.25), adults' singing (in a high or low pitch) (1.10.7), and adults' singing (without picking up the tune) (1.2.15, 2.9.18, 3.0.29).

The child tried to imitate sounds of some animals and those emitted by some inanimate objects. He imitated the croaking of a crow ("kar-kar") (1.6.4), squeaking of guinea pigs (1.6.4), barking of a dog ("how-how") (3.0.0), sound of an airplane ("gkh") (1.5.7), ticking of a clock ("tick-tick-tick") (1.5.16), rattling of the chain while lifting a clock weight ("drr") (1.6.22), sound of a squeaking door (1.7.23), rattling of a curtain ("drr") (1.6.11), and sound of a trumpet (2.0.18).

I observed Roody (1.6.13) picking up and reproducing intonations; he repeated "bye-bye" after me in a crooning manner. Speech flavored with intonation was observed in Roody as early as 21 months. The child understands intonation very early in his life. We have noted that Roody was offended and started crying when his father reprimanded him for something in a strict voice.

At the age of 10 months, Roody slapped me on the cheek. "You cannot do that," we said to him. The child immediately skewed his eyebrows, turned down his lower lip, narrowed his eyes, whimpered, and was ready to cry. Later, Roody (1.1.12) cried inconsolably in response to a loud voice or even more so in response to shouting.

It is well known that the child willingly croons a song after adults, but his musical ear (in most cases) is yet undeveloped and does not pick up the melody. Nevertheless, children like to sing in a group, apparently not caring much about the cacophony they produce.

As soon as the child learns how to reproduce simple tunes, he willingly sings, enlivening this singing with the silent procedures of taking off and putting on his clothes. I never observed Joni singing or even uttering long sounds, except for modulated hooting.

It has been noted that Roody tried to imitate a human reading. He would take a book or newspaper (1 year 6 months to 2 years 7 months), open his mouth wide, and utter a series of inarticulate sounds [plate 69(4)]; later, he would take a sheet from the calendar and mutter various words. He often would stop on the street and look at posters as though he (1.4.27) was reading them.

Roody (1.4.9) easily reproduced the familiar words daddy, mother, grandma, and nanny; he enthusiastically shouted these words with all his strength. Roody (1.5.22) sometimes was unable to reproduce less familiar words. For instance, I heard him trying to pronounce the word "Kathy" for 2 days in a row. Urging him to repeat the word, I said, "Say, Kathy." He pronounced certain words resembling Kathy, but still not exactly this word.

Or, for instance, I shouted, "Alya!" He (1.5.24) repeated after me, "Alo" and then, "Ava." Roody (1.8.14) reproduced certain words correctly, for instance, Kolya and Boria (male names), grandma, and Nadya (female name). This does not mean, of course, that from this moment on the child will start pronouncing all the words correctly; much later, we still can encounter distorted verbal imitations.

Roody (1.10.26) often tried to pronounce words immediately after adults, but he distorted them immensely. Despite the distortion of the words and substitution of the vowels, he put the accents correctly.

The child's tendency to reproduce everything adults have said becomes more and more visible. Roody (2.3.25) reproduced entire sentences. I said, "Lena, bring the ham, please," and the boy immediately repeated this after me. Walking in the garden with his nanny, Roody (2.2.22) repeated almost everything she said, for instance, "pitty" (pretty) flower; "markible" (remarkable) flower, "Gasha [female name], open the gate, please."

He (2.6.24) repeated fully or partially almost every sentence coming from adults, sometimes completely unaware of what they actually meant. Once, the boy pushed a big football in my presence, and his father said to me, "Why are you allowing him to do it, Semashko [Secretary of Health Services] says it's harmful." In a little while, the boy forgot the ban and started pushing the ball, but he suddenly remembered and admonished himself, "Semashko does not allow."

When rhymes were read to the child, he listened intently and asked for them to be repeated because he liked them very much. Once, Roody (2.4.13) made me repeat the same rhyme 14 times in a row:

> The squirrel is singing songs
> While he is cracking nuts;
> These nuts are not plain,
> The shells are pure gold,
> Kernels—pure emerald.

The last two stanzas he knew by heart. Having learned several rhymed lines, the child tried to use them in speech, often not out of place. Rhymes fascinated Roody so much that he (2.3.6) muttered them even half-asleep. Roody (2.4.8) already had learned long rhymes by heart. The child, of course, distorted the words and produced childish guttural sounds that we, adults, like so much.

Roody (2.6.3) read by heart, distorting the words, a poem consisting of 23 stanzas. He (2.7.7) recited the entire "Railway" poem by Durov. He (2.9.17) recited with proper intonation the poem "Telephone" by Marshak, distorting only a few lines.

It is remarkable that, if you change or miss some words in the text or add something while reading it to the boy, he (3.0.2) immediately corrects you even if the book was read long ago. If you keep the poem in rhyme, the child only smiles; if the change has been significant, he laughs loudly. The child does not miss even the finest inaccuracies.

Roody (2.7.25) asked me to read him a book, *Elephant Vambo*, many times in a row, trying to remember some lines from it by heart. He even tried to reproduce three lines from that book, but it was yet difficult for him at this age; he certainly failed and started crying. Roody (2.8.10) also tried to tell, in his own words, a book, *A Fire in the Cat's House*. He told in his own words the captions in books about animals, accompanying his narration with expressive gestures.

When Roody was telling about a hedgehog that found a worm and ate it, he said, "She ate it," grabbed something, opened his mouth, and seemed to put "the worm" there.

Sound is the first and the most powerful way of the child's self-expression early in his life. The child is born with crying; he often keeps crying for the first several hours of his life.

Development of Roody's sound language (from his birthday to 1 year of age). Even this first cry of a child consists of many sounds; Roody's first cry was lilting, including more vowels ("a," "e") than consonants ("v" was the most frequent among consonants, "a" among vowels). In 5-day-old Roody, I could observe a changing, roaring, moaning, squealing cry that included vowels "u," "a," and "e" and consonants "m," "v," and "r."[31]

At 1 to 2 months, Roody uttered a singing sound with the vowel "a" after his satiation. At this time, you can discern easily an angry, demanding, monotonous, quacking cry of the hungry child, which stops immediately after the child's satiation. Sometimes, you can hear "agi" and long "a-a" after the satiation (0.1.1), to which consonants "g" and "r" are often added (0.1.14). Later, Roody uttered "ma" (0.1.17), very long "a-a-a" (0.1.27), "b," "gkh," "egu," "ekhg," "eai," "gkhe," "ekhu," "ekhe" (0.2.3), "gli," "gai," "agi," "kkh," long "m," "eo," "bau," "ki," "igy," "gkh," dull "sh" (0.3.2), loud "ai" (0.3.4), "uie," "g-a-a-a-a," "g-e-e-e-e," "kh-u-a-a-a," "m-a-a-a," "miam," "gi-gau," "ey-mue," and "meu" (0.3.9).

Every new day brings the enrichment of sounds; exercises in sound become a dominant entertainment for the child. Here are my chronological records (Roody's age 0.3.10 to 1.1.9) of his two-element, three-element, and eventually four-element sounds.

"bue," "ge," "ee," "gi," "eee," "ea," "ei," "gaa"	0.3.10
"gau," "gl," "kl," "kl-kl"	0.3.13
"bu-bu-bu," "gi," "gli," "gy," "ka-lia"	0.3.21
"gkha"	0.3.22
"a-a-a," "ma"	0.3.27
"m-m-m-a," "gai," "mm-e"	0.4.6
"gi," "eu-m-n-n-ei," "gei," "gl-y," "gl-ym"	0.4.21

"eu–e," "uai," "e–e–e–u," "nm," "m–kkh," "mmm," "mmai," "gkkh"	0.4.29
"biu–biu–biu–i–aa–kh"	0.6.30
"pa"	0.7.10
"pu–bu"	0.7.26
"ba–eba–ebe–ebiu–ebia–be–bia," "ama," "dia–ekhe"	0.8.19
"bava–ma–m–m–m"	0.8.20
"eba," "ba–aba–ma–ma"	0.8.21
"ba–ba–ba–be–be–be–va–bia–ba–ma–ma–m"	0.8.22
"baba," "ababa," "biama," "ebab," "beba," "miamia"	0.8.23
"a–baba," "baba"	0.8.24
"ma–ma–ga–baba"	0.9.0
"ga–ge–ga–ka–gaka–kkh"	0.9.1
"baba," "kaga"	0.9.3
"ga–ka–gaga–aba–ababa–pa"	0.9.7
"ba–pa–apu–pu"	0.9.8
"k," "kkk"	0.9.10
"kokh," "ke," "baba"	0.9.11
"aba" (persistently)	0.9.13
"d–de," "adia," "daba," "diaba," "baba," "dai," "dedia," "de," "bu"	0.9.17
"dedia," "abia," "ede," "de"	0.9.23
"mapa," "dedia"	0.9.24
"dedia," "tiodia," "detia," "diatia," "diadia," "diadia," "adia," "baba"	0.10.5
"tateta," "baba," "dai," "dedia"	0.10.6
"dedia," "ta–ta–ta"	0.10.8
"bu–ta–ta," "da–te–te," "dia–ta–ta," "papa"	0.10.10
"bapa," "baba," "tata," "da–data," "da," "tatata," "papa," "ptfu"	0.10.11
"natia–br"	0.10.15
"biu," "da," "diadia," "adia," "baba," "khi," "ki," "kkhi," "k–kh"	0.11.7
"maba," "aba," "bu"	0.11.16
"gage," "bu–br"	0.11.16
"bua"	1.0.5
"ke–ge"	1.1.2
"dlia," "glia," "dl–gl"	1.1.8
"mmama," "ppappa"	1.1.9

[Translator's note: Some sounds among those above have a meaning. These are "mama" (mother), "papa" (daddy), "diadia" (uncle), "baba" (grandma).]

For a child (1.1.5), a word means the name of an object, and it can be associated easily with the object. For instance, Roody gave me the toy I named from a group of five different toys: rattle, young pioneer, chimpanzee, lamb, and soldier. He never made a mistake in distinguishing them.

As soon as the child learns the words, he starts to apply them correctly. Roody (1.2.20) gave the names of members of our family and our domestic animals when we pointed at each of them: "Mama," "Papa," "Uncle," "kor" (crow), "kkh" (cat). Roody (1.3.8) correctly called Daddy, Uncle, Nanny (more seldom),

Baba (grandmother), and Mama (me). Roody (1.4.26) repeated familiar words the way he earlier repeated separate sounds and syllables.

The boy (1.5.26) understood many sentences and responded to them with individual words. Answering my question, "Where have you been?" he said, "Bua" (walked). "With whom did you walk?" "Mama." "What did you see?" "Gaga" (aunt in his terms) or "Din-din," a streetcar.

Roody (1.7.6) already used the verbal affirmative "da" (yes); he (1.10.5) later used the negative "nyet" (no) (before that he expressed his negative attitude by saying "bia," which is what he used to call unpleasant food, or by the long "a-a-a," pronounced with a special intonation. Later, the child clearly said "nyet" and repeated it many times, "Nyet-nyet-nyet," to express a persistent protest, for instance, during a doctor's examination (1.10.16).

Conditional Language of Gestures and Sounds

As in Joni's case, the initial language of the human child is the language of facial expressions and gestures. When a 9-month-old child wants to be taken by his mother, he extends both hands, cries, and stares at her, thus eloquently, unequivocally expressing his wish. When a 9-month-old child wants to change his position from lying to sitting, he extends his hands forward, bends his back, and makes shaking movements with his hands. When the child wants to take some object, he reaches for it, bends his head and the entire body in that direction, and often (as early as 11 months of age) uses his index finger to point at a desired object [plate 48(1)], groaning intensely, blushing, opening his mouth, sticking out his lips, and uttering a long, drawn-out, persistent sound "e-e." For a stronger request, he makes impatient shaking movements with his head. If his request is not fulfilled, the child stamps his feet, makes the drawn-out sound louder and cries. Verbal expression of the same wishes comes considerably later. To reject an object, the child makes a thrusting, pushing-aside gesture (7 months); when the child does not want to give an object to somebody, he (1.4.21) turns his back to that person. Roody (1.5.10) expressed his strong wish by the word "dai" (give); he (2.3.7) later used "dai-dai" (give-give).

When rejecting an unpleasant food and after satiation, the child (1.12.7) turned his head to the side, shook his head (2.5.21), and covered his face with his hands as though he were reluctant even to look at it. If the food found its way to his mouth, Roody (1.0.10), like Joni, often dumped it out. Roody (1.6.15) asked to drink by staring at a cup, putting his palm to his mouth, and slapping his lips.[32]

Roody (1.9.3) expressed his demand to give him something to smell by putting his hand to his nose and uttering an inhaling sound. To attract the attention of an adult, the child (9 months to 1½ years) tugged at the adult's clothes and staring at the adult with his eyes wide open. One time, Roody (2.2.2) wanted to gain my attention while I was talking with someone else; Roody took my face in his hands and turned my head to him.

The child begins very early to point with his eyes at different objects and persons. Later, he uses his finger in such situations (in Roody's case, at the age of 10 months).

The child (1.0.6) points at the door with his hand; he separates his index finger from the other fingers and points in the direction he wants to walk [plate 48(3)] and in the direction of everything that interests him and to which he wants to call other people's attention [plates 48(1)–48(4), 48(6)]. Roody (2.2.25) did not yet look in the direction pointed by a finger, but if we threw a stone in this direction, he followed its flight. When the child seeks the help of an adult, he usually tugs at the person's clothes and points at the thing he wants to have. Roody (1.6.1) urged me in such a way to get a little ball for him from under the bed.

Sometimes, the child fulfills our complex requests using such gesticulation. Once, Roody (1.6.19) was asked to request of his grandmother that she sit on a chair. He came over to the grandmother, tugged at her dress and pointed at the chair with his finger. The child keeps using gestures for a long time not only during a common conversation, but when he recites especially expressive poems.

Very early in the child's life, his expressive gestures start to be accompanied by certain sounds. A 6-month-old child, while reaching for a mirror, utters a drawn-out mooing-quacking sound "m." Often, when crying, he says "ma" (as if calling his mother), or he growls "m" if we do not allow him (0.7.9) to touch something.

I often heard Roody (0.8.12) saying "a-a-a" accompanied by a deep breath when we were showing him something. Such an aspirate sound of astonishment, a kind of exclamation "a-a-a," escaped Roody's mouth when he saw a freshly washed floor (darker than it used to be) in his room and in response to the uncommon sight of a doll sitting on a horse. We mentioned the frequent choking sighs of the boy during his visit to a museum, where he looked at new unusual things (1.0.3).

In taking some highly desired thing, the child (younger than 1 year of age) utters a mooing growling sound; I heard a similar sound from Roody (2.6.9) when he was eating something he liked very much.

We noted, in the corresponding sections, the earsplitting cries of the child when he (1.8.19) was forbidden to do something, and the squealing joyous cries when he (1.9.5) managed to realize his dreams (for example, to destroy a toy house).

Roody called "khkh" something warm or hot (in imitation of steam and hissing); he called "kkh" a piece of cotton wool resembling cat's hair in its softness (he also called cat "kkh").

When 6-month-old Roody was catching things, he usually emitted a groaning sound. He spread out his arms and made shaking movements with them.

Chapter 15

Memory and Habits of the Human Child
(Conditioned Reflex Acts)

Let us discuss the conditioned reflexes of human child. The child develops firm visual-motor reflexes after only one demonstration. Once, I showed a brush to 9-month-old Roody. The child grabbed it very quickly and pricked his finger. When I brought him the same brush in a week [plate 49(3)], he did not reach for it immediately, but first stared at it, tentatively extended his hand, touched the central metal wire of the brush, and only then touched the prickly hair. In contrast, when I bought him a smooth, shiny metal pen (which was familiar to him from previous experience), he grabbed it and immediately put its end in his mouth. One time, 8-month-old Roody played with the rings of a wooden pyramid, accidentally hurt his forehead, and started to cry. After that he flatly rejected our attempts to offer him the same toys; he turned his back to us every time we brought the toys closer to him. Only in a couple of hours (after we offered them to him four times in row) did he finally take the toys.

Once, Roody (2.9.17) scalded himself with hot tea; several days after the incident, he still started crying "hot" when he saw tea coming until we convinced him to try it. When Roody (1.4.22) saw a finger covered with iodine, he said, "hurt," because he associated it with a cut finger.

At 11 months, Roody (like Joni) was covered with a blanket when he was carried around the apartment; after that, every time he was covered, he jerked his arms and legs joyously because he knew that soon he would be brought to a new entertaining environment.

Quick and steady development of other conditioned reflexes in the human child (i.e., auditory and visual-motor reflexes) is conducive to his mutual understanding with adults, which appears well before he starts to speak. As in Joni's case, in the human child, this conditioned language initially is mechanical, stereotypical, and inflexible; nevertheless, when accumulated and enriched in associations, it serves its purpose as a means of communication with people around him and promotes his mental development.

At 7 months, Roody reacted to somebody's calling his name by turning his head in the direction of the call.

Below are some of the spontaneously developed auditory-motor visual reflexes.

Auditory-motor conditioned reflexes. When 7-month-old Roody was asked to "take this pillow," he took the pillow. When asked to show how mice scratch, the child (8 months old) scratched with his fingers over a hard substrate. When asked to "feel sorry for me," Roody (1.2.1) stroked my face. Asked to "clap," he clapped his hands. To the request, "Give me your hand," he gave me his hand. "Bring me," "Bring there," "Come over," "Pick up," "Give," "Help," "Wash your hands," "Wash your feet," and other requests were all operations he performed (1.6.23). The 1-year-old child reacted to the words "not allowed" by stopping his actions, understanding clearly what the words meant. At my request, "Show me how Roody cries," and "Show me how Roody laughs," he made crying and laughing expressions, respectively.

Auditory conditioned reflexes. When asked, "How does Roody shout, sing?" the child yelled and uttered a long drawn-out sound. When asked, "How does this lady cry?" Roody made a semblance of sobbing. To the question, "How do you call your dog?" Roody (1.3.3) made a smacking sound with his lips. When asked, "How does the bell ring?" Roody (1.3.20) muttered "d-r-r." To the questions, "How does the [factory] whistle hoot?" and "How does the crow croak?" the child (1.1.21) responded with "u-u-u" and "kar-kar" sounds, respectively.

Auditory-visual-motor conditioned reflexes. A 7-month-old child, responding to questions, points at different objects with his eyes. When asked how the soldiers walk, Roody (1.4.3) raised his feet high and stamped loudly. In answering questions, Roody directed his eyes to the clock and turned to the pictures, the baby carriage, the light, and other objects in the room. Roody (9–10 months), when asked where various family members (mother, father grandmother, nanny) were, pointed correctly at the certain members of the family. At 11 months, Roody used his index finger and extended his hand in the direction of the person whose name had been called.

The 10-month-old child with his finger at the eyes, legs, and other body parts of people or toy animals.

In response to our requests, he found the family member we named and threw a ball at that person; he was not confused by the fact that these people were in different parts of the room. He brought things to different people according to our request.

It is remarkable that some new reflexes of the visual-auditory type develop from a one- or two-time action. Roody (1.5.20) was shown the moon and the stars; from that moment, he always pointed at the moon and the stars with his eyes or his finger when we asked him about them. From one demonstration, the child (1.6.15) remembered and showed such fine things as a red mark that indicated a zero point on the thermometer's scale.

Roody (1.4.21) was shown an airplane that flew so low the sound of the propeller could be heard; the next time the child heard the propeller sound, he raised his head and started looking for an airplane.

Visual-auditory and visual-motor associations. Once, I took a cookie out of a cabinet and gave it to Roody (1.4.0); after that, every time Roody saw that cabinet, he extended his arms toward it in the hope of getting a cookie. Once, Roody (1.8.13) saw a cat near a fence. After that, every time he passed that fence, he said, "kkh" (his name for a cat). One time, I brought him to a yard to show him chickens; after that, every time we passed that yard, he demanded I take him in the yard and show him the chickens. At the circus, after seeing a monkey that waved his hands and jumped, he also waved his hands when asked how the monkey performed.

Usually, before crossing the street, I held the boy firmly by the hand and waited until the street was absolutely clear of traffic; in turn, he (2.9.3) became so careful that he did not want to cross the street if he saw a streetcar or automobile at even a great distance, although I urged him to start moving. He was adamant and waited until there was nothing more in sight, saying, "Mommy, wait, autobile is coming."

Characteristically, the child (2.9.3) sometimes relied exclusively on visual signals; for instance, instead of asking whether or not his uncle was coming, he asked, "Is uncle wearing a clean or a dirty apron?" For him, it meant the following: If the clothes were clean, the uncle would come to visit; if the clothes were dirty, he was not coming because he was working.

Quick and firm establishing of reflexes sometimes leads to behavioral conservatism in the human child, as well as in the chimpanzee Joni. We all know that even small babies acquire firm habits, and if these habits are bad, they can become a problem.

Roody (3.0.18) found a pacifier. Although he had not been using it for a long time, he asked, "Mommy, may I suck the pacifier?" He made a grimace of dissatisfaction after chewing it.

For instance, 9-month-old Roody acquired the habit of being carried around in somebody's arms, then he was given a small bottle of milk, after which he went to sleep. It so happened one time that the inner surfaces of the rubber nipple on the bottle stuck together, and the bottle had to be taken from him to straighten out the nipple. It only took a few seconds; nevertheless, the baby started crying and did not want to take the bottle back. Then, it occurred to me that the familiar sequence of events must be restored. I took the baby in my arms, carried him around for awhile, and then gave him the bottle, which he took immediately and soon fell asleep as usual.

Later, Roody (2.0.4) saw me lying on the bed with my head at the "wrong" end of the bed. The child insisted I change the position of my body by pointing with his finger at the usual place of my pillow and saying, "There," although this change clearly did not matter much for him.

Visual-gustatory-motor reflexes are also very firm. After Roody (1.1.27) tasted an orange, apple, and egg, he recognized these things by appearance. He joyously reaches for the fruit and turns his back to the egg.

When the process of "recognition" was described, we gave examples to illustrate how a baby less than 1 year of age established visual–emotional reflexes with his mother, with people familiar and unfamiliar to him, and with objects associated with feeding. The appearance of pleasant stimuli brings about a joyous mood in the child; elimination of these stimuli throws him into a sullen mood that usually ends in crying.

The conditioned reflexes of the human child and the chimpanzee can be characterized by the quickness of their formation and firmness of their retention. Obviously, the conditioned reflexes of the human child are more diverse than those of the chimpanzee. These reflexes establish in the child as in the chimpanzee, spontaneously, but in the case of the child, they do not require so much repetition for their fixation.

Child's Speech Evolution

In one respect, there is a tremendous difference between the chimpanzee and the human child; while the chimpanzee shows stereotypical conditioned reflexes that are predictable and are based on the chimpanzee's experience, the human child often surprises us with a new conditioned reflex reaction that reveals to us that, within the pattern of these simple associations, real comprehension of events and genuine work of thinking become involved. We were amazed to discover that Roody (1.6.1–1.6.26) memorized the first names and patronymics of all people in our household and answered the corresponding questions correctly.

"Who is Alexander Fiodorovich?" "Daddy."

"Nadezhda Nikolaevna?" "Mother."

"Yevgenia Alexandrovna?" "Grandmother."

"Philip Yevtikh'evich?" "Uncle."

"Lena?" "Nanny."

"Ein liebes Kind hat viele Namen" ["a lovely child has many names"]. It was actually realized in the case of our boy; it turned out that the child (1.8.7) remembered how he was called by different people.

"Who calls you Alfik?" "Mother."

"Rudochek?" "Uncle and Nanny."

"Apa?" "Daddy."

"Lulin'ka?" "Grandmother."

"Little boy?" "Gaga" (technician).

The chimpanzee Joni knew only my name; as soon as he heard it in my absence, he uttered a joyous grunt. However, he did not react to the names of other people in the household.

Roody (1.5.10) composed his first two–word phrases as follows: "Grandma, take", "Uncle bye-bye," he said when he saw men lying on the grass. Roody (1.8.2) used a phrase consisting of three words: "Mother, apple, there," inviting me to go with him to the next room, where he usually ate an apple.

The child (1.5.10) exercised for a long time to pronounce words that he had learned previously: Gelia, Mania, Polia, Mother, Grandma, Kolya, Katia, Gaga, and so on. In 3 days I heard Roody say 23 words in a row; 3 of these words (Mother, Daddy, Nanny), he repeated several times.

The child kept enriching his vocabulary. Roody (1.8.30) when taking his toys one by one out of a box, asked "Bu?" until we gave him the name of the toy; then, he said "yes," threw the toy aside, and took another one. Sometimes, he asked about the names of the things he already knew. When we asked him about that, he told us its real, distorted, or conditional name of his own choosing.

In a month (2.3.0), during a similar sorting-out procedure, Roody asked questions almost about every object. "How do they go?" he asked about a few pieces of coral. "How can I put it on?" he asked about a brooch. "What do you do with this brush?" and "Why does it smell?" were asked about the odor of a medicine. "Where to paint?" he wanted to know about watercolors.

The child (2.3.4) added some comments to the name of a thing. For example, when he took out a stick, he asked, "How much does it cost?" When taking out a fork, he said, "This is a little fork to eat." About a small piece of cardboard, he said, "Mommy, this is good, take it home." As he took things out of a box, he counted them, "One, second, third, fourth." When he found a small ball of hair, he said, "Hair is bad, I am crying—bad." When he found a hat, he said, "Mommy, to put on, please, put on," and so on.

Roody (1.10.13) gave the names of some domestic animals when he saw pictures in a book: dogs were "am," a cat was "ks," goat was "bu-bu-bu," horse was "tpru," cow was "moo," and sheep, donkey, and camel were "tpru" (like a horse). He called every ball "bliam" (as well as an apple).

Roody (1.11.12) names a series of pictures from a new book. He called

A horse "am" (confusing it with a dog)

A child "katiura" (generalized children's name)

A dog "amura" (endearing derivative of "am," dog)

A colt "gebel"

A bone "kukula"

A bird "gagura" (derivative of "gaga," goose)

A cat "ks"

A little mouse "mika" [in Russian "myshka"]

A goat "tu-tuk" (for the sound of butting horns)

A little goat "mu-u"

A hen "ariaba" [hen Riaba is a character in Russian fairy tales]

Chickens or ducklings "ti-ti-ti"

A hare or rabbit "liai"

A carrot "murop" [in Russian "morkov"]

A hedgehog "mukha" (Roody makes a pricking movement with his hand)

A pig "mu-u" (confused with a cow)

A girl "aunt"

In 7 months, we discovered that Roody (2.6.13) recognized from pictures and could name 45 wild animals (1) brown bear, (2) moose, (3) fox, (4) wolf, (5) hare, (6) marten, (7) hedgehog, (8) squirrel, (9) heron, (10) crane, (11) owl, (12) heathcock, (13) giraffe, (14) polar bear, (15) African elephant, (16) Indian elephant, (17) hippopotamus, (18) rhinoceros, (19) gorilla, (20) elk, (21) aurochs, (22) buffalo, (23) camel, (24) walrus, (25) tiger, (26) wild cat, (27) bird-of-paradise, (28) bat, (29) zebra, (30) monkey, (31) kangaroo, (32) parrot, (33) anteater, (34) lion, (35) pheasant, (36) yak, (37) bison, (38) baby fox, (39) sea lions, (40) leopard, (41) condor, (42) eagle-owl, (43) antelope, (44) Northern deer, (45) white duck.

Roody (1.10.13) characterized pictures of animals. He called pictures of donkey, dog, cat, horse, cow, pig, goose, and cock "ntsa" (he liked these pictures). He called pictures of a goat, hen, and duck "byaka" (he disliked them for some reason).

When sorting out things from a drawer, Roody (1.11.4) took each thing, named it, and threw it aside. He often stated to whom this thing belonged, "This belt is Apa's," "This plate is grandma's," and so on.

The child's visual memory was highly developed. At my request, he found, from 10 books, a book with a picture of a hen inside. Interestingly, Roody sometimes qualified books by characteristics that revealed the part of the text or pictures that had attracted his attention. For example, "little stick" was what he called the book, "*We Are Riding a Little Stick*"; a book, "*Fire in Cat's House*," he called "Don-don," for the first line of the rhyme. A book, "*Who is Faster*" he called "atabil" because the first picture and the first rhyme about an automobile had captured his imagination.

Two-word or three-word phrases became usual for the child: "Daddy, napkin," "Nanny, wash," Grandma, drink," "Mommy, dig," "Mommy, put on," "Mommy, run," "Mommy, my hand" (to take his hand), and so on. I heard him (1.11.1) using the preposition "with" for the first time in response to my question about whom he was going to go for a walk with, he said, "With Mother, Daddy, Nanny."

Roody (1.11.12) used phrases consisting of four words. Telling us about what he had just seen on the street, he said, "Atabil nanny boom in snow.[1] The boy (1.11.15) sorting toys one by one from a box, named 34 different toys.

Now, all his play was accompanied by incessant babbling. For instance, during a game of bowling, I managed to write down the following: Roody (taking a light bowling pin) said, "White"; he took another one and said, "Another"; he took the third one and said, "Three"; he set up a pin and said, "Stand"; the pin fell, and he said, "Old." He set four pins in a row, knocked them down with a ball, and said, "Tuk"; one pin fell down, he laughed and stood it up again, saying, "Stand, white"; he threw the ball again, saying "rrr"; he knocked down one pin, laughed, and said "one"; he carried all the pins to another spot and tried to stand them there; he saw that they also did not stand firmly at this new spot, and he said, "Eh, stand," and "Eh, two," as two pins fell down.

Roody (2.0.27) used a phrase of five words: "Here caught fly, show Mommy." At that time two- or three-word phrases were a common occurrence. Four-word phrases were significantly rarer; five-word ones were extremely rare. I witnessed Roody (2.1.25) using a subordinate clause for the first time. He said to me, "The house in which cats were."

At the age of almost 28 months, Roody used a personal pronoun for the first time. The boy was covered with sheets after a bath; I touched him through the sheets and asked, "Who is it?" Instead of his usual answer, "Roody," this time he said, "me."

While walking outside, Roody (2.2.29) actively reacted to everything around him; he was the one who did most of the talking. "Clean path," "Such a carriage to buy," "This church don-don-don," "Such a huge pit," "Bus," "He won't catch me," "Ant" (he saw an ant on the sidewalk), "After the sparrow" (he wanted to catch a sparrow), "Milk is coming" (he saw cans with milk on a cart), "Mice live here" (a hole in a fence), "What smells?" (the odor of lighting gas), "I am afraid of this hill" (he came down from a steep hill), "Much water" (a river), "Horse is eating, take off" (a horse was eating from a sack), "Why did they throw out the can?" (a can was on the ground), "Burr, I won't take it any more" (previously he pricked his finger with a burr), "Frog is lying" (a leaf on the water surface), "Forehead is bad" (large drops of sweat on a man's forehead), "Both are going, one behind the other" (two boats at a pond), "Paper wanted to swim" (a piece of paper in the river), "Where it come from?" (a log on the river bank), "Where road clean" (he wanted to go by the road, not by a muddy shortcut), "Two seagulls, can I catch," "Engine," "Hear, engine" (an engine was hooting), "Mother, duck, babies" (a duck with ducklings), and "Where is daddy?"

The child gave logical answers to various questions, and he was willing to lead long dialogues. Roody (2.3.22) said, "I'll paint the porch."

"You'd rather paint your nose," Nanny said.

"I can't."

"Why?"

"It will be dirty."

"I'll bring a chair, and speak to my mother" (2.3.22).

"Mother, how are you doing? How are you running? How are you sitting?"

I showed him the new moon and said, "Here is the young moon." He (2.3.28) said, "And where is the old moon? Where did it go?"

"Daddy's throat is aching, he needs iodine," Roody said, seeing his father's throat bandaged and remembering previous manipulations of throat treatment with iodine. "I have already applied iodine," his father said. "Is it funny to put iodine in your mouth?" "No," said his father. "Can father still speak?" Roody asked.

At the age of almost 33 months, when arranging toy animals around him, Roody talked incessantly, apparently not uncomfortable with the repetitions of predicates.

"The tiger went to hunt, and the camel also ran to hunt, and the elephant also ran to hunt, and the cat also ran to hunt, and the polar bear also ran to hunt, and the lion also ran to hunt, and the seal also ran to hunt, and the little monkey also ran to hunt, and the horse also ran to hunt, and the fox also ran to hunt, and the bear also ran to hunt, and the cow also ran to hunt, and the goose also ran to hunt, and the deer also ran to hunt."

"And what is it?" He said when he took the tusk of the toy elephant, and answered his own question, "This is of the big elephant."

"And the black bear also ran to hunt, and the baby goat also ran to hunt, and the owl was sitting and screwing up its eyes, and watching them running to hunt. And the hare, too, and the doggie, too."

At this moment he dropped the baby goat and said, "Excuse me, baby goat." He moved the animals around, saying, "And this way too." He took a squirrel and said, "I love this squirrel," "Let him also run to hunt" (he pressed the squirrel to himself with all his strength). "The elephant went to hunt and ran again" (he took the elephant).

Roody (3.0.6) told me in his own words a story told to him by a man about how he was bitten by a horse when he was in the army during the war. "When our man was a soldier, he had a good horse, only the horse bit him, but he never flogged her, neither with a whip, nor with a rope, nor with a belt, maybe she was stupid and she bit the man's ear, sure she was quite stupid."

This quick language development and enrichment of his vocabulary is no surprise if we consider his powerful inner stimulus, which urged him to acquire new words. This stimulus was his insatiable desire for perceiving objects of the surrounding world: their names, properties, and applications and their complex relationships with the environment.

Intellectual Features of the Human Child

The human child's language development is the main factor in his mental development. Speech, curiosity, and the inquisitiveness of the child's mind are three horses that powerfully haul the child's soul in his difficult climb to the intellectual mountain in his difficult road to become a human.

1. Inquisitiveness of the Mind.   The child at the age of 2 years burns with one desire: "I want to know everything." The questions, "What?" and, "Why?" are constantly on his lips. In the beginning of the third year of his life, the child limits himself to the question, "What?" and shows only superficial curiosity, at the age of about 3 years, we can notice his growing intention to penetrate the depth of things.

While looking through a new book, "*Wunder der Natur*," Roody (2.4.12) pointed at hundreds of pictures and asked me about every one of them, "What is it?" He turned the pages so quickly that he did not have enough time to see the pictures; he did not keep his attention an extra second on one picture. Even when I wanted to direct his attention to especially interesting pictures, he rushed forward until he reached the end of the book.

Roody (2.6.18) asked about every familiar thing: "Mother, tell me about a pan, cake, porridge, ham, roll, grapes, potato, Vaseline, pin, piece of paper," and he listened intently to your stories. Coming to a new environment (a kitchen), Roody (2.10.20) asked literally about each and every thing. He noticed different properties of things and wanted to know why they were different. For example, Roody (2.6.22) said, "Mother, which color are your lips?" (I said, "Red.") He said, "And which color are Apa's?" (I said, "Red.") He said, "Why red?" "Why is the sink pink?" "Why is the bowl green?" "Why are the hands white?" "Why don't you have a beard?" (2.0.26) "Why does father have a beard? Why doesn't Apa have a beard?"

When we read a book to Roody (2.7.3), he asked the meanings of the words he did not know. For example, when he listened to a book, *Railway*, by Durov, he asked, "What is Durov? What is a fine? What is a switchman?" When a book, *Our River*, was read to him, Roody (3.02) asked, "What are rafts, rudders, flow, bonfire?" He (3.1.3) was intrigued even more with everything unusual: "Why the horse's eyes are covered?" or "Why that man's eye is black?" (he saw an eye covered with a black band).

Roody (2.9.10) arranged animal toys and bombarded everybody with questions: "Where does a zebra live?" (answer: in Africa). "Where does a giraffe live?" (answer: in Africa). "Why does a giraffe live in Africa? He is not striped" (like a zebra). "And where does a lion live?" When he learned that a lion also lives in Africa, he looked totally confused and said, "It [the lion] is yellow." (In other words, it lives in Africa despite the fact it is yellow; he apparently was convinced that the animal had to have stripes to live in Africa.)

Roody (2.9.27) arranged all his animal toys in a long row; as can be seen in plate 99(1), he made up groups according to a certain system. That is, he put a giraffe in the front row, a camel in the second row, three cats of different sizes in the third row, a tiger and a lion in the fourth row, a brown and a white bear in the fifth row, two dogs in the sixth row, a squirrel and two seals in the seventh row, and a duck in the eighth row.

The child showed keen observation ability. "Why does it glisten from far away?" Roody (3.0.9) asked, seeing that only at a certain distance and at a certain angle, a mark made by the ink pencil reflected light.

Nature awakened the vivid mind of the child (2.11.11) and caused countless questions: "What is the sky?" "What are clouds?" "Can clouds be lower?" "Where are they flying?" "Can they be caught?" "Why not?" "And can the stars be lower?" "Can the sky be brought down?" (3.05), and so on with no end.

The child tried to understand things deeper, in the figurative as well as literal sense of the word. Roody (2.9.14) asked, "What is inside the bear?" (a toy bear), "What is inside Rebecca?" (a doll). He wants by all means to break the toys apart and see what is inside.

We all know that one "Why?" starts the child on a series of logically connected "Why?"s; The child tries to find the primary cause of the phenomenon and get an exhaustive explanation, sometimes putting adults in a quandary.

2. Thinking, Comprehension, Comparison, Memory. The child's verbal expressions are bright illustrations of his conscious thinking activity. Looking at an automobile on the street, the child (3 years old) compared it loudly with the toy

automobile at home. He said, "Daddy, it doesn't have a steering wheel." Then, after looking more closely, "Daddy, it has such a funny steering wheel, mine doesn't have it, this has a spare tire, we don't have it," and so on.

Once, Roody (2.0.23) saw three white enameled nails in the back of his bed (one big nail, two small nails) and noticed a hardly visible black dot on one of the small nails. He said, "Mother, this no coal" (the big nail) (does not have a black dot, like coal), "baby coal" (the small one), "[the] other not" (that is, one of the little nails has a black dot, the other does not).

The boy (2.6.21) listened when I read from a book: "Two boys are standing and holding a black, shaggy, spooky cat." There was a drawing in the text that depicted one boy holding the cat and the other just standing next to the first one. Roody noticed the inconsistency and said, "But this one is not holding."

The child had a well-developed memory. Once, Roody (3.1.13) broke a porcelain toy (a boy in purple pants and a green hat and with a flower in his hand) and started to cry. The same day we bought him another one that differed from the first toy in one small detail. I thought the child would not notice the difference and said, "We bought you the same toy." Roody only took one look at the new toy (the new boy wore a blue hat) and said, "Green hat!" I asked, "What about the pants?" He said, "Pants and here are alike" (referring to the flower on his knees). I asked, "What color is his hat?" (Assuming he could not tell the colors). He said, "Blue."

Once, the child (2.9.20) noticed, "There is something hanging out of Lisichka's [a dog] mouth" (it was a ball of dog hair). Next day, he saw the same dog in perfect order; Roody remembered how the dog looked yesterday and asked, "Why is nothing hanging out Lisichka's mouth?"

Roody had a topographical memory and could (2.7.13) find a way to a newsstand four blocks away from us in two directions (however, this was after we had walked along that route many times).

Conditioned reflexes spontaneously developing in Roody numbered in the hundreds or thousands; the human child was ahead of the chimpanzee in visual-auditory and especially in auditory reflexes (for instance, his reciting of rhymes). The child used a gesture language associated with mental processes and an unparalleled advantage—an ability to form words and speak. He started making assumptions and drawing practical conclusions, logical inferences that bore witness to his genuine work of thinking and not only to mechanical accumulation of words and speech associations.

Roody (3.0.25) saw a man sitting on a bench (apparently a dozing drunk) and asked, "Mother, why is this man bending his head?" and added, "Maybe his head aches?"

Roody (2.10.29) asked, "Of what are these dishes [toy dishes] made?" "Of wood," his grandmother answered. "What about the duck [celluloid]?" "I don't know, to tell you the truth," the grandmother said. "I know," said the boy and added, "It's made of eggshell."

The child tried to understand fully every word that adults said. For instance, I told Roody (2.7.19), touching his legs, "How thin Apa has become!" [The word "thin" may also mean "worn out" in Russian.] "Then where are the holes?"

Roody said. Or, for another example, Roody (3.0.1) asked, "Are these photos?" "No," he was told, "These are postcards" [literal translation from Russian is "open cards"]. "And how do you open them?" asked the boy (revealing the syncretism of children's thinking).

I said to Roody (2.4.9), "Let's go and see who is outside?" He came out to the yard and said: "Where is Outside?" One time I was talking the boy into eating something and was careless enough to say, "Eat this, and the little car will be looking at while you are eating." The boy's immediate response was "It has no eyes."

One of the eloquent dialogues with the child shows how he tries to make sense of proper names: There were three men among our colleagues: Alexey Pavlovich Svirin, Ivan Pavlovich Svirin, and Nikolay Alexeevich Bobrinski. Roody (2.11.0), who often heard their names and saw the people, asked, "What is Svirin?"

"It is a man's last name," we told him.

"What is Alexey Pavlovich?"

"This is the man's name."

"What is Nikolay Alexeevich?"

"This is Mr. Bobrinski's name."

"What about Ivan Pavlovich?"

"The name of the other Svirin."

"Can there be Ivan Alexeevich?"

"Yes."

"Can there be Svirin-Bobrinski?"

3. Practical Deduction.   Sometimes, the child draws a practical conclusion right in front of your eyes, experimenting with objects. Roody (2.3.25) took a photographic tripod and tried to put it up in different positions, extending its legs and talking constantly while he was doing it. "Would it stand or not?" (the tripod fell down). "It won't stand," he added. "Now I see that the earth is turning," Roody said (4½ years of age) after he had been going around a pole in circles and became dizzy [plate 116(3)].

The child (2.11.15) used his conclusions in his behavior, sometimes noticing the logical inconsistencies of adults. We said, "Roody you cannot throw your galoshes, you'll break them." Roddy said, "What are they made of?" "Rubber." "You cannot break rubber, can you?" We said, "You, Stepka-Disheveled, let me tie your belt." Roody (2.9.16) said, "Stepka-Disheveled is not like that." "Like what?" "He is red." (Stepka is described in a book as having a red face.)

Roody (2.11.15) ate an egg and asked, "May I eat the eggshell?" His father replied, "No, only birds eat that." After a while, the child asked, "Father, am I a little bird?" His father, who completely forgot what he had said previously, answered, "Yes, you are my nice little bird." The boy immediately exclaimed, "I'm going to eat the eggshell," and brought it up to his mouth.

4. Logic and Wit.   Roody (3.0.0) heard that I used the phrase, "This is not a joke." The boy asked me immediately, "Mother, what is a joke?" I said, "This is when somebody says something funny on purpose, to make everybody laugh."

In a minute, the boy said with a smile, "Bandeagle." I said, "Maybe you wanted to say bandit." "No, bandeagle." "He wanted to say eagle," the nanny said. Roody said, "Yes," laughed and added in the form of a question, "Mother, is this a joke?" I said, "Yes, you said a joke." He uttered another unintelligibly distorted word and laughed again. Another time, Roody named a picture of an eagle incorrectly, apparently meaning it as a joke. Answering the question, "What do you see in this picture?" He (1.11.1) said, "Mother" and repeated it several times. I said, "Apa doesn't know, but Kolya [a boy of the same age] knows." Roody immediately named the bird correctly. I showed him another picture, of a turtle; he said "Mother" again. But, when I said, "Vitya knows, but Apa doesn't know," he said, "Turtle."

5. Counting.   Roody knew how to count as early as at the age of 3 years. Touching loops in the screen of his bed one by one, Roody (2.9.14) said, "1, 2, 3, 4, 5, 6, 7, 11, 12, 14, 15, 17, 21, 22, 24, 25, 27, 28." A month later, while playing the buy-and-sell game, he understands what quantity means and performed simple addition and subtraction based on quantitative identification.

The game goes as follows. I say, "Sell me one stick, please." He says, "Which one?" I say, "The red, small one." He (giving me the stick) says, "Here you are." I say, "How much?" He says, "Seven." I give him one piece of paper as money.

I say, "Sell me two sausages, please." He puts two sticks on the balance and says, "This and this." I give him two pieces of paper.

I ask, "Sell me three sausages, please." He picks three sticks. I give him three pieces of paper. He says, "Three sticks and three monies."

Now, I ask, "Give me one sausage, please." He gives me one bowling pin. I pay him not with one, but with two pieces of paper. He says, "I don't need the other one, I gave you only one sausage."

I ask, "Give me three sausages, please." He puts three bowling pins in my bag. I give him two pieces of paper. He says, "For three sausages, I need three monies." He waits for my reaction and seems to be relieved when I give him another piece of paper.

I ask, "How much for two apples?" He says, "Seven," and gives me two balls. I give him three pieces of paper. He says, "It's two apples, I don't need a kopeck" (puts one of the pieces aside).

I ask for two cookies and give one piece of paper as payment. He says, "For two cookies, I need two kopecks." He waits until I add one more. I ask, "One piece of meat, please," and give him three pieces of paper. He says, "It's only one meat, I don't need these kopecks" (throws two pieces of paper aside). I ask for two bowling pins and give him three pieces of paper. He says, "It's only two pins, I don't need this kopeck" (gives me one piece of paper back). I ask for three cookies and give him one piece of paper. He says, "But it's three, and there should be three kopecks," and takes from me two more pieces of paper. I ask for four sticks; he throws me three sticks saying, "Three, four, five," gets confused and is not willing to play any more.

He picked the correct number of objects if there were fewer than four; other-wise, he could not pick objects correctly.

Accordingly to my data, Joni at the age of 3½ to 4 years could not understand quantity even when the number of objects was less than three, my 3-month efforts in this direction notwithstanding. (This is described in detail in volume III of my research on the chimpanzee.)

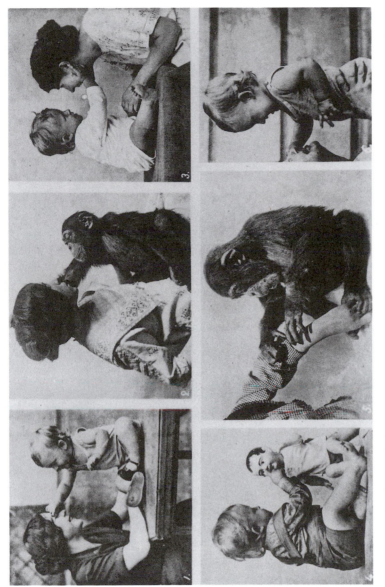

Plate 41. Roody and Joni examining and touching various intriguing objects: (1) Roody (1 year 2 months) examines and touches my eyes; (2) Joni examines and touches human hair; (3) Roody (1 year 4 months) examines and touches human ears; (4) Roody (1 year 4 months) examines and touches face of doll; (5) Joni examines and touches a scratch on human hand; (6) Roody (1 year 2 months) examines and touches human face.

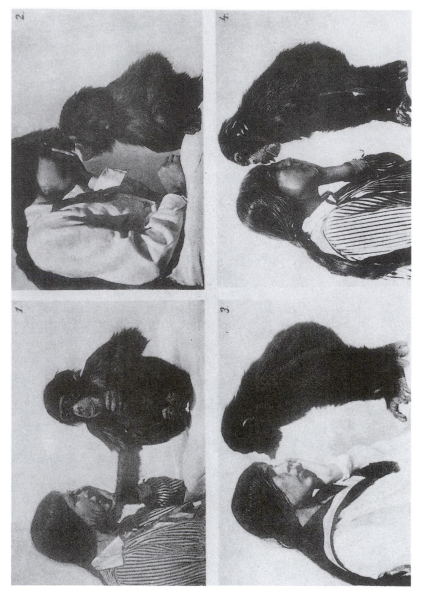

Plate 42. Expressions of tender feelings—caresses and sympathy of the chimpanzee: (1) carefully touching face of weeping human; (2) putting hand on head, protrusion of tightly closed lips; (3) ready to touch the face of the weeping human with his lips; (4) ready to touch the face of the weeping human with his tongue.

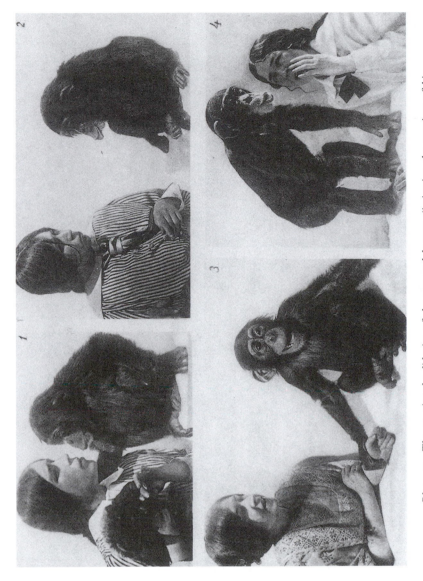

Plate 43. The emotional solidarity of the ape and human (imitating the emotions of his human friend): (1) induced sorrow; (2) induced angry excitation; (3) the chimpanzee's joy; (4) anxiety and excitation mingled with sadness.

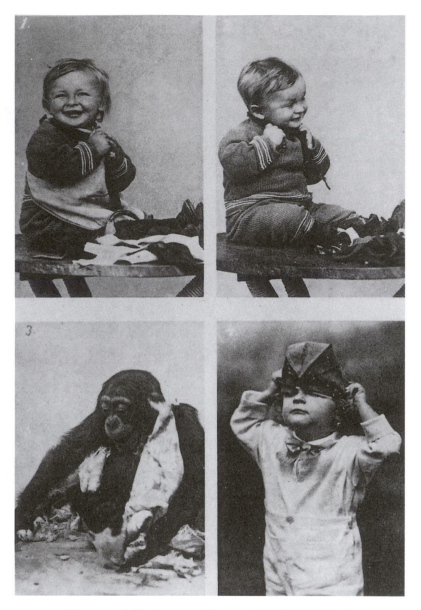

Plate 44. Self-adornment of ape and child: (1) Roody
(1 year 5 months) dons a boa of yellow feathers; (2) Roody
adorning himself; (3) Joni puts a piece of tulle around his
neck; (4) Roody (3 years 4 months) dons a peculiar kind
of headgear.

Plate 45. The chimpanzee's specific kinds of play: (1) moving a crumb of bread in the mouth; (2) passing head through a rubber ring; (3) setting a "mouth extender" between teeth.

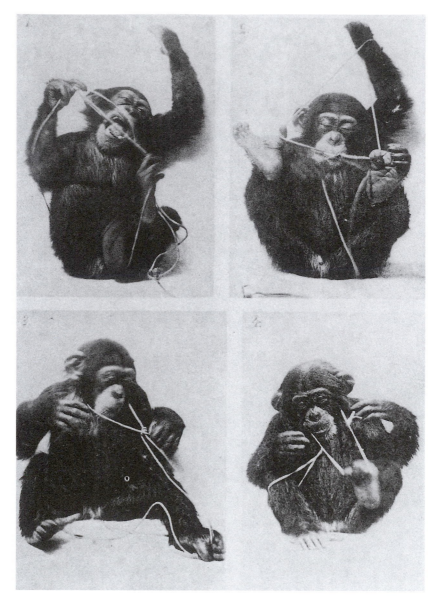

Plate 46. The chimpanzee playing with elastic objects (with tape and thread): (1) and (2) winding and unwinding a thread; (3) and (4) climbing through an improvised loop.

Plate 47. Mirror reactions of child and ape: (1) Roody
(2 years 5 months) looks into mirror, touches mirror image with
face; (2) Joni looks into mirror, examines his image; (3) Roody
touches mirror image; (4) Joni draws mirror to himself; (5)
Roody grimaces before mirror; (6) Joni grimaces before mirror.

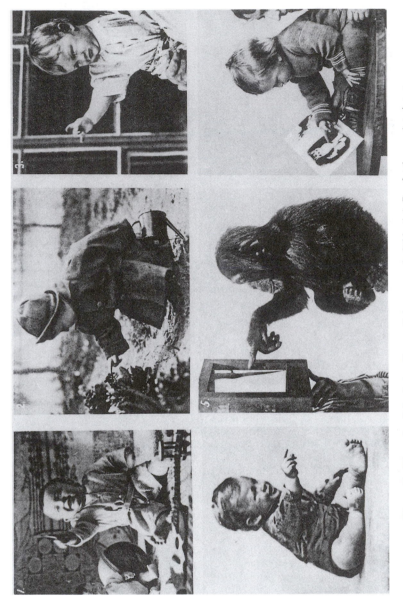

Plate 48. The role of the index finger of the ape and child: (1) Roody (1 year) points to intriguing object; (2) Roody (2 years 1 month) points to intriguing object with index finger; (3) Roody (1 year 2 months) points to direction in which he wants to go; (4) Roody (1 year 3 months) points to object he wants to acquire; (5) Joni touches intriguing object with index finger; (6) Roody (1 year 5 months) touches intriguing object with index finger.

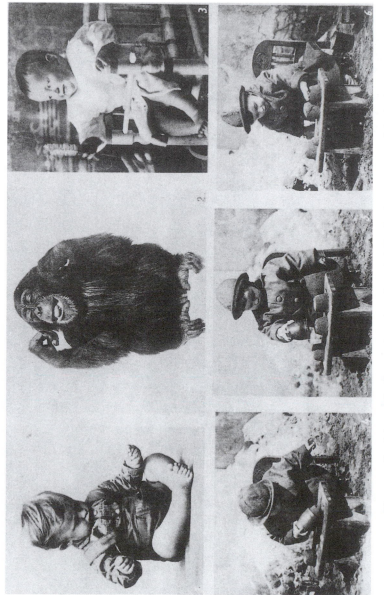

Plate 49. Roody's and Joni's experimenting play: (1) Roody (1 year 4 months) playing with a hair; (2) Joni playing with a hair; (3) gets acquainted with a brush (Roody 9 months); (4) making sand cakes (Roody 2 years 1 month); (5) the sand cakes ready (Roody 2 years 1 month); (6) flattening out sand cakes (Roody 2 years 1 month).

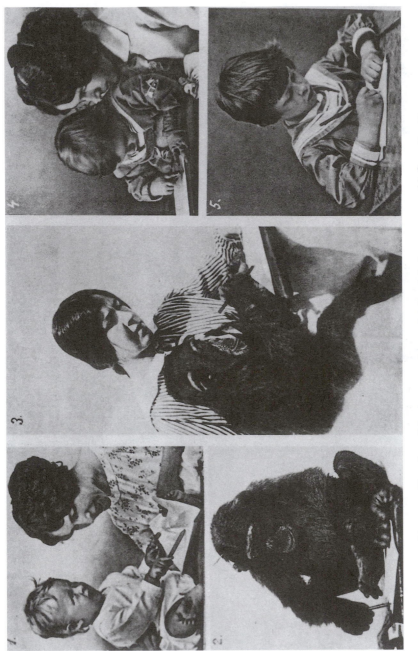

Plate 50. Use of pencil by human and ape: (1) Roody's (1 year 4 month) first attempts at using pencil; (2) and (3) Joni "draws" lines with pencil; (4) Roody (1 year 9 months) "draws"; (5) Roody's (7 years) regular pencil work.

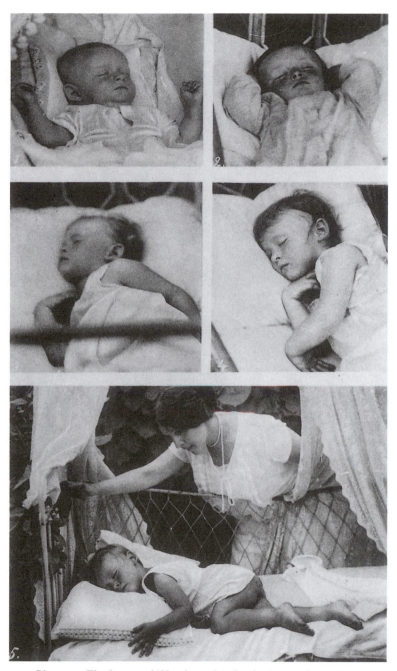

Plate 51. The human child asleep: (1) Roody (3 months 21 days) sleeping on his back (symmetrical sidewise extension of arms); (2) Roody (2 years 1 month) sleeping, arms behind head; (3) Roody (2 years 4 months) sleeping on his side; (4) Roody (2 years 3 months) sleeping on his side; (5) most typical posture of sleeping child.

Plate 52. Expression of affection of the ape and child:
(1) tenderness of Joni, pressing closer to observer; (2) tenderness of Roody (2 years 8 months) pressing close to his mother;
(3) tenderness of Joni, hugging; (4) tenderness of Roody
(1 year 9 months), pressing himself against his mother's breast;
(5) Joni kissing; (6) Roody (2 years 4 months) kissing.

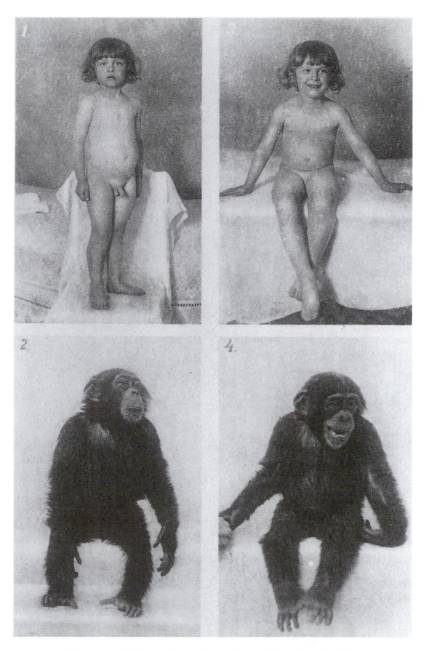

Plate 53. Sitting and standing postures of human and chimpanzee: (1) Roody (5 years) standing; (2) Joni (4 years) standing; (3) sitting with legs freely hanging down (Roody 5 years); (4) sitting with legs freely hanging down (Joni, 4 years).

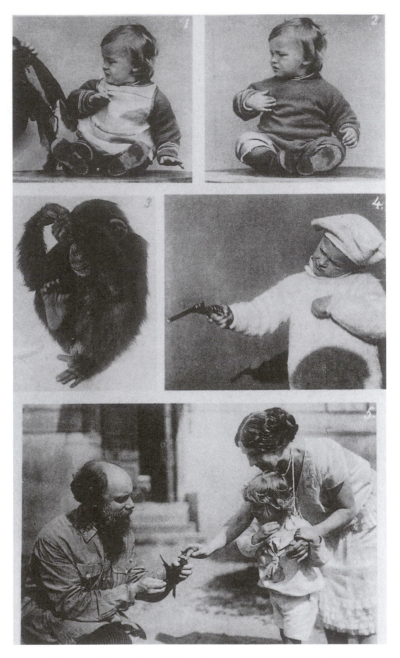

Plate 54. Expression of fear of ape and child: (1) Roody (1 year 6 months) afraid of a piece of dark fabric; (2) Roody (1 year 6 months) afraid of a piece of dark fabric; (3) the gesticulation of frightened Joni; (4) the gesticulation of frightened Roody (5 years), self-protecting gesture with revolver; (5) Roody (3 years 3 months) frightened of a live fluttering bird.

Plate 55. Destructive play of the child: (1) demolishing part of a brick house (Roody 1 year 5 months); (2) demolishing almost all the brick house (Roody 1 year 5 months); (3) destroying a self-made pillar (Roody 4 years 1 month); (4) throwing pebbles into a pool of water (Roody 1 year 7 months); (5) Kicking up a twig with stick (Roody 3 years 3 months); (6) directional throwing of stick.

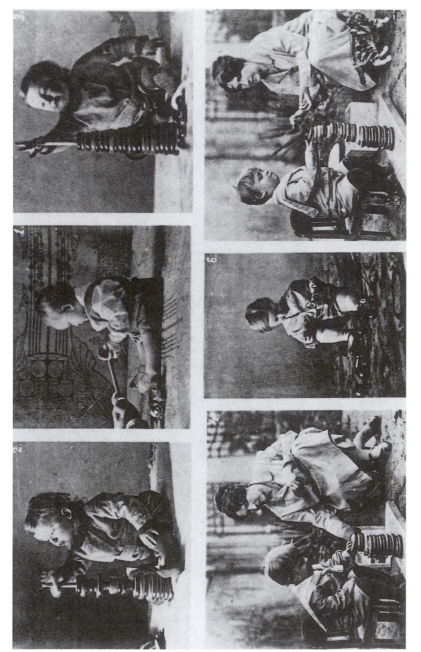

Plate 56. Deconstruction and reconstruction in the child's play: (1) Roody (9 months) removing ring from pyramid; (2) and (3) Roody (1 year 9 months) putting ring onto pyramid; (4) and (5) Roody (2 years 6 months) putting rings of pyramid into place (no vertical spindle available); (6) Roody (1 year 4 months) assembling ninepins after having disjoined them.

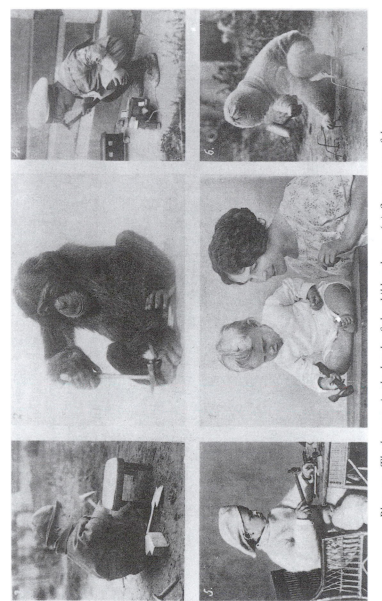

Plate 57. The hammer in the hands of the child and ape: (1) first unsuccessful attempts to use hammer (Roody 1 year 4 months); (2) ineffective use of hammer by Joni; (3) attempts attacking in nails (Roody 2 years 1 month); (4) hammering-in defective wheel (Roody 2 years 5 months); (5) Roody turns hammer to good account (4 years 7 months); (6) proper use of hammer (Roody 5 years).

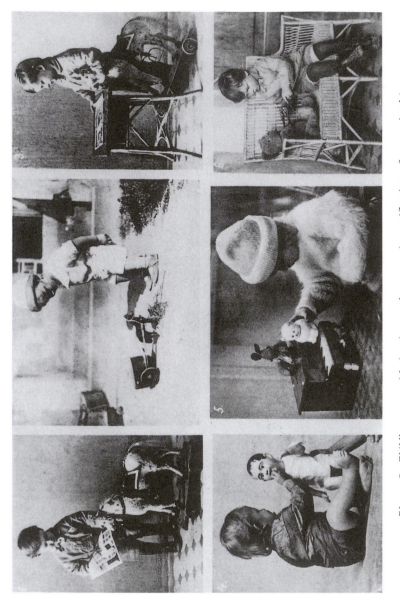

Plate 58. Child's games with inanimate playmates (personification of toy animals): (1) *Roody (2 years 9 months) showing pictures to horse;* (2) *Roody (2½ years) feeding toy horse;* (3) *Roody (2 years 9 months) reading for horse;* (4) *Roody (1 year 4 months) tries to feed doll;* (5) *Roody (4 years) playing the piano for cat;* (6) *Roody (2 years 9 months) auscultation of teddy bear.*

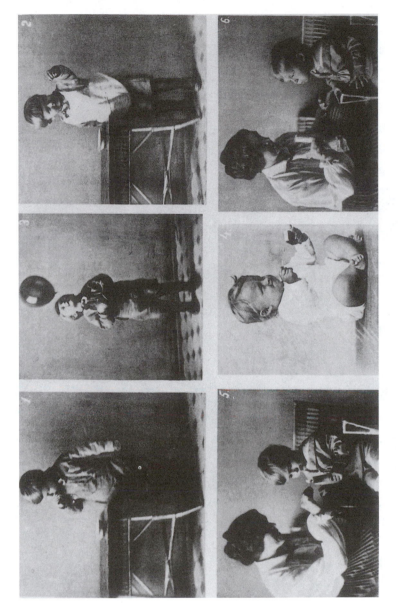

Plate 59. *The human child playing with water, fire, and lustrous objects: (1) blowing soap bubbles (Roody 3 years); (2) admiring a bubble; (3) looking at the play of colors of a toy balloon; (4) striking a match (Roody 1 year 2 months); (5) looking at the flame of a candle; (6) extinguishing flame (Roody 3 years).*

Plate 60. Washing and toilet of human child: (1) Roody (2 years 4 months) washing face; (2) Roody (3 years 2 months) washing hands; (3) Roody (2 years 2 months) putting on cap; (4) Roody's (2 years 2 months) attempt to put on shoes; (5) Roody (2 years 6 months) preparing to blow nose; (6) Roody (2 years 6 months) using handkerchief.

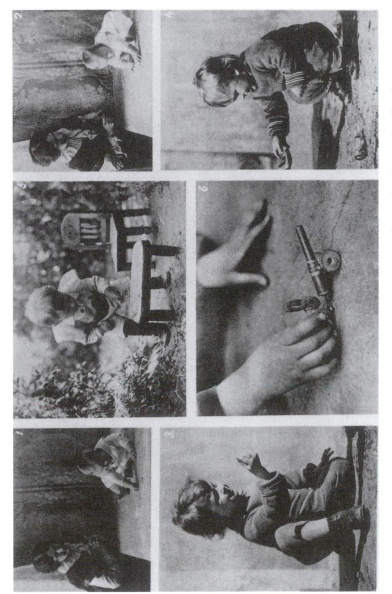

Plate 61. The child likes miniature objects: (1) gathering bread crumbs (Roody 9 months); (2) gathering bread crumbs (Roody 9 months); (3) "I like all that's very small" (Roody 2 years 3 months); (4) gathering little stones (Roody 1 year 5 months); (5) examining minute snails; (6) the smallest hare gets the gun (as a special distinction).

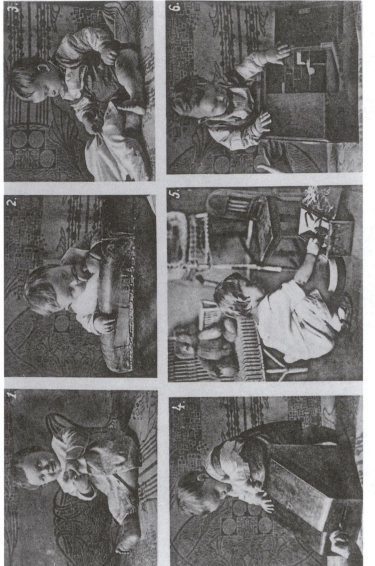

Plate 62. Investigation of holes and cavities: (1) Roody (1 year) puts index finger into hole; (2) Roody (1 year) sucks finger; (3) Roody (9 months) tears hole open; (4) Roody (9 months) opens table lid with index finger; (5) Roody (2 years 2 months 8 days) takes out different things; (6) Roody's (9 months) exploration of the depth of boxes.

Plate 63. Child's imitative play with water: (1) navigating ships in a puddle (Roody 4 years); (2) pushing a raft; (3) fishing in a puddle; (4) "a real fisherman"; (5) shipbuilding (Roody 3 years 10 months).

Plate 64. Expression of social instinct—intercourse and friendly attitude of child toward animals: (1) Roody (1 year 2 months) tries to make friends with cat; (2) feeding dog (Roody 2 years 7 months); (3) Roody (2½ years) expresses tender feelings toward cat; (4) hugging a toy elephant; (5) Roody's reaction to Joni's stuffed representation (examining chimpanzee's face); (6) Roody's (2 years 5 months) tenderness toward Joni ("hugging the little ape").

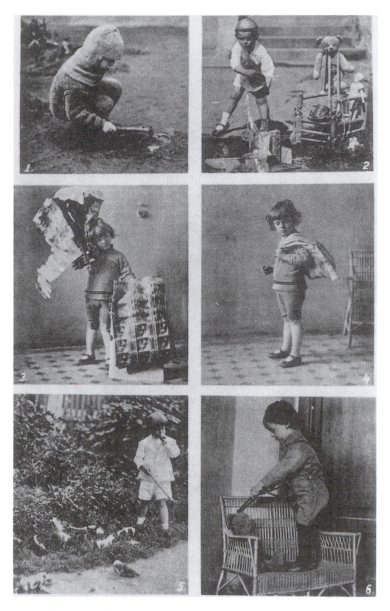

Plate 65. *The child imitating various trades and professions: (1) road building, paving a path (Roody 4 years 7 months); (2) posing as a fireman (Roody 4 years 4 months); (3) posing as a newspaper vendor (Roody 3 years 10 months); (4) posing as a second-hand dealer (Roody 3 years 10 months); (5) posing as a shepherd, playing the horn (Roody 4 years 1 month); (6) posing as a doctor, examining a teddy bear patient (Roody 2 years 9 months).*

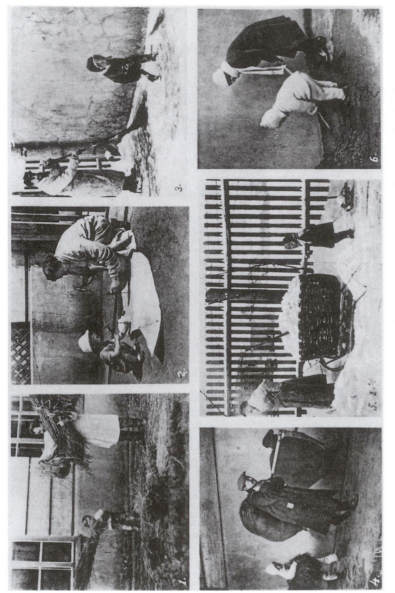

Plate 66. Roody's actions of imitation (immediately following performance of action by adult): (1) shaking straw (Roody 2 years 2 months); (2) tacking in nails (Roody 2 years 2 months); (3) cleaning snow (Roody 2 years 6 months); (4) carrying hay (Roody 4 years); (5) hauling snow (Roody 2 years 6 months); (6) shoveling sand (Roody 3 years).

Plate 67. Child's constructional play: (1) collecting material for proposed structure (Roody 1 year 11 months); (2) first reproduction of airplane (made of twigs; Roody 1 year 11 months); (3) two-dimensional reproduction of airplane (Roody 2 years 3 months); (4) ballooning (Roody 3 years 3 months); (5) flying in self-made airplane (Roody 3 years 6 months).

Plate 68. Child's constructional play: (1) Roody (2 years 1 month) builds house by putting a number of blocks beside each other; (2) Roody (2 years 6 months) places bars one on the other; (3) Roody (3 years 3 months) combines horizontal and vertical bricklaying; (4) Roody (3 years 4 months) making a kind of tower; (5) Roody (3 years 4 months) making a gate out of round bars; (6) Roody (6 years 1 month) building a fortress of many stories.

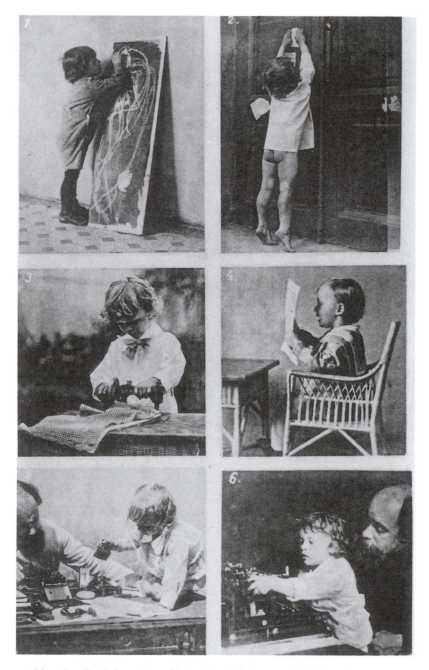

Plate 69. Roody's actions of imitation: (1) draws on blackboard with piece of chalk (Roody 2 years 11 months); (2) pasting on an advertisement: "Who will get me back my lost cat?" (Roody 3 years 3 months); (3) sewing (Roody 3 years 4 months); (4) reading (Roody 2 years 7 months); (5) stamping (Roody 3 years); (6) typing (Roody 3 years 4 months).

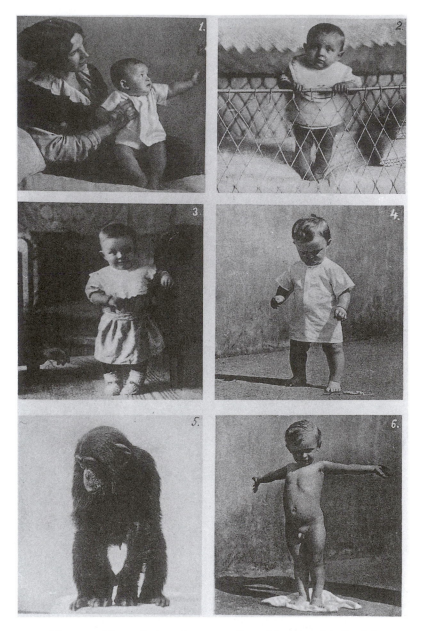

Plate 70. The evolution of erect standing of human: (1) Roody's (5 months) first attempts to stand; (2) Roody's (6 months) standing posture, strongly leaning on arms; (3) Roody's (9 months) standing posture, slightly leaning on arms; (4) Roody (1 year) standing erect, but with poor stability; (5) Joni's typical erect standing; (6) Roody (4 years) standing with full equilibrium.

Plate 71. Walking, running, and using hands for carrying objects, human versus ape: (1) Joni (3½ years) running; (2) Roody (2 years) running; (3) Roody (1 year 1 month) carrying ball; (4) Roody (2 years 1 month) carrying chair; (5) Roody (2 years 5 months) carrying armchair; (6) Roody (2 years 2 months) carrying ninepins.

Plate 72. The evolution of walking of human and ape: (1) The chimpanzee's typical running posture; (2) the crawling human infant (9 months); (3) Roody's (1 year 1 month) first attempts at walking (profile); (4) Roody's (1 year) first attempt at walking (en face); (5) the slow gait of Joni; (6) Roody (3 years) running.

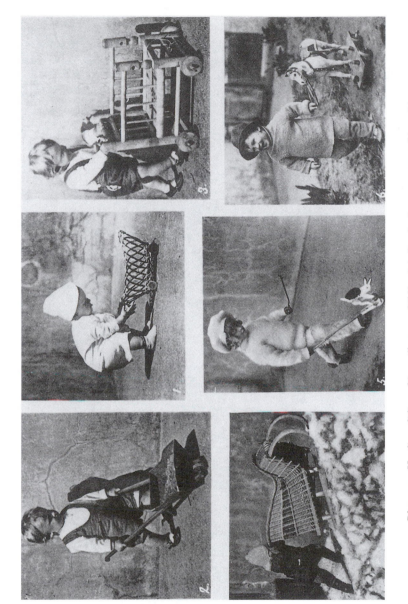

Plate 73. Hauling sliding objects (human child): (1) *Roody (1 year 1 month) pushing basket car;* (2) *Roody (1 year 4 months) pushing wheelbarrow with sand;* (3) *Roody (1 year 4 months) pushing cart;* (4) *Roody (2 years 11 months) pushing sledge up hill;* (5) *Roody (4 years) pushing toy hare;* (6) *Roody (2 years 1 month) dragging toy horse.*

Plate 74. Obstructed walking, running, and jumping (impeded and self-complicated motion of the child): (1) Roody (2 years 4 months) walking over improvised bridge; (2) Roody (2 years 5 months) walking on plank; (3) Roody (2 years 7 months) walking through snow; (4) Roody (3 years 6 months) jumping over barrier; (5) Roody 3 years 6 months) jumping over teddy bear.

Plate 75. The human child drawing diverse objects: (1) Roody's (3 years) train for the locomotion of his toy playmates; (2) Roody's (2 year 4 month) train passes over a self-made bridge; (3) an improvised accident of the train.

Plate 76. Running, sliding, climbing, and drawing various objects (by the child): (1) Roody (2 years 11 months) running down a snow hill; (2) Roody (2 years 11 months) sliding down hill; (3) Roody climbing up steps of snow hill; (4) Roody sliding down hill without sleigh; (5) dragging sleigh up hill.

Plate 77. The evolution of climbing of the human child: (1) Roody (1 year 1 month) climbing up stairs on all fours; (2) Roody (2 years 1 month) negotiating stairs on all fours; (3) Roody (2 years 1 month) has recourse to help in descending stairs; (4) Roody (1 year 1 month) has recourse to help in ascending stairs.

Plate 78. The evolution of climbing of the human child: (1) Roody (1 year 4 months) climbing into armchair; (2) Roody (1 year 4 months) climbing down from armchair; (3) Roody (1 year 11 months) climbing into seat of sleigh; (4) Roody (1 year 11 months) climbing down from seat of sleigh; (5) Roody (2 years 4 months) climbing onto steps of car; (6) Roody (2 years 4 months) climbing down from steps of car.

*Plate 79. The evolution of climbing of the human child:
(1) Roody (2 years 1 months) negotiating stairs in vertical posture for the first time; (2) Roody (2 years 7 months) ascending
stairs unaided in vertical postures; (3) Roody (2 years 1 month)
carefully descending stairs in vertical posture for the first time;
(4) Roody (2 years 7 months) descending stairs in vertical posture.*

Plate 80. Mobile games of human and ape (climbing, swing-ing): (1) Joni on a rope ladder; (2) Roody (6 years) on a rope ladder; (3) Joni on a tree; (4) Roody (4 years 7 months) on a tree.

Biopsychological Similarities and Dissimilarities in the Behavior of Infant Chimpanzee and Human Child

Chapter 16

Similarities in Behavior of
Human and Chimpanzee

Comparison of the infant chimpanzee and the human child of the same age reveals extensive similarities in many essential instances of their behavior. This does not come as a surprise; any superficial and brief observation of a chimpanzee and a human easily demonstrates the "humanness" of an anthropoid ape and is quick to conclude that "humans descend from apes" because an ape is "almost human" (as noted by Professor R. M. Yerkes).

Indeed, if we look closely at natural motor habits of the infant chimpanzee (his lying and sitting poses; his ability to stand up vertically, walk, climb, and jump), we see that all of this is also characteristic of the human child of the same age.

If we consider the instinctive behavior of the chimpanzee and the human governed by the self-sustenance instinct (food, water, sleep, self-maintenance), we again find a number of coincidences in both infants. Self-preservation, self-defense, and attacking instincts are strongly developed in both infants. Joni and Roody were extreme cowards; if we were to indicate their frightening stimuli, it would be much easier to name what they did not fear.

All new, sudden, abrupt stimuli (luminous, auditory, tactile) and large objects frightened both of them. Unfamiliar faces, unknown surroundings, unseen or big animals (horses, cows), a large number of people, tall people, a loud cry, noise, thunder, lightning, the sound of a gunshot, a snapping sound, photographic flash, brightly illuminated rooms, darkness, and dark objects could also be considered frightening stimuli.

Both Joni and Roody were most afraid of self-propelling inanimate objects, live animals with the ability for abrupt movements, dummies of large animals, and birds and their pictures.

The fear emotion was expressed very similarly in both infants. While frightened by an object, both infants opened their eyes tensely and wide, fixed their eyes on the object and remained motionless. Periodically, their bodies might shake, and their hearts might beat more rapidly; they might sweat, try to hide in a remote place, ask for somebody's protection, or run away as far as they can from the object. A loud roar or yell also caused fear.

We can observe many similar features in the chimpanzee and human child while they act in self-defense. They use threatening hand gestures, stamp their feet, pound a frightening object with their fists or palms, pinch, scratch, or bite; they even wave a rag or stick at it (if they are afraid to come in close contact with it).

The facial expressions of anger were basically the same for Roody and Joni. These expressions manifested themselves in wrinkling of the upper part of their faces and baring of their teeth and gums.

Roody and Joni burned with an aggressive feeling not only during their self-defense, but also they became irritated when their instinctive demands (for food, water, or sleep) were not fulfilled in time. They were angry when their unreasonable desires and unbridled behavior were thwarted or when their actions were met with strong resistance on the part of any live creature.

Roody and Joni were inflamed easily with an angry feeling out of compassion toward allegedly offended people (especially their relatives); both were ready for revenge against the offender. Both infants could be pretty fierce when they claimed or protected their property or when freedom of their movement was limited.

The ownership instinct was developed strongly in both infants; it was expressed not only in their reluctance to share their favorite food, but also in fervent protection of their belongings, in passionate collection of various objects, and even in claiming somebody else's property. A thorough study of the objects (especially of the valued ones) appropriated by the infants allows us to determine their tastes and preferences and perhaps their primitive aesthetic tendencies.

Characteristically, the infant chimpanzee and human child, with respect to their visual perception, were drawn to shiny, transparent, and bright, intensely colored objects (predominant colors were red, blue, yellow, and white). Both infants disliked black objects (which apparently frightened them).

With respect to size, both of them showed their interest and attention (even tenderness in the case of the human child) toward tiny objects; with respect to form, spheres were preferred. With respect to tactile perceptions, they had a tendency for soft, smooth, plastic, netlike objects. With respect to temperature perceptions, they liked everything warm. With respect to olfactory perceptions, pleasant odors (fragrant fruit) were preferred. With respect to gustatory perceptions, they enjoyed sweet or sour-sweet things. Both infants took shiny, intensely colored plates, small balls, bright colored velvet scraps of cloth, pieces of transparent oilcloth, pieces of glass, vials, laces, nets, and pieces of rubber; the infants played with these objects and did not want to part with them under any circumstances.

Both infants showed the tendency for self-embellishment, especially with attractive bright pieces of cloth, which they hung (in quite a similar fashion) on their neck like a scarf.

Both the infant chimpanzee and the human child were extremely freedom-loving creatures. They were reluctant to get dressed or covered, they suffered and cried when they were confined to one room, and they were happy to get access to the broad space of a yard, field, or forest. They were ready to broaden the arena of their actions and movements even farther.

Both infants had a highly developed social instinct, especially (according to their age) the tendency for providing themselves with the care and protection of adults—women who fed them. If a good contact with such women was lost, the children looked for another guardian, as if they were afraid to remain without care.

Both children distinguished their "nannies" among other people; they rejoiced when she was coming, and they were upset when she was leaving. They wanted to be with her all the time, day and night, passionately insisting on it when they were ill. She was the first person to whom they ran if they got in trouble. They expressed their tenderness and trust, surrendering only to her for any kind of hygienic manipulations or medical treatment and resisting anybody who tried to assume this role. Both infants sympathized with their respective nannies, each in his own way, if they noticed in her any signs of sickness; they made attempts to "treat" or comfort her and were always there for her if they thought she had been mistreated by other people.

External forms of their sympathy were especially similar. Both Joni and Roody, enveloped in tender feelings, ran up to their guardian, pressed their body against hers, embraced her, tenderly touched her face, and pressed an open mouth to her cheeks or to her hands, breathing rapidly. Both infants kissed her after they learned how to kiss (in the case of the human child, this was a later acquisition; for the chimpanzee, this was rather an artificial act that he used very rarely).

The sense of love and affection clearly was egocentric in both infants and was combined with jealousy. Both Roody and Joni had a tendency for monopolization of their power over their favorite people; if these people broadened their tender feelings to include other children or adults, both infants demonstrated dislike for such rivals.

Both infants differentiated between "their own" and "strange" people, being relaxed with the former and apprehensive with the latter. Both were ready for friendly enthusiastic play with a human or animal. There was nothing the infants disliked more than being left alone; there was nothing they liked better than the process of communication.

The infants differed in their reaction to people with different temperaments. They behaved delicately and carefully with calm, quiet people, apprehensively with the gloomy and wary, and vividly with the cheerful.

The infant chimpanzee, as well as the human child, possesses a strongly developed imitative instinct, getting easily infected from a human by the emotions of fear, sorrow, joy, or even anger. Infants often reproduce the human's unintelligible sounds, gestures, and body movements associated with these emotions.

There is partial similarity between natural facial expressions that reflect the main emotions of the human child and those of the infant chimpanzee: They extended their lips in the form of a tube in anxiety; they opened their mouths wide, cried, and sobbed in sorrow; the smile was narrow or broad, they moved their hands haphazardly, and they produced sounds in joy; they pressed in the corners of the mouth, and the mouth assumed the form of a rectangle in surprise; they tightly closed their extended lips and touched their lips with the index finger in attention; and they turned their backs in indignation. Also, there was similarity (mentioned above) in the expression of fear, anger, and tenderness.[1]

In both chimpanzee and human, we observe tight closing of the lips and pulling to the sides during cautious movement that requires fine coordination of the fingers. Both infants also showed a series of similar imitative actions: sweeping the floor with a broom, wiping a puddle with a rag, hammering a nail, opening a lock with a key, reaching for objects with a stick, turning an electrical switch, drawing with a pen or pencil. Thus, both infants used the following objects, tools, and utensils: cup, spoon, knife, handkerchief, napkin, blankets, broom, rag, stick, hammer, key, pencil, pen.

Both infants sometimes used substitutes: For lack of a pen, Roody and Joni tried to scratch paper with a nail, stick, or even fingernail. With no ink around, both dipped their fingers in milk, jelly, water, or saliva and tried to leave stains on paper. When there was no hammer, both tried to use their fists, stones, or any other heavy object; they substituted a piece of paper for a napkin and a handkerchief. Sometimes, the imitative actions of both infants were not very effective; for instance, they could not assess the relative dimensions of locks and keys and tried to fit a tiny key into a large keyhole.

Both infants imitated some sounds after adults: hand clapping, lip rattling, or knocking with their knuckles on solid substrates. Both had the tendency for voice imitation; however, they limited themselves to certain sounds that they were able to reproduce.

Seeing a hole in the bed screen, Roody welcomed his luck; from that moment, he refused to get into the bed in the usual way and, like Joni, squeezed in through this widened hole, grunting, wheezing, and spending considerably more time and effort than was otherwise needed.

Moving objects were the most attractive for both infants. A wooden ball was the most entertaining and desirable toy for both Roody and Joni. Both infants could be engaged for a long time in throwing and catching balls, rolling round baskets over the floor, and pushing baby carriages or chairs on rollers. Both could not help moving any movable object within their reach. They would several times in a row open or close doors, windows, cabinets; slam a piano lid; turn the wheel of a sewing machine; spin swivel chairs; or simply throw various objects around the room, in a matter of minutes transforming a tidy place into a chaotic pile of half-broken, mutilated things.

Both infants liked to watch street traffic from a window, pressing their faces to the windowpane and following the course of pedestrians, processions, or carriages; they became so totally immersed in this that they did not respond when called by name. While in the country, both infants watched passing droves of cattle or playing children with great interest. Even slow moving creatures such as caterpillars, ants, and roaches intrigued both infants. When, for instance, Joni saw roaches sitting in the cracks of his cage, he sometimes drove them out with a straw, watched their commotion, and renewed their fear when they seemed to calm down. Roody liked to watch running ants for a long time.

Play with animals presented a special attraction for both children: chasing dogs or pigs; scaring hens, sparrows, or crows; finding and attacking cats were for Roody and Joni, deprived of companionship with infants of their age, extremely exciting. Both infants were less enthusiastic about playing a game of hide-and-seek; they both preferred hiding to seeking.

Roody and Joni were even happier if play was accompanied by sounds; the more noise, rattle, and thunder, the more they were elated. While running around the room, both infants grabbed thundering metal chains or baby carriages; they liked to shake keys or rattles; they repeatedly knocked on the walls or floor with metal rods or wooden balls; both loved to beat the drums, producing deafening sounds. Both infants, for the lack of other entertainment, often entertained themselves by clapping their hands, rattling their lips, pounding their fists, or stamping their feet. Both were willing to thump on the piano.

The infants demonstrated many similar features during experimenting play with fire, water, and sand as well as with solid, elastic, transparent, or sharp objects. Both infants were so drawn to fire that we had to restrain them to prevent them from being scorched. With great joy, they watched a burning candle or stove or the lighting of a match. They liked to turn on electrical switches. They were no less attracted to water. It is hard to name all of the pranks these little heads could think of when they saw water. Joni always took water in his mouth and made repetitive rinsing movements. Both did not miss the opportunity to splash water from a puddle. Sitting in front of a bowl of water, Roody, and Joni, scooped water with his hands or with a mug, poured it and scooped again. Roody liked to watch things floating on the surface of a water reservoir and their immersion in water. He tried to spread spilled water over the floor, dipping his fingers in it and splashing it around. Both children were intrigued with tap water; they caught the running water with their mouths, stopped water with their fingers pressed to the faucet opening, and splashed cascades of water on themselves. Both infants willingly dug in the sand and mud, picking small stones and looking closely at them; they shifted sand from place to place, raked it in piles, scattered it around, filled and emptied vessels, and found endless material to touch and examine.

Objects found in the mud or in the sand (small stones, vials, nails, chips, straws, matches) were used immediately for playing. Joni looked closely at larger stones, took them in his mouth, gnawed and scratched them, and threw them at some target. He shuffled small stones, nails, and vials in his mouth, turned them with his tongue, moved them from one cheek to the other, sucked them with a smacking sound, but never swallowed them. Roody also tended to put small stones in his mouth. Sharp sticks, vials, and straws were especially valued by both infants, who either pricked their hands with them or used them as props. Joni put a vertical prop between his lips, tried to run around with it in his mouth, tickled himself, and fidgeted, but managed not to drop the prop from his mouth. Roody set a similar prop in his palm and tried to keep it there while running fast.

Both infants shared a passion for various kinds of sticks. Finding a stick, Joni dug in the ground with it, beat on the walls and on the floor, reached for things otherwise unreachable, and waved it in a threatening gesture or attack. Roody drew on the ground with a stick, threw it over fences, destroyed things, drove away dogs, knocked down bowling pins, and used it to fling objects up in the air. Both infants also shared a passion for balls and balloons.

Transparent things, such as pieces of colored glass, vials, yellow oilcloth, and combs, brought joy to both children. Roody, as well as Joni, looked through them at the light, pressed them to his eyes, and shifted the glance from the sky to the ground to flowers to human faces. They kept even the smallest pieces of some oilcloth or colored glass as treasures.

Long elastic objects, such as rubber tubes, rope, shoelaces, even hair, were for both infants of no less value than any real toy. I cannot describe all their manipulations with a rubber tube. Ropes, laces, and bands were wound up in balls. Loops were pulled out of the balls and served as invitations for squeezing through; these invitations were accepted immediately. Both children use a long human hair for similar play.

Roody and Joni revealed many similarities in their tendencies for destructive play. Throwing, scattering, rupturing, and breaking various objects presented an overwhelming pleasure for them. All the things, all the toys that had been in their possession usually bore the results of the destructive effect of their hands and teeth; these objects were conglomerates of broken or mangled parts of what could be called a particular item in the past. In Joni, this passion was so strong that it targeted practically everything: He ripped wallpaper from the walls, gnawed plaster and the bars of the cage, broke glass, and tore apart all pieces of cloth he could. A tendency for rending pillowcases and pillows, for breaking glass, and for destroying the screen of the cage was absolutely inevitable; he did this despite instances of severe punishment. Joni demonstrated a strong will, especially in the case of his destructive actions; resistance of certain things to destruction only enhanced his excitement and his passion to finish the thing off. While achieving his goals despite our counteractions or during his attempts to free himself from the cage, Joni developed such energy that he was ready to withstand strong language, inconvenience, pain, punishment, or quarrel with someone he loved. I observed in both infants that a ban on something could cause an effect quite contrary to the intended one; both Roody and Joni were especially persistent in doing something forbidden.

Goal-oriented actions sometimes were erratic in both infants and were performed only out of stubbornness. After persistent solicitations, yelling, and crying for something, Roody, as well as Joni, could show total indifference toward the desired object they finally had gotten in their possession. Thus, it is clear that the very process of getting this object, and not the object itself, governed the passion for acquiring it.

We observed in Roody and in Joni the sense of being offended when their desires were not fulfilled. In response to an abrupt, severe affront, they both turned their backs to the offender, did not look the person in the eye, and did not want to take even their favorite food from the person.

Both Roody and Joni, doing forbidden things, sometimes tried to deceive us and conceal the traces of their trespassing; being too naive, they did not know how to do it.

Sometimes their "prudent" tactics had all the features of "Straussenpolitik" (the ostrich sticking its head in the sand). For instance, Joni took some forbidden thing, looking not at it, but at the person who did not allow him to take it. Not seeing the thing himself, he evidently assumed that it also was invisible to others. Roody's tricks of that nature were even more naive. Roody (1.5.22–2.2.8) showed similar tactics during a hide-and-seek game when he tried to hide by covering himself with a kerchief, by standing behind a wicker chair, by squatting quickly near the steps of the porch, by covering his eyes with his hands, by turning his back to the seeker, or by some other way of preventing himself from seeing the outside world.

The sense of guilt was expressed clearly in both infants. Joni tried not to look at anybody after he committed an offense and stoically endured reprimand and punishment. Roody (2.1.5) reacted to a reprimand by blushing; he was ready to cry and cried even when the punishment was light.

Both infants, Joni and Roody, were extremely mobile creatures not only physically, but also mentally. From the moment of their morning awakening until going to sleep in the evening; half-asleep; half-sick; during their meals, dressing, washing; even when they sat on a chamber pot, they managed to think how to entertain themselves. When left to their own devices and having many objects in their possession, they changed their actions and methods of entertainment all the time. Their attention was extremely scattered; it shifted from one thing to the next and went back to the first one.

The free play of both human and chimpanzee not only reflects their unconcentrated attention and inconsistency of desire, but also their intense curiosity toward everything new. New people, new surroundings, and new things were very intriguing for both Roody and Joni. By showing something unusual and exciting their curiosity, you could stop any emotional manifestation, any play. Roody once expressed his desire for newness in the following words: "I could go all the time if I saw new things all the time!"

Joni's entire behavior confirmed that idea. Without new impressions, both children showed clear signs of being bored. Joni would lie on his back, stare vacantly at one point in front of him, pick his teeth, or rattle his lips. Roody would whimper, "Show me something new!" Both of them were delighted to be exposed to some new experience; a strange yard or house, forest, or new sight from the windows were inexhaustible sources of curious contemplation. They stared greedily through the windows of automobiles and asked persistently for permission to peek in the low windows of strange apartments. They got into sideboards, cabinets, and baskets filled with junk, every time finding new interesting things to examine. Semiopen cavities, such as ovens, pockets, even deep vessels, vials, and human nostrils and ears, excited their curiosity; they directed their curious eyes, exploring hands, and testing index finger everywhere. Both vision and touch helped them get acquainted with new things. To look was not enough; they had to touch the object, take it in their mouths, taste it with their tongue. My Roody was not behind in this respect. Despite our categorical objections, he put in his mouth almost all things that interested him. Convex objects intrigued both infants no less than concave; in this case, their touching was especially thorough. Both infants sometimes became prey to visual deception. For example, seeing stereometric images in books, they tried to take the objects off the paper. Joni, in his diligence to do that, made holes in the paper. Roody (2.0.23) said, after a number of unsuccessful attempts, "I can't," with disappointment in his voice.

Shiny, bright, moving things aroused their curiosity; both opened their mouths in amazement or stared attentively, extending their lips forward and employing the index finger for exploration.

Both infants liked to look through books with pictures; in their craving for new visual experiences, they quickly turned the pages, emotionally qualifying the pictures. Joni would pass some pictures, but suddenly would start pounding other ones with his fist, for instance, pictures of wild animals with brightly glimmering eyes or pictures of monkeys. Looking at the pictures of two birds of different size, Roody (2.1.6) called the small one "ntsa" (good) and kissed it; he called the bigger one "byaka" (bad). When Roody (2.1.25) looked through a calendar, he called the dates written in red ink "good" and the dates in black ink "not good."

Both children transferred their curious attention to people around them or to themselves. Their reaction to a mirror was remarkably similar. Here are seven sequential stages of the reaction of Roody and Joni to a mirror: (1) staring at their image in the mirror; (2) smiling; (3) touching the image with their fingers; (4) striking the image and looking for somebody in the mirror; (5) spitting at the image; (6) lip rattling, grimacing, gesticulation; (7) aggressive waving at the image with a weapon.

Both infants showed fine observation ability. Joni, as well as Roody, immediately noticed any new object against the background of "old" objects and started staring at it. They both noticed a new dress, shoes, or decoration; every scratch, pimple, or ink stain on somebody's hands; tiny specks of dust on the floor; stains on the wallpaper; pins, thin needles, or nails—they looked at everything, touched it, and picked it up.

Both infants also recognized similarities. Watching the chimpanzee in his free play, I observed his tendency for combining objects by color (breaking down the groups of 35 plates of seven colors) and for picking out only light blue plates for his play. Another time, Joni picked out only small, white, round plates from the group of different plates by shape and color. Sometimes, he ignored dissimilarity of forms and sizes: From the group of sticks, acorns, plates that were rectangular, round, large, or small, he picked out things exclusively by color. The tendency for finding similarities and for analogization was also highly developed in Roody. After he found his own eyes and ears, he pointed at the eyes and ears of everybody around him. Having discovered legs in a picture of a horse, he touched his own legs.

Both infants had traces of the general ideas of objects they discovered when changing their tools of action: a stick, a finger, or a nail instead of a pencil; milk jelly, water, or urine instead of ink; a stone or a fist instead of a hammer; a piece of paper instead of a napkin or handkerchief. Joni apparently had a general idea of what a key and a lock were like because he tried to fit different keys to different keyholes (in doors, suitcases, etc.).

Roody, at the age of 10 months, easily mastered the general idea of a button. When asked, he pointed at buttons differing by shape, size, and material and attached to different things in different places. He (1.5.15) pointed at the hats on his toys despite dissimilarities in their appearance.

All this indicates the infants' ability for elementary abstraction and development of memory. This development is visible clearly when we consider their conditioned reflex activity. Both human child and infant chimpanzee developed, without great difficulty and special training, the motor skills associated with their acts of self-maintenance when using a spoon, knife, napkin, blanket, and the like.

Both children spontaneously developed almost similar gesture languages for expressing some of their desires. For instance, a request was expressed by extending a hand, rejection of food by turning the face away, thirst by pressing a finger to the lips, and attempts to attract somebody's attention by tugging at that person's dress.

Here are some other types of conditioned reflexes easily observed in both infants:

1. Visual-gustatory-motor. For instance, when we show them familiar tasty food (oranges), they joyfully run over and grab these things.
2. Visual-pain-motor. For example, when they feel heat coming from an oven or fire, they both move away from the source.[2]
3. Auditory-motor. They properly respond to requests such as take it, sit, lie down, give me your hand, and so on.[3]
4. Auditory-visual-motor. Hearing the word "fly," Joni looks around for a fly. Responding to the question, "Where is the tick-tack?" Roody turns his head and stops his eyes at the clock.[4]
5. Visual-motor. I put my books back on the shelves, which is a direct sign of my imminent departure. Joni runs over to the door and blocks my way out. If Roody spots a faraway bus before crossing the street, he stops and does not budge, while previously in a similar case we needed to hold him firmly.
6. Visual-emotional-auditory. The sounds of the elevator and our voices are the signs of our homecoming; Joni joyfully grunts behind the door. The sound of the elevator and subsequent silence are the signs of our departure; Joni cries. I say, "Roody, let's go for a walk"; he kisses my hand and squeals from joy. I say, "Roody, let's go to the park"; his face dims, and he is ready to cry because he does not like to go there.

Comparing natural sounds of the infant chimpanzee and the human child, we found the following similar sounds: "e," "u-au,"[5] "ym,"[6] "khru-u,"[7] "u-khu,"[8] "o,"[9] and "yu."[10] Apart from this, the following sounds were very similar in both infants: rapid breathing, sneezing, coughing, grunting, snoring, deep yawning, and partially the sounds of crying.[11]

There are only a few sound-imitating examples that we can cite. Both of them imitated, with different degrees of perfection, the barking of a dog. Joni did it better because barking was his natural sound. Also, both infants reproduced the sounds of foot stamping, hand clapping, and lip rattling.

Chapter 17

Differences in the Behavior of
Human and Chimpanzee

Thus, the similarity between the infant chimpanzee and the human child is beyond any doubt: it is vast and multifaceted. But if we make a deeper, broader, and more accurate analysis of every aspect of their behavior, we will find that each has his own highly developed specific features (see Table I at the end of the text for a summary).

When we considered similarities in the sitting poses of the chimpanzee and human, we predominantly had in mind some untypical and quite artificial ways of sitting (on a podium, on a bench, or in a human's lap). A natural, typical sitting pose of the chimpanzee, which shows many times in our photographs, is with his legs bent and pressed tightly to the body while he is resting firmly on his hands. This particular pose is uncharacteristic of the human child of the same age; it can be observed only in a 5-month-old baby learning how to sit. On the other hand, I never observed the chimpanzee in a squatting pose[1] or resting on his knees, the poses so typical of the human child.

The same dissimilarity pertains to standing poses. It is true that the chimpanzee, like the human child, can stand vertically, but to keep his balance, he must always rest on the outer edge of at least one foot. He stands with his legs positioned widely. His stance is extremely unsteady; he can retain it for a very short time and is ready at any moment to seek additional support to remain standing. At that time, he is like a 6-month-old human child who is only learning how to stand. A typical standing pose of the chimpanzee is his standing on all fours with the axis of his body inclined; he can remain in such a position for a fairly long time. A typical standing pose of a 3-year-old child is vertical standing with closely positioned legs while the child rests on his spread soles.

The vertical walk of the chimpanzee also raises many doubts. It can be called vertical with certain reservation: The infant chimpanzee can walk vertically for only as many as three or four steps, but he must use his arms or have some sort of support to keep his balance. He walks vertically only when he wants to see a large area of terrain or when he must turn around constantly in unfamiliar and threatening circumstances. He uses this uncanny way of walking with difficulty and is ready any moment to change it back to his typical way of walking (or running) on all fours with an inclined or horizontal position of his body. Using this quick, confident, and resilient gait, he easily can outrun not only a human child of his age, but also an adult human.[2]

Even when the chimpanzee is led by the hand, he is inclined from time to time to use his free hand as a support to ease his movement. While in the child from 9 to 11 months we still can observe a somewhat inclined walking position and balancing with the arms, a 3-year-old child can walk and run vertically easily and confidently for a couple of hours without getting tired.

The fact that child's hand is free from the necessity to participate in moving on the ground gives him an opportunity to take an object and play with it while he runs or walks. If the chimpanzee wants to take something along, he is not willing to use his hands for that purpose; he takes the object in his mouth or drags it behind him with his leg.

We have mentioned the ability of both infants for climbing the stairs or trees. In the ways of doing this and in the degree of perfection we can see the most conspicuous differences between the chimpanzee and the human. While Joni could climb the stairs only on all fours and could come down in the same manner (head first), Roody used this method only at the age of 1½ to 2 years (he always descended the stairs backward, legs first). But, at the age of 3 years the child went up and down the stairs, walking vertically, in the usual human way, without the help of his hands.

The human child at the age of 3 to 5 years can climb the high smooth trunk of a tree with difficulty. He hardly can hold on to a tree with his weak hands; his face is contorted in a grimace, and he is about to fall off the tree any minute. The infant chimpanzee easily and slyly hitches to a tree, using all four extremities; he gets on high fences and buildings up to the very ridges of the roofs and fearlessly strolls there, quickly coming down vertical poles or inclined planes and surpassing most skillful gymnasts in this climbing ability.

Not only is the chimpanzee's foot, compared with the human foot, much more flexible in the fingers, but his leg is also more mobile in the hip joint. For instance, Joni could raise his leg so high that it formed an obtuse angle with his other leg. Roody was unable to do it; only acrobats can achieve this after special training.

The infant chimpanzee and human child frequently and joyously jump on both legs, but only the human child is able to jump on one leg. Only in the chimpanzee can we observe jumping from feet to hands and back while his body stays in a horizontal position.

Summarizing the anatomical and physiological features of the human child and the infant chimpanzee (during walking, running, jumping, and climbing), the firmness and endurance of their bodies, power of their hands and teeth, and keenness of their senses, we can draw a conclusion about the greater adaptedness of the chimpanzee compared with that of the human. Let us discuss the divergence of the chimpanzee's behavior associated with his self-supporting instinct.

While the human child performs all the mundane procedures (such as eating, drinking, dressing, washing, or combing) hastily and sloppily and always tries to get it done as soon as possible because these things are boring for him, the infant chimpanzee eats, drinks, examines, and cleans himself with great attention, thoroughness, and alertness.

Before taking even a piece of familiar food in his mouth, Joni smelled it, tasted little bits of it, and then slowly ate it with apparent pleasure. If Joni was given a bowl of milk that was cooler than usual, he did not want to drink it: He held it in his mouth until it warmed, and only after that did he swallow it. Joni never swallowed solid objects (fruit stones or inedible things) as Roody usually did. Joni detested meat (especially chicken) and even small amounts of butter added to his food, but he willingly devoured some parasites, even those found on his body. Roody ate butter and meat with great pleasure and was so squeamish toward insects that if he spotted a tiny one in his plate, his face wrinkled in disgust, and he rejected the entire meal. While eating especially tasty food, Joni grunted loudly, this grunt evolving into a dull cough. Roody, under similar circumstances, not only grunted, but also sometimes made a mooing sound that resembled that of baby bears sucking milk.

While the human child during eating and drinking does not like to be helped by adults and tries to learn how to use such everyday things and utensils as a cup, spoon, soap, towel, clothes, and the like, the infant chimpanzee is entirely comfortable with outside help and is not inclined to improve his skills. Joni drank from a cup or saucer awkwardly and unwillingly; he often used his foot, instead of his hand, to hold a dish. Roody used dishes and acquired skills associated with his selfmaintenance much more efficiently than Joni.

We noted the reluctance of both infants to share their favorite food with anybody. However, while Joni totally refused to share his food even at the request of his favorite person, Roody might spare, in response to a request, at least tiny bits; he might easily give up one piece when he had two. All this despite the fact that Joni generally wasted much more food than Roody and in his tasting attempts spoiled and scattered a lot of edible material.

The infant chimpanzee, during his self-cleaning and self-examination, also tries to treat himself, taking out splinters, licking out his wounds and scratches, wetting the wounds with saliva, or sucking out the blood. He may twitch from pain, but he would not stop these manipulations. When the human child hurts himself, gets a splinter, or sees blood, he becomes scared, cries, runs for an adult's help, and cannot bring himself to touch the sore spot. While Joni liked being treated by a person close to him, Roody either objected to such manipulations or met them with tears.

Observing both infants go to bed and fall asleep, we can conclude that the infant chimpanzee takes time to prepare his bed. Joni made his bed soft and in the form of a nest; he shifted the sheets around in a businesslike manner (although his bed had already been prepared for him), put softer sheets closer to the periphery, and made a semblance of a pillow. The human child uses his bed without any intention to rearrange it according to his taste. While Roody always covered himself completely, up to his neck, with a blanket and even tried to "dive" under it, covering his head (despite our persistent objections), Joni never (even during cold seasons) allowed us to cover him up to his neck; he always freed his arms and left his chest uncovered. Roody often talked, cried, and gesticulated in his sleep; Joni did not utter any sounds during his sleep and made no gestures.[3]

Maybe Joni's objections to covering his arms was caused by his desire to have his hands free in case he needed them to defend himself during his most helpless state—sleep. Joni resisted, in any conceivable way, being dressed, even in loose shirts; he ripped off compresses or bandages, unwilling to tolerate even one of his fingers being wrapped. Roody did not put up special resistance in such cases. He tried to learn how to put on his clothes; although with much difficulty, he finally succeeded. Joni, apart from covering himself with a blanket, did not want to use any other clothes.

Perhaps we are dealing here with Joni's more developed freedom instinct. Characteristically, when Roody was let out of the house, he was inclined to run without any apparent direction. Joni was drawn by heights; he got on fences or roofs, where he could stroll for hours (incidentally, Joni invariably climbed on high places before defecation and urination).

Joni was more sensitive to freedom restrictions than Roody, probably because the chimpanzee, deprived of the freedom of movement, was doomed to solitary confinement in his room or cage, while the human child spent all his time among people.

Also, there are differences in the manifestation of their defense and attack instincts, particularly in the forms of fear emotion and in the related stimuli. While Roody's face usually reddened as a result of fear and he pressed his hands to his chest, Joni's face became pale; he became fluffed up, made a defensive gesture, and usually raised his hand in front of his face or forehead to protect his eyes. In the state of awe (for instance, from a gunshot, a blast, or a photographic flash), Joni lay flat on the ground face down, crossed his arms over his head, and defecated involuntarily (I did not have the opportunity to observe Roody under similar circumstances).

I never observed Roody in such a panic caused by the light or sound stimuli, as in the case of Joni, who by the way, showed special fear toward such reptiles as a tiny turtle or a small grass snake and toward animal hides, especially the hide of a spotted panther. The fear of any olfactory stimuli was also uncharacteristic of Roody; such fear was often observed in Joni. I am absolutely certain that Joni was more apprehensive than Roody; therefore, it is remarkable that Joni used to overcome his fear by repeatedly challenging the frightening stimulus and was generally braver and more courageous than his human counterpart. Roody, while afraid of something, tried to involve an adult to help him overcome the fear.

Joni compared to Roody, showed a stronger ability to defend himself. Threatening poses can be observed frequently in a scared chimpanzee. Fluffed up, enlarged almost to twice his size, standing on all fours, and staring at the scary object, he jumps from hands to feet and back, choosing the right moment to assault his enemy. He bares his teeth and gums, pulls up his upper lip, emits a loud sound, assumes the vertical position, fiercely bites, and if possible, rips his victim apart; when he cannot do this exactly, in fits of a helpless rage, he even starts biting himself.

Obviously, Roody did not present me with such a colorful picture of rage toward the same scary objects (dummies or animal hides) that caused similar anger in Joni. Carried away with an angry feeling, Joni often clenched his fingers and toes; Roody only clenched his fingers.[4]

Interestingly, Joni was enraged the most when I showed him the dummy of a small (6-month-old) chimpanzee; the dummy of Joni (showed in 13 years later to 2½-year-old Roody) aroused in my baby the most affectionate feelings. Obviously, Roody did not have even traces of that hateful, fierce attitude toward small helpless animals or insects that was so characteristic of Joni, who used to torture, beat, and rip apart frogs, crawfish, bugs, and the like.

Roody felt special affection for insects or small animals (live animals and toys). He showed toward them his sympathy, protection, and compassion when they got into trouble; he tried to defend them and to help them if they were wounded; he always resisted catching mice. Also, Roody helped to care for children smaller than he was and did them various favors.[5]

We wrote about the sympathy of both infants toward the people closest to them and about certain interference of the infants on behalf of these people. However, only Roody, seeing signs of illness in the ones he loved, expressed his sympathy by crying, and he could not hold back his sad feelings.

I never observed in Joni an intention to sacrifice his own interests, at least to share tasty food with a person he loved. Even in response to a request, Joni did not want to treat you with tasty food; even when he tried to defend somebody, this did not go beyond a point at which this might be dangerous for him. As soon as the danger increased, Joni preferred to step back, leaving the one he defended to the mercy of fate.

Roody burned with a genuine feeling of revenge when he was defending small animals from larger ones; for example, in his desire to break up two fighting dogs, he was so busy pulling away the big one that he usually forgot about the danger to himself. I observed Roody many times stop his noisy pranks and games when he was told that somebody was not feeling well.

In the behavior of a 3-year-old child, we can see spontaneous elements of fairness, justice, morality, and altruism; the chimpanzee does not have even traces of that. A rare example was an instant when he held back anger toward a person he loved: One time, when applying an ointment to Joni's nose, I accidentally hurt him; he grabbed my hand with his teeth, but at the last moment, he recoiled and pressed his jaws together only weakly.

There is a substantial difference in how both infants express their affection. Roody's were articulate forms of verbal expressions that revealed the power and depth of his feelings (particularly, his feelings toward his mother).

We observed in Joni, compared with Roody, an easier contact with unfamiliar adults, more officious treatment of a strange person, and certain despotic traits in his intention to retain the leadership role in his mobile play with a human. The smaller and the more helpless the creature with which Joni communicated, the more visible his despotic tendencies. The human child is entirely different: He easily submits himself to adult initiative and joyfully runs errands. In company of his peers, he shows elements of companionship; among younger children or among harmless animals and toys, Roody most often set up a familylike, patronizing form of communication.

Let us discuss the differences between the infant chimpanzee and the human child in the field of emotions. We wrote about the similarities in Roody's and Joni's gestures when they were experiencing basic emotions (anxiety, sorrow, joy, anger, fear, disgust, curiosity, surprise, tenderness) and about similarities in most of the external stimuli corresponding to these emotions. To make this comparison more accurate, we must emphasize that objective representations of emotions in sounds and facial expressions are much more vivid in the chimpanzee than in the human. Power, expressiveness, and duration of emotional representation in the chimpanzee allow us to follow sequential stages of emotional development easily, which is very difficult to do with respect to the relatively weak external manifestations of human emotions.

For example, when worried, the chimpanzee gets fluffed up, often assumes the vertical position, emits a modulated hooting sound, clenches his toes into fists, and gesticulates with his arms. When Roody was worried, his face blushed, and he breathed rapidly.

Experiencing emotions of sorrow and despair, Joni roared; he spread, raised, and wrung his arms and tumbled over his head; his face darkened somewhat, but there were no tears in his eyes. The preliminary stage of Joni's sorrow was accompanied by moaning and extension of his lips forward.

When Roody was very upset, he cried. The tears might be profuse; they dropped straight to the floor, and his face blushed; he often pressed his hands to his face, rubbing his eyes and wiping his tears (Joni did not use this gesture because there were no tears in his eyes). A weak stage of sorrow preceding Roody's cry was expressed by his lower lip, which inside out (the human child cried without tears, like Joni, only when he was younger then 1½ months).

The joyful experiences of the human child are accompanied by loud laughter and other squealing sounds (which appear at the age of 3 to 4 months and grow later). The chimpanzee does not have loud laughter even at the moments of light tickling: His face is lit with a broad smile, his eyes shine, and he is happy and energetic, but we do not hear his laughter, only rapid breathing.

We mentioned that, during fear, the human child's face blushes; he yells and presses his hands to his chest. The chimpanzee, experiencing a similar emotion, pales, gets fluffed up, and utters a short dull "u." When raging, the chimpanzee extends forward his upper lip, bares his gums, opens his mouth, and pounds with his knuckles. The human child under similar circumstances presses his teeth tightly, clenches his fists, and stamps his feet.

The human child is inclined to express his tender feelings by pressing his face to a human's face or kissing it; the chimpanzee expresses such feelings by touching a human's face with his [chimpanzee's] open mouth (this is characteristic of the child only when he is younger than 2½ months) and with his [chimpanzee's] tongue (which is not found in the child at all).

We observe in the human child verbal forms of expressing tenderness very early in his life (at the age of 2 years).

The emotion of surprise is accompanied in the human by deep breathing; disappointment is accompanied by a grunting dull sound, disgust by a weak, coughing, cracking sound. In the chimpanzee, the surprise and disgust emotions occur without any sound, and disappointment is accompanied by a hoarse sound. In the chimpanzee, surprise, anxiety, and fear often go with fluffing up of his facial and body hair. In the chimpanzee, curiosity and attention go together with his smelling the intriguing stimulus; this is not observed in the human child.

The human child has a peculiar facial expression that is also partially characteristic of the chimpanzee. The human pulls out his tongue during intense movements of his arms; in similar cases, the chimpanzee pulls out his tongue and turns it.

There is a significant difference in the range of conditions under which emotions appear. Comparing three basic emotions (anxiety, sorrow, and joy), we must emphasize that, while the anxiety emotion is a frequent and essential event in the chimpanzee's psychic life and precedes the development of his basic affects, in the human child, this emotion is relatively rare and weak. Compared to Roody, Joni often reacted to some events vigorously, but indefinitely and indiscriminately. It seemed he could not at once understand the nature of the stimulus encountered (favorable or unfavorable) and how (joyfully, sadly, defensively, or offensively) he should react.

The human child's response in most cases is more precise emotionally. He is glad, upset, afraid, or angry, with the apparent exclusion of a long intermediate stage of anxiety. He undoubtedly is quicker to take into account the biological significance of the stimulus and to react to it in a more definitive way.

Furthermore, in the human child, the range of joyful, sad, tender emotions and emotions with cognitive overtones (for instance, curiosity or surprise) is usually much wider. The human child is sensitive to physical pain and is drowned in tears when he is even slightly hurt. He often cries out of compassion toward his relatives or, for instance, when he fails to carry out his plans; this can be observed in the chimpanzee only as an exception. In the human child, there is also a large space for joyful excitement: a sense of humor in response to sudden extraordinary combinations of usual stimuli. This feeling reveals to us that the child notices any divergence from the norm and is amused by the novelties. He is as strongly upset when his desires are not fulfilled as he rejoices when they are; this expands enormously the manifestation range of his joyous feelings. The child's curiosity, attention, and surprise develop as the most powerful and deep among other feelings, and his verbal reactions do not leave any doubts about that.

To summarize the human child's and infant chimpanzee's emotional manifestations, we must state that external forms of almost all of their emotions are more expressive in the chimpanzee than in the human. Even the chimpanzee's silent laughter is offset by his earsplitting sounds. His vigorous joy begins with a vivid hooting gamut and ends with a loud barking; it is accompanied by unrestrained intense movements, not to mention the extreme expressiveness of his anxiety, sadness, and fear emotions and the like. Only with respect to tender feelings do we see almost identical external manifestations in both infants.[6]

In terms of expressiveness of the infant chimpanzee's external manifestations of his emotions, he can be compared to mentally ill people, who also may have exaggerated facial expressions.

As for the range of the emotion manifestations (i.e., the variety of stimuli that elicit them), we must point out major differences between the infants. While the chimpanzee shows whole-scale development of the anxiety, fear, and anger emotions, in the human child, we have a vast space for manifestation of joyful, sad, and tender feelings and emotions with cognitive overtones, such as curiosity and surprise.

Roody was inclined much more than Joni to imitate a human; his imitative actions were versatile and effective. While Joni could imitate only individual actions, Roody was able to reproduce a whole series of them. He imitated the professions of adults, their apparel, their facial expressions, and gestures, and their intonations. He imitated voices, wheezing, snoring, singing, and laughing. He tried to reproduce sounds of some animals and birds (for instance, the croaking of a crow, squealing of guinea pigs) and sounds emitted by inanimate objects (the ticking of a clock, rattling of a curtain, squeaking of a door, noise of a propeller, etc.). In essence, the entire process of mastering speech by a child initially is based on subconscious and then conscious imitation.

At the age of 2½ years, Roody, imitated, partially or fully, three-word phrases; he remembered rhymes consisting of 86 stanzas. Obviously, Joni was considerably behind his counterpart; his imitations were limited to those of the barking of a dog or sounds of a chimpanzee, intentionally reproduced by a human.

The human child considerably exceeds the chimpanzee in imitations involving tools or construction play. Let us take, for example, the actions with a hammer and a pencil.

Joni, despite all his efforts, did not drive in a single nail. Roody (as early as at the age of 2.1.10) did it very aptly. Joni's drawing, in spite of his repeated use of a pencil, did not go beyond the first drawing stage, in which crossed lines were rare occasions, while Roody tried to make a complete drawing. Analysis of these drawings, as well as of verbal expressions accompanying them, reveals the child's intention to differentiate between the drawings, identify the reality with the image, and qualify the images according to the characteristic features of the object. It also knows the existence of percepts of real objects in the child's consciousness, the ability to compare the drawing with nature, and the desire to make it look closer to nature. When the child is unable to make the image look similar to the object, he adds eloquent verbal comments that attest to the work of his imagination. All of these high-quality human attributes observed in his creative activity did not bear even a slight resemblance to the components of the chimpanzee's behavior.

Analyzing the differences in egocentric instincts of the infant chimpanzee and the human child (particularly the ownership instinct), we must state that the chimpanzee guards his property more aggressively, more passionately claims things he likes, and more aptly and frequently applies a deceitful tactic to acquire forbidden objects. Joni vigorously asserted his ownership, but he rarely used the things he had acquired if they were outside his day-to-day needs (as, for instance, bed sheets that he used to appropriate). Sometimes, you could see clearly that he was stimulated by the acquisition process and not by the object itself. Joni usually showed complete indifference toward most of his things; he did not even care to hide them until the danger to lose them arose. The moment one of his things attracted somebody's attention, it acquired special significance for him, and he energetically fought for it. Roody, compared with Joni, accumulated his property more greedily (e.g., sticks or stones). He was more inclined to acquire things in a peaceful way; he guarded them more zealously and hid them more thoroughly, but he defended them with less passion; he easily gave more neutral things, to which he was not so attached, to other children, saying, "I don't need them." He often did not want to part with an old, mutilated toy and gave away a new one instead, one that had not yet participated in his games.

Obviously, Roody used "raw materials" much more creatively than Joni. Even such neutral objects as stones, sticks, or pieces of metal were utilized in his construction play (the child did not always know for what purpose he needed this or that stick, but the time would come, and he would use it in his play). Most often, the child estimated the value of material and its future application right away.

We observed certain differences in the primitive-aesthetic sense of Roody and Joni, namely, in their color preferences. While Joni most often took blue plates, Roody preferred red ones. Roody, more than Joni, developed a tendency for self-adorning, self-admiration: He liked to wear his favorite outfits.

The human child transposes his aesthetic tendencies from the real world to the field of cognitive entertainment: Not only does he distinguish between different drawings, but also he qualifies them. When choosing his books, he rejects some of them and takes others. He likes to listen to certain pages from a book and does not want to listen to other pages. Among his preferred books usually are those with dynamic, merry, expressive, fantastic stories. The attraction to fantastic elements can be observed only in the human child, which reflects the functioning of his imagination.

The chimpanzee lives exclusively in the world of concrete things and relationships. Perhaps only in his play, in which he creates obstacles, we could assume that he substitutes imaginary for real objects, but even in this case, such an assumption seems too far-fetched.

The sexual instinct was more developed in Joni: all of his affects invariably were accompanied by sexual excitement. This excitement clearly was visible in his play with a football and soft objects; we did not see even a hint of that in Roody under similar circumstances.

Thus, almost all kinds of instinctive activity (self-sustenance and self-preservation instincts; instincts of defense, offense, freedom, and ownership; as well as social and sexual instincts) were expressed more strongly in Joni; only in the development of the imitation instinct was the chimpanzee behind.

The smallest difference between the infant chimpanzee and his human counterpart was observed in play, especially in motor play. Roody liked to entertain himself with imaginary movement and often set up chains of carriages to imitate trains; he used to sit in such a train and pretended to be moving. While in his room, he often imitated skating or skiing: He stood on two planks, put his feet in big shoes, and made sliding movements. Such ways of movement never were observed in Joni because they apparently did not amuse him.

While 3-year-old Joni liked to ride on doors, holding the doorknobs with his strong hands, Roody was able to do that only at the age of 4 years, but he became tired very quickly and had to let go of the knobs. All kinds of gymnastic play (climbing, hanging, rocking, jumping) were performed more easily and agilely by the chimpanzee due to the strength and tenacity of his hands. Roody could only hang for 2–5 seconds on his weak arms. Joni could hang for several minutes. Joni could hang on his legs with his head down for 2 or 3 seconds; Roody could not do it. Joni easily and fearlessly jumped from 2 meters. Roody was ready to jump only from a height of 0.25 meter. Joni climbed bravely and quickly onto a one-story building; Roody climbed without fear only to a height of 2 meters.

Due to his more passionate temperament, Joni was drawn more strongly to all types of sports contests than Roody; Joni was more aggressive, quick, and agile in the breaking-free play and in games of catching or grabbing things.

In the case of failure, Joni almost never cried, but most often was angry at his partner; Roody, in contrast, cried loudly when he lost. Of all sports activities, Roody liked the wrestling and racing play the most.

While Joni during movement put objects in his mouth and his feet, Roody loaded only his hands. Joni exercised in overcoming obstacles associated with carrying objects in his mouth and his feet, in stealing things, and in squeezing through narrow passages. Roody excelled in jumping over barriers, walking, and riding a bicycle over uneven surfaces (such as wooden bridges). While walking, Roody exercised in carrying things or tugging them over the ground.

In his games with creating obstacles, Joni often hurt himself and endured the pain stoically (he used to carry such heavy loads that he could hardly breathe). Roody, in contrast, avoided pain and exercised vigorously. The human child trains to a greater extent his psychic resourcefulness; the infant chimpanzee trains the endurance of his body.

Observing the hide-and-seek games of both infants, we must say that Joni hid better and more ingeniously and slyly than Roody.

In his play with live animals Joni demonstrated his despotic nature and his tendency for chasing, torturing, fighting, and even killing these creatures. Roody persistently tried to involve all living creatures (and sometimes inanimate ones) in the circle of his human interests. He arranged play with a succession of logical actions that reproduced episodes from adult life (fire, demonstration, trip, hunting).

In this animation play, the child's imagination works more than anywhere else; he starts to ascribe human feelings, thoughts, and words to his inanimate (or live) animal partners. Interestingly, Roody did not take these attributions seriously. He was aware that his toys and animals were not quite human, not equal partners, but "make-believe humans." For example, when he brought a book in front of his toy horse's eyes and "made the horse read," he read for it using a special "horse" language, uttering a series of incomprehensible syllables. In another case, Roody made his doll give an utterly incorrect statement. This convinces us that, in his play with animals and inanimate objects, the child does not substitute one reality for another, but seems to rise above real life and the field of imagination, fantasizing creatively. He transforms reality without completely departing from it, but not completely substituting for it, either.

If we assume that the infant chimpanzee also demonstrated some kinds of an "imaginary" fight with an enemy, imaginary resistance against somebody, and imaginary obstacle (in play when Joni was overcoming barriers he had erected intentionally), we still have to emphasize the following: The way he conducted this play (passionately, emotionally) made it clear to us that here we were either dealing with full substitution of the inanimate for the real objects (i.e., his full identification with the substituted reality) or the chimpanzee did not make any mental substitution at all perceiving these obstacles as such (i.e., as no more than concrete objects).

In his play with animate and inanimate partners, the human child takes on the role of a leader, but in this role, he tries to demonstrate his most noble and ethical qualities: courage, which he lacks in his day-to-day life, mercy, and compassion toward the offended or oppressed, in this respect being the exact opposite of the chimpanzee.

As for their play with sounds, we observed that, while Roody (at the age of 3 to 4 months and especially later) was inclined to entertain himself with the sounds of his own voice (initially unintelligible, then intelligible sounds, muttering or yelling words, singing, or reciting self-rhymed unclear syllables or poems), Joni did not even make attempts to do that despite his strong proclivity for producing sounds (for instance, rattling his lips, clapping his eyelids, knocking on a metal substrate with his hands or with solid objects, strumming on a tight rubber string, etc.).

I could observe the following differences between the infants in their play with fire, water, and sand as well as solid, soft, transparent, elastic, or pricking objects. Roody genuinely experimented with things, trying not only to discover their concrete properties, but also to find the cause, and even the primary cause, of the appearance and disappearance of these properties. For example, after blowing out a candle light, amazed by its disappearance, he (2.5.27) asked, "Where is the light? Where did it go?" and started looking for the light under the furniture.[7]

When Joni blew out a candle, he did not display any external signs of surprise or bafflement; he took the fact of disappearance of the light as such. Roody's similar experimenting behavior was observed when he played with water and with solid objects (a photographic tripod, a watch, etc.).

It goes without saying that Roody used most of the objects he found in his construction play; Joni did not do this altogether. Playing with sand or soil, Joni only poured it into the vessels or out of the vessels, and he made piles or small pits. Roody, at the age of 2 years, did not limit himself to it: He tried to construct something out of sand, for example, making a zoo (a series of sand cages) and burying his toy animals in the sand.

In his play with solid, sharp objects, Joni was more reckless than Roody; Roody never dared to take a bunch of nails into his mouth and move them around as Joni used to do. Also, Roody never put sharp props between his lips, only in the palms of his hands; the human child patently avoided painful sensations.

While Joni limited himself to collecting twigs and small sticks, using the stick as an auxiliary tool for reaching otherwise unreachable objects at best (for instance, scaring roaches out of the floor cracks or reaching a chandelier with a long stick), Roody used sticks to build an airplane, a well, a boat, and more.

In Joni's case, the destructive play was naturally more common and conducted with greater enthusiasm and effectiveness. Joni was amused with overwhelming destruction; he easily ripped apart and destroyed with his strong hands and teeth the objects that defied Roody's strength. Roody also was carried away by destruction, but if he failed to achieve a desired effect with his natural means and sparing his weak hands, he took a tool (a stick or a stone) very early in his life. He tried to expand his sphere of destructive activity by throwing sticks or stones at certain targets. Trying to achieve greater effect, he combined his destructive and creative actions. He made a boy's catapult, a bow, a sword. He joyfully fired from a toy gun or pistol.

While Roody's imitations were most productive in the field of constructive actions, Joni was most effective in his destruction. Roody took a tool to make something (an airplane, a boat, or a train); he arranged a zoo, built a bridge, a house, a telephone, a cage, and a well. And how delighted he was when he succeeded!

Roody made quick progress, for instance, in building a house (see plate 68, which represents his consecutive attempts to build a house at the age of 2½ and 3 years 3 months).

Joni, on the other hand, was more successful in taking out nails than hammering them in, taking off rather than putting on the trapezes, opening rather than closing latches, unraveling rather than tying knots.

Both infants had in their possession similar collapsible bowling pins. While Joni could take apart these pins for hours and expected me to put them together only to break them apart again, in contrast, Roody took the pins apart, put them together, and did this again and again.

The inclination for constructive play in the human is far ahead of his destructive tendencies. I observed Joni trying to make a semblance of two things: a stringed instrument and a rattle. One time, he poured some sawdust in a vial and tried to use this for rattling. Another time, he stretched a rubber band on his head and produced a rattling sound; then, he hitched the band by his canine, stretched the rubber even more, and started strumming on it as on a string.

In 3-year-old Roody, we observe active development of constructive activity expressed in his reproduction of day-to-day things, different ways of movement, machines or institutions, and imitation of real objects more or less accurately (see also Figure 2 regarding constructing an airplane). At this age, construction projects of the child were primitive and represent a poor likeness of reality, but this did not upset him. He realized what he had wanted to do and what he had failed to achieve; on the wings of fantasy, he rose to an unreal world and with a legislative act, with omnipotent words, "Let it be!" he ordered his ships and boats to sail, his trains to move, and his airplanes to fly. Such an amazing imagination, such a metamorphic ability to achieve bright results on the basis of a bleak, poor pattern of reality, plain materials, and the weak, inexperienced hands of a child is a specifically human feature.

Analysis of the chimpanzee's and human's play shows that, in the case of the chimpanzee, mobile, gymnastic play, play to overcome self-inflicted obstacles, destructive play, and play with live animals are more developed, and the infant chimpanzee is involved most passionately in them. In the case of the human child, the more developed play is imitative and constructive. Both infants could engage in play with moving objects or experimenting play almost to the same extent; the latter play was more meaningful and important in the case of the chimpanzee. It is in this play that the child presents himself not as a passive watcher of phenomena around him, but as an active naturalist who uses the method of natural experiment to examine properties of things and search for a cause for their appearance. In imitative, constructive, and experimenting play, the human child at the age of 3 years reveals such qualities that will make him the reformer of the world later in his life.

Mental initiative clearly manifests itself in an unhampered tendency of the chimpanzee and the human for entertainment; this tendency is greater in the child who dares to imitate even adult behavior that exceeds the limited strength and abilities.

In spontaneous play of the chimpanzee and the human, it clearly is visible that their attention is scattered; this is especially characteristic of the chimpanzee, whose play, even during a very small period of time, represents a mosaic of shreds of different unassociated and unstipulated actions that can stop at any point and suddenly resume. In the human child, we often observe ordered forms of play, including a series of sequentially evolving and targeted actions (see, for instance, his play with dolls and live animals).

Both Roody and Joni often showed their curiosity, but only in the human child did this curiosity broaden to include his cognitive interests and transform into a craving for knowledge.

Comparing the intellectual features of the chimpanzee and the human (their observation ability; their ability for recognition, identification, generalization, abstract comparison, logical conclusion, imagination, and memory), we must state the following. If we based our conclusions only on direct observations, without taking their verbal expressions into account, we should have limited ourselves to what we cited in the chapters dedicated to Joni's and Roody's similarities; consequently, in this respect, we should have placed both infants almost at the same level. But would it be in accordance with the real situation? Do we have a right not to mention the child's words; the child's speech, which reveals the treasures accumulated in the nooks of his soul? We must dig out this treasure and account for it; only in this case can we look at a child as a whole and complete individual.

In his speech, the human child makes comparisons, statements, practical generalizations, logical conclusions that reflect the genuine work of his mind, his comprehension of words, and his operations with concepts. The child's speech helps us understand the degree to which he has mastered and transformed reality.

The absence of speech in the chimpanzee is the reason why, as a result of his direct observation (outside the framework of experiment), we hardly can tell to what degree these higher intellectual processes are characteristic of him, the processes easily observed in the child's incessant, lovely, naive, but so meaningful, babble. The child's words brightly reflect the hidden psychic processes that occur in the depths of his soul.

The child's words are like rays coming out of real diamonds, which collect the scattered light from everywhere, refract it through their thinnest facets, and direct cascades of blinding glare into our eyes. The strength and intricacy of this play help us judge the quality of the natural stone and the exquisiteness of its polish.

The child's bright, imaginative, and colorful speech reveals how complex, multifaceted, playful, inexpressibly beautiful, and, in its own way, progressive the human soul is. It conquers and enraptures our mind and heart. We do not see such an intricate, fine, and diverse play, particularly a play of mental strength and abilities, in the chimpanzee.

Expanding our comparison, we are inclined to liken the intellectual pattern of the chimpanzee and its dim, unclear, and gray manifestations obviously not even to a fake diamond, with its blinding but scattered glimmer, and not to a natural unpolished diamond, which can give sparkling glitter after a certain treatment, but to a variation of the transparent, radiant diamond, to its closest kin—a dull, gray, uniform graphite.

And, as soon as we take the child's verbal expressions into consideration we immediately have to change these "equals" signs we were about to put between the intellectual abilities of a 4-year-old human child and his chimpanzee counterpart. We change the equals signs into the sign >, which still does not look expressive enough; you want to say, or rather yell, not only more, but better, qualitatively higher, and incomparably, inexpressibly more perfect!

Table 1. Summary of biological and psychological similarities and differences between behavior of chimpanzee and human child

	Features Characteristic Predominantly or Exclusively of Chimpanzee	Similarities in Behavior of Chimpanzee and Human of the Same Age	Behavioral Features That Are Exclusively or Predominantly Human
1. Comparison of Poses and Body Movements of Chimpanzee and Human			
§1 Sitting	Sitting with rest on arms[1]	Artificial sitting poses (sitting on elevated places) and certain poses untypical for chimpanzee and human	Sitting on bent knees and squatting[2]
§2 Standing	Brief vertical standing supported by the outer edge of the foot; widely spread legs[3]; standing on four extremities[4]	Vertical standing	Prolonged vertical standing supported by metatarsus, with legs closed and knees unbent; typical standing on two extremities
§3 Walking	Brief vertical walking by worried chimpanzee (typical walking on four extremities with the body in inclined position)[5]	Vertical walking	Prolonged long-distance vertical walking with straight body position
§4 Climbing	Climbs stairs in horizontal position or on all fours[6]; skillful climbing on trees, fences, and roofs	Climbing stairs and trees	Climbing the stairs in vertical position without support of arms; imperfect and brief climbing on trees
§5 Gripping	Gripping with foot (big toe)[7]; carrying objects in feet and in mouth while walking (foot tenacity); hanging on feet	Hanging on hands[8]; gripping with hands	Carrying in hands while walking; lack of tenacity of big toe
§6 Jumping	Jumping from feet to hands and back	Jumping on feet at one spot	Jumping on one foot
§7 Lying	Lying with legs pulled to head (during sleep); mobility of hip joints	Lying pose on back, side, stomach; tucking hands under head during sleep; pulling legs to body	Putting both hands under head

(continued)

Table 1. (*continued*)

	Features Characteristic Predominantly or Exclusively of Chimpanzee	Similarities in Behavior of Chimpanzee and Human of the Same Age	Behavioral Features That Are Exclusively or Predominantly Human
II. Comparison of External Expressions of Emotions[9] in Chimpanzee and Human			
§8 Emotion of anxiety	Getting fluffed up in anxiety, modulated hooting sound, clenching toes, erect penis; gesticulation with hands; tentative standing in vertical position	Expression of anxiety: extending lips forward in the shape of a tube	Sound of breathing in anxiety; blushing of face
§9 Emotion of sadness	Very loud roaring, expressive gesticulation; darkening of face during crying[10], falling face down, tumbling; extending lips (sound "uu") in sudden sadness	Expression of sadness: crying, opening mouth, roaring, falling on the floor; convulsive hand movements	Crying with tears; blushing of face; pressing hands to face; wiping tears; turning lower lip inside out in sudden sadness; sad sighing
§10 Emotion of joy	Rapid breathing during laughing[11]; broad smile (soundless)[12]; during joyful excitement there is hooting sound ending in loud barking	Expression of joy: narrow and broad smile, making sounds with different objects; haphazard hand movements; abrupt movements of extremities and whole body	Loud laughter; squealing; shouting
§11 Emotion of fear	Getting fluffed up; dull sound "u"; paling of face; falling on the floor, raising hands or pressing them to eyes	Expression of fear: dilation of eyes, frozen look and pose, shaking, excessive heartbeat, sweating of face, hiding from fearful stimulus, fleeing, roaring, crying	Blushing of face in fear, then paling, pressing hands to chest; sounds "okh," "akh," "oy"
§12 Emotion of anger	Fluffing up; jumping threatening poses; barking sound "a"; clenching toes in fists; turning lower lip inside out; knocking with knuckles; rapid breathing in angry excitement; sounds "ukh," "khru" (grunting when upset)	Expression of anger: wrinkling of the upper part of face, baring teeth and gums, aggressive hand gestures, biting, foot stamping, waving with an object or with fist,[13] pinching, scratching, pounding, self-biting; scattering of things; tendency for scaring	Stamping with both feet, pounding with fist; use of finished or self-made tool or weapon; shouting (grumbling when upset)

§13 Emotion of tenderness	Touching with open mouth or with extended tongue; shaking with entire body; kiss as artificial way of expressing affection: help out of compassion not to the point of sacrifice	Expression of tenderness: holding hand, touching with hands, touching with wide open mouth,[14] rapid breathing, pinching with lips, sucking with mouth, hugging, clinging with body, kissing; expression of jealousy and compassion to the loved ones	Kissing as a rule, verbal expressions of affection, crying out of compassion for the loved ones, disinterested help out of compassion; expression of affection toward inanimate toys, dummies, and pictures of animals
§14 Emotion of disgust	Silent expression of gustatory repulsion	Expression of disgust: pressing in corners of mouth, rectangular form of mouth, wrinkling of the upper part of face, jerking the nose up	Excessive salivation during gustatory repulsion; weak coughing or grunting sound
§15 Emotion of surprise	Fluffing up during surprise; no sound	Expression of surprise: opening mouth wide, fixed staring	Uttering sound "a" and taking deep breath; sometimes spreading hands in a characteristic gesture
§16 Emotion of curiosity and attention	Smelling of intriguing objects, touching with hands and at the same time making pinching movements with lips	Expression of curiosity and attention: extending closed lips forward, sound "m," touching with index finger; touching intriguing objects with hands and mouth; similar optical illusions while looking at stereometric images and distance evaluations	
§17 Specific facial expressions	Turning extended tongue during poorly coordinated movements	Closing mouth tightly during alert movements of fingers	Extending tongue forward during poorly coordinated movements

(continued)

Table 1. (*continued*)

	Features Characteristic Predominantly or Exclusively of Chimpanzee	Similarities in Behavior of Chimpanzee and Human of the Same Age	Behavioral Features That Are Exclusively or Predominantly Human
III. Comparison of the Stimuli[15] Triggering the Main Emotions			
§18 Stimuli triggering main emotions	Fear caused by olfactory stimuli; mistreatment of helpless creatures; no crying during physical suffering; overcoming of fear without outside help	Similarity of most of the stimuli triggering main emotions: anxiety, sadness, joy, fear, anger (revenge), tenderness (compassion), surprise, curiosity, repulsion	Sadness: crying from physical pain or out of compassion; crying[16] and laughing[17] for "principle" reasons; sense of humor; affection toward helpless creatures; helps the weak; overcoming fear with outside help
IV. Comparison of Instinctive Actions[18]			
§19 Self-support—food	Loud grunting in response to tasty food; thorough preliminary smelling and tasting of food; slow eating; unwillingness to share food even after satiation; eating of insects; repulsion toward butter and meat; scattering food around; discarding unsuitable pieces of food from mouth	Similarity in eating habits (providing food and water, ways of obtaining, treating and using them, greediness in eating, using utensils, ignoring spoons and forks during eating); similarities in specific gustatory preferences (eating chalk, lime, coal, lemon, nasal mucus); apprehensiveness toward new food; eating with greater appetite during entertainment; grumbling with a smacking sound while eating tasty food	Grumbling and mooing sound while eating tasty food; swallowing fruit seeds, vials; tendency to learn how to use utensils during meals; sharing food with others after satiation; repulsion toward food touched by insects; willingness to eat butter and meat; hastiness in eating

§20 Self-supporting instinct—sleep	Prepares his "nest" and "pillows" without assistance; protests if we tried to cover his arms before he goes to sleep	Similarities in the state of sleep; using pillows, blanket, putting hands under head, holding to something if sleeping in an unenclosed space; tendency to have a company of human during sleep; shaking during sleep; snoring	Tendency to cover self fully; talking, crying, gesticulating in sleep; lack of tendency for arranging sheets and pillows
§21 Self-supporting instinct—self-maintenance	Stoical endurance of pain; readiness for medical treatment; keeps clean without assistance	Self-examination; self-cleaning (washing, biting fingernails, skin cleaning, wiping of nose); self-treatment (taking out splinters, examining of sore spots on skin)	Resisting treatment; sensitivity to pain; tendency for hasty, sloppy cleaning procedure without assistance; untidiness
§22 Ownership instinct	Indiscriminate accumulation of property and its inadequate utilization; aggressive claiming; poor guarding; unwillingness to share unclaimed property	Accumulation of property; hiding, guarding, taking away someone else's property	Effective utilization of property, collecting with a sense of purpose; greedy accumulation; utilization in constructive play, indecisive claiming; willingness to share unclaimed property
§23 Primitive aesthetic sympathies	Acquiring of blue objects; undeveloped tendency for self-embellishment	Similarities of tastes in choosing attractive objects Preferred qualities: a) In color perception (bright shiny colors, predominantly from first half of spectrum) b) In size (miniature things) c) In shapes (round, spherical) d) Sense of touch (soft, smooth, netlike, elastic) e) In temperature (warm) f) In olfaction (odor of fragrant fruit) g) In gustatory perception (sweet, sour) Tendency for self-embellishment; fear of black color; predominant use of blue plates	Preference for red objects; strong tendency for self-embellishment; certain forms of principle preferences: choosing pictures, books, their contents; principle qualitative criteria for drawings (his own and somebody else's); imposing aesthetic judgments on creative activity (his own and somebody else's)

(continued)

Table 1. (*continued*)

	Features Characteristic Predominantly or Exclusively of Chimpanzee	Similarities in Behavior of Chimpanzee and Human of the Same Age	Behavioral Features That Are Exclusively or Predominantly Human
§24 Freedom instinct	Tendency for excursions to heights	Tendency for freedom and broadening space for movement; protesting against being dressed	Tendency to travel far away
§25 Social (communication) instinct	Despotic tendencies in treatment of lower and weaker creatures (human children, small animals, insects)	Objecting to being left alone; intention to communicate with people and animals; dominant forms of social communication; communication with a person taking care of them and with their peers	Peaceful intentions in communication with lower creatures; orderly forms of communication with inanimate "friends"
§26 Instinct of imitation	Greater expressiveness of emotionally involved chimpanzee compared with human	Picking up moods and emotions of people around: imitation of certain sounds, body movements, gestures associated with emotions	Imitation of facial expressions and apparel of children and adults; imitation of images of movements and facial expressions
§26a Imitation of actions	Imitation of individual human actions, predominantly destructive (see §39 for more details)	Imitation of human actions associated with using dishes, utensils, and tools, as well as with self-maintenance procedures (see §39 and §40 for more detail)	Reproducing of logically connected actions that he has seen or read about (see the section on constructive play and §40 for more detail)
§26b Imitation of sounds	Immediate imitation of chimpanzee sounds intentionally reproduced by human; most accurate imitation of dog barking	Similar imitative sounds: foot stamping, rhythmic knocking with knuckles, lip rattling, imitation of clapping and dog barking	Imitation of animal sounds and specifically human sounds (wheezing, snoring, yelling, talking, laughing, singing); imitation of sounds produced by inanimate objects (rattling, squeaking, ticking of clock, etc.); reproducing of syllables, words, phrases, poems, prose

§27 Natural sounds of chimpanzee and human	Specific natural sounds uttered by chimpanzee: 1) Modulated hooting 2) Squealing sound (tiredness) 3) Five vocal stages preceding roaring 4) Hoarse rapid breathing 5) Barking angry sound "a" 6) Sound of dissatisfaction—grunting 7) Sounds "khru" and "ukh" of angry excitement	Similar natural sounds of chimpanzee and human child: "e, u–a–u" in child at age of 0.0.1; "m" (0.0.5); "kh–r–u" (0.1.27); "u–khu" (0.2.3); "o" (0.3.5); "yu" (0.6.30); roaring in sadness; grunting, sneezing, coughing, snoring, rapid breathing, deep yawning	Ability to master intelligible speech; speech evolution

V. Comparison of Play[19] of Chimpanzee and Human

§28 Mobile play	Entertainment with moving at a height: roofs and trees	Mobile play (running, riding, pulling, jumping)	Entertainment with imaginary movement
§29 Gymnastic play	Hanging on feet, jumping from heights, and other tricks more elaborate and agile than those of human	Gymnastic play (climbing, swinging, spinning, squeezing through, hanging)	Less diversity and agility in gymnastic tricks
§30 Sports play	Anger in response to failure	Sports play (contests in running, catching, taking away, wrestling; a preference for fleeing strong rival and chasing weak one	Crying in case of poor result
§31 Play with obstacles	Carrying things in mouth and feet; self-training to endure pain, steal objects, squeeze through obstacles	Play with obstacles infants themselves have erected; hindered running, riding, swinging, climbing	Carrying objects in hands; training legs and jumping; walking over uneven ground; running; training boldness and agility; psychic ingenuity
§32 Play with moving objects		Play with easily moved objects (balls, balloons, baby carriages); throwing; knocking down; rolling; spinning; catching	

Table 1. (*continued*)

	Features Characteristic Predominantly or Exclusively of Chimpanzee	Similarities in Behavior of Chimpanzee and Human of the Same Age	Behavioral Features That Are Exclusively or Predominantly Human
§33 Watching moving objects		Entertainment by watching movement of inanimate and animate objects (animals, people, cars)	Preference for listening to dynamic stories
§34 Hide-and-seek game	More elaborate hiding	Preference for hiding as opposed to seeking	Imaginary hiding
§35 Play with live animals	Despotic, torturing play with live creatures	Play with live animals	Orderly play with animate and inanimate objects; humanizing of animals; animating toys; sympathizing with toys
§36 Entertainment with sounds	Clapping with his own eyelid; strumming on a taut rubber string; cracking with teeth; clanging with jaws; earsplitting noises	Entertainment with objects that make sounds and reproducing various sounds (foot stamping, hand clapping, lip rattling, knocking, making noises, chains, rattles)	Entertainment with his own voice: unintelligible and intelligible murmuring, yelling, reproducing rhymes, singing, whistling, clicking with tongue
§37 Experimenting play	Bold playing with sharp objects; putting props between lips	Similar forms of experiments with water, sand, fire, and solid, elastic, transparent, sharp objects; strong tendency for playing with flexible objects (hair, rubber tube, ball)	Cautious playing with sharp object; putting props in palms of hands; verbal statements; intention to understand phenomena and what causes them; real experiments and verbal generalization
§38 Destructive play	Entertainment with destructive activity without the tendency to restoration (taking and ripping apart, taking out, throwing away, untying, opening)	Destructive play: throwing, ripping apart, breaking[20]	Connection between destructive and constructive activities; destruction with tools and weapons; military play

§39 Imitative entertainments	Ineffectiveness of imitative constructive activities	Imitation of individual actions (sweeping, wiping, opening, waving, taking out, digging, drawing, hammering) associated with the following tools: broom, rag, stick, key, pencil, hammer, tongs; tendency to substitute semblance of tools for real ones	Effective imitation of a sequence of logically connected actions, professions; artistic activity (drawing); play with sand and solid materials
§40 Constructive play	Making semblance of rattle and string instrument	Primitive edifices; "instrumental thinking"	Complex constructive actions; reproducing cars, boats, ships, trains, airplanes, buildings
VI. Volitional Qualities of Chimpanzee and Human			
§41 Manifestations of will	High energy and endurance of pain in achievement of desired goal; anger in response to failure	Persistence; tendency for counteractions (disobedience); stubbornness; whims; sensitivity to offense[21]; sense of guilt	Great patience in achieving goals; crying in case of failure
VII. Intelligence of Chimpanzee and Human			
§42 Intelligence			
§43 Intelligence		"Einsicht"—premeditated actions, deception, naive slyness; mental activity; indomitable tendency for entertainment and new impressions (scattered attention, inconsistency of interests)	
§44 Intelligence		Curiosity toward similar stimuli: new, bright, shiny, moving, convex, concave	Curiosity; "naive teleologism"; "forecausative thinking"

(continued)

Table 1. (continued)

	Features Characteristic Predominantly or Exclusively of Chimpanzee	Similarities in Behavior of Chimpanzee and Human of the Same Age	Behavioral Features That Are Exclusively or Predominantly Human
§45 Thinking	Exceptionally aggressive emotional qualification of images	Ability for discerning, recognizing, and subjective emotional qualification	Ability for comparison, generalization, logical conclusion, humor; prevailing sympathy in emotional qualification of images
§46 Thinking	Negative attitude toward his own image in mirror; no identification of self with this image	Fine observation ability (to discern tiny objects), interest toward self-examination and contemplation; seven similar stages of reaction to mirror	Finding principle inconsistencies; comparison of object with its image; identification of self with the mirror image; favorable attitude toward image of self
§47 Thinking		Ability for identification, assimilation, analogizing, and elementary abstraction[22]	Strongly developed tendency for analogizing, assimilation ("syncretical" thinking)
§48 Intelligence	Possibility of imagination activity in play of fighting with pseudoenemies and intentionally placed obstacles		Imagination (during, drawing, imaginary actions, and animation of objects); love for fantasies (autistic thinking)

VIII. Skills—Conditioned Reflexes of Human[23] and Chimpanzee

	Features Characteristic Predominantly or Exclusively of Chimpanzee	Similarities in Behavior of Chimpanzee and Human of the Same Age	Behavioral Features That Are Exclusively or Predominantly Human
§49 Memory (conditioned reflexes)	Dodging from using dishes, utensils, clothes	Similar motor skills associated with day-to-day procedures (using dishes, cups, spoons, napkins, handkerchiefs, blankets, etc.)	Tendency for independent use of day-to-day things, putting on clothes, washing, and cleaning self
§50 Memory (conditioned reflexes)	Gesture language is associated only with physiological needs and emotions	Similarities of gesture languages in expressing request, desire, rejection, denial	Gesture language is associated with thinking, combined with sounds and speech; speech conditioned reflexes (from 1½ years of age)

§51 Memory (conditioned reflexes)	Similar types of conditioned reflex association 1) Visual-pain-motor 2) Visual-gustatory-motor 3) Auditory-motor 4) Auditory-visual-motor 5) Visual-motor 6) Visual-emotional-sound 7) Auditory-emotional-sound	In Roody, enormous development of all types of conditioned reflexes, particularly auditory-intellectual-sound and visual-intellectual-sound; speech reflexes include new associations acquired without assistance

[1] Observed only in a 5-month-old child learning how to sit.

[2] The latter is observed in the chimpanzee sometimes during fear.

[3] Resting on the outer edge of the foot (6-month-old child standing with support); standing on widely spread legs (child at the age of 1 to 1½ years).

[4] Resting on arms (6-month-old child learning to stand with support).

[5] Movement with resting on arms is observed in a human child crawling (8 months old) and learning to walk (9 to 11 months old). (In the chimpanzee, walking on three extremities is observed even when he is led by the hand; under similar conditions, the human walks on two extremities.)

[6] Climbing stairs on all fours is observed in the human child only between 1½ and 2 years of age.

[7] A special deviation of Roody's: gripping with the small toe of his left foot.

[8] Hanging is more protracted in the chimpanzee than in the human.

[9] All kinds of emotions in the chimpanzee assume more expressive external forms than in the human.

[10] Crying without tears is observed in the human child until 2½ to 3 months of age.

[11] Rapid breathing during laughter in the child until 4 months of age.

[12] Silent broad smile in the human child until 4 months of age.

[13] Clenching his hand, the human child puts his thumb outside the fist; the chimpanzee puts his thumb inside the fist.

14Touching with open mouth as a sign of affection; rapid breathing (child until 2½ years old).

15A greater range of anxiety, fear, and anger emotions in the chimpanzee; in the human, a greater range of sadness, joy, tenderness, curiosity, and surprise emotions.

16Crying because of failure in creative activities.

17Laughing in response to stimuli different from what he has expected.

18All kinds of instinctive activity, with the exception of imitation, are more developed in the chimpanzee than in man.

19In the case of the chimpanzee, the following types of play are more developed: mobile, gymnastic play, play with obstacles, play with live animals, destructive play; in the case of the human, imitative play and constructive play are prevalent.

20Entertainment with destruction (human child until 1½ years of age; later, constructive plays dominate).

21Similarities in taking offense: turning back to offender, turning face away, ignoring his requests and actions.

22All the intellectual features are immeasurably more developed in the human child than in the chimpanzee.

23All the types of day-to-day instincts develop faster in the human child than in the chimpanzee.

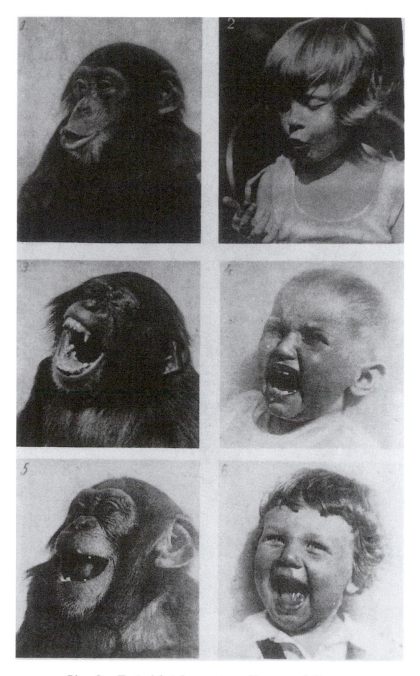

Plate 81. *Typical facial expressions of human and chimpan-zee: (1) Joni excited; (2) Roody (3 years 4 months) ex-cited; (3) Joni crying; (4) the human child crying (with tears); (5) Joni laughing; (6) the human child laughing.*

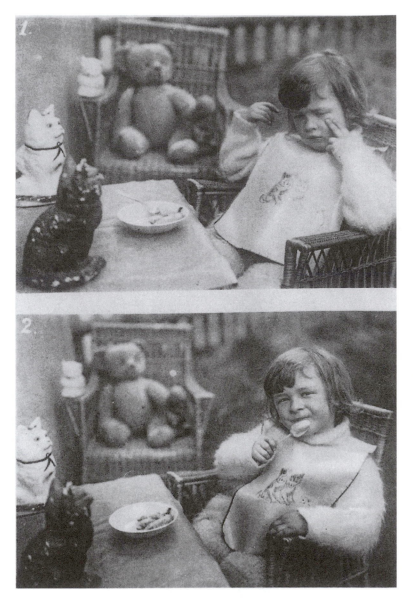

Plate 82. Child's reaction to palatable and nonpalatable food: (1) displeasure at porridge without sugar (Roody 4 years 1 month); (2) pleasure at porridge with sugar (Roody 4 years 1 month).

Plate 83. Crying of human and ape: (1) the initial or first stage of Roody's (7 years) crying; (2) the subsequent or second stage of Roody's crying; (3) restrained crying from physical pain; (4) wiping eyes; (5) Joni crying; (6) Roody crying (third stage).

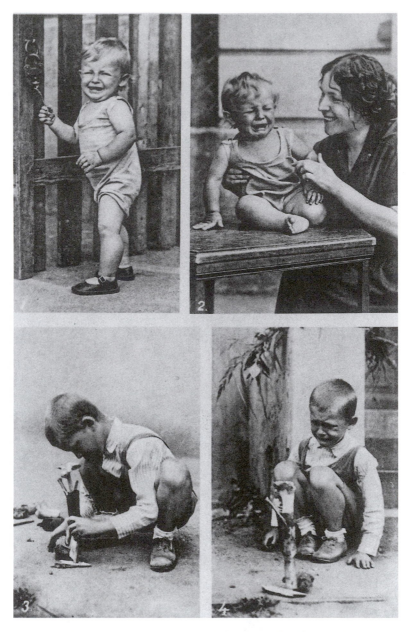

Plate 84. Weeping (caused by different stimuli) of the child: (1) crying (fourth stage) because of moral hurt (Roody 1 year 2 months); (2) crying (fifth state) because of physical pain (Roody 1 year 2 months); (3) finding amusement in structural activity—Roody's (6 years) attempts at making "cave man"; (4) Roody's (6 years) restrained crying because of psychic hurt (cave-man has gone all to pieces).

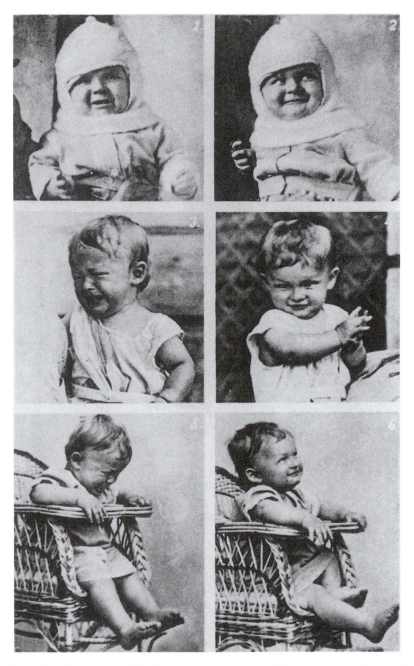

Plate 85. The human child's laughter and crying at different ages: (1) Roody (6 months) crying (third stage); (2) Roody (6 months) smiling (first stage); (3) Roody (1 year 2 months) crying (fourth stage); (4) Roody (1 year 2 months) smiling (first stage); (5) the posture of Roody (1 year 2 months) crying (fifth stage); (6) The posture of Roody (1 year 2 months) smiling (first stage).

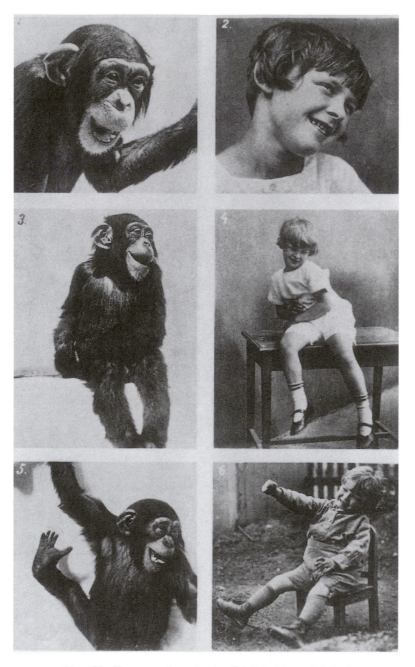

Plate 86. Human and ape in cheerful dispositions: (1) Joni frolicsome; (2) Roody frolicsome; (3) Joni's reaction to tickling; (4) Roody's reaction to tickling; (5) Joni in a playful mood; (6) Roody (3 years 1 month) in a playful mood.

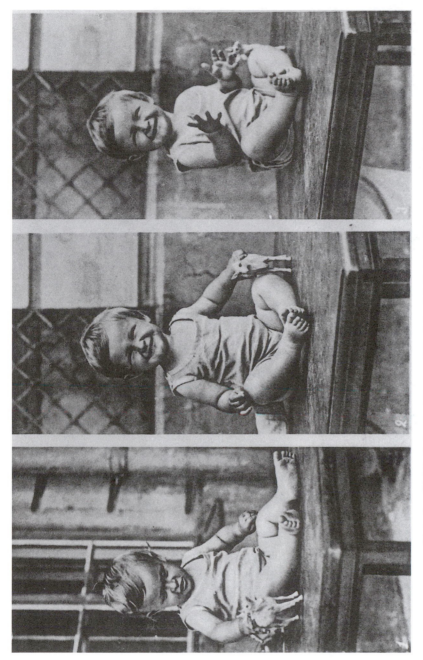

Plate 87. Child's first attempts at construction: (1) first attempt to put object in right position (Roody 1 year 2 months); (2) ready; (3) the joy of achievement.

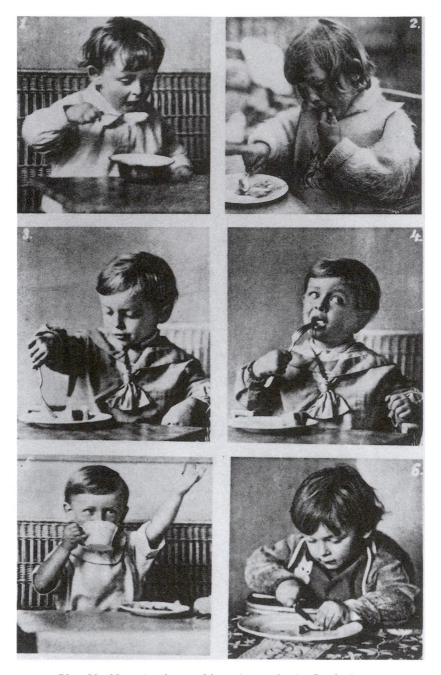

Plate 88. Mastering the use of domestic utensils: (1) Roody (2 years 1 month) using a spoon; (2) Roody (4 years 1 month) taking food with a spoon, tasting bread with finger; (3) and (4) Roody (2 years 7 months) using a fork; (5) Roody (2 years 3 months) drinking from a cup; (6) Roody (2 years 11 months) using a knife.

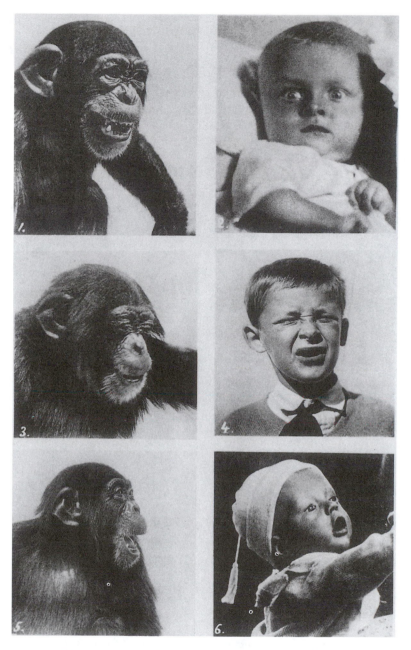

Plate 89. Typical facial expressions of human and chimpanzee: (1) Joni afraid; (2) Roody (4 months) afraid; (3) Joni disgusted; (4) Roody (7 years) disgusted; (5) Joni astonished; (6) the human child astonished.

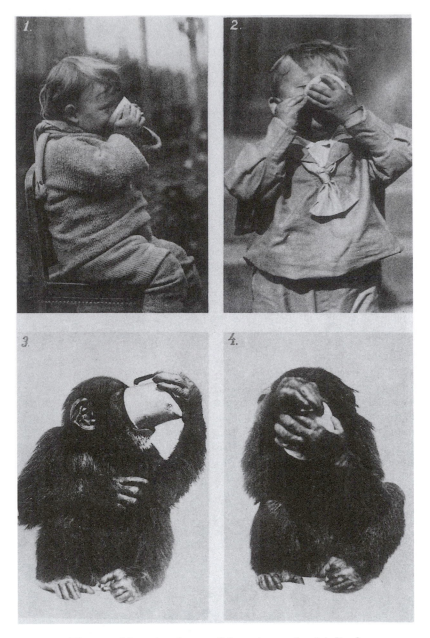

Plate 90. Mastering the use of domestic utensils: (1) Roody (2 years 2 months) uses a cup; (2) Roody (2 years 5 months) drinking from a cup; (3) Joni's use of a cup; (4) Joni drinking from a cup.

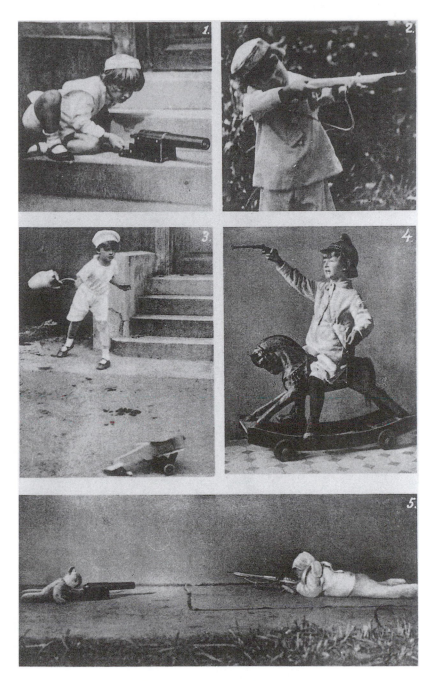

Plate 91. Military games of the child: (1) shooting a gun;
(2) shooting a rifle; (3) "gas attack"; (4) shooting a re-
volver, attack of enemy; (5) two adversaries.

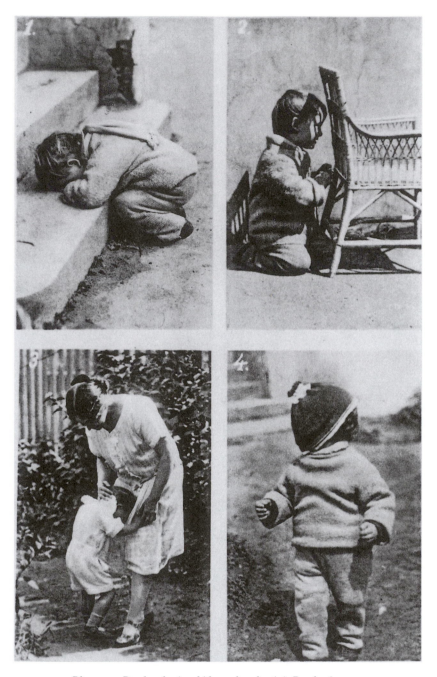

Plate 92. Roody playing hide-and-seek: (1) Roody (2 years 2 months) has hidden; (2) Roody (2 years 5 months) has hidden behind his chair; (3) Roody (2 years 4 months) hiding in the knees of his mother; (4) Roody (2 years 2 months) "invisible" (has covered head with cap).

Plate 93. Child's organized play with teddy bear: (1) swinging with teddy bear and pushing off teddy bear and catching bear on swing (Roody 4 years); (2) Roody has caught teddy bear and they now swing together; (3) collision with bear in an improvised accident; (4) climbing ladder with bear; (5) teddy bear driven in sled; (6) teddy bear tobogganing.

Plate 94. Roody's (3 years) intercourse with other children: (1) weighing; (2) playing horses; (3) joint watering of flowers; (4) demonstration of museum; (5) driving; (6) cycling.

Plate 95. Child's mobile play—swinging: (1) swinging on the wooden horse (Roody 3 years 4 months); (2) on the rocking chair (Roody 3 years); (3) on wooden horse (Roody 4 years 4 months); (4) purposeful falling down (Roody 4 years 4 months); (5) little girl friend in tow (Roody 4 years 7 months).

Plate 96. Roody's mobile play: (1) "Car does not go" and Roody (2 years 2 months) feels depressed; (2) attired as a chauffeur—Roody (3 years 11 months) in big chauffeur's gloves; (3) tinkering with the steering gear (Roody 2 years 2 months); (4) "I'll take the car!" (Roody 3 years 2 months); (5) "I'll repair the wheel!" (Roody 3 years 6 months); (6) "I'll repair the wheel!" (Roody 3 years).

Plate 97. Roody's organized play with teddy bear: (1) Roody (4½ years) setting the bait; (2) Roody preparing sling knot; (3) Roody awaiting prey; (4) Roody catching bear; (5) Roody carrying bear away; (6) Roody tying bear to tree.

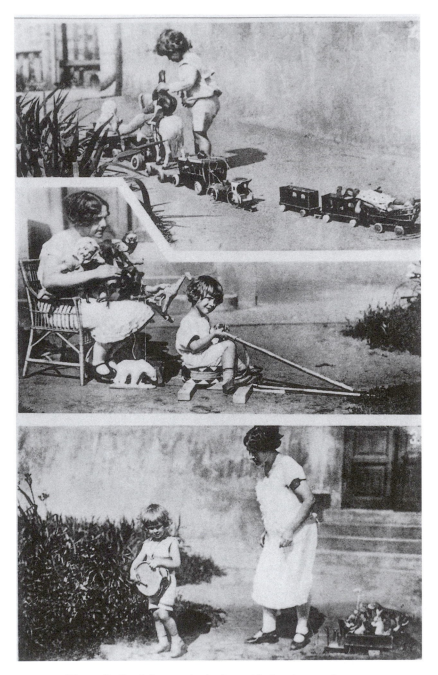

Plate 98. Roody's organized play with inanimate playmates: (1) Roody (3 years 3 months) "travels"; (2) Roody (3 years 4 months) makes an airplane flight; (3) Roody (3 years 4 months) goes to demonstration.

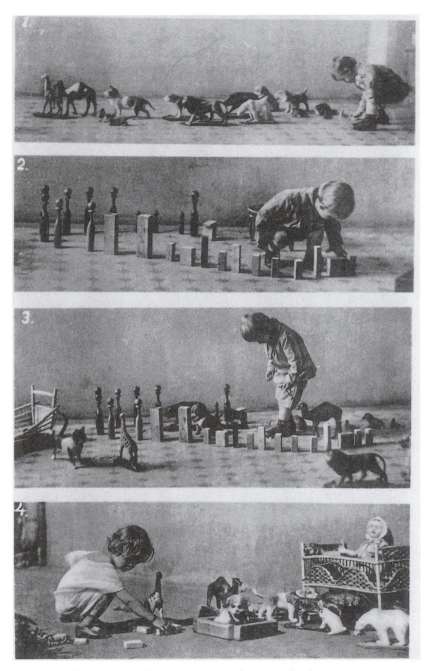

Plate 99. Roody's construction play: (1) distributing toy animals by groups (Roody 2 years 9 months); (2) making the fence of an improvised zoo; (3) bringing animals into the zoo; (4) making enclosures for animals (Roody 3 years 7 months).

Plate 100. Child using a stick: (1) touching intriguing object with stick (Roody 2 years 10 months); (2) drawing lines with stick (Roody 2 years 2 months); (3) waving stick (Roody 2 years 1 month); (4) beating off leaves with stick (Roody 1 year 1 month) (characteristic protrusion of tongue); (5) riding a stick (Roody 4 years 5 months).

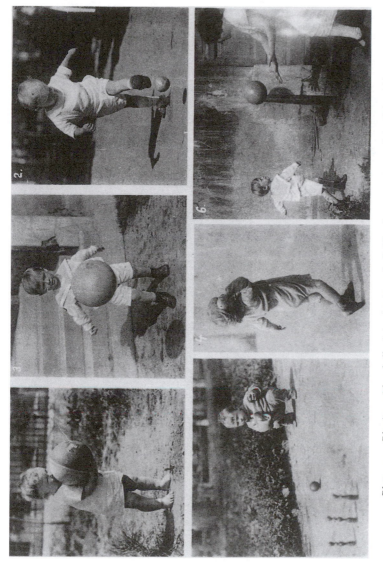

Plate 101. Playing with a ball: (1) carrying a ball (Roody 1 year 2 months); (2) pushing a ball with foot (Roody 2 years 3 months); (3) catching a ball (Roody 2 years 4 months); (4) throwing a ball with hand (Roody 2 years 1 month); (5) throwing ball at ninepins (Roody 2 years 1 month); (6) playing with ball (Roody 2 years 4 months).

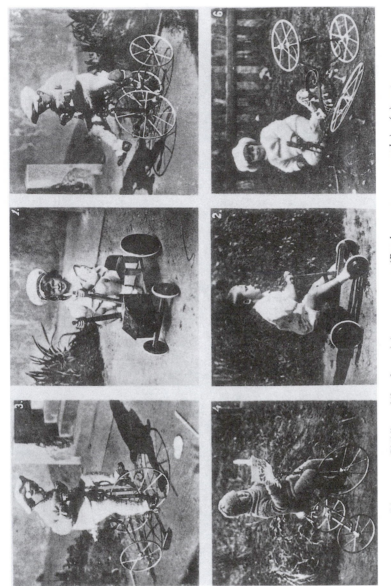

Plate 102. Child's mobile play: (1) auto-go-cart (Roody 3 years 3 months); (2) auto-go-cart (Roody 4 years 3 months); (3) cycling with improvised obstacles (Roody 4 years 5 months); (4) demonstrating boldness in cycling (without holding steering gear) (Roody 4 years 7 months); (5) cycling without feet on pedals; (6) purposeful falling down from bicycle.

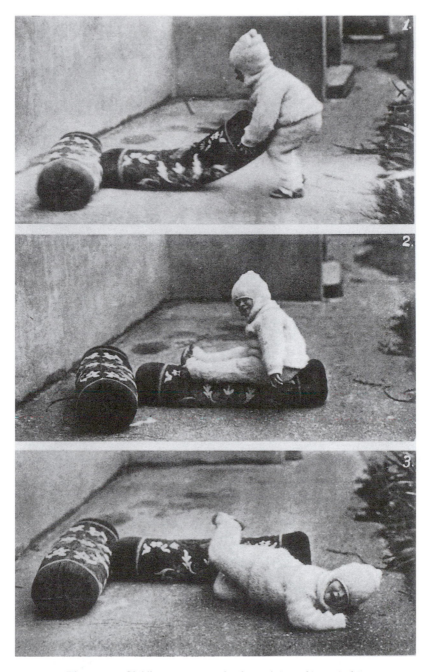

Plate 103. Child's constructional play: (1) making airplane (Roody 3 years 6 months); (2) flying in airplane (Roody 3 years 6 months); (3) deliberate wreck of airplane (Roody 3 years 6 months).

Plate 104. Child's constructional play: (1) "making boat" (Roody 2 years 3 months); (2) "boating" (Roody 2 years 3 months); (3) "boating," holds "oars" (Roody 2 years 3 months 25 days); (4) uses oars (Roody 2 years 3 months 21 days); (5) "travels by steamer" (Roody 2 years 4 months).

Plate 105. Roody playing with easily movable objects: (1) turning a pianoforte stool (Roody 2 years 2 months); (2) turning the wheel of a sewing machine (Roody 2 years 9 months); (3) driving a hoop (Roody 3 years 3 months); (4) driving a small wheel (Roody 4 years 1 month); (5) rolling eggs (Roody 3 years); (6) spinning a humming top (Roody 4 years 4 months).

Plate 106. Winter sports—genuine and imitated: (1) pseudoskis (Roody 3 years 11 months); (2) pseudoskates (Roody 3 years 11 months); (3) skiing on real large skis (Roody 2 years 11 months); (4) skiing on children's skis (Roody 4 years).

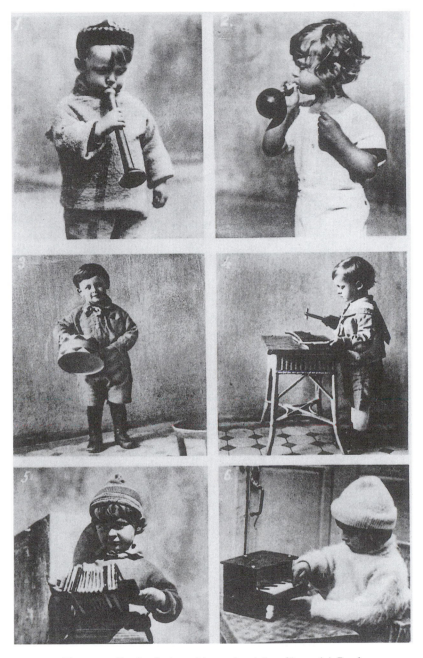

Plate 107. Roody playing with sound-emitting objects: (1) Roody (2 years 2 months) blowing the trumpet; (2) Roody (2 years 3 months) inflating a squeaking "devil"; (3) Roody (2 years 9 months) beating the drum; (4) Roody (2 years 6 months) playing the cymbals; (5) Roody (3 years) playing an accordion; (6) Roody (4 years 1 month) playing a toy-piano.

Plate 108. Roody's expression of astonishment: (1) Roody (1 year 2 months) opens mouth while contemplating intriguing object; (2) tightly compressed lips and eager movement of hands (Roody 1 year 2 months); (3) drawing arms aside while astonished (Roody 1 year 1 month); (4) feeling intriguing object with mouth (Roody 1 year 1 month).

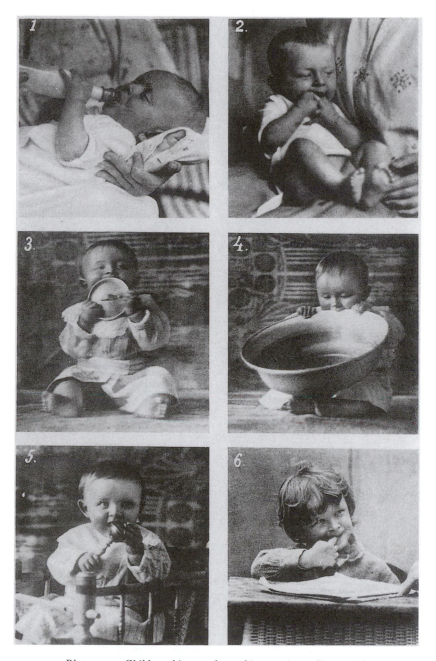

Plate 109. Child sucking and touching various objects with mouth: (1) Roody (4 months) in the act of sucking; (2) Roody (5 months) sucking his hand; (3) Roody (9 months) touching a metallic object with mouth; (4) Roody (9 months) touching a metallic rattle with mouth; (6) Roody (3 years) sucking his thumb while listening to someone read.

Plate 110. Roody's experimenting play: (1) splashing water (Roody 1 year 1 month); (2) spreading water about (Roody 1 year 4 months); (3) pouring water (Roody 2 years 1 month); (4) playing with water—launching a steamer (Roody 2 years 4 months); (5) playing with water—washing linen (Roody 2½ years); (6) making a water fountain (Roody 2 years 6 months).

Plate 111. Child's games of imitation: (1) first attempts at using watering can are unsuccessful (Roody 1 year 5 months); (2) Roody's (1 year 10 months) first, but vain, attempts at making use of spade; (3) first attempts at using scythe (Roody 4 years); (4) effective use of watering can (Roody 2 years 1 month); (5) effective use of spade (Roody 1 year 10 months).

Plate 112. Roody's and Joni's experimenting play with transparent objects: (1) Roody (2 years 4 months) looking through transparent objects; (2) looking into stereoscope (Roody 4 years); (3) examining objects through magnifying glass; (4) looking into self-made "field-glass."

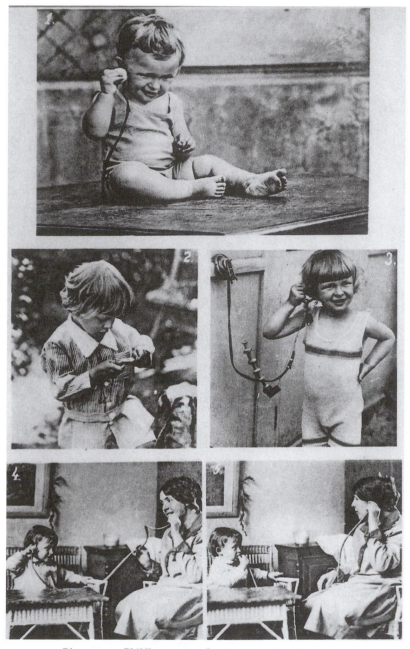

Plate 113. Child's games of imitation: (1) listening to watch (Roody 1 year 2 months); (2) pretending to find the time of day (Roody 3 years 4 months); (3) imitating conversation by self-made telephone (Roody 3 years 4 months); (4) mimicking a talk over the telephone; (5) imitating a telephone conversation (Roody 2½ years).

Plate 114. Child's directional throwing: (1) ready to hit at target (Roody 2 years 2 months); (2) preparing to throw stick; (3) "off the stick goes."

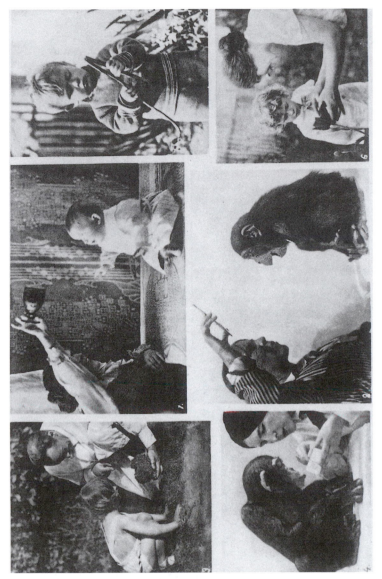

Plate 115. Expression of astonishment and attention of human and ape: (1) astonishment — opening of mouth by child on seeing novel object (Roody 9 months); (2) astonishment—Joni opens mouth while looking into mirror; (3) typical posture of fixed attention—Roody (4 years) examining hedgehog; (4) typical posture of fixed attention—Joni looks at box; (5) Roody (1 year 5 months) shows attention; (6) Roody (3 years 4 months) shows attention.

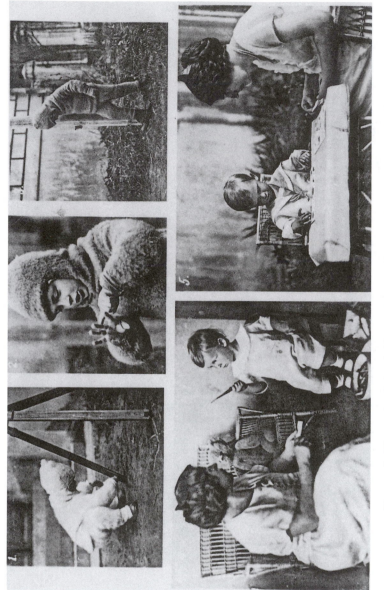

Plate 116. The work of the child's mind: (1) will it stand or tumble down? (Roody 2 years 4 months); (2) "one, two, three!" finger counting (Roody 4 years); (3) "Now I see that the earth turns round!" Roody (4 years 7 months) related his impressions after turning around a post; (4) "What is this? (Roody 2 years 3 months); (5) identification of representation with object (Roody playing lotto at 2 years 4 months).

Plate 117. Child's games of imitation: (1) imitating house painters (Roody 2 years 3 months); (2) imitation of greasers with the help of brush (Roody 3 years 11 months); (3) "smoking" (Roody 1 year 5 months); (4) pretending to smoke by "lighting cigarette" (Roody 2 years 3 months).

Plate 118. The child imitating a photographer: (1) ready to take a snapshot (Roody 2 years 2 months); (2) "photograph ready" (Roody 2 years 2 months); (3) ready to take a snapshot (Roody 3 years 6 months); (4) photographing—focusing objective (Roody 4 years 1 month); (5) photographing—snapshotting (Roody 2 years 1 month); (6) photographing—setting the camera (Roody 4 years).

Plate 119. *The human child's use of tools:* (1) "*sawing*" (*Roody 2 years 1 month*); (2) *cutting with scissors* (*Roody 4 years*); (3) *sawing* (*Roody 4 years*); (4) *lifting hay with fork* (*Roody 4 years*); (5) *raking* (*Roody 3 years*); (6) *turning hay* (*Roody 4 years*).

Plate 120. Child's structural activity: (1)–(3) making a "worm," three successive stages (Roody 2 years 1 month).

CONCLUSION

Now, based on the materials studied, we can turn our attention in another direction. What is the modern chimpanzee like? We can state with all certainty that, not only isn't he "almost human," as he usually is called, but also he is "by no means human." The following arguments prove this.

Similarities in the infant chimpanzee and human child can be found in many aspects, but only in the case of superficial observations of both infants during their instinctive and emotional behavior. These similarities are especially evident when their behavior is compared in relatively neutral spheres and circumstances: in certain kinds of play (mobile, destructive, and experimenting play), in external expression of their main emotions, in volitional actions, in certain conditioned reflex skills, in elementary intellectual processes (curiosity, observation ability, recognition, identification), in neutral sounds (grunting, snoring, rapid breathing, moaning, crying). As soon as we deepen our analysis and try to equate similar forms of behavior of both infants, we become convinced of our inability to do that, and that we have to use an inequality sign ($<$ or $>$) instead. As a result, we observe the divergence of these creatures. It turns out that the more vital the biological feature we take for comparison, the more frequently the chimpanzee has the advantage over man. The higher and more refined the mental qualities in the center of our analytical attention, the more frequently the chimpanzee is surpassed by man.

Eventually, we find in humans the specific features that cannot be found in the chimpanzee and that are not included in the sphere of our comparison. In the anatomical and physiological sphere, these are his vertical walking and carrying objects in his hands; in the sphere of instincts, these are his imitation of the human voice (laughing, singing, reproducing individual words, intelligible speech);

in the emotional sphere, these are his moral, altruistic feelings and his sense of humor; in the sphere of egocentric instincts, these are his easy parting with his belongings; in the sphere of social instincts, these are his peaceful, orderly communication with lower creatures; in the sphere of play, these are his creative and constructive games; in the sphere of intelligence, these are his imagination, his intelligible and logical speech, and his ability to count; in the sphere of habits, these are his ability to improve his vital day-to-day skills and to develop conditioned reflexes, such as hearing-intelligence-sound and vision-intelligence-sound reflexes.

On the other hand, it is worth noting that the chimpanzee does not have a single mental feature that would not have been characteristic of the human at certain stages of his development.

Even the most specific characteristics of the chimpanzee, such as his walking on all fours (the motor sphere); his smelling of new objects and food and his touching with lips and tongue (the sphere of senses); his threatening poses and biting (the sphere of instinctive actions); his anxiety that accompanies affects and peculiar facial expressions associated with general excitability (expression of emotions); his play with sharp and prickly objects and with obstacles he has created (the sphere of play); his language of gestures (the sphere of conditioned reflex actions); and his tendency for counteraction (the sphere of volitional actions) are typical of the human child.

The only feature that is exceptionally the chimpanzee's is his ability to utter a hooting-grunting modulated sound followed by barking (accompanying the emotion of general excitability), as well as a barking angry sound and a hoarse sound of resentment. However, I am sure that, with the exception of the modulated sound, the child could reproduce these sounds easily and accurately if he were asked to do so (maybe after certain practicing).

If we project certain bodily and mental features of the chimpanzee on ontogenetic age periods of the human, we will come to the following conclusions.

> The face of a chimpanzee 3 to 4 years old, furrowed with deep wrinkles, reminds us of the face of a decrepit old man approximately 70 years of age.

> With regard to his senses (olfaction and vision according to our observations[1]) and their fineness and sharpness, the chimpanzee exceeds the adult human in his prime (24–35 years).

> With regard to the strength of his hands and teeth and by motor characteristics, the infant chimpanzee exceeds a physically developed teenager (16–18 years old).

> With regard to the development of his vitally important instincts (self-support, self-preservation, ownership, social, etc.), the chimpanzee may be compared to a child who has just turned 7 years old.

> With regard to the strength of his emotions, the infant chimpanzee can be likened to a mentally ill man who has caricature-like facial expressions, exaggerated theatrical body movements, and reckless affects.

> With regard to his destructive and mobile play, the infant chimpanzee can be equated with the human child of the same age (i.e., from 1½ to 4 years). And this is where the similarity stops.

With regard to his creative and constructive play, the chimpanzee is behind the human child of the same age and can be compared to a child of 1 to 1½ years of age.

With regard to his ability to form conditioned reflexes, the infant chimpanzee can be equated with a child of 6 months to 1½ or 2 years of age.

With regard to the way he conducts communication using his gesture and body language, the infant chimpanzee can be likened to a child of 9 months to 1½ years of age.

With regard to certain sounds,[2] the chimpanzee can be compared to a child in the first 2 to 3 months of his life.

And, with regard to intelligible speech (loud laughter, singing, rhymed and prosaic speech), the chimpanzee obviously cannot be compared to the human child.

Of Joni's 25 sounds that I logged and that pertained to his different emotional states, all were included in the lexicon of 7-month-old Roody. At 8 months, Roody already could reproduce a word consisting of four letters; at the age of 1 year, he persistently exercised word formation. Later, he (1.2.2) used words as names of objects. He greedily acquired new words, tried to understand their meaning, demanded to be told the names of things, and compiled initially (1.5.10) two-word phrases, and soon after that (1.8.2) three-word and then (1.11.2) four-word phrases. From that time, the child accompanied all his actions by incessant babbling, unequivocally and directly revealing to us the entire complex mechanism of his soul: his refined observation ability, his accurate recognition, his bold digressions, his elementary logic.

Now, we must emphasize strongly that the infant chimpanzee, possessing rudiments of human traits and abilities, is not inclined to perfect them even if they give him certain benefits. For example, the chimpanzee can walk two or three steps in a vertical stance; he always straightens his body when he walks over an open space and looks around (he feels the need for that). But, I have never observed him trying to exercise in this vertical walking.

The chimpanzee often wants to carry something along, but he takes this thing in his foot because he does not have a free hand. He drags the thing instead of carrying it, loses it, and hampers his own walking; all this does not stimulate him to train himself in vertical walking[3] and in carrying objects in his strong, tenacious hands.

The infant chimpanzee hears human sounds all the time, reacts correctly to certain verbal orders of a human, uses diverse natural sounds for expressing his feelings, and learns complex facial expressions and signs to illustrate his desires; nevertheless, he does not show even the slightest tendency for imitation of the human voice or for understanding words to make communication with humans easier and to deepen their relationship. Moreover, as the experiments of Professor Yerkes[4] have shown, the chimpanzee does not comprehend this vocal language even after persistent special training.

The chimpanzee desires passionately to experiment with various objects and to acquire something for himself, and he heatedly claims his property, but he does not employ these objects in any constructive way and most often destroys them. It is not accidental that, in the chimpanzee, destructive, as opposed to constructive, imitation is developed more strongly.

All Joni's drawings that I have in my possession failed to show clear progress in his artistic skills; apparently, the very action of drawing with a pencil and making lines was for the chimpanzee an entertainment.

The infant chimpanzee living among humans does not want to learn how to use day-to-day objects (dishes, utensils), although he is completely capable of doing it. Joni, unlike the human child, never objected to being helped by a human in this matter.

The infant chimpanzee craves companionship with animals, but his behavior toward weaker creatures evolves into chasing and killing them, while they could be his true companions and entertain, if not serve, him.

Thus, the chimpanzee acts as follows: (1) in the functional biological field, he does not intend to improve his vertical walking and to free his hand from its function as a support; (2) in the field of instinctive activity, he ignores exercising sound imitation and broadening his imitative actions; (3) in the field of altruistic and social emotions, he underestimates the joy and advantage of agreeable contact and peaceful communication with lower creatures; (4) in the field of skills, he does not improve his motor skills associated with various tools and day-to-day objects; (5) in the field of play, he does not engage in creative constructive play.

It is hard to predict how far the chimpanzee could have advanced in these specifically human features, but it is important that he lacks a strong, sanguine, active, organic desire, a tendency for development in those directions, especially in relation to qualities with which nature has not endowed him sufficiently. How sharply he differs from his human counterpart, who dares to overcome his physical and psychic imperfections!

The human child, with his weak little hands, is yet unable to climb, but tries to climb anyway, falls down, hurts himself, and cries until he eventually learns how to do it (although not to such perfection as the chimpanzee). A number of years will pass; the human will put on grapplers and will climb to greater heights than the chimpanzee is capable. The human child is behind the chimpanzee in running, but by 3 years of age, he puts on skis or skates and accelerates his movement; then, in some 40 years, he will be breaking speed records in an automobile or train. The child is afraid of heights and is unable to fly, but he is anxious to fly; at approximately the age of 3 years, he runs around the room, spreading and waving his arms, and yells, "Let's fly!" He will grow up, and he will dash into the airstreams in an airplane or balloon.

And this is so in everything. The human child complements his teeth and hands, which are weak compared to those of the chimpanzee, with tools, taking a stone, hammer, or tongs when the chimpanzee uses his strong teeth and hands. It turns out eventually that the human's strength is in his weakness, in his indomitable tendency to overcome his natural deficiencies. Perhaps it was this process of constant surmounting of the human's physical imperfections that enabled his hypothetical ancestor to use his mental ingenuity and to become a real human.

Overcoming the natural weakness of the human's own body, arms, and teeth made him work, take a tool in his hand, and become an inventor.

The weakness of vitally important egoistic instincts (self-support, self-preservation) had to be compensated by his enhanced altruistic and social instincts in the form of orderly peaceful and agreeable communication with other people.

The human's emotional vulnerability, broadened by the sphere of physical suffering, led on one hand to development of his compassion toward other creatures, which laid the foundation for his moral feelings, and on the other hand to a counterbalance for his sad feelings and brought about a special form of his joyous experience: a sense of humor.

The insufficient strength of the human's facial expressions urged him to use additional vocal attributes and to exercise in uttering sounds. An organized communication inevitably required more differentiated methods of mutual comprehension than stereotypical facial expressions might provide, and this was achieved by development of intelligible speech.

Perhaps this particular human ancestor, living in scant external conditions, had to exercise inevitably and constantly in work, in using tools, in creative conquering of the environment to obtain everything he needed to survive. Absence or unavailability of goods made him substitute a world of his fantasy and imagination for concrete things and real relationships, which gave rise to art.

Hence, seven specifically human attributes have appeared: (1) work, (2) invention, (3) organized communication, (4) ethics, (5) sense of humor, (6) speech, (7) art.

The words *forward* and *higher* are instinctive mottos of the human's life long before he is aware that he is a thinking creature. The human seems to try to acquire the seven talents or more to add to what he was given by nature. In contrast, the infant chimpanzee, in the aspects we observed him, appears to be a creature who has buried the talents he once had.

Thus, our description, our analysis, and our synthesis brings us to qualification of the chimpanzee as a creature stagnant in his own narrow-mindedness, regressive compared with the human, and a creature lacking the desire or ability to progress in his development.

"Was ein Häckchen wird krümt sich bei Zeiten" [a hook must be bent in advance]. If the infant chimpanzee in the earliest years most suitable for his self-perfection does not show a tendency to improve himself in the qualities that rank him behind man, chances are that he will not succeed in his mature years, when the entire mental and physical structure is set and, therefore, even more stagnant.

And now, at the end of our research, it appears that the bridge with which I have tried to span the mental chasm separating the chimpanzee from the human is cracking. This complex gigantic bridge that I, like a conscientious engineer, have been building so persistently, so thoroughly, and so patiently, checking the applicability and strength of every fact contributing to the research to obtain a monumental edifice that could be used to join the unjoinable, this bridge has collapsed with a bang. This has been such a surprise for me.

It seemed to me that, if I led the two infants from different sides of this bridge and made them go in each other's direction, then after a prolonged and difficult journey, by all means I would see them in the middle of the bridge with their hands extended to each other and their bodies psychically connected.

But what has actually happened?

While at the beginning of the construction (during morphological and biological comparison of the chimpanzee and the human) my connecting path had separate small holes that could be ignored and over which both infants could only stumble, at the central and most crucial point—at the boundary of intelligence and a tendency for progress—at a point at which the chasm appeared to have no bottom, my bridge caved in; to my surprise, it was the chimpanzee, with his characteristic effusiveness, who fell into the hole formed, leaving his human counterpart way above, baffled by what seemed an incomprehensible disappearance of the one who had just stood beside him and to whom he had been ready to extend his hand.

Whether or not I succeed in raising the chimpanzee to the human level and in discovering, by a new form of experiment in higher spheres of the chimpanzee's behavior, intellectual features that this most ancient "Proantropos" had to possess to become "Homo Sapience," the third volume of my research, *Chimpanzee's Abilities to Distinguish Shapes, Size, Quantity, to Counting, Analysis and Synthesis*, will show.

Only then, having encompassed by universal comparison the entire psychic structures of the infant chimpanzee and the human child, will it be possible to venture in their genealogy and determine their relation more accurately.

Perhaps the concluding lines of this book, as though departing from the final solution of the genealogical problem, will disappoint the readers.

To justify myself in their eyes, I will remind them of the analogy cited in the preface.

Could anybody come up to the base of a high mountain and climb straight to its top? The summit may seem very clear and close to that man, but trying to climb higher and higher, he may lose his wind and stop, unable to make the entire climb at a stretch. Coming closer to the apex, the man has to reject the straight and shortest way and elect a steep, winding, zigzagging path instead.

Sometimes it appears to an observer that the traveler goes in circles or away from the destination, but the traveler knows that every step brings him closer to his goal. Instead of reproaching the tired traveler, it would be fair to wish him the strength and energy to complete his journey.

NOTES

Preface

1. Illustrative material (particularly pertaining to the child's facial expression) sometimes encompasses a later age period.

2. See his article, "Das Heranwachsen des Schimpansen," in *Der Zoologische Garten, 11*, B.4, 3/V (1931).

3. See my article, "A Study of Cognitive Abilities of Chimpanzee," Moscow, Gosizdat, 1924, preface, p. 5.

4. I needed to reread, make notes, and analyze 3,040 pages from a 4-year diary on my son.

5. The study of the chimpanzee has been completed; the results of almost a 2½-year observation of Joni have been processed for the last 3 years, hundreds of photographs reviewed, and 16 printer's sheets of text prepared for the publisher.

6 The second comparative psychology part of the study has been completed—"Behavior of Human Child" (16 printer's sheets).

7. The manuscript, which contains unprocessed experimental results, is due to be published in the third volume, dedicated to the intellect of the infant chimpanzee and human child.

8. Partially published in my first paper.

Chapter 1

1. This tendency increases as the animal grows.

2. These forehead wrinkles develop and grow as the animal grows.

3. The deepest pit formed by the median furrow as it cuts across the nose.

4. In the nose area, 17 frames the nasal mound, 18 frames the cartilaginous nose, 19 cuts across the nose, and 20 is the under-the-nose furrow.

5. Center-of-the-lip 21 is unsymmetrical.

6. Group 6b is noticeable only when the mouth corners are pulled aside [figures 8(3), 8(4)].

7. When the lips are pulled forward strongly, the third upper cheek furrow becomes visible.

8. Center-of-the-lip 37 is unsymmetrical.

9. On the bridge of the nose, the three lower furrows merge in the center; two upper furrows are tilted and merge only when the face wrinkles.

10. The width of the palm was determined across its widest part, along line aa₁.

11. See the book, *Die körpermasse und der aussere Habitus eines jungen weiblichen Schimpansen*, by O. Schlaginhaufen, Abhandlungen und Berichten d.k. Zoologie u: Anthrop., Ethn. Museum zu Dresden, 1907, Bd. XI. 4.

12. Joni's palm lines look more complicated if the relief of the palm is compared with that published in a book by J. Lecierc *Le Caractere et la Main*, F. Juven, Ed., Paris), although pictures in that book were taken from plaster models of chimpanzee's hands and therefore were relatively less distinct.

13. Digital marks of the lines are ours; literal marks are Schlaginhaufen's. Digital marks obey the following rule: Arabic numbers pertain to horizontal lines; Roman numbers pertain to vertical lines.

14. This view is supported by Schlaginhaufen's diagrams and descriptions; he believes that the cc₁ line consists of two parts.

15. We have to point out that this analysis becomes more difficult when one deals with a wax cast of the hand of a dead animal; in this case, the boldness of the lines changes with the conditions of lighting. Therefore, to orient and mark the lines, we had to follow every line at different lighting conditions and looking at them from every possible point of view to determine their true course, including the points of departure and arrival and all its possible ties with the nearest contacting lineal components.

16. All the drawings of animal hands were done on my suggestions and with my participation by artist V. A. Vatagin, in the second case, they were from a dead animal, and the third and fourth cases from live animals.

17. I express my gratitude for the help extended to us (to me and to artist Vatagin) by M. A. Velichkovky with the drawings of the hands and feet of live chimpanzees.

18. Showing the soles of the chimpanzee Mimoza (the chimpanzee at the Moscow Zoo).

19. As in figures 4(4), 4(7).

20. As in Joni's case.

21. Tragically died in 1933.

22. On the right and on the left sides of the cheeks [figure 8(2)].

23. Furrows 16 and 17.

24. Only shorter.

25. When crying hard.

26. The ninth group of wrinkles.

27. Almost a circle [plate 16(6)].

28. Particularly the upper gums.

29. Unfortunately, in our photograph, the chimpanzee's eyes are closed to protect them from the bright sun.

30. The 16th and 17th wrinkles under the eyelids and their extensions downward.

Chapter 2

1. For a detailed description of the direction and location of the wrinkles, see discussion in chapter 1 [plate 3(2)].

2. This has a shorter duration than for anxiety with sadness overtones.

3. Even in a book by such a great scientist as Charles Darwin, the chimpanzee facial expression pertaining to general excitability is ascribed mistakenly to facial expressions of rage or discontent.

4. Canines and molars.

5. The latter when the chimpanzee is in active play.

6. The chimpanzee's usual place.

7. The prototype of greeting.

8. This stage of sadness usually is accompanied by steady erect penis.

9. Once, in similar circumstances (but when we were absent) it was noted that the chimpanzee had been crying for three hours without pause.

Chapter 3

1. Particularly when he had pneumonia.

2. Two or three times a day, he gets additional meals (e.g., fruit, nuts, berries, vegetables, depending on the season).

3. I want to underline that the chimpanzee enjoyed eating raw lemons, not in the least objecting to their sourness.

4. I heard similar grunting when I saw a gelada baboon eating at the zoo.

5. The unusual propensity of the chimpanzee for eating lime can be explained easily by his need for the material to build his bones. We should not forget that our chimpanzee was an infant living through a period preceding the change of his baby teeth to permanent ones. I tried, in vain, to exchange lime for limewater or chalk; he was not inclined to eat them. Afterward, the autopsy revealed certain signs of rachitis caused by the lack of the building material for bones. We were late to regret our strictness, which only could have been justified by our fear of his possible intestinal disorders in the form of constipation associated with eating lime, which is hard to digest.

6. Joni shows the same fastidiousness toward boiled chicken. Only after intensive smelling does he make up his mind to try it; having tried it, he eats it with reluctance, smelling it again from time to time.

7. Clenching his fist, Joni puts his thumb inside the fist, unlike a human, whose thumb remains outside the fist.

8. The album of the Berlin Zoo.

9. See in more detail in the section, Imitation, in this chapter.

10. Thus, he had not eaten for 7 hours.

11. If Joni lived in the wild at his current age, he would probably sleep with his mother.

12. How strikingly such love differs from the true love that strives to give rather than to take.

13. Here, I cannot help describing the recently observed instance of jealousy in a dog. Jim (that was the name of my temporary little admirer), a mongrel that had two rather inert ladies who were indifferent to his canine life as his masters, found me (during our 2-month sojourn in the summer house) as his guardian. I made sure that he was fed on time, arranged his sheets for the night, and took him for long walks in the forest, where he could run around, frolic, and play the field-mouse hunt. The animal became so attached to me that he followed me everywhere. To the distress and surprise of his real masters, he "didn't care a thing about them," as they put it; he did not pay any attention to their calls and signs of affection. But, as we have

noted already, love and jealousy are two inseparable sisters. If in Jim's presence I happened to cuddle a big dog, a close friend of his, Jim was enraged; he jumped up to my face, as if trying to shift my attention to him; attacked his rival, barking angrily; ran around us; shrilled; and did not calm down until I drove the other dog away, and he could count on my total affection and sympathy.

Chapter 4

1. I notice that, during catching games, the chimpanzee can evaluate the distance from him to whoever is chasing him. For instance, if Joni is running near the ceiling of the cage and I am dashing around at the bottom of the cage at a distance no less than 5 meters, he shifts from place to place very calmly. But, as soon as I move under the cage bottom so that I am right opposite him and look straight at him, he begins breathing distressfully, as in the cases when he was caught. All this happens because the distance between us becomes somewhat less, although he is as unreachable for me now as he was in the past.

2. He can be occupied with the belt endlessly, doing various manipulations with it.

3. The term was coined by K. Groos; see his book, *Spiele der Tiere*, Fischer, Jena, 1907.

4. Possibly an act of imitation.

5. K. Groos, *Spiele der Tiere*.

6. I observed similar tendencies for producing artificial difficulties and overcoming intentionally created obstacles in the orangutan Frina in the Moscow Zoo.

7. The same as in human children (see part II of this book).

8. I observed similar lip plays in a young female orangutan, Frina, in the Moscow Zoo.

9. The term is by Groos.

10. As do small children.

11. The similar action of the tongue in performing movements that are difficult to coordinate is also observed in children.

12. I noticed many times that Joni, tossing up some hard object (for example, a wooden ball), falls on the floor face down, shielding the face from a possible strike.

13. Joni saw this vial on the window sill in the next room; he was eager to take it many times, but he was not allowed. Eventually, the chimpanzee's desire was fulfilled.

14. Joni showed a similar form of behavior in dealing with collapsible bowling pins.

15. It is worth mentioning that this jug, usually filled with water, was always the special object of Joni's attention, but was guarded against the chimpanzee's devices.

Chapter 6

1. Once, I observed my son (3 years old) in similar circumstances. He put the end of the stick into a crack in the stone floor, bent the stick, and finally broke it.

Chapter 7

1. It is natural that the chimpanzee learns how to open a padlock (closed, with the key hanging on it) quicker, better, and more accurately than the macaque Daisy did it. See more details in my work, *Adaptive Movement Skills of a Macaque Under*

Experimental Conditions, Publishing House of the State Darwin Museum, Moscow, 1928.

Chapter 8

1. We did this in order not to expose Joni to smoke, which usually filled the room due to some defects in the stove structure.

Chapter 10

1. Little children are known to kiss in the same way, by pressing an open mouth to people's faces.
2. R. M. Yerkes and B. W. Learned, Williams & Wilkins, Co. Baltimore, Md.

Chapter 11

1. According to his body proportions.
2. Without leaning on his musculus gluteus.
3. Already at 2½ years of age.
4. Incidentally, I noticed this way of leaning on the outer edge of the foot in little children [9 months; plate 70(3)] who could not stand and tried to find some support and as an atavistic habit in adults and teenagers when their mental brakes were activated, for example, in an embarrassed young speaker in front of an audience of strangers or during the shy moments of a child.
5. The rut was 10–15 cm wide.
6. At this age (1 to 2 years), the human child can come up the stairs in a vertical position only with someone else's help [plate 77(4)].
7. It is well known that kittens climb high trees very easily, but they may be afraid of coming down and may sit on the top of the tree for a long time, mewing and asking for help.
8. Pedologist N. I. Kasatkin.

Chapter 12

1. Once, Joni showed the same expression under similar circumstances.
2. Although C. Darwin observed tears in lower monkeys.
3. In his book, *The Expression of the Emotions in Man and Animals*. (1872). London: Murray.
4. *Der Gesichtsausdruck des Mensches*, Ferdinand Enke, Stuttgart, 1923, p. 175, figure 50.
5. Yerkes, R. M. and A. W. Yerkes. (1929). *The Great Apes: A Study of Anthropoid Life*. New Haven: Yale University Press.
6. Sounds "g," "k," and "kh" in this baby talk are not accidental: All these sounds later participate, as important elements, in the formation of the child's loud laughter.
7. This smile was rendered masterfully by Leonardo da Vinci in *La Gioconda* (Mona Lisa).
8. As is well known, in the best Moscow theaters (for example, in the former Khudozhestvenny Theater), the viewers were not allowed to applaud, according to a sublime logical and psychological notion that truly beautiful experiences, as well as other deep emotions, ought to be reflected inward, remaining invisible and silent.

Chapter 13

1. When nobody was around to keep him company, he liked to be with his toy companions—a teddy bear, a cat, and the like [plates 82(1), 82(2)].

2. My 3-year-old boy had a tendency, when eating especially tasty things, to make mooing/growling sounds; it took a lot of effort on my part to break his nonimitative and obviously atavistic habit. These sounds reminded me of the growling of a baby bear (which once was in my care) when he was drinking milk; similar growling can be observed in dogs when they gnaw bones.

3. By Yu. A. Polyakova.

4. Although Joni could also drink from a cup or mug, holding them in one hand [plate 35(2)], he usually took the utensil by the edge; when he tilted it, he had to put his hand under its bottom [plate 35(1)].

5. When the food was semiliquid.

6. The human child opens his mouth wide when he swallows from a spoon or eats with a fork [plates 88(1), 88(4)].

7. Observations of the orangutan Frina (at the Moscow Zoo) confirmed the same pattern. Every time she was urged (by M. A. Velichkovski) to use a spoon, she was very uncooperative and ate slowly. He had to remind her about the spoon again and again. But, she willingly and energetically opened her mouth when she was fed with a spoon. I have never seen her take the spoon from a person feeding her and try to eat independently.

8. The maid.

9. In the human child, the initial stage of fear usually is accompanied by reddening of his face; a more advanced stage of fear, according to my observations, causes reddening at first and then paling of his face. The face turns deadly pale only as a result of maximum fear.

10. Even fairly recently (when Roody was 9½ years old), at the Timiryazev Biology Museum, Roody was unwilling to look at a freak calf that had one huge eye.

11. A colorful illustration of this expression can be found in Professor Krukenberg's book (p. 310, figure 273, "Rage").

12. This is even more visible in predatory mammals (for example, angry wolves or dogs).

13. For instance, when I put a cold compress on his leg.

14. A friend of mine, an extremely reserved and rational woman, constantly jerked up her nose in response to any unpleasant or irritating impression.

15. As mentioned, I intentionally provoked this expression and photographed it under the conditions of a natural experiment. It appeared when I added a bitter rhubarb powder to the sugar he had requested.

16. It is known that dogs and other animals express their tenderness toward a human by licking him with their tongues; in Russian common parlance, kissing or other signs of affection between loving spouses or between lovers are pejoratively called "licking."

17. In his military games, Roody always armed that particular hare as if trying to help him out [plate 61(6)].

Chapter 14

1. The definition was made under the following circumstances: Once, I asked Roody, "What do you think the plush teddy bear is, a man or an animal? He said,

"Half human, half animal," and added, "His face is human, his arms and legs are bear's."

2. Roody arranged similar artificial accidents to accelerate the play, when riding a bicycle [plate 102(6)] or flying a so called "airplane" [plate 102(3)]. When Roody could not move himself while sitting in his toy car, he managed to bring some dynamism into this static situation: He created a semblance of an impending accident by sitting up his teddy bear right in front of the car [plate 93(3)].

3. The chimpanzee Missy, observed at the Berlin Zoo in 1913, also could ride a bicycle very fast, without holding the handlebars and turning only with her feet.

4. A little later, Roody found another way of riding: He pushed against the ground with his feet. At frequent and inevitable stops, he pretended to restart the car by "turning the ignition key." In one more year, Roody managed to push against the ground so hard that, with his legs on the car, he could coast for a short time, assuming the pose of a passenger [plate 3(4)].

5. Joni usually took a small ball with him.

6. One boy I knew fell from the second floor into the staircase; he was lucky enough to survive, only bruising his head and knees, which had to be treated for a long period of time. Nevertheless, when I asked him whether he would ever be sliding down the banisters, he very firmly answered "yes," which revealed the enjoyment this had given him.

7. As we noted, the human child is even more inferior to the chimpanzee in the speed, agility, and strength needed to climb and hang on trees.

8. Obviously, when he is not so ill that he is bedridden, and his entire organism restructures itself to fight the disease.

9. In my book, *Cognitive Abilities of Chimpanzee*, Gosizdat, 1924.

10. For example, Tolstoy or Aksakov.

11. Especially in museums of natural history.

12. This is permitted in some museums and is recommended for children's museums, for instance, in the former toy museum.

13. Old people told me that once there was a preposterous fashion in Moscow to wear shoes that had been produced to make the screeching sound. Screeching high boots were always a highly valued item in villages.

14. The term by K. Groos.

15. I observed a similar form of behavior in the young orang Frina in the Moscow Zoo when she was given a bowl filled with water.

16. Similar to plate 47(3); in Roody's case, 17 months of age.

17. Similar to plate 47(5); in Roody's case, 17 months of age.

18. Similar to plate 47(1); in Roody's case, 17 months of age.

19. Roody's hair in the picture was much lighter than his hair in real.

20. In the child even as early as before 1 year of age, the index finger becomes prevalent compared with other fingers [plates 48(1)–48(6)]. A similar exclusive role of the index finger was also found in Joni [plate 48(5)].

21. Although, during play Joni preferred blue.

22. The colors were red, orange, pink, purple, blue, brown, white, and yellow.

23. Not counting the abominable odor of some medicines, such as cod-liver and castor oils.

24. Joni was also afraid of some pictures.

25. Roody usually was not allowed to take any sharp or prickly things, particularly knives.

26. By this play, I mean the overwhelming entertainment by destruction itself, that is not associated with curious watching of the thing being destroyed.

27. In L. Huxley, *Life and Letters of Thomas Henry Huxley*, MCM, London, 1900, p. 435.

28. *A Cat That Was Lost*, Publ. Mirimanov, 1927.

29. Although the famous trainer V. L. Durov showed me drawings of rings that his chimpanzee Mimosa allegedly had done in imitation of Durov's actions, unfortunately, as was usually the case in all Durov's interpretations, it was hard to discern where the guiding, or rather leading, human influence ended and the animal's independent activity started.

30. I observed a baboon with similar behavior; in a state of irritation, he took a piece of wood that lay on the floor of his cage and threw it at the offender.

31. During Roody's first week, I also observed the dull sound of moving lips, and sounds of swallowing, sucking, a hiccup, smacking of the lips, quacking, and snoring.

32. Roody (2.0.27) expressed this wish verbally.

Chapter 15

1. The event was that, suddenly, an automobile skidded onto the sidewalk where they were walking; the nanny with the child rushed aside and fell on the pile of snow.

Chapter 16

1. An overwhelming majority of external stimuli causing all of these emotional reactions were also quite similar for both infants.

2. Observed in 5-month-old Roody.

3. Observed in Roody at the age of 8 to 9 months.

4. Observed in 7-month-old Roody.

5. Observed in Roody at the age of 1 day.

6. Observed in Roody at the age of 5 days.

7. Observed in Roody at the age of 1 month and 27 days.

8. Observed in Roody at the age of 2 months.

9. Observed in Roody at the age of 3 months.

10. Observed in Roody at the age of 7 months.

11. These sounds are given in the order of declining similarity.

Chapter 17

1. I observed brief squatting in the chimpanzee only very rarely, when he was frightened or worried.

2. I did not observe walking on all fours in my child at any stage of his development; even during crawling (8–9 months of age), the child rested on his spread palms and on his knees, but not on his feet or on his knuckles, as we can see in the chimpanzee.

3. With the exclusion of the jerking movement common for both infants.

4. To make a fist, Roody positioned his thumb over other fingers; Joni, in contrast, covered his thumb with other fingers. As Fritz Kan observes in his remarkable book, *Man*, mentally retarded people, as well as babies, form their fists in the same manner as the chimpanzee.

5. As we mentioned, the conditions in which both infants were brought up were almost identical in this respect: Not only was torturing animals discouraged, but also forbidden. I tried to nurture in them sympathy for mistreated animals, but only

Roody reached a state of compassion; with respect to Joni, I was unsuccessful. It is difficult even to assume that, by certain pedagogical methods, real expressions of sympathy toward lower animals could be reinforced in Joni.

6. In the case of the human, they are complemented by deep verbal expressions of love and affection.

7. Only one time I observed something slightly resembling this in Joni's behavior, when I applied turpentine to his body. When Joni started feeling the sting, like Roody, he looked around to find the invisible enemy, but in this case, his curiosity was prompted exclusively by the real stimulus of the painful sting, not by cognitive intention.

Conclusion

1. And all the other senses, according to other authors (for instance, L. A. Kellog).

2. Associated with various physiological states (for example, snoring, grunting, coughing, etc.).

3. Which is possible in principle, for instances are known of anthropoids that lived in zoos that could move by vertical walking (for example, the orangutan Jacob in Hamburg, Germany).

4. See, for instance, R. M. Yerkes and A. Petrunkevitch, *Chimpanzee Intelligence and Its Vocal Expressions*, Williams & Wilkins, Baltimore, Md., 1925.

ACKNOWLEDGMENTS

The authors have been supported by grants RR-00165 from the National Institutes of Health/National Center for Research Resources to the Yerkes Regional Primate Research Center, National Institute of Neurological Disorders and Stroke 29574 to W. D. Hopkins, and R01-RR09797 and a grant from the John Templeton Foundation to F. B. M. de Waal. The authors would like to thank Paul Ekman for his tenacious efforts in seeing this manuscript published and Phillip Laughlin for his editorial assistance. Special thanks to the many students and collaborators who made this work possible and to W. D. Hopkins for his help in developing the computerized paradigm used in these experiments. We also thank Frank Kiernan for photographic assistance and the animal care staff at the Yerkes Regional Primate Research Center. Helpful comments on earlier versions of this manuscript were provided by A. Lacreuse.

The APA standards for the ethical treatment of animals were adhered to during these studies. The Yerkes Primate Center is fully accredited by the American Association for Accreditation of Laboratory Animal Care. Correspondence should be addressed to L. A. Parr, Living Links, 954 North Gatewood Road, Atlanta, GA 30329 or parr@rmy.emory.edu. Visit www.emory.edu/LIVING_LINKS/ for additional information.

AFTERWORD: RESEARCH ON FACIAL EMOTION IN CHIMPANZEES, 75 YEARS SINCE KOHTS

Lisa A. Parr
Signe Preuschoft
Frans B. M. de Waal

The long-awaited publication in English of the observations by Nadezhda (Nadie) Ladygina-Kohts of her chimpanzee Joni arrives at an opportune moment. Years of intensive study of chimpanzees have nibbled away at cherished human-animal distinctions, including insights into chimpanzee culture, moral tendencies, politics, intergroup violence, emotional life, and mental capacities. Not that there no longer are distinctions, but given the customary assumption of human uniqueness, it is the similarities that have been most shocking. Now, we can flashback to one of the pioneers of chimpanzee research, who writes with great love for, and appreciation of, the species and shows few prejudices as to whether her "little creature" is supposed to be similar to or different from us.

Most experts on primate behavior have caught glimpses of the original Russian version of this book in other publications though reproductions of some of the photographs, especially those concerning facial expressions. However, the scope and detail of Kohts's observations have never been realized fully. Since the Living Links Center is devoted to the study of human and ape evolution, we are pleased to sponsor the publication of this classic not only for obvious scholarly reasons, but also because it contains many insights still relevant today. In historic importance, the work ranks with Köhler's (1925) provocative analyses of chimpanzee mental life, which first challenged behaviorist assumptions, and Yerkes' studies of chimpanzee tool use and temperament. There are frequent references to Kohts in Yerkes and Yerkes (1929), and Yerkes was instrumental in the translation of one of her studies (Yerkes & Petrunkevitch, 1925).

Kohts was an ethologist avant la lettre in that she believed that thorough description should come first in science. In her entries, she hit on themes that sound familiar to anyone who has followed current debates about the cognitive abilities of the chimpanzee. She investigated Joni's reactions to pictures of chimpanzees and other animals, animal furs, as well as to his mirror image. Even though Joni was probably still too young for mirror self-recognition (Custance & Bard, 1994), Kohts describes how, once he had gotten used to the mirror, he "entertained himself by sitting in front of it and making cracking sounds with his lips" (p. 108).

The book also provides numerous observations relevant to ape art, imitation, deception, and theory of mind. Here is a typical example of tool use imitation without a clear understanding of the goal:

> Left to his own devices, Joni often takes a broom or brush and tries to sweep the floor, raking the trash into a pile. However, he does it so awkwardly and inefficiently, due to a lack of direction, that he spreads the trash over the floor rather than gathers it, and the floor is never clean as a result. Joni even moves the furniture, as is done during a cleanup, although he often does not sweep the floor at the freed spot. (p. 183)

Joni is said to show a wide range of emotional responses, from jealousy and guilt to sympathy and fierce protection of those he loves. The following passage relates, with the attention to detail typical of Kohts, the extreme concern and compassion Joni feels for his mistress:

> If I pretend to be crying, close my eyes and weep, Joni immediately stops his play or any other activities, quickly runs over to me, all excited and shagged, from the most remote places in the house, such as the roof or the ceiling of his cage, from where I could not drive him down despite my persistent calls and entreaties. He hastily runs around me as if looking for the offender; looking at my face, he tenderly takes my chin in his palm, lightly touches my face with his finger as though trying to understand what is happening, and turns around, clenching his toes into firm fists. (p. 121)

Kohts's accounts are entirely in line with what we know now about the sociality and complex emotional life of the chimpanzee, in both the field and captivity (for example, see de Waal, 1996; Goodall, 1986). That her descriptions occasionally include unqualified anthropomorphism is not necessarily problematic. Her descriptions clearly reflect intimate knowledge of the ape and hence rarely represent naive assumptions about the meaning of his behavior. In addition, if a species is related so closely to us (chimpanzees and humans belong to the same Hominoid family and split from their common ancestor only about 5 million years ago), anthropomorphism is inevitable. The safest assumption about the species is that, if it acts similar to us, the underlying motivations and psychology probably also are similar. Rather, it is the opposite assumption—that humans and apes are fundamentally different—that needs defense (de Waal, 1997).

The great limitations of Kohts's work, of course, are that she had only a single subject, and that the subject was so young (estimated between 1 and 4 years of age). This means that she had little insight into the usual social organization among chimpanzees and never got to see the full-blown cognitive capacities and mature psychology of the species. Similarly, a psychologist who studied and tested a single boy of around 5 years of age would be unable to draw many conclusions about the human species. On the other hand, the facts that Kohts was Joni's mother figure, was in close physical contact with him every day, and systematically collected all possible information in the greatest detail made her see a chimpanzee up close in a way that very few people can or ever will match. She looked hard and deep into the ape's soul and made no secret that she was impressed by what she saw.

This chapter serves to place Kohts's monograph in the context of current knowledge of emotional behavior of nonhuman primates to highlight the importance of her work. We compare Kohts's observations of chimpanzee facial expressions with those made independently by researchers of nonhuman primates in the field and captivity during the rest of the 20th century. In doing so, we illuminate similarities and differences in the phylogenetic continuity of facial expressions across several species of nonhuman primates and provide an interpretation of how social organization and ecological variability may have shaped such behavior. Finally, we present the results of recent experiments conducted at the Living Links Center that employ a computerized paradigm to assess how chimpanzees respond when presented with discriminations involving their facial expressions.

Part 1. Historical Overview of Emotion Research in Primates   The role of facial expressions in the social and emotional lives of animals has been of interest in the area of comparative psychology since Darwin's seminal book *Expression of the Emotions in Man and Animals* (1998). Although Darwin was not the first to speculate about the role of faces in emotional behavior, he is credited for bringing these ideas into the focus of scientific thinking (Fridlund, 1994; Izard, 1971; Russell & Fernández-Dols, 1997). Darwin acknowledged remarkable similarities in the behavior and sensitivity of nonhuman species to emotional situations and was fascinated particularly by similarities in the facial expressions of humans and great apes. He was among the first to speculate that these expressions may correspond to basic emotional states in animals as they do in humans.

A similar theme runs throughout Kohts's manuscript: The continuity between nonverbal expressive behavior in apes and humans is unmistakable. The facial expressions of apes function to communicate a range of complex emotional messages that most likely are homologous evolutionarily to our own nonverbal communication. In fact, the majority of Kohts's manuscript is devoted to highly detailed descriptions of the expressive and emotional behavior of one young male chimpanzee, Joni. While Kohts did engage in a variety of experimental procedures with Joni, some of which foreshadowed research that would dominate animal cognition in the next half century, it is her detailed descriptions of behavior that make the monograph special. She describes both the expressive composition of Joni's behavior (chapter 1, Chimpanzee's Face in Dynamics) and how this behavior is illustrative of eight states of emotion—excitement, sadness, joy, anger, fear, repulsion, surprise, and attention (see also chapter 2). This list of emotions is not far removed from the six basic human emotions identified more than 30 years later by Ekman and his colleagues—sadness, happiness, anger, fear, disgust, and surprise (Ekman, 1972, 1992; Ekman, Friesen, & Ellsworth, 1972). Together, these descriptions seem to provide an exhaustively complete profile of the mental and emotion life of a young chimpanzee.

While studies of emotional behavior in nonhuman primates have had a long history in ethology and comparative psychology, attempts to isolate and measure the emotional experience of nonverbal species remain limited even today, thus emphasizing the importance of preserving Kohts's contributions. Her descriptions of a young chimpanzee's behavior are a refreshing contrast in a field that increasingly is becoming more focused on experimental techniques that examine exactly how behavior is elicited and maintained rather than the attention to detailed descriptions characteristic of earlier approaches, particularly those of European ethologists. Before the rise of behaviorism in America, research on nonhuman primates, particularly chimpanzees, focused very strongly on how these creatures were similar to humans, both intellectually and emotionally. Several groups attempted to raise infant chimpanzees like human children (Gardner & Gardner, 1969; Hayes, 1951; Kellogg & Kellogg, 1933), teaching them languages, including American Sign Language, and a series of complex lexigrams associated with objects and actions (Fouts, Chown, & Goodin, 1976; Savage-Rumbaugh, McDonald, Hopkins, & Rubert, 1986; Terrace, 1979).

Under the influence of behaviorism, however, those studying nonhuman primates were left to concentrate on behavior while ignoring the emotional aspects. Emotion was part of the black box that was considered inaccessible, especially in nonverbal species. Animals may well have emotions, but how would we ever know?

It is clear here why detailed descriptions of behavior are so important. How can one study the causes and consequences of emotional behavior without a clear understanding of what this behavior looks like and the situations in which it is elicited? The template for emotional behavior in any species therefore became our own human behavior, and discussions of an inner, emotional life in animals were met promptly with criticisms of anthropomorphism and human projection (de Waal, 1997). This dilemma has slowed comparative research greatly even to this day, while in humans, tracking emotions in various situations has become a major field and has added substantially to our interpretation of how humans process stimuli and make decisions during cognitive tasks. In contrast, few studies have addressed the relationship between facial displays and emotion in nonhuman primates or have done so rather indirectly. These few studies, however, have contributed significantly to our understanding of how nonhuman primates perceive emotional signals and interact in social situations. These studies are reviewed in part 4.

Part 2. Chimpanzee Facial Expressions   Primates use a vast number of ritualized displays in all their sensory channels to gain information about nonsocial events and the social attitudes and emotional states of their interaction partners (Preuschoft & Preuschoft, 1994). Among these, the visual and vocal channels figure most prominently. Facial displays either can be silent, functioning as visual input alone, or can be combined with vocalizations, forming compound visual-acoustic messages. Research over the past half century has produced considerable advancement in our understanding of facial behavior, particularly emotional behavior, in nonhuman primates. This research includes detailed descriptive reports of facial expressions in a variety of primate and nonprimate species, including chimpanzees, bonobos, rhesus monkeys, capuchin monkeys, and canids (Andrew, 1963b; Bolwig, 1962; de Waal, 1988; Fox, 1969; van Lawick-Goodall, 1968; Goodall, 1986; Hinde & Rowell, 1962; Preuschoft & van Hooff, 1995, 1997; Redican, 1975; van Hooff, 1962, 1967, 1973; Weigel, 1979).

Almost any behavior is capable of conveying information and therefore can act as a signal (Moynihan, 1998). Even if communication is not the primary function, an observant animal may derive information from behaviors performed by other individuals. Behavior patterns that convey information in such "incidental" fashion are termed *unritualized*. These are contrasted with patterns that are specialized through evolution to function as communication signals, that is, to convey information in specific ways. This evolutionary process is called *ritualization* (Tinbergen, 1952). Ritualization, defined as "the adaptive formalization of emotionally motivated behavior to promote better, more unambiguous signal function" (Huxley, 1966, cited in Redican, 1982), is the process by which a formerly adaptive, unspecialized behavior becomes divorced from this original function to take on a different meaning. The raw material from which signals are derived are elements, or by-products, of such behavioral acts as intention movements that precede an action (Lorenz, 1941), protective responses [Andrew, 1963b; see also figure A1(a)], autonomic responses (Rinn, 1984), and displacement or of redirected activities that occur when an animal experiences conflicting motivations (Tinbergen, 1952). As a result of this ritualization, communication signals become more readily recognizable: They appear more stereotypical and conspicuous and less ambiguous [figure A1(b)]. This can be achieved by elaborating some aspects of the display and reducing others, while elements of the original unspecialized behavior pattern may be repeated or omitted. The result of ritualization is a "typical intensity" with which the signal behavior usually is displayed (Morris, 1957).

The distinction between ritualized and unritualized displays, however, is not always clear or neat (see figure A2). Evolution never leads to a final stage. This, in turn, implies that the evolution of signals is a dynamic process by which, at any point in time, we happen on a snapshot that includes signals that are changing their shape or their meaning, falling into disuse or adopting new functions, thus becoming "emancipated" from their previous motivational backgrounds. This dynamic process makes it acceptable to refer to different degrees of ritualization (Moynihan, 1998). Virtually all species of animals, from insects to mammals, display ritualized expressions. The chimpanzee has a set of such ritualized facial displays, but at the same time is capable of subtle variations of and grading between expressions. This makes the study of facial expressions in this species particularly fascinating. In chimpanzees, for example, variable expressions may represent emotional responses that are subject to as much individual variability

(a)

Figure A1. The evolutionary process of ritualization turns an instrumental action into a communication signal by increasing its stereotypy and exaggerating its movements. Andrew (1963a, 1963b) speculated that the bared-teeth expression, commonly seen in many primates, derives from the retraction of lips in response to noxious stimuli. Shown are the instrumental act (a) in a baboon eating a cactus and (b, facing page) in a rhesus monkey who is being approached by a more dominant group member. In this species, teeth baring reliably signals submission. (Photographs by Frans de Waal)

and as many idiosyncrasies as in humans. Chimpanzees are one of the only species apart from humans in which facial expressions are so variable and in which individual differences are so prominent.

There are two dimensions on which signals may be graded: They may vary quantitatively via intensity and qualitatively by forming intermediates with other signals. As with human speech (Studdert-Kennedy, Liberman, Harris, & Cooper, 1970), however, these intermediates are not necessarily appreciated, or accurately perceived, by the receiver. Instead, perception of graded signals may be categorical (Green, 1975). As a result, the somewhat paradoxical relationship between ritualization and typical intensity, as opposed to graded communicative signals, probably is conceived best in terms of fuzzy set theory (Gouzoules, Gouzoules, &

 (b)

Figure A2. An example of an unritualized facial expression is frowning, an action accomplished by contraction of the corrugator muscle. This face is not common in most primates, but they are certainly capable of producing it. Here, a juvenile bonobo male charges two others, staring at them with piercing eyes and a frown. (Photograph by Frans de Waal)

Tomaszycki, 1998; Massaro, 1987). Fuzzy sets consist of a central prototype that is constructed actively by the receiver, either a conspecific or a human observer, although different species do not necessarily form the same constructions. The implication of this is that, however refined our signal recording techniques become, we always need to validate our understanding of what constitutes a signal, and what is just irrelevant variation, by referring to the natural communicative repertoire of the species under investigation (Janik, 1999).

Here, we concentrate on the pioneering descriptions and classifications of chimpanzee facial expressions provided by van Hooff (1962, 1967, 1971) and Goodall (1968, 1986) and compare these with the observations made by Lady-gina-Kohts (chapter 1, Chimpanzees Face in Dynamics). Both van Hooff and Goodall worked with chimpanzees in large mixed communities in which the full range of social behavior could be expressed; the former worked with chimpanzees in captivity, and the latter worked with them in the field. The goal of this comparison is to formulate a list of facial actions that are prominent and typical elements of chimpanzee emotional communication, provide a taxonomy for these expressions (something that Kohts did not attempt in her own descriptions), and briefly describe their context of use and possible social function. Following this, we compare the phylogenetic continuity of facial expressions across humans and different species of monkeys and apes and discuss the constraints that ecological factors may have played in the evolution of these displays.

There are between 20 and 30 different facial expressions and vocalizations described for chimpanzees. These expressions predominantly are graded; they may intermix and blend into one another with the same kind of physical and contextual flexibility that characterizes human expressions (Marler, 1976). It is difficult to convey the elaborateness with which chimpanzee use facial and vocal signals to communicate intentions and emotions without the use of visual materials, such as video, in which all shades and transitions between the different expressions can be seen. We estimate that it takes students at the Yerkes Primate Center at least 1 year of simply watching chimpanzees before they are able to make "sense" of what they see, that is, recognize all the relevant elements of these often-confusing and complex interactions. Yet, despite the graded nature of chimpanzee signals, there is considerable agreement among researchers as to which expressions are prominent social signals and what their social function may be.

Chimpanzee facial expressions can be grouped into several major categories, including those associated with aggression and submission, play, long-range communication, food announcements, and expressions of affection and consolation. Van Hooff's (1973) analysis stands as the most systematic attempt to date to provide an empirical, quantitative foundation for such a classification on the basis of observed behavioral sequences. This work had its roots in earlier ethological studies on so-called motivational systems (reviewed in Hinde, 1970), a now all-but-forgotten field that nevertheless remains relevant to the study of emotion. While Kohts did not attempt to develop specific labels for Joni's expressive behavior, although it is clear that she was aware of its graded nature, she did choose to associate the behaviors with emotional motivations, perhaps foreshadowing the later work of the ethologists. She said the expressions are "fleeting and correspond to the maximum point in the emotion development." They are, therefore, only peak examples of each emotional state that, due to its graded nature, changes constantly with intensity and context. Table A1 provides a list of several major classes of facial expressions that have been described for the chimpanzee and cross-references these to similar descriptions in the existing literature.

Chimpanzee Facial Expressions

BULGING-LIPS FACE

Description. In this expression, the individual stares ahead with eyes open, mouth closed, and lips pressed together. Characteristic of this expression are the lips that bulge out as though the individual is blowing air, although the lips remain locked together. No vocalizations occur. Goodall (1968) has described this expression as the glare, or compressed lips face, given by individuals prior to attack or copulation or when two rivals stare at one another. Van Hooff (1973) has called it the bulging lips face, attack face (van Hooff, 1962), or tense mouth face. The lips are pressed tightly together; the upper lip arches, but the jaws remain clenched firmly (Marler & Tenaza, 1976; Redican, 1982; van Hooff, 1967).

Table A1. Chimpanzee facial expressions cross-referenced to previous literature descriptions

Expression	Other Names
Bulging-lips face	Glare or compressed lips face[a]; attack face[b]
Relaxed open mouth display	Play face[a,c–f]; relaxed open-mouth display[g]
Silent bared-teeth display	Grin[a,b,f]; horizontal bared-teeth expression[h,i]; bared-teeth yelp face; silent bared-teeth display[g]
Staring bared-teeth scream face	Rough scream[j]; roar, growl, scream[k]; double-tone scream[l]
Stretched pout-whimper	Stretch pout-whimper[b,h,m]; whimper face, hoo-whimper, pout-moan[a]
Silent pout	Pout[c]
Pant-hoot	Pant-hoot[a,m]
Pant-grunt	Pant-grunts[m]; panting, bobbing pants[a]; rapid oh-oh[c]
Teeth clacking	Lip smack and teeth clack[m]; lip smacking[j]
Splutter	See van Hooff, 1971

[a]Goodall, 1968; [b]van Hooff, 1962; [c]van Hooff, 1973; [d]Andrew, 1963a & b; [e]Bolwig, 1962; [f]Redican, 1982; [g]van Hooff, 1972; [h]van Hooff, 1967; [i]van Hooff, 1971; [j]Marler, 1969; [k]Reynolds & Reynolds, 1965; [l]Yerkes & Learned, 1925; [m]Marler & Tenaza, 1976.

Context and function. The context of use of this expression is most often aggression, but it can also be associated with sexual activity (Goodall, 1968). Adult male chimpanzees often make this face prior to an attack, while chasing another, or during a bluff display (figure A3). It may also be used prior to copulation. When given by an adult male during a bluff display, it probably functions in combination with other components of the display (i.e., bipedal charge, piloerection, and swinging arms) to make that individual appear larger and more powerful.

RELAXED OPEN MOUTH FACE

Description. This expression commonly is referred to as the *play face* due to its clear context of use. The face generally is relaxed with the eyes and mouth open. The mouth corners are in their usual position, or slightly retracted, but the lower jaw is open to expose the bottom teeth. The upper lip may be raised slightly, exposing the upper teeth, otherwise, it folds over them, creating whiskerlike wrinkles along the mouth corners and the sides of the nose. These wrinkles can become more intense if the play gets rough or in combination with other expressions [see plates 2, 3(8), 25]. Vocalizations may include a fast, rhythmic staccato breathing similar to panting in which the voiced exhalation can be quite strong. This expression has been described similarly by a number of other researchers and represents a very common, easy-to-identify facial expression in the chimpanzee and other primates (Andrew, 1963b; Bolwig, 1962; Goodall, 1968; Marler, 1976; Redican, 1982; van Hooff, 1973).

Figure A3. The bulging-lips face. The bulging-lips face occurs during intimidation displays. Here, it is made by a male chimpanzee who has sat down after such a display. (Photograph by Frans de Waal)

Figure A4. The relaxed open mouth face (or play face) in a juvenile bonobo. (Photograph by Frans de Waal)

Context and function. This expression typically is given in the context of play or during a friendly approach (figure A4). It is believed to be a ritualized form of biting (van Hooff, 1972). The ritualization has resulted in a highly stereotypical form of mock biting, or gnawing, characteristic of play wrestling. One individual may put their fingers or toes into the mouth of their play partner, allowing themselves to be led around this way and producing a frenzy of laughter in the partner. The play face is a very widespread expression in the animal world, observed in many genera of primates, dogs, lizards, and a variety of other animals, suggesting that it has very old phylogenetic origins (Fagen, 1981).

SILENT BARED-TEETH DISPLAY

Description. The silent bared-teeth display may be one of the more widespread and easily identifiable facial expressions among mammals. The mouth can be partially open or closed, with the mouth corners and lips retracted laterally, fully exposing the top and bottom teeth (van Hooff, 1972). The eyes usually are open, but in extreme forms, they may become squinted. The lack of vocalizations helps to define this from the other bared-teeth expressions, which are accompanied by a variety of screams. It is sometimes referred to as a grin due to its resemblance to the smile (Goodall, 1968; Redican, 1982; van Hooff, 1962). Van Hooff (1967, 1972) describes three different forms of bared-teeth expression that differ subtly in terms of their function: the horizontal bared-teeth face, bared-teeth yelp face, and silent bared-teeth face. The silent bared-teeth display is distinct in that no vocalizations occur.

Context and function. The silent bared-teeth display is performed most often by individuals who experience a combination of nervousness, fear, and submission. It may be given when a more dominant individual is approaching, after being attacked, during reconciliations, or when soliciting support from another individual (figure A5). The open mouth version of this expression is also given in a more affiliative context, such as a blend with the play face in which the mouth is more open and the top lip is stretched flat over the upper teeth, or by females during sexual intercourse (figure A6). A wide variety of vocalizations may accompany this expression, depending on the context in which it is elicited. These can range from screaming or screeching in the context of a fight, to pant-barks or pant-screams during the excitement of feeding, to an absence of vocalization in the submissive context.

Figure A5. The silent bared-teeth display. A male chimpanzee shows a bared-teeth expression and penile erection at a tense moment related to the complex politics among males of the species (de Waal, 1982). The male is of top rank, but his position depends on the male on the right. He is trying to reconcile with this male after a fight between them. Both males extend their hands toward one another, and moments after this scene, the two males embraced. (Photograph by Frans de Waal)

Figure A6. Even though bonobos are close relatives of chimpanzees, they use the bared-teeth expression in slightly different ways. In bonobos, the expression occurs regularly in affectionate and sexual contexts, as seen here between two adult females engaged in genito-genital rubbing, a common occurrence in this species. Because of this context, it has also been called a "pleasure grin" (de Waal, 1988). (Photograph by Frans de Waal)

STARING BARED-TEETH SCREAM FACE

Description. This form of the bared-teeth face is accompanied by raised eyebrows that wrinkle the forehead. The eyes are partially open, as may be the mouth. The lips are withdrawn fully, completely exposing the teeth. Individuals may have piloerection and exhibit forward movements. Unlike the silent bared-teeth display, this expression includes loud, harsh, rasping screams. Van Hooff (1973) describes this version as the most intense vocal display of the chimpanzee, and the calls are high-pitched intermittent screams with sharp timbre. Van Hooff divides screaming into further categories based on sonographic analysis: pulsed scream–rasping scream (see also rough scream, Marler, 1969; roars, growls, and screams, Reynolds & Reynolds, 1965); double-tone scream, which has a higher frequency and the rasping disappears; and pant scream. Kohts describes a similar expression and associates it with sadness [plates 3(7), 16(4), 19(5)], although in her examples, the expression appears to be an extreme version of the pout-whimper (described below), which Joni must have done when his attempts for contact were ignored.

Context and function. The staring scream expression is seen most often in individuals who have been the focus of an attack, although the tone of this expression is one of protest rather than submission. One common eliciting situation is when a dominant male performs a bluff display and haphazardly slaps a nearby individual (figure A7). The forward movements, piloerection, and loud vocalization by the victim of this attack are indicative of protest rather than submission. This expression is extremely variable from individual to individual and is highly context dependent. It is reasonable to assume that it serves a social function by recruiting support in the conflict (de Waal & van Hooff, 1981) or eliciting consolation from bystanders after an attack (de Waal & Aureli, 1996) (figure A8). In Kohts's examples, this expression may be part of a temper tantrum given by Joni when he is denied access to something or if he is ignored for too long.

STRETCHED POUT-WHIMPER

Description. This expression resembles a combination between a bared-teeth display and a pout. The lips are slightly puckered, pushing the mouth corners forward. The mouth corners are withdrawn partially to expose the teeth; at the same time, the upper lip is curled up and protruded as in the pout face, two movements human beings cannot combine easily. The bottom teeth usually are exposed. The eyes and mouth may be open only partially. Kohts provides several nice photographs of this expression in association with the emotional context of general excitement [plates 16(1)–18(13)]. The vocalizations that accompany the whimper are very distinctive and resemble

Figure A7. This figure shows two entirely different responses to a displaying adult male chimpanzee (on the right). The juvenile on the left screams in fear, whereas the adult female in the middle bobs her head while uttering soft pant-grunts in an attempt to appease the intimidating male. (Photograph by Frans de Waal)

Figure A8. Staring bared-teeth scream face. The male chimpanzee in front shows the staring scream face, while his supporter, who is mounting him, screams along with him. Both males are together confronting a political rival, thus forming a coalition against him. (Photograph by Frans de Waal)

a low wailing or moaning. Goodall (1968) refers to this as the whimper face, a more intense form of "hoo-whimper," often used when an infant is ignored. Van Hooff (1973) also describes it as intermediate between a bared-teeth yelp and a pout-moan, for which the lips curl outward and protrude slightly, especially the upper lip. In adults, this may resemble a bared-teeth display more than a pout.

Context and function. The context of use for the stretched pout-whimper may vary. It can occur in an individual when frightened by strange things, begging for food, clinging to its mother, or searching for the nipple (Marler & Tenaza, 1976). It is a good example of a blended expression (i.e., bared teeth and pout), both morphologically and motivationally (figure A9).

SILENT POUT

Description. This expression is characterized most obviously by a slightly open mouth, in which the lips are pushed forward, pursed, and rounded, creating

Figure A9. Stretched pout whimper made by an adult male chimpanzee after he has been defeated in a confrontation with another male. These confrontations are usually non-physical, involving ritualized forms of displays. The male walked away from the intimidating scene, and when he encountered a young female, showed his nervousness by whimpering at her with this half-pouting, half-teeth-baring expression. (Photograph by Frans de Waal)

a small aperture. It is mostly the upper lip that takes the curled posture. The teeth do not show, and no vocalizations are apparent (van Hooff, 1973). This expression can blend into a pout-moan [see plate 29(6)], hooting, or the stretched pout-whimper, depending on the context and intensity of the situation (van Hooff, 1962, 1967). This expression was described frequently by Kohts since it is more common among infant than adult chimpanzees [plates 3(1), 5(3)–5(6), 23, 24].

Context and function. This expression is given most often by young infants when they are distressed. It is believed to have served originally as a function of succor to draw attention to an infant rejected from the nipple by its mother (i.e., the puckered lips are reminiscent of a nursing form) [figures A10(a) and A10(b)]. If the mother does not comfort the infant, the expression may turn into the more extreme form, the stretched pout-whimper, or a temper tantrum, behaviors that Joni appeared to display with some frequency. Over time, this expression became emancipated from its original function to regain nipple access and instead took on a more general function of expressing desire for comfort or food in a variety of situations.

PANT-HOOT

Description. In the pant-hoot, the lips are pushed forward and puckered, resulting in a round opening in the mouth. The mouth can open more widely as louder and more rhythmic vocalizations "hoo-hoo" occur. These vocalizations are voiced on inhalation and exhalation and are used most often in long-distance communication, as others approach, when greeting group members after separation, or to signal the presence of food (Clark & Wrangham, 1993; Goodall, 1968; Marler & Tenaza, 1976; Mitani & Brandt, 1994). Hooting is most often seen in animals older than Joni, which may explain the absence of a hooting expression in Kohts's descriptions. She does, however, describe a kind of vocalization that can accompany a pout, which may be the infant form of the hoot, and that shares many similarities with hooting (chapter 2, General Excitability).

Context and function. Individuals hoot in many different contexts: when engaged in long-range communication, in the presence of desirable food, when greeting individuals who have been absent for a period of time, or when performing the very stereotypical and highly ritualized bluff displays [figure A11(a)]. In the latter situation, these long displays are often followed by a climax scream [figure A11(b)]. The ritualization of this expression is thought to stem from an infant's rejection from the nipple and is therefore a variation of the pout seen in adults as opposed to infants. Rather than pout, adult chimpanzees may begin to hoot when they are denied access to a desired object. Pant-hooting is probably the most ritualized of all chimpanzee facial expressions, taking a highly stereotyped and predictable form.

PANT-GRUNT

Description. Pant-grunts are soft rhythmic vocalizations "uugh-uugh" that can be accompanied by bobbing and/or bowing, often resulting in the vocalizer looking up to the recipient from the crouched position. The upper lip is curled slightly, and the mouth is open. Goodall (1968) describes panting, or bobbing pants, that are given when kissing or bowing to another more domi-

(a)

Figure A10. The pout face is shown in its the most natural
context, (a) and (b, following page) when a juvenile's nursing attempts are rejected. This juvenile is almost 5 years old,
the age at which mothers stop nursing them. (Photographs
by Frans de Waal)

(b)

nant individual. No teeth typically are exposed during the grunts. Van Hooff
(1973) describes the vocalizations as a "rapid oh-oh" series consisting of
breathy rough grunts delivered in rhythmic succession, with both the inhalations and exhalations vocalized. The mouth is open fairly wide, with lips
covering the teeth and the mouth corners drawn slightly forward, creating a
round aperture. This may develop into pant-screams, yelps, or shrill barks
if proximity is reduced or if the individual loses his or her nerve and flees
(Marler & Tenaza, 1976).

Context and function. Pant-grunts are given in the context of dominance. Subordinate individuals greet more dominant ones with a series of pant-grunts,
which can be quite variable across individuals, particularly among different

(a)

(b)

Figure A11. (a, top) Panting-hooting adult male chimpanzee. (b) Pant hooting may end in a higher pitched screaming climax, as made here by an adult male chimpanzee, whose elaborate charging display culminated in a climax while he rhythmically stamped on a resounding metal plate hidden in the grass. (Photographs by Frans de Waal) age and sex classes. Young males pant-grunt more than females in an attempt to strengthen their relationship with high-ranking males. The pant-grunt, or submissive greeting, is considered the chimpanzee's *formal* signal of subordination: It is the only status indicator of the species that is fully consistent in its direction between any two individuals; that is, if A pant-grunts to B, B will never do so to A during the same period (figure A12). For systematic data on the use of this important display, see Bygott (1979), de Waal (1986), Hayaki (1990), and Noë, de Waal, and van Hooff (1980).

TEETH CLACKING

Description. The mouth is opened and vigorously closed in repetitive fashion, making the teeth clack or clap together audibly. The lips remain relaxed and cover the teeth.

Context and function. Teeth clacking typically accompanies social grooming, especially at the beginning or after breaks. It may also be shown by an individual sitting next to others already involved in grooming; in this case, it seems to indicate a readiness to groom.

Figure A12. The pant-grunt. In the chimpanzee, the pant-grunt is a formal status indicator. It is doubtful that Kohts's infant chimpanzee Joni uttered this sound as it is more typical of adults, particularly the males. In this photograph, the dominant male (on the left) raises his hair and walks bipedally, whereas the subordinate (to the right) avoids him with a crouched posture while giving pant-grunts. In reality, these two males are the same size. (Photograph by Frans de Waal)

SPLUTTER

Description. Air is blown out through compressed lips. The facial configuration looks relaxed so that it is sometimes hard to detect which individual is producing the spluttering sounds.

Context and function. Spluttering accompanies social grooming or may be performed before grooming begins, quite analogous to teeth clacking. It can also accompany begging from group mates or human beings. Spluttering has not been described by Goodall (1968; but see Marler, 1976; Marler & Tenaza, 1976), but has been observed in a variety of different captive groups (Burgers Zoo, The Netherlands; Yerkes Field Station; and various zoo populations; Marshall, Wrangham & Clark, in press).

Part 3. Evolutionary Aspects of Primate Facial Communication   Emerging from comparative studies is the idea that facial expressions, as well as other communicative displays, are related phylogenetically across species, an assumption commonly referred to as *homology* (as opposed to *analogy*) in the scientific literature (Andrew, 1965; Chadwick-Jones, 1998; Chevalier-Skolnikoff, 1973; Darwin, 1872; Preuschoft & van Hooff, 1995, 1997; van Hooff, 1972, 1976). Not all of the facial expressions listed in Table A1, for example, are found exclusively in chimpanzees. Also assumed by this is that other species may have expressions not found in the chimpanzee (i.e., see figure A13). The scream, silent bared teeth, relaxed open mouth displays, staccato breathing or laughter, and tooth clacking (also referred to as lip smacking) are prominent displays shared by species of Old World monkeys (Preuschoft & van Hooff, 1997; Redican, 1975; van Hooff, 1962), as well as other Homonoids (i.e., bonobos, orangutans, gorillas, and humans). Evidence for these expressions in gibbons and siamangs, the "lesser" apes, is not so clear at present (Preuschoft & van Hooff, 1995, 1997).

The fact that primates share facial expressions should not come as a surprise, given their common evolutionary ancestry. Yet, the continuity of facial expression from one species to another has been a matter of stormy debate over the last few decades, especially as it relates to the universality of facial expressions in humans (Ekman, 1998). The debate is complicated by the fact that similarly looking displays may have quite different functions in different species. This is particularly troubling for psychologists, who do not always realize that function is a poor criterion of homology (Lorenz, 1941; Preuschoft & van Hooff, 1995; Remane, 1952; Wickler, 1961). As a result, inquiries into the evolutionary origin of facial expressions must build on precise descriptions of their form, or morphology, that are devoid of functional criteria. The appearance of a facial display is characterized best as a gestalt or recurring configuration of various facial elements, that is, the lips, jaw, eyes, and brow are in their specific states, such as retracted, open, squinted, lifted, and so on (Ekman & Friesen, 1978; Hjortsjo, 1970; van Hooff, 1967). This can be extended to include vocalizations, gaze, and posture. Phylogenetic relationships can only be reconstructed among facial displays that are performed regularly in the same fashion and hence are ritualized

Figure A13. An open mouth staring threat face, an expression not known in apes and humans, but common in cercopithecine monkeys, is shown here by two adult female rhesus monkeys. This display is highly ritualized and rarely leads to a fight. (Photograph by Frans de Waal)

in the sense described in part 2. The next step is to investigate how the morpho-logically defined displays are distributed across the phylogenetic tree.

Displays that are shared by a number of closely related species are likely to be ancient signals already present in the last common ancestor of these species. The fact that a relaxed open mouth expression is found in Old World monkeys, great apes, humans, New World monkeys, and prosimians suggests that this dis-play already was present in the insectivorelike common ancestor of all the pri-mates, which would make this display at least 54 million years old. Using the same deductive rationale based on similarities in facial morphology, the human expressions "smile" and "laughter" can be traced back to two distinct displays in nonhuman primates, the silent bared-teeth display and the relaxed open mouth display, respectively (Preuschoft & van Hooff, 1995, 1997). The distribution of these displays in the phylogenetic tree suggests that they are ancient features of our lineage and potentially are universal to all primates.

The comparative approach shows that facial expressions that resemble each other in appearance may fulfill different social functions in closely related species, while they may be used in the same context by species that are related only distantly (Preuschoft, 1995; Preuschoft & van Hooff, 1995, 1997). For instance, the silent bared-teeth display, assumed to be homologous to the human smile, is used by rhesus monkeys, long-tailed macaques, Barbary macaques, and chimpan-zees to signal submission, de-escalating aggression, and appeasement (Angst, 1975; de Waal & Luttrell, 1985; Preuschoft, 1995; van Hooff, 1972). Virtually the same expression is used to communicate social attraction, affection, pleasure, and amusement in humans (Frank, Ekman, & Friesen, 1993), bonobos (de Waal, 1988), and Tonkean and lion-tailed macaques (Preuschoft, 1995).

Remarkably, some species, including humans, gelada baboons, and Tonkean and lion-tailed macaques, have derived an open mouth bared-teeth display that replaces the relaxed open mouth display and appears to be a blend between the bared-teeth smile and the relaxed open mouth display. Evidently, even signals that represent a common evolutionary heritage are exposed to different selection pressures in different species. The principal factors of selection are rooted in the perceptual disposition of the receiver and the degree to which the interests of the sender and receiver conflict or converge (Krebs & Davies, 1984).

Current evidence suggests that the precise function of displays such as smiling or laughter can be predicted on the basis of the dominance style that species. *Dominance style* refers to covarying features of social behavior that are indicative of the degree of power asymmetry among the members of a social group (de Waal, 1989; Thierry, 1985). These include the intensity and direction of ag-gression, the rate of reconciliation, the distribution of profits from competition, the degree of favoritism toward kin, and the ratio of aggression to affiliation. When the dominance style of a species is relaxed, the social functions of submis-sion, affiliation, and playfulness are not able to be separated clearly, and the appearance of the displays used in these contexts tends to blend as well (Preu-schoft & van Hooff, 1997). If, however, dominance is enforced strictly, the dis-plays of submission, socially positive tendencies, and playful interactions are dis-cerned clearly.

The concept of dominance style also ties in with socioecological theories of resource distribution and their influence on the way in which competition is executed (van Schaik, 1989; Vehrencamp, 1983; Wrangham, 1980). A similarly strong socioecological influence on the shape and the graded nature of displays is evident when the vocalizations of common chimpanzees and mountain gorillas are compared (Marler, 1976). Chimpanzees use more graded signals, while gorillas employ more discrete versions of what are believed to be homologous vocalizations. Gorillas live in a forested habitat where dense foliage constrains how far sounds can travel naturally without distortion. In contrast, chimpanzees live in more open terrain where subtle gradations in acoustic parameters are less likely to be lost. Yet, despite this acoustically favorable environment, Marler showed that chimpanzees often uttered vocalizations that were prototypical and unblended (i.e., the laughter and cough vocalization). On the other end of the continuum were vocalizations, such as the woaow-bark and scream, that were expressed concurrently as blends, or intermediates, between call types.

It is unclear why chimpanzees produce expressions and vocalizations in both blended and stereotypical forms, although the preceding paragraph discussed several influential variables. Even less is known, however, about how chimpanzees perceive their own communicative displays, whether they recognize that vocalizations may be in a pure or blended form, and what social or environmental cues they use to help interpret these signals. In the domain of cognitive psychology, the importance of both the sender's message and the factors that influence the perception of the receiver is receiving more attention. In a species that has a communicative repertoire as complex and variable as the chimpanzee, there is no necessarily straightforward correspondence between message and meaning. For this reason, research at the Living Links Center has focused on how chimpanzees perceive their own signals and on what basis they organize these signals into discrete categories.

Part 4. Experiments on the Chimpanzee's Perception of Facial Expressions

Communication of Emotions in Monkeys. One of the first psychologists to speculate about the relationship between emotion and social behavior was Harlow at the Wisconsin Regional Primate Research Center in Madison. His research focused on the emotional needs of young rhesus monkeys (*Macaca mulatta*) during their early stages of development (Suomi & Leroy, 1982). While these studies did not emphasize the importance of facial expressions or emotional communication explicitly, this research was pivotal in documenting the severe consequences that resulted when young rhesus monkeys were deprived of social contact with their mother from birth and were raised in isolation (Harlow, Harlow, & Suomi, 1971). The attachment of an infant to its mother was not based simply on the need for food, but on the even more fundamental need for what Harlow termed "love" and "affection." These studies raised awareness within the primatological community, and far beyond, as to the importance of early social experience for normal social and emotional development (Harlow & Mears, 1983).

During this period, Sackett, a student of Harlow, demonstrated that socially naive monkeys raised in isolation produced threatening facial expressions in response to photographs of threatening conspecifics (Sackett, 1966). This was interpreted as evidence that social experience was not necessary for the appropriate production of facial expressions, but that these expressions were the product of an innate releasing mechanism (Sackett, 1965). When the isolation-reared monkeys were reintroduced into a social group, however, they were unable to respond appropriately to the range of facial displays and complex emotional messages produced by their conspecifics. It seemed that, while the appropriate production of facial expressions may be the result of innate mechanisms, the comprehension of facial displays made by other individuals in a social context requires a period of normal social and emotional development.

Sackett's studies suggest that the subtle context-dependent meaning of facial expressions can be learned only from direct social experience. A direct gaze, for example, is considered threatening to rhesus monkeys; hence, an important component of a threat face is to stare directly at the opponent (Hinde & Rowell, 1962; van Hooff, 1962). But, do monkeys understand this cue from birth, or must they see others engage in these interactions before they understand their meaning? To study the development of how infant rhesus monkeys respond to a direct stare threat, 1-, 3-, and 7-week-old rhesus infants were tested for their responses to faces of conspecifics with direct or averted gaze (Mendelson, Haith, & Goldman-Rakic, 1982). These faces were presented using projection slides. No differences in viewing were detected in the 1-week-old infants. By 3 weeks of age, however, infant monkeys showed social responses and looked for a longer time at the faces with the direct gaze than at those with the averted gaze. This increased viewing was accompanied by negative emotional responses, that is, the individuals squealed submissively, grinned, and lip-smacked at the staring faces. By 7 weeks of age, the infants avoided looking at the direct-gaze faces altogether. Instead, these infants preferred to look at the faces with the averted gaze and no longer produced emotional responses in their presence. These results suggest that, by 3 weeks of age, rhesus monkey infants understand the social implications of the direct stare and respond submissively to this threatening stimulus. These infants, however, do not as yet have the social competence or knowledge to avoid looking at these faces. By 7 weeks of age, the infants' behavior changed dramatically. They now learned to respond to this socially threatening stimulus by looking away, a response that is critically important for integration in a rhesus macaque social group. It is important to note that these monkeys lived in social groups when they were not participating in the experiments. Unlike the experiments of Harlow and Sackett, these studies are unable to provide any evidence that these responses developed innately because the monkeys were able to learn them within a social group.

A series of intriguing studies was performed in the 1960s by Miller and colleagues. They demonstrated that rhesus monkeys could communicate affective information to one another using facial expressions. Specifically, Miller, Murphy, and Mirsky (1959) exposed monkeys to a photograph of a familiar group mate with a neutral face. The presentation of this conditioned stimulus (CS) was paired with a shock that subjects could only avoid by pressing a lever in the test cage. Subjects quickly learned to press the lever when this face appeared, thereby avoiding the shock (i.e., standard operant conditioning). Next, subjects were shown another photograph of the same CS individual, this time making a fearful facial expression. When subjects were shown this new stimulus, they spontaneously produced significantly more avoidance responses than during the acquisition of the conditioned response when the CS was the same monkey with a neutral face. This suggests that a negative facial expression is much more effective in communicating an upcoming aversive event (i.e., the shock) than a neutral face, illustrating that at some level, the monkeys understood its inherent negative emotional meaning.

In a second study, Miller, Banks, and Ogawa (1963) tested whether a stimulus monkey and a responder monkey were able to cooperate in producing an avoidance response. In this paradigm; only the stimulus individual could see the CS that signaled the upcoming shock, but unlike the previous situation, this individual had no access to the avoidance lever. In another room, the responder monkey could access the lever, but could not see the projection screen that displayed the CS. The responder monkey could only see the face of the stimulus monkey on a small black-and-white video monitor, which provided live video feedback of the stimulus monkey in the other room. The idea was that changes in the facial expression of the stimulus individual as it viewed the CS would communicate the upcoming danger to the responder monkey, who in turn would press the lever so that they both would avoid the shock. The result of this cooperative avoidance paradigm was that the responder monkey produced more lever presses during the period when the CS was visible to the stimulus monkey, indicating that the stimulus monkey's facial expression was sufficient in communicating the appropriate affective information to the responder monkey.

These studies demonstrate several important characteristics of facial expressions in nonhuman primates. First, these expressions have an innate component. This includes both the production of specific facial expressions, such as the bared-teeth display, and the type of information that these expressions communicate. Thus, threatening facial expressions are better for communicating negative events than neutral faces. They also provide an excellent example of the importance of normal social development and social experience for learning the appropriate meaning of these expressions and to be able to use them with sufficient skill.

These results in monkeys are similar to what is known about the development of facial expressions in human infants. Human infants, for example, produce a variety of facial displays from birth, illustrating their reflexive nature (Meltzoff & Moore, 1977; Steiner, 1974). Other studies have shown that even very young human infants are able to discriminate between some facial expressions, indicating that the recognition of some facial expressions may be more adaptive than others; this skill emerges before the recognition of other expressions. Nelson

(1987) concluded that human infants as young as 4 months are able to discriminate between some facial expressions (i.e., happiness and fear), but it is not until the infants reach 2 years of age that they begin to understand the emotional meaning that these expressions communicate. These basic findings are similar to the way in which vocalizations develop in both human and nonhuman primates, an area of research that has received considerably more attention than the development of facial expressions. This developmental course is one in which the ability to produce the expression precedes its comprehension (Cheney & Seyfarth, 1990; Seyfarth & Cheney, 1986). In addition, knowing when it is socially appropriate to make certain facial expressions appears also to be regulated by a long period of normal social development and social experience (Rinn, 1984). These *display rules* as Ekman and Friesen (1971) have termed them, provide evidence that, although facial expressions are determined biologically (meaning that they develop under the influence of a strong genetic predisposition), they are subject to considerable social shaping and cultural rules of display.

Kohts's descriptions seem to support these general principles, albeit not with systematic experimentation. Her descriptions of Joni's response to a large drawing of a chimpanzee suggest that he was both curious and a little fearful of this image. It is unclear to what extent Joni had experience with other chimpanzees in his young life, but his curiosity seemed to indicate the inherent salience of the face of a conspecific compared with other animals or inanimate objects. Since Joni died very early in his life, it is unclear how his responses to this and other types of social images would have changed as he matured.

Research on the early developmental life of the chimpanzee has been pioneered at the Yerkes Primate Center by Bard and coworkers. Bard (1994a) demonstrated highly similar patterns of development between humans and chimpanzees in the first month of life. Neonatal chimpanzees show strong orienting responses to social stimuli (i.e., an experimenter's face) and exhibit motor and attentional reflexes comparable with human neonates (Bard, 1994b). Bard's studies also demonstrate that chimpanzees reared with humans exhibit many of the same emotional facial expressions and vocalizations present in adult chimpanzees and human infants, suggesting that, while Joni's behavior may not have been identical to that of a mother-reared chimpanzee, it is reasonable to conclude that it was at least comparable (Bard, 1998).

Currently, the Living Links Center is involved in several long-term projects investigating the interaction between cognitive processes and social perception in chimpanzees. The goal of this research is to gain a more complete understanding of the way in which chimpanzees perceive their social environment, including their facial expressions, and the extent to which they associate a social/emotional meaning to these signals.

Specifically, one series of experiments examined different aspects of face perception, including how chimpanzees identify one another, and whether face recognition is disrupted when the faces are altered in various ways, such as changing their normal orientation and masking certain features. These studies demonstrate that chimpanzees recognize unfamiliar individuals using facial cues, and that the eyes appear to be the most important feature (Parr, Winslow, Hopkins & de

Waal, 2000). Also, inverted faces are more difficult to recognize than faces in their normal upright orientation (Parr, Dove, & Hopkins, 1998). Together, these results strongly suggest that chimpanzees process faces differently from other stimuli (i.e., photos of automobiles or abstract shapes), but not any differently from the way in which humans process facial information.

This research involved the use of a computerized joystick testing paradigm (figure A14). Subjects were trained to manipulate a cursor displayed on a computer screen by controlling the movements of a joystick. Training was performed using the Language Research Center Training Software (LTS) developed by Rumbaugh and colleagues at the Language Research Center, Georgia State University, Atlanta (Washburn & Rumbaugh, 1992). This software includes a variety of tasks that assess psychomotor performance, as well as other cognitive skills such as short-term memory and spatial navigation.

Recently, two experiments were conducted to assess the ability of five adult chimpanzees to categorize facial expressions typical of their own species (for a more detailed description, see Parr, Hopkins, & de Waal, 1998). These tests were performed using a variation of the well-known short-term memory task—sequential matching to sample (SMTS). Subjects were presented first with a sample image on the computer monitor. This represented the stimulus to be matched and consisted of an unfamiliar chimpanzee showing one of five facial expressions: bared-teeth display, hoot face, scream face, relaxed-lip face, and relaxed open mouth display (see figure A15). These stimuli were all taken from the unique Living Links Stimulus Set®, a large collection of high-quality black-and-white photographs of chimpanzees and other species of nonhuman primates. The pho-

Figure A14. The computerized joystick testing paradigm. (Photograph by Frans de Waal)

Figure A15. An example of six facial expressions (left to right): neutral, hoot face, bared-teeth display, play face, scream, and relaxed-lip face. (Photographs by Frans de Waal)

tographs, taken mostly by Frans de Waal, are archived at the Living Links Center, Emory University (Atlanta, Ga.), and are available to researchers on request (www.emory.edu/LIVING_LINKS).

Subjects were required to move the joystick-controlled cursor to contact the sample image. When this was done, the sample was cleared from the screen, and two comparison stimuli were presented. These always depicted individuals who were different from the one in the sample (i.e., each trial showed three different chimpanzees). The correct comparison matched the sample in that it showed another example of the same facial expression made by a different individual. The other nonmatching comparison showed either a neutral portrait or one of the remaining four facial expressions. Therefore, these experiments required subjects to recognize the same facial expression made by two different unfamiliar individuals and to indicate this by selecting that facial expression using the joystick-controlled cursor.

Experiment 1: Facial expression matching. Five adult chimpanzees, three males and two females approximately 10 years of age, participated in these experiments. They all were raised by humans in the Yerkes Primate Center nursery and then were moved into social groups when they were around 4 years of age. They had been trained previously to use the joystick paradigm and had some experience with the matching-to-sample format, although they had never been exposed to photographs of facial expressions before these experiments. All of the stimuli used depicted individuals unfamiliar to the experimental subjects. These photographs were taken mostly from groups living outside the United States, precluding any recognition of the individuals on the basis of familiarity or the recognition of genetic similarities and the like. The first experiment tested whether these subjects could match photographs of two unfamiliar chimpanzees displaying the same facial expression. For the purpose of this experiment, the nonmatching comparison always showed a third individual's neutral portrait.

There were 22 matching trials involving the five different facial expressions. These included discrimination of six play faces, five scream faces, five hoot faces, four silent bared-teeth displays, and two relaxed-lip facial expressions. It should be emphasized that there was no overlap in the individuals depicted in any of the stimulus photographs. Therefore, the photographs of three different individuals were combined in each stimulus set to prevent subjects from matching the facial expressions on the basis of photographic similarity or by selecting the same individual within a trial. A total of 66 different photographs were needed to make this task. According to this arrangement, subjects had to attend to the expressions themselves to make the discriminations accurately. Figure A16 gives an example of a typical stimulus set that depicts the bared-teeth display as the correct comparison on the left, paired with a neutral portrait on the right as the nonmatching comparison.

The acquisition data for this experiment after 5 days of testing can be seen in figure A17. Performance is plotted across the five testing sessions, for which subjects received approximately 50 trials per session. This clearly shows that three of the five facial expressions (the bared-teeth display, scream, and relaxed open mouth display) were discriminated above chance levels on the first session, that is, Day 1. This makes any explanation of performance based on subjects learning which stimulus was correct from their prior history of reinforcement extremely unlikely. More interesting, however, is the fact that the relaxed-lip face was never discriminated above chance levels from the nonmatching neutral face. In fact, by Day 5 subjects more reliably selected the plain neutral comparison, even though they were never rewarded for doing so. This finding was very interesting and seemed to indicate that, although subjects were able to use the distinctive feature of the relaxed-lip face, the droopy lower lip, to discriminate it from the plain neutral face, they did not. Considering the fact that both of these

Figure A16. An example of a stimulus set presented in the ex-pression discrimination task. The correct choice is the bared-teeth display on the lower left. (Photograph by Frans de Waal)

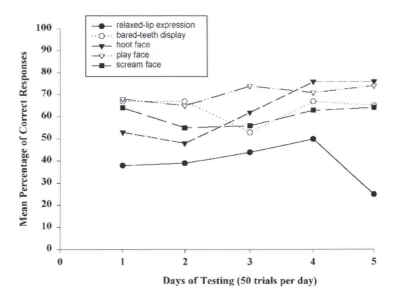

Figure A17. Performance on the Expression Discrimination Task over 5 days of testing.

expressions are functionally neutral in a social context and communicate no overt emotional or motivational predisposition (see part 2), these results raise the possibility that subjects may have relied on the emotional content or intensity of the faces to perform the discriminations instead of using distinctive facial features. Determining the extent to which the chimpanzees relied on distinctive facial features to discriminate their facial expressions was the focus of the next experiment.

Experiment 2: Dyadic expression discrimination. In this task, stimuli were arranged systematically so that each of the five expressions was combined with every other expression as the nonmatching comparison. The result was 20 directional, dyadic stimulus sets in which every expression was paired with every other expression. Due to the limited number of expressions that were available, the pair of photographs used as the sample and correct comparison stimulus was repeated for each unique nonmatching comparison expression. Therefore, it was possible in principle for subjects simply to learn the correct stimulus in a pair through repeated exposures over the course of the experiment.

The stimulus sets that resulted from the dyadic combination technique could be divided into two main types, those in which the nonmatching comparison shared features in common with the sample expression and those in which the two expressions were distinct featurally. A dyad with similar features is shown in figure A18(a); a scream face sample and a nonmatching relaxed open mouth face are combined. Figure A18(b) shows an example of a dyad with distinct features. This shows the same sample and correct comparison expressions as in

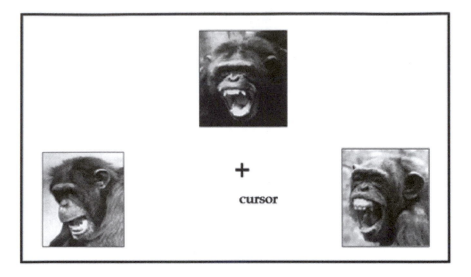

(a)

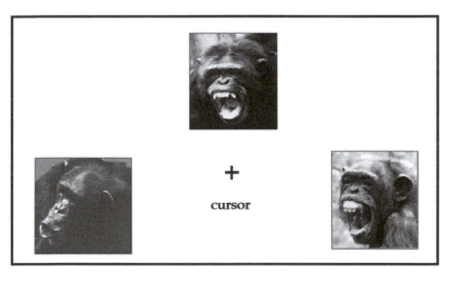

(b)

Figure A18. (a, top) A stimulus set in which the sample (a scream face) and nonmatching comparison expression (a play face) share similar features. (b) A stimulus set in which the sample (a scream face) and nonmatching comparison expression (a hoot face) are distinct in their features. (Photographs by Frans de Waal)

figure 18a but with a hoot face as the nonmatching comparison, an expression that does not share any features with the scream face. The hypothesis under investigation was that, when the correct and nonmatching comparison expressions were similar in appearance (i.e., shared similar facial features), the discrimination would become more difficult, and performance would deteriorate compared to trials in which the correct and nonmatching comparison expressions appeared very distinct.

Data from the dyadic expression matching task were analyzed by comparing performance on those stimulus sets in which the sample and nonmatching comparison expressions shared similar features with performance on those dyads in which the two expressions appeared distinct. According to this analysis, subjects performed significantly better on those dyads in which the features of the expressions were distinct, as opposed to similar, supporting the hypothesis that chimpanzees rely on the distinctiveness of facial features to categorize facial expressions. This finding was inconsistent with the results of the expression discrimination task, which showed subjects were unable to use the distinctive feature of the relaxed-lip face to discriminate it from the neutral face.

If distinctive features were used exclusively in this experiment, one would predict that the overall correlation between performance and the number of features that the expressions in a trial shared would be significantly negative: The more features the expressions shared, the worse performance should be. This, however, was supported only weakly, mainly due to the fact that the trend was true only for some of the expression types (i.e., the bared-teeth display, the hoot face, and the relaxed-lip face). Two expressions showed the reverse pattern, with performance better when the sample and nonmatching comparison expressions shared similar features (i.e., the scream face and the relaxed open mouth display). So, while there may be a trend for features to matter when categorizing facial expressions, it does not seem to be true for all expressions and does not appear to be the exclusive feature used.

In summary, our research showed that chimpanzees do not seem to rely on the distinctiveness of facial features to discriminate facial expressions, although there was a trend for distinctiveness of facial features to be important in the dyadic expression-matching task. One might speculate that subjects relied on one type of feature more than another (i.e., emotional content and intensity), and only if this feature was not easily discerned did they move on to another discriminative cue. In addition, subjects may have perceived the facial expressions differently due to morphological variability in the way different expressions look when made by different individuals (i.e., a correct pair in these experiments showed two different individuals making the same facial expression). It also raises the possibility that the expression types that did not seem to be affected by an overlap in facial features (i.e., the scream face and the relaxed open mouth display) may be perceived as examples from a fuzzy category. Within a social group, these expressions may be interpreted differently based on the social context in which the expression was used, environmental conditions, the individual who performed the expression, and so on. Future studies should try to isolate these possible cues and to test each of them systematically. In addition, the presentation of facial expressions as static black-and-white stimuli, as done in our experiments, is hardly analogous to the way in which these dynamic social displays are encountered normally. It may have been that, in the absence of ecologically valid cues like vocalizations and/or dynamic motion, subjects attended to perceptual similarities in the stimulus array that they may have focused on in a more naturalistic setting. Our ongoing research utilizes new digital technology to make dynamic presentations that will allow facial expressions to be displayed in short video clips complete with an audio track. This will provide a more ecologically relevant stimulus for categorization and will allow subjects to use cues that may be closer to the way in which facial expressions are processed in their social environment.

Conclusions and Future Directions.   Research on facial expressions and emotion in nonhuman primates illustrates several important findings. First, chimpanzees show a rich variety of highly graded facial expressions and vocalizations. Their expressions show ritualized elements, but are among the most diverse in the primate world. Second, the phylogenetic continuity of facial expressions across the primate order should not be surprising. Some of these expressions are homologous to expressions in other Old World monkeys and even humans. Differences and similarities in the message and meaning of these signals can be explained by considering the social and ecological constraints of that species, such as habitat, social organization, and dominance style. Third, facial expressions are perceived as salient and discriminable stimuli by chimpanzees even when they are presented as black-and-white two-dimensional images. The ability to categorize facial expressions most likely involves more than just the recognition of distinctive facial features and highlights the versatility in the perceptual organization of chimpanzees as a species and the importance of social context for the accurate interpretation of emotional signals.

Finally, while researchers have speculated about the relationship between behavior and emotional states in primates for centuries, a single accepted method for measuring emotional perception in nonverbal organisms has yet to be established, despite the relatively well-established relationship between facial expressions and emotional responses in humans. The Living Links Center focuses its study of chimpanzee communication by combining naturalistic observations of behavior with controlled experimentation in a modern, computerized setting. Kohts's pioneering observations serve as a continuing inspiration for these studies.

REFERENCES

Andrew, R. J. (1963a). Evolution of facial expression. *Science*, *142*, 1034–1041.

Andrew, R. J. (1963b). The origin and evolution of the calls and facial expressions of the primates. *Behaviour*, *20*, 1–109.

Andrew, R. J. (1965). The origins of facial expressions. *Scientific American*, *213*, 88–94.

Angst, W. (1975). Basic data and concepts on the social organization of *Macaca fascicularis*. In L. A. Rosenblum (Ed.), *Primate Behavior* (pp. 325–388). New York: Academic Press.

Bard, K. A. (1994a). Similarities and differences in the neonatal behavior of chimpanzee and human infants. In G. Eder, E. Kaiser, & F. A. King (Eds.), *The Role of the Chimpanzee in Research, Symposium, Vienna 1992* (pp. 43–55). Basel: Karger.

Bard, K. A. (1994b). Very early social learning: The effect of neonatal environment on chimpanzees' social responsiveness. In J. J. Roeder, B. Thierry, J. R. Anderson, & N. Herrenschmidt (Eds.), *Current Primatology: Volume 2, Social Development, Learning and Behavior*, pp. 339–346). Strasbourg: Université Louis Pasteur.

Bard, K. A. (1998). Socialexperiental contributions to imitation and emotion in chimpanzees. In S. Braten (Ed.), *Intersubjective Communication and Emotion in Early Ontogeny* (pp. 208–227). Cambridge: Cambridge University Press.

Bowlig, N. (1962). Facial expression in primates with remarks on a parallel development in certain carnivores (a preliminary report on work in progress). *Behaviour*, *22*, 167–192.

Bygott, J. D. (1979). Agonistic behavior, dominance, and social structure in wild chimpanzees of the Gombe National Park. In D. Hamburg & E. McCown (Eds.), *The Great Apes* (pp. 405–428). Menlo Park, Calif.: Benjamin/Cummings.

Chadwick-Jones, J. (1998). *Developing a Social Psychology of Monkeys and Apes*. East Sussex, UK: Psychology Press.

Cheney, D., & Seyfarth, R. M. (1990). *How Monkeys See the World*. Chicago: Chicago University Press.

Chevalier-Skolnikoff, S. (1973). Facial expression of emotion in nonhuman primates. In P. Ekman (Ed.), *Darwin and Facial Expressions* (pp. 11–89). New York: Academic Press.

Clark, A., & Wrangham, R. (1993). Acoustic analysis of wild chimpanzee pant hoots: Do Kibale chimpanzees have an acoustically distinct arrival pant hoot? *American Journal of Primatology*, *31*, 99–109.

Custance, D., & Bard, K. A. (1994). The comparative and developmental study of self-recognition and imitation: The importance of social factors. In S. T. Parker, R. W. Mitchell, & M. L. Boccia (Eds.), *Self-Awareness in Animals and Humans: Developmental Perspectives* (pp. 207–226). Cambridge: Cambridge University Press.

Darwin, C. (1998 [1872]). *The Expression of the Emotions in Man and Animals* (3rd ed.). New York: Oxford University Press.

de Waal, F. B. M. (1986). Integration of dominance and social bonding in primates. *Quarterly Review of Biology*, *61*, 459–479.

de Waal, F. B. M. (1988). The communicative repertoire of captive bonobos (*Pan paniscus*) compared to that of chimpanzees. *Behaviour*, *106*, 183–251.

de Waal, F. B. M. (1989). Dominance "style" and social organization. In V. Standon & R. A. Foley (Eds.), *Comparative Socioecology* (pp. 243–264). Oxford: Blackwell.

de Waal, F. B. M. (1996). *Good Natured: The Origin of Right and Wrong in Humans and Other Animals*. Cambridge: Harvard University Press.

de Waal, F. B. M. (1997). Are we in anthropodenial? *Discover*, *18*, 50–53.

de Waal, F. B. M., & Aureli, F. (1996). Consolation, reconciliation, and a possible cognitive difference between macaque and chimpanzee. In K. A. Bard, A. E. Russon, & S. T. Parker (Eds.), *Reaching Into Thought: The Minds of the Great Apes* (pp. 1–34). Cambridge: Cambridge University Press.

de Waal, F. B. M., & Luttrell, L. M. (1985). The formal heirarchy of rhesus macaques: An investigation of the bared-teeth display. *American Journal of Primatology*, *9*, 73–85.

de Waal, F. B. M., & van Hooff, J. A. R. A. M. (1981). Side-directed communication and agonistic interactions in chimpanzees. *Behaviour*, *77*, 164–198.

Ekman, P. (1972). Universal and cultural differences in facial expressions of emotion. In J. K. Cole (Ed.), *Nebraska Symposium on Motivation* (Vol. 19, pp. 207–293). Lincoln: University of Nebraska Press.

Ekman, P. (1992). Facial expressions of emotion: An old controversy and new findings. *Philosophical Transaction of the Royal Society of London*, *335*, 63–69.

Ekman, P. (1998). Afterword. In C. Darwin, *The Expression of the Emotions in Man and Animals* (3rd ed., pp. 363–394). New York: Oxford University Press.

Ekman, P., & Friesen, W. V. (1971). Constants across cultures in the face and emotion. *Journal of Personality and Social Psychology*, *17*, 124–129.

Ekman, P., & Friesen, W. V. (1978). *Facial Action Coding System*. Palo Alto, Calif.: Consulting Psychologists Press.

Ekman, P., Friesen, W. V., & Ellsworth, P. (1972). *Emotion in the Human Face: Guidelines for Research and an Integration of Findings*. New York: Pergamon Press.

Fagen, R. (1981). *Animal Play Behavior*. New York: Oxford University Press.

Fouts, R. S., Chown, B., & Goodin, L. (1976). Transfer of signed responses in American Sign Language from vocal English stimuli to physical object stimuli by a chimpanzee (*Pan*). *Learning and Motivation*, *7*, 458–475.

Fox, M. W. (1969). A comparative study of the development of facial expressions in canids; wolf, coyote, and foxes. *Behaviour, 36*, 4–73.

Frank, M. G., Ekman, P., & Friesen, W. V. (1993). Behavioral markers and recognizability of the smile of enjoyment. *Journal of Personality and Social Psychology, 64*, 83–93.

Fridlund, A. J. (1994). *Human Facial Expression*. New York: Academic Press.

Gardner, R. A., & Gardner, B. T. (1969). Teaching sign language to a chimpanzee. *Science, 165*, 664–672.

Goodall, J. (1986). *The Chimpanzees of Gombe: Patterns of Behavior*. Cambridge: Belknap Press of Harvard University Press.

Goodall, J. van Lawick (1968). A preliminary report on expressive movements and communication in the Gombe Stream chimpanzees. In P. C. Jay (Ed.), *Primates: Studies in Adaptation and Variability* (pp. 313–519). New York: Holt, Rinehart & Winston.

Gouzoules, H., Gouzoules, S., & Tomaszycki, M. (1998). Agonistic screams and the classification of dominance relationships: Are monkeys fuzzy logicians? *Animal Behaviour, 55*, 51–60.

Green, S. (1975). Communication by a graded vocal system in Japanese monkeys. In L. A. Rosenblum (Ed.), *Primate Behavior* (Vol. 4, pp. 1–102). New York: Academic Press.

Harlow, H. F., Harlow, M. K., & Suomi, S. J. (1971). From thought to therapy: Lessons from a primate laboratory. *American Scientist, 59*, 538–549.

Harlow, H. F., & Mears, C. E. (1983). Emotional sequences and consequences. In R. Plutchik & H. Kellerman (Eds.), *Emotion: Theory, Research, and Experience* (pp. 171–197). New York: Academic Press.

Hayaki, H. (1990). Social context of pant-grunting in young chimpanzees. In T. Nishida (Ed.), *The Chimpanzees of the Mahale Mountains* (pp. 189–206). Tokyo: University of Tokyo Press.

Hayes, C. (1951). *The Ape in Our House*. New York: Harper and Brothers.

Hinde, R. A. (1970). *Animal Behaviour: A Synthesis of Ethology and Comparative Psychology* (2nd ed.). New York: McGraw-Hill.

Hinde, R. A., & Rowell, T. E. (1962). Communication by postures and facial expressions in the rhesus monkey (*Macaca mulatta*). *Proceedings of the Zoological Society of London, 138*, 1–21.

Hjortsjo, C. H. (1970). *Man's Face and Mimic Language*. Lund, Sweden: Studentlitteratur.

Izard, C. E. (1971). *The Face of Emotion*. New York: Appleton-Century-Crofts.

Janik, V. M. (1999). Pitfalls in the categorization of behaviour: A comparison of dolphin whistle classification methods. *Animal Behaviour, 57*, 133–143.

Kellogg, W. N., & Kellogg, L. A. (1933). *The Ape and the Child*. New York: Whittlesey House.

Köhler, W. (1925). *The Mentality of Apes*. New York: Vintage Books.

Krebs, J. R., & Davies, N. B. (1984). *Behavioural Ecology: An Evolutionary Approach*. Oxford: Oxford University Press.

Lorenz, K. (1941). Vergleichende Bewegungsstudien an Anatinen. *Journal für Ornithologie, 89*, Sonderheft, 194–294.

Marler, P. (1969). Vocalizations of wild chimpanzees, an introduction. *Proceedings of the Second International Congress of Primatology, Atlanta, 1*, 94–100.

Marler, P. (1976). Social organization, communication and graded signals: The chimpanzee and the gorilla. In P. P. Bateson & R. A. Hinde (Eds.), *Growing Points in Ethology* (pp. 239–279). Cambridge: Cambridge University Press.

Marler, P., & Tenaza, R. (1976). Signaling behavior of apes with special reference to vocalization. In T. Sebeok (Ed.), *How Animals Communicate* (pp. 965–1033). Bloomington: Indiana University Press.

Marshall, A. J., Wrangham, R. W., & Clark, A. (1999). Does learning affect the structure of vocalizations in chimpanzees? *Animal Behaviour, 58,* 825–830.

Massaro, D. W. (1987). Categorical partition: A fuzzy-logical model of categorization behavior. In S. Harnad (Ed.), *Categorical Perception* (pp. 254–283). Cambridge: Cambridge University Press.

Meltzoff, A. N., & Moore, M. K. (1977). Imitation of facial and manual gestures by human neonates. *Science, 198,* 75–78.

Mendelson, M. J., Haith, M. M., & Goldman-Rakic, P. S. (1982). Face scanning and responsiveness to social cues in infant rhesus monkeys. *Developmental Psychology, 18,* 222–228.

Miller, R. E., Banks, J. H. J., & Ogawa, N. (1963). Role of facial expressions in "cooperative avoidance conditioning" in monkeys. *Journal of Abnormal and Social Psychology, 67,* 24–30.

Miller, R. E., Murphy, J. V., & Mirsky, I. A. (1959). Relevance of facial expression and posture as cues in communication of affect between monkeys. *Archives of General Psychiatry, 1,* 480–488.

Mitani, J. C., & Brandt, K. L. (1994). Social factors influence the acoustic variability in the long-distance calls of male chimpanzees. *Ethology, 96,* 233–252.

Morris, D. (1957). "Typical intensity" and its relation to the problem of ritualization. *Behaviour, 11,* 1–12.

Moynihan, M. (1998). *The Social Regulation of Competition and Aggression in Animals.* Washington: Smithsonian Institution Press.

Nelson, C. A. (1987). The recognition of facial expressions in the first two years of life: Mechanisms and development. *Child Development, 58,* 889–909.

Noë, R., de Waal, F. B. M., & van Hooff, J. A. R. A. M. (1980). Types of dominance in a chimpanzee colony. *Folia Primatologica, 34,* 90–110.

Parr, L. A., Dove, T. A., & Hopkins, W. D. (1998). Why faces may be special: Evidence of the inversion effect in chimpanzees (*Pan troglodytes*). *Journal of Cognitive Neuroscience, 10,* 615–622.

Parr, L. A., Hopkins, W. D., & de Waal, F. B. M. (1998). The perception of facial expressions in chimpanzees (*Pan troglodytes*). *Evolution of Communication, 2,* 1–23.

Parr, L. A., Winslow, J. T., Hopkins, W. D., & de Waal, F. B. M. (2000). Recognizing facial cues: Individual recognition in chimpanzees (*Pan troglodytes*) and rhesus monkeys (*Macaca mulatta*). *Journal of Comparative Psychology, 114,* 47–60.

Preuschoft, S. (1995). *"Laughter" and "Smiling" in Macaques: An Evolutionary Perspective.* Utrecht, The Netherlands: University of Utrecht.

Preuschoft, S., & Preuschoft, H. (1994). Primate nonverbal communication: Our communicative heritage. In W. Noeth (Ed.), *Origins of Semiosis* (pp. 61–100). Berlin: Mouton de Gruyter.

Preuschoft, S., & van Hooff, J. A. R. A. M. (1995). Homologizing primate facial displays: A critical review of methods. *Folia Primatologica, 65,* 121–137.

Preuschoft, S., & van Hooff, J. A. R. A. M. (1997). The social function of "smile" and "laughter": Variations across primate species and societies. In U. Segerstrale & P. Molnar (Eds.), *Nonverbal Communication: Where Nature Meets Culture* (pp. 171–189). Mahwah, N.J.: Erlbaum.

Redican, W. K. (1975). Facial expressions in nonhuman primates. In L. A. Rosenblum (Ed.), *Primate Behavior* (pp. 103–194). New York: Academic Press.

Redican, W. K. (1982). An evolutionary perspective on human facial displays. In P. Ekman (Ed.), *Emotion in the Human Face* (pp. 212–280). Cambridge: Cambridge University Press.

Remane, A. (1952). *Die Grundlagen des natürlichen Sytems der vergleichenden Anatomie und der Phylogenetik.* Leipzig, Germany: Geest & Portig.

Reynolds, V., & Reynolds, F. (1965). Chimpanzees of the Budongo Forest. In I. De Vore (Ed.), *Primate Behavior* (pp. 368–424. New York: Holt, Rinehart & Winston.

Rinn, W. E. (1984). The neuropsychology of facial expression: A review of the neurological and psychological mechanisms for producing facial expressions. *Psychological Bulletin, 95,* 52–77.

Russell, J. A., & Fernández-Dols, J. M. (1997). *The Psychology of Facial Expression.* Cambridge: Cambridge University Press.

Sackett, G. P. (1965). Response of rhesus monkeys to social stimulation presented by means of colored slides. *Perceptual and Motor Skills, 20,* 1027–1028.

Sackett, G. P. (1966). Monkeys reared in isolation with pictures as visual input: Evidence for an innate releasing mechanism. *Science, 154,* 1468–1473.

Savage-Rumbaugh, E. S., McDonald, K., Hopkins, W. D., & Rubert, E. (1986). Spontaneous symbol acquisition and communicative use by pygmy chimpanzees (*Pan paniscus*). *Journal of Experimental Psychology, 115,* 1–25.

Seyfarth, R. M., & Cheney, D. (1986). Vocal development in vervet monkeys. *Animal Behaviour, 34,* 1640–1658.

Steiner, J. E. (1974). Innate, disciminative human facial expressions to taste and smell stimulation. *Annals of the New York Academy of Sciences, 237,* 229–233.

Studdert-Kennedy, M., Liberman, A. M., Harris, K. S., & Cooper, F. S. (1970). Motor theory of speech perception: A reply to Lane's critical review. *Psychological Review, 77,* 234–249.

Suomi, S. J., & Leroy, H. A. (1982). In memoriam: Harry F. Harlow (1905–1981). *American Journal of Primatology, 2,* 319–342.

Terrace, H. S. (1979). Can apes create a sentence? *Science, 206,* 891–895.

Thierry, B. (1985). Coadaptation des variables sociales: L'example des sytèmes sociaux des Macaques. *Les Colloques de l'INRA, 38,* 91–100.

Tinbergen, N. (1952). Derived activities: Their causation, biological significance, origin and emancipation during evolution. *Quarterly Review of Biology, 27,* 1–32.

van Hooff, J. A. R. A. M. (1962). Facial expressions in higher primates. *Symposia of the Zoological Society of London, 8,* 97–125.

van Hooff, J. A. R. A. M. (1967). The facial displays of the Catarrhine monkeys and apes. In D. Morris (Ed.), *Primate Ethology* (pp. 7–68). Chicago: Aldine.

van Hooff, J. A. R. A. M. (1971). *Aspecten van het social gedrag en de communicatie bij humane en hogere niet-humane primaten.* Unpublished doctoral dissertation, University of Utrecht, The Netherlands.

van Hooff, J. A. R. A. M. (1972). A comparative approach to the phylogeny of laughter and smiling. In R. Hinde (Ed.), *Non-Verbal Communication* (pp. 209–241). Cambridge: Cambridge University Press.

van Hooff, J. A. R. A. M. (1973). A structural analysis of the social behaviour of a semi-captive group of chimpanzees. In M. von Cranach & I. Vine (Eds.), *Expressive Movement and Non-Verbal Communication* (pp. 75–162). London: Academic Press.

van Hooff, J. A. R. A. M. (1976). The comparison of facial expressions in man and higher primates. In M. von Cranach (Ed.), *Methods of Inference From Animal to Human Behavior* (pp. 165–196). Chicago: Aldine.

van Schaik, C. P. (1989). The ecology of social relationships amongst female primates. In V. Standon & R. A. Foley (Eds.), *Comparative Socioecology* (pp. 195–218). Oxford: Blackwell.

Vehrencamp, S. (1983). A model for the evolution of despotic versus egalitarian societies. *Animal Behaviour, 31*, 667–682.

Washburn, D. A., & Rumbaugh, D. M. (1992). Testing primates with a joystick-based automated apparatus: Lessons from the Language Research Center's computerized test system. *Behavioral Research Methods, Instruments and Computers, 24*, 157–164.

Weigel, R. M. (1979). The facial expressions of the brown capuchin monkey (*Cebus apella*). *Behaviour, 68*, 250–276.

Wickler, W. (1961). Ökologie und Stammesgeschichte von Verhaltensweisen. *Fortschritte der Zoologie, 13*, 303–365.

Wrangham, R. W. (1980). An ecological model of female-bonded primate groups. *Behaviour, 75*, 262–300.

Yerkes, R. M., & Learned, B. (1925). *Chimpanzee Intelligence and Its Vocal Expressions*. Baltimore, Md.: Williams & Wilkins.

Yerkes, R. M., & Petrunkevitch, A. (1925). Studies of chimpanzee vision by Ladygina-Kohts. *Journal of Comparative Psychology, 5*, 99–108.

Yerkes, R. M., & Yerkes, A. W. (1929). *The Great Apes: A Study of Anthropoid Life*. New Haven, Conn.: Yale University Press.